AF531358

PRINCIPLES
OF
PLANT PHYSIOLOGY

PRINCIPLES
OF
PLANT PHYSIOLOGY

By
Walter Stiles
M.A., Sc.D., F.L.S., F.R.S.

DISCOVERY PUBLISHING HOUSE
NEW DELHI-110002

Edition - 2016

ISBN: 978-81-7141-247-1

Principles of Plant Physiology

Published by:
DISCOVERY PUBLISHING HOUSE PVT. LTD.
4383/4B, Ansari Road, Darya Ganj
New Delhi-110 002 (India)
Phone: +91-11-23279245, 43596064-65
Fax: +91-11-23253475
E-mail: discoverypublishinghouse@gmail.com
sales@discoverypublishinggroup.com
web: www.discoverypublishinggroup.com

Printed at:
Infinity Imaging Systems
Delhi

PREFACE

THE writing of this introduction to the principles of plant physiology was begun ten years ago. Even then our knowledge of the physiology of the plant had been greatly increased since the publication of the last general statement of the principles of plant physiology by an Englishman, the late S. H. Vines, in 1886, and an attempt at a re-statement of these principles appeared eminently desirable. Further work in the subject which has been carried out during the last decade has resulted in further changes in view of certain aspects of plant physiology, and the need for an introduction to the principles of plant physiology appears to be greater than ever.

In attempting to make these principles clear it has been necessary to illustrate them by reference to the results of many older and more recent researches, but no attempt has been made to mention all the important researches of even the last ten years. This would indeed be impossible in any case in a book of this size and would, moreover, be out of place in a book of this scope, for it would only tend to obscure the general principles in a mass of details. Failure to refer to any particular piece of work is not therefore to be taken by the reader to indicate either that the author was unaware of that work or was inappreciative of its value.

Purely biochemical details have been introduced only so far as reference to them has appeared essential to an understanding of the physiology of the plant. But inasmuch as the physiology of the plant is largely involved with chemical changes taking place in the plant body, a certain acquaintance with the chemistry of the essential substances which make up this body is necessary for a proper appreciation of the principles of plant physiology. For those who find the biochemical treatment in the present volume inadequate, there are a number of works to which reference may be made. The late Mrs. Onslow's *Practical Plant Biochemistry*, although essentially a text-book for practical work forms an excellent introduction to the subject for botanical students. *The Chemistry of Plant Products*, by P. Haas and T. G. Hill, is a work of wider scope and greater detail, while Czapek's monumental

Biochemie der Pflanzen remains the standard work of reference in the subject.

It is hoped this introduction to the principles of plant physiology will be found particularly useful to University students reading for pass or honours degrees. For the benefit of the latter and of others who desire to read further on the subjects dealt with, references to monographs on the respective subjects are given at the ends of some of the chapters. In addition a bibliography is appended of works cited in the text. It is hoped this will be found useful to those students who wish to obtain detailed information from the original sources.

I have pleasure in acknowledging the assistance I have received from Mrs. H. I. Coombs (Dr. F. M. Carter) for reading the whole of the manuscript, and from my colleagues Dr. W. Leach and Mr. C. G. C. Chesters for reading certain chapters, and for making many suggestions which have been incorporated in the text. I would also take this opportunity of thanking the following botanists for permission to use figures from their respective publications: Dr. G. F. Asprey, Mr. G. E. Briggs, Mr. K. W. Dent, Dr. L. E. Hawker, Dr. W. O. James, Dr. F. Kidd, Dr. W. Leach, Professor J. H. Priestley, Mr. R. Snow and Dr. C. West. Acknowledgements are also due to the Council of the Royal Society for permission to use the blocks of figures 9, 10, 18 and 47, the *New Phytologist* and Cambridge University Press for figures 1, 2, 6 and 24, *Annals of Botany* and the Clarendon Press for figures 7, 55 and 56, and *Annals of Applied Biology* and Cambridge University Press for figure 44.

WALTER STILES

CONTENTS

INTRODUCTION

BOOK I—THE GENERAL PHYSIOLOGY OF THE PLANT CELL

BOOK II—METABOLISM

BOOK III—THE PHYSIOLOGY OF DEVELOPMENT

BOOK IV—IRRITABILITY AND MOVEMENT

INTRODUCTION

CHAPTER I

THE SCOPE OF PLANT PHYSIOLOGY

To the ordinary non-botanical observer there appear to be three obvious ways in which living things differ from non-living: in their capacity for taking in materials from outside themselves and building them up into their own bodies, in undergoing changes in form, structure and composition which usually involve increase in complexity, and in their capacity for detaching from their bodies fragments which develop into fresh bodies similar to those from which they arise. These three obvious characteristics which distinguish the living from the non-living are respectively metabolism, development and reproduction. But there is a further characteristic of living matter which is not so apparent at first sight. When subjected to certain changes in the conditions external to it, the organism responds by undergoing changes frequently involving transferences of energy of a magnitude out of all proportion to those involved in the changes in the external medium. Thus tilting the axis of a bean seedling out of the perpendicular results in curvatures in the stem and root, so that the tips of these again come to be orientated vertically. Bringing the tendril of a pea in contact with a woody twig or some other kind of solid rod results in the tendril coiling round this. In so responding to changes in its environment the living organism is said to be irritable, and to possess the quality of irritability.

Whereas it is the scope of morphology, in its widest sense, to determine the structure, including the minute structure, of the plant, it is the aim of physiology to obtain an understanding of the way in which the plant lives and keeps alive; that is, to understand the working of the processes involved in metabolism, development, reproduction and irritability. These form four aspects of physiology which have developed to a great extent independently of one another, and it is convenient to consider these various aspects of the physiology of the plant separately. But it must always be remembered that these four aspects of the life of the plant are closely related to one another. For example, normal development can only proceed as a result of metabolism while reproduction can be regarded as a particular stage of development.

As a preliminary to an understanding of the physiology of the plant, it is necessary to have a knowledge of the component parts of the plant. The plant is a machine, a particularly complicated one, working in a particular way, and like any other machine, its working can only be understood if knowledge of its structure and composition is adequate. In the case of the plant two methods of analysis have been employed for acquiring the requisite information, the morphological and the physico-chemical. Plant morphology, as already noted, has for its aim the determination of the form and structure of the plant. The study of external form of the plant early revealed the existence of different organs which were very naturally assumed to possess different functions. While it was also realized that the plant had an internal structure, it was only with the discovery of the compound microscope that this structure was at all adequately understood, and it came to be known that the plant was built up of small compartments or cells of extremely varied form and with very varied contents. The study of the internal structure of plants constitutes the branch of morphology called histology. With the elaboration of the microscope and the introduction of special methods of treatment of cells, the examination of the minute structure of the cell-wall and the cell contents was rendered possible, and the study of cytology came to rank as an important branch of morphology.

But while morphology provides the necessary information relative to the structure of the plant, it does not provide other information which is extremely necessary to the plant physiologist, information with regard to the chemical composition of the various structures met with in the plant. We cannot hope to understand much about the physiological processes by which the plant is able to absorb material from the outside world and build it up into itself, unless we know something of the physical and chemical condition of the plant. The study of the physics and chemistry of the plant, biophysics and biochemistry, is of much more recent growth than plant morphology, and has revealed the fact that the physics and chemistry of the plant are exceedingly complex.

Theoretically these two methods of analysis, the morphological and the physico-chemical, should become ultimately identical, for morphological analysis will only be complete when it has succeeded in laying bare the molecular structure of the plant, while the complete physical and chemical analysis of the plant should yield the same result. Indeed a meeting-place between these two essentially different methods has already taken place in the investigation of the plant by means of the ultra-microscope, by which it has been established that in plants we have to deal very largely with colloidal systems, so that a knowledge of colloid chemistry, a subject itself yet very imperfectly developed, is a requisite for a proper understanding of the structure of the living plant.

It has been mentioned that the different aspects of plant physiology have developed very largely in independence of one another.

The study of plant metabolism, involving the determination of the substances taken into the plant and their subsequent fate, has naturally developed largely as a chemical study, and as in all but lowly plants there is considerable differentiation of function throughout the plant body, it is not surprising that metabolism has developed largely as a biochemistry of plant organs. The other sides of plant physiology have not so far lent themselves as readily to chemical methods of experimentation, although, as will be seen in the sequel, the application of chemistry and physics to various problems of growth, reproduction and irritability, has yielded results of great promise. Whether it will ever be possible to describe the whole of the life activities of the plant in terms of chemistry and physics no one can say. What is called the mechanistic view of the organism holds that all the processes and activities going on in it, and together constituting what is understood as life in biology, are completely subjected to the same physical and chemical laws as those which govern the non-living world, so that if our analysis were carried far enough the organism could be described as a physical and chemical mechanism. The outlook of the plant physiologist is essentially mechanistic and there is no positive evidence to suggest that the mechanistic conception of the plant is inadequate to explain the metabolic processes of plants. On the other hand it has been urged that development, reproduction and irritability are not in the least understood, and it must certainly be admitted that physico-chemical explanations of these characteristics of living things appear to be exceedingly remote; it does not follow, however, that such phenomena are not physico-chemical phenomena rendered difficult of comprehension on account of the very great complexity of the system in which they are involved. Nevertheless, there are no grounds for asserting that a complete mechanistic explanation of these phenomena is indubitably possible. In any case an attempt at the present time to explain most of the activities of the organism in terms of physics and chemistry is out of the question, and J. S. Haldane has even enunciated the view that 'the problem of physiology is not to obtain piecemeal physico-chemical explanations of physiological processes but to discover by observation and experiment the relatedness to one another of all the details of structure and activity in each organism as expressions of its nature as an organism'. At the same time it is true that, as Hans Fitting has pointed out, for the conception of the plant as a unit it was necessary to have a proper understanding of the physiology of organs in order to obtain an idea of the physiological relations of the different parts to one another.

We may, indeed, go further than this. The separate organs, each of which has a special function, are themselves built up of cells, and to have an understanding of the physiology of the individual organs it is necessary to understand the structure and activities of the individual cell. While cells differ largely among themselves in regard to structure and function, there are, nevertheless, certain characteristics common to all cells displaying vital activities. We are thus justified in speaking of a general physiology of the cell a knowledge of which is a necessary preliminary to a study of the physiology of organs, just as this is necessary for a proper appreciation of the activities of the plant as a whole which are summed up in the terms development, reproduction and irritability.

In this introduction to plant physiology it will therefore be convenient to review our knowledge under the four different heads: general physiology of the cell, metabolism, development, including reproduction, and irritability. While such a division of the subject is in some respects arbitrary it has the merit of convenience as these divisions correspond very largely to aspects of the subject which have developed to a considerable extent independently of one another.

REFERENCES

FITTING, H., *Die Pflanze als lebender Organismus*. Jena, 1917.
HALDANE, J. S., *The New Physiology*. London, 1919.

BOOK I

THE GENERAL PHYSIOLOGY OF THE PLANT CELL

CHAPTER II

THE PLANT CELL

It is a fundamental fact of plant anatomy that the bodies of all plants are built up of elements called cells. These elements are typically very small, only the largest being individually visible to the naked eye, so that while the smallest plants consist of a single cell, the larger ones are composed of millions of them, of various shapes and fulfilling various functions, but having been nevertheless derived by successive divisions from one original cell. In the unicellular organism the one cell that composes the whole plant body must of necessity fulfil all the vital activities of the plant. In the multicellular plants, on the other hand, there is nearly always a so-called division of labour, different cells having different forms and carrying out different activities.

The multicellular plant, in contrast to the multicellular animal, generally retains throughout its life a number of cells, apart from germ cells, which continue to divide actively. These meristematic cells are localized at the apices of stem and root and in the cambium and phellogen of plants with secondary growth in thickness. While some of the products of division of meristematic cells thus remain meristematic, others lose their power of division, undergo various changes in size and shape and are modified into cells of various permanent tissues. All the same, it appears that in very many cases the cells of permanent tissues, so long as they remain alive, have a power of rejuvenation under certain conditions and can again undergo active division.

The ordinary plant cell, in general contrast with those of animals, is provided with a more or less rigid envelope called the cell-wall, within which is the living substance, the protoplasm. Except in meristematic cells the protoplasm forms only a part of the cell contents, a vacuole or vacuoles consisting of an aqueous solution of various substances forming an inclusion or inclusions within the protoplasm. In most adult cells, indeed, the vacuole comprises by far the greater part of the cell interior, the protoplasm being limited to a thin layer lining the inside of the cell-wall. To this lining of protoplasm the name primordial utricle was at one time given.

Although plant cells may vary tremendously in size, shape and function, yet, so long as they are living they exhibit certain characteristics and behaviour in common; the characteristics and behaviour, in short, of living matter. It is with these properties of living matter in general and the living plant cell in particular that we are concerned in this first book, for an understanding of the general physiology of the plant cell is a necessary preliminary to a study of the functional and growth activities of the plant as a whole.

PROTOPLASM

It has been stated above that all living cells contain protoplasm. While cells devoid of protoplasm do occur, sometimes in very considerable numbers, in the plant body, such cells are dead and no longer possess the characteristics of living material. Such cells play a part, and often a most important part, in the life of the plant as a whole, but only cells containing protoplasm have the power of division or, having lost it, can regain this power of division. Once the protoplasm disappears from the cell this power can never be regained, and other characteristics of living cells are permanently lost, for these are bound up with the protoplasm. It is thus true that the characteristics of living organisms which differentiate them from non-living material must be due to the protoplasm. It is therefore reasonable that our first inquiry into the mode of living of the plant organism should be into the nature, chemical and physical, of the living substance.

It so happens that the first cells to be figured, those of cork observed by Robert Hooke in 1665, were dead, and it was not until much later that the importance of the cell contents was recognized, particularly as a result of the work of von Mohl and Schleiden in the middle years of the nineteenth century. The name protoplasm, coined by Purkinje in 1839, was first used for the living substance of plants by von Mohl. The discovery of protoplasmic masses without any apparent envelope helped to the view that the protoplasmic contents are the essential part of the cell and the wall itself of less importance.

Examination of living cells under the microscope shows that the protoplasm is not homogeneous. It contains, with very few exceptions, a spherical or ovoid body, the nucleus, and in the very few cases where such a nucleus is not present, as in bacteria, a number of granules with the staining properties of a typical nucleus have been observed, and possibly function as a nucleus. The nucleus may sometimes occupy one-third of the total volume of the cell, but generally is smaller than this and may occupy less than one-hundredth of the cell volume.

Other granules present in plant cells are the plastids. These are, as a rule, smaller than the nucleus, and a number of them generally occur in each cell. They are usually classified as chloroplasts, chromoplasts and leucoplasts according to their colour. Chloroplasts are green, chromoplasts red, orange or yellow, and leucoplasts colourless. No doubt these, by the loss or development of pigment, can change from one into the other. In the higher plants they are most usually shaped like a biconvex lens, although other forms exist. In the filamentous green algae, however, the green pigment is often contained in a comparatively large structure, as in the case of the well-known spiral band of *Spirogyra*, and there may be only one such chromatophore in each cell. It was at one time thought that plastids never arise *de novo* but only by division of pre-existing plastids, but the work of Guilliermond and others has shown that in the higher plants the plastids may arise from small rod-shaped granules called chondriosomes. In various algae, liverworts and mosses, on the other hand, chloroplasts have been traced through the whole life-cycle and have been found to arise only by division of pre-existing plastids.

The nucleus and plastids are always embedded in the general ground mass of the protoplasm. To this general ground mass the distinguishing term cytoplasm is applied. Thus the protoplasm consists of cytoplasm, nucleus and plastids.[1]

Under the ordinary microscope cytoplasm often has the appearance of a clear liquid, but sometimes a large number of small granules are clearly visible in it. In animal cells some of these granules are the mitochondria of Benda ; these are said to be characterized by staining with Janus green, by containing a lipoid substance and by dissolving in fat solvents. It is possible that the mitochondria of animal cells are of similar nature to the chondriosomes of plant cells which are said to develop into plastids. In other cases the granules may be small non-living inclusions consisting of fat, glycogen, protein or some other substance.

It has been stated above that the cytoplasm has the appearance of a clear liquid sometimes containing a number of tiny granules. There is considerable evidence that the cytoplasm usually has the properties of a liquid, although some observers have thought it to be of a more rigid nature ; it is highly probable that the consistency of cytoplasm varies. But that cytoplasm generally behaves as a liquid is shown by (1) the assumption of a spherical form by drops of water enclosed in it, (2) the movement of small particles in it : the so-called 'Brownian movement' first observed by Robert Brown in 1827, (3) the assumption of a spherical form by the cytoplasm itself under the action of an electric shock, and (4) the streaming movement of the cytoplasm observable in many cells and the somewhat similar 'amoeboid' movements of Myxomycetes. There are, however, some cells in which these characteristics are not observable ; in which, for example, Brownian movement has not been seen. In such cases it is likely that the cytoplasm is more rigid.

There can be no doubt that cytoplasm is a colloidal system. The characteristic of a colloidal solution or sol is that the particles dispersed through it are not of small molecular dimensions, but either consist of aggregates of small molecules or are very large molecules :

[1] This is the ordinary usage of the terms protoplasm and cytoplasm, but it is not universal. Thus Lepeschkin writes : 'I call the mostly colourless kind of living matter which surrounds the nucleus (which is also a kind of living matter), protoplasm.' Seifriz writes of the plastids as cell inclusions, which perhaps suggests that he does not regard them as part of the protoplasm.

so large that their size is of the same order as that of an aggregate of a considerable number of small molecules. Colloidal solutions are, in fact, very fine suspensions (suspensoids) or emulsions (emulsoids) according as the disperse phase, that is, the substance dispersed through the liquid (dispersion medium) is solid or liquid.[1] The granules to which reference has already been made, and which are visible under the ordinary microscope, can be regarded as constituting a coarse disperse phase of protoplasm, but it is generally held that the essential colloidal character of protoplasm is related to the presence of much finer particles. Thus, Heilbrunn states that ' it is certain that the protoplasm of practically all cells contain vast numbers of tiny granules ', some as much as 3 or 4 μ in diameter, but most of them much smaller and almost at the limit of microscopic visibility. On the other hand, Lepeschkin states that the only particles observable in protoplasm under the ultra-microscope are the coarser granules visible under the ordinary microscope, and he concludes that on account of their character the smaller particles of the disperse phase are not visible under the microscope or ultra-microscope. He speaks of coarse disperse phases and colloidal disperse phases.

The substances of which these disperse phase particles are composed, as far as we have knowledge of them, are such as are known to form colloidal systems with water, and the physical properties of cytoplasm in general are those of a colloidal system. It should be noted that the old idea that cytoplasm possesses a reticulate structure was based on observations of fixed (killed) and stained material. Such a reticulate structure appears to be produced by the process of fixing and not to exist in living protoplasm. Hardy showed that the structure of cytoplasm observed in dead material could be varied by varying the method of fixation.

It has been mentioned above that cytoplasm usually exhibits the characters of a liquid, but that in some cases its consistency is more rigid, or, at any rate, more viscous. Having regard to its colloidal character, this means that the cytoplasm is usually a sol, but in some cases it may approximate more to a jelly or gel, just as a solution of gelatin, a colloidal system, may have a liquid or gel consistency according to the concentration of the gelatin, temperature and other conditions. Of late years a number of attempts have been made to obtain relative or absolute values of the viscosity of cytoplasm, and various ingenious methods have been developed for this purpose. These have recently been critically examined by Heilbrunn, who has pointed out

[1] It is more usual now to classify colloids into lyophobe and lyophile systems. In the former, which correspond largely to suspensoids, the disperse phase has no affinity for the dispersion medium, whereas in the case of lyophile colloids the particles of the disperse phase take up, and thus contain, a certain amount of the dispersion medium. The lyophile colloids correspond roughly with emulsoids and include those which form jellies or gels. For further information on the classification and nature of colloids, works on colloid chemistry must be consulted.

that it is possible to measure both the viscosity of the cytoplasm as a whole, including the granules, and that of the intergranular material. Where the granules are very numerous the two values of the viscosity may be very different. Heilbronn measured the rate at which movable starch grains in the endodermal cells of the broad bean, *Vicia faba*, fall under the influence of gravity and compared this rate with that at which they fall in water. He found that the time taken for the grains to fall through a certain distance in the cell was about 8 times that taken for them to fall through the same distance in water, and concluded from this that the viscosity of the protoplasm of the cells in question is about 8 times that of water [1]—considerably less, that is, than the viscosity of sulphuric acid, which is 23 times that of water, or of olive oil and glycerol, which possess respectively viscosities 99 and 1069 times that of water. Heilbrunn adduces evidence to show that Heilbronn's results are probably too high, and he places the viscosity of the protoplasm of the endodermal cells of *Vicia faba* at probably not more than 4 times that of water and certainly not more than 8.

A second method used by Heilbronn for determining the viscosity of the cytoplasm consists in the introduction of small iron particles into the protoplasm of Myxomycetes which are then placed in a magnetic field. The magnetic force necessary to produce a certain movement of the particle was determined and from this the viscosity, on which the value of the required force would depend, was calculated. While results obtained by this method are not precise, Heilbrunn, after considering them critically, considers that they show that the protoplasm of the interior of the plasmodia of Myxomycetes (the endoplasm), possesses a viscosity not more than 15 to 25 times that of water and probably one much less than this. The outer layer of the plasmodium, the ectoplasm, appears to possess a very considerably higher viscosity, in some cases approaching infinity. This suggests that the ectoplasm is of gel consistency.

The values thus obtained for the viscosity of plant protoplasm are surprisingly low, but they agree with observations made by Ewart on protoplasmic streaming and with measurements of the viscosity of the protoplasm of sea-urchin eggs made by Heilbrunn. Indeed, here values in the neighbourhood of three times that of water were obtained for the viscosity of the intergranular fluid and of less than 11 times that of water for the whole protoplasm. At 18° C. the viscosity of the protoplasm of *Amoeba dubia* according to Heilbrunn is only about twice that of water. But the viscosity of some other animal cells, such as muscle cells and Protozoa, is probably very much higher.

Information with regard to the viscosity of the protoplasm but of a less quantitative character has been sought by microscopic observations on the dissection of living cells, the procedure known as micro-

[1] It must, however, be pointed out that it is not certain whether the starch grains move through the cytoplasm or whether the starch grains, with a covering layer of cytoplasm, move through the vacuolar sap.

dissection. The micro-dissection technique has been largely employed by two American investigators, Chambers and Seifriz. Their results suggest that protoplasm may vary greatly in viscosity, not only from one species to another, but also in the same cell at different times. Thus ' the consistency of Myxomycete protoplasm when in the active vegetative state is liquid ', while in the mould *Rhizopus* the protoplasm ' is of gel consistency, more usually that of a soft gel, is sticky, quite elastic, and very extensile, closely resembling bread dough in these physical properties '. A Myxomycete plasmodium increases in viscosity as it prepares to fruit. Changes in viscosity also appear associated with cell division and during the ripening and fertilization of ova, those of *Fucus*, for example. The results obtained by micro-dissection have been criticized by Heilbrunn, chiefly on the ground that the viscosity values of protoplasm obtained by this method are generally much higher than those obtained by other methods of measurement. That variations in viscosity do occur in protoplasm has, however, been observed by other methods. Thus Price, from observations on Brownian movement by means of the ultra-microscope, decided that in resting spores of *Mucor*, for example, the protoplasm is in the gel condition, but that it becomes a sol on the germination of the spore. Seifriz observed that in a quiescent *Amoeba* the number of microscopically visible particles is small and the amplitude of their movement short, while in an active *Amoeba* all the microscopically visible particles except the largest exhibit movement, the amplitude of which is long. These results suggest the general conclusion that active protoplasm is more mobile than quiescent protoplasm, an increase in viscosity taking place as a cell becomes less active and passes into a resting condition.

Seifriz has produced evidence in support of the view that protoplasm is elastic, that is, that it will return to its original form after deformation. The evidence for this is based on experiments in which a particle of magnetic metal was suspended in the protoplasm of the egg of the sea-urchin *Echinarachnius* and subjected to the attraction of an electromagnet ; after release from the magnetic attraction the particle returned to its original position. Were the protoplasm indeed elastic, the fact would be of the greatest importance in affording information with regard to its structure, and Seifriz himself pictures protoplasm as forming either an entangled mass of fibres or an orderly arrangement of chain molecules, but in view of Heilbrunn's criticism of the work just described and his own observations on the plasmodia of Myxomycetes, which he finds are quite inelastic, it seems hardly possible to suppose that protoplasm is generally elastic, although it may be in certain cases, as when it is in the gel state.

There is little definite information with regard to the consistency of the nucleus and plastids. While Gaidukov and Price, from observations with the ultra-microscope, concluded that the nucleus is in the gel condition, Kite and Chambers concluded that the resting nucleus of the ovum is a sol, while Lepeschkin observed Brownian movement of the granules within the nucleus and Heilbrunn records the movement of the nucleolus within the nucleus under the influence of centrifugal force. Indeed, from the observations of Gray, Heilbrunn concludes that the viscosity of the nucleus of the egg of *Echinus* is only about twice that of water, in which case the nucleus is very fluid.

If the conclusions of Heilbrunn with regard to the low viscosity of protoplasm should be well founded, it would appear that protoplasm belongs rather to the suspensoid or lyophobe colloids than to the emulsoid or lyophile colloids. Certain other evidence points in the same direction.

The stability of colloidal systems depends largely on the fact that the particles of the disperse phase carry an electric charge. There is a certain amount of evidence that in the case of protoplasm this charge is positive. Thus Hardy showed that when an electric current is passed through the tip of the onion root the protoplasm migrates to that side of each cell nearer the negative pole. The movement of colloidal particles in this way is known as cataphoresis and the movement of the particles towards the negative pole indicates that they carry a positive charge. A number of similar, though less definite, examples of this cataphoretic effect in protoplasm have been noted.

There are, however, reasons for supposing that the definition of protoplasm as a suspensoid or lyophobe colloid is insufficient. Chief among these are the results of chemical analyses of protoplasm. In this connexion it has to be remembered that the ordinary methods of chemical analysis bring about changes in the protoplasm so that it no longer has the attributes of living matter; it is, in fact, dead. It is also difficult to separate protoplasm from other cell constituents. Hence, analyses of plant protoplasm have chiefly been of the plasmodia of Myxomycetes which consist of protoplasm to a greater extent than any other plant material available. Three such analyses are given in Tables I–III, the first the classic analysis of Reinke and Rodewald, the results of which were published in 1881, the second and third the more recent analyses of Lepeschkin and Kiesel, published in 1924 and 1925 respectively.

Table I

Analysis of the Plasmodium of *Aethalium septicum* (= *Fuligo varians*)

(Reinke and Rodewald)

Substance	Percentage of dry weight
Proteins	40·0
Albumins and enzymes	15·0
Other nitrogenous compounds	2·0
Fats	12·0
Carbohydrates	12·0
Cholesterol	2·0
Resins	1·2
Calcium salts (except calcium carbonate)	0·5
Other salts	6·5
Undetermined matter	6·5

Table II

Analysis of the Plasmodium of a Myxomycete resembling *Fuligo varians*

(Lepeschkin)

Substance	Percentage of dry weight
A. Water-soluble organic substances chiefly contained in the vacuoles :	
Monosaccharides	14·2
Proteins	2·2
Amino-acids, purine bases, asparagine, &c.	24·3
B. Insoluble organic substances which principally form the ground mass of the protoplasm :	
Nucleo-proteins	32·3
Free nucleic acids	2·5
Globulin	0·5
Lipo-proteins (plasmatin)	4·8
Neutral fats	6·8
Phytosterol	3·2
Phosphatides	1·3
Other organic matter (polysaccharides, pigments, resins)	3·5
C. Mineral matter, of which about half is extractable with water	4·4

Table III

Analysis of the Plasmodium of *Reticularia lycoperdon*

(Kiesel)

Substance	Percentage of dry weight
Fat	17·85
Lecithin	4·67
Cholesterol	0·58
Reducing carbohydrates	2·74
Non-reducing soluble carbohydrates (exclusive of glycogen)	5·32
Glycogen	15·24
Polysaccharides, hydrolysable with difficulty	1·78
Extractives (containing nitrogen)	12·00
Proteins, including plastin	29·07
Nucleic acid	3·68
Oil of lecithoproteins (?)	1·2
Unknown substances	5·87

Analyses of the cytoplasm of the cells of cabbage-leaves were published by Chibnall and Channon in 1927. The average of values given by six analyses are shown in Table IV.

Table IV

Composition of Cytoplasm of cabbage-leaves

Substance	Percentage of dry weight
Protein	63·1
Ether-soluble substances (fats, lipoids, &c.)	20·75
Ash	6·45
Undetermined	9·5

These analyses indicate that the chief constituents of the protoplasm are water, proteins and fatty (lipoid) substances, while microchemical tests also show the presence of proteins and lipoid substances in the protoplasm, and these are substances which can form lyophile or emulsoid systems. According to Lepeschkin, the dispersion medium of the protoplasm is not pure, or almost pure, water, but a combination of water, lipoids and proteins. It cannot be water because the protoplasm does not mix with water. It does, however, contain much water and Lepeschkin thinks of it as taking up water to a certain maximum amount, any water taken up by a cell above this amount presumably separating as a vacuole. Also it appears that water passes readily through the protoplasm, another indication, were one needed, that the dispersion medium of protoplasm contains water. Further, most substances dissolved in water penetrate through protoplasm very slowly, but, as Overton showed, substances soluble in lipoid substances pass through the protoplasm readily. This, according to Lepeschkin, indicates that the dispersion medium contains lipoid substances, while the presence of proteins is supposed to be indicated by observations of Pfeffer that weak acids, salts of heavy metals and other substances which exert no influence on lipoid substances, but which modify the properties of proteins, also enter cells readily. But the proteins, lipoids and water must be so intimately associated that particles are not observable, and Lepeschkin thinks the lipoids and proteins are combined in some form which is easily broken down. Into this combination water is taken in some way up to a limiting amount. It is suggested that in different species there are probably different combinations.

The various particles forming the disperse phases of protoplasm are, according to Lepeschkin, roughly of three kinds, (*a*) microsomes, (*b*) granula, and (*c*) mitochondria, chondriosomes, or chondriokonts. The microsomes appear to be protein, sometimes denatured, sometimes not. The granula also appear to be proteins and are often of a solid character. Some appear to dissolve in water, others appear to swell up and to be jelly-like. The smaller particles, on the other hand, which appear to constitute the greater part of the disperse phase material, are, judging from their behaviour towards dyes and other reagents, probably composed of a protein-lipoid combination which is certainly different from that composing the dispersion medium.

Lepeschkin thus thinks of protoplasm as a polyphase system containing a protein-lipoid-water dispersion medium, and disperse phases which are partly suspensoid or lyophobe, partly emulsoid or lyophile. One of the reasons for supposing that protoplasm, at any rate to some extent, is an emulsoid system, namely, that it contains substances which form such systems, has already been stated (cf. p. 14). Another reason given by Lepeschkin is that measured values of the viscosity of protoplasm are too high for a suspensoid sol. But we have already noted that Heilbrunn concludes that protoplasm is a suspensoid very largely on account of the low viscosity of the protoplasm.

A third view of the structure of protoplasm has been put forward during recent years by Seifriz. It has been already mentioned that Seifriz regards protoplasm as elastic. It must be remembered, however, that this view of the elasticity of protoplasm as distinct from that of membranes surrounding the protoplasm, is by no means generally held. There is no doubt that naked cells and nuclei are extensible and more or less elastic, but it is one view that the cells are surrounded by a membrane which is elastic, the protoplasm itself being no more elastic than water contained in an indiarubber bladder. This can be deformed by pressure and the water along with it, and recovery of the original form will take place after removal of the pressure, but it is the indiarubber which is elastic, not the water. Similarly with the protoplasm, it may only be the surrounding membrane which is elastic, not the liquid protoplasm within. However, Seifriz regards this supposed elastic property of protoplasm as of the greatest importance in affording information with regard to its structure. There is certain work of Freundlich and others which indicates that colloid systems with rod-shaped particles are elastic while those with spherical particles are not. Since, then, elastic colloid systems have a fibrous structure and protoplasm is also elastic, it is held that protoplasm has a fibrous structure. Moreover, its high elasticity, its rigidity and its capacity for imbibitional swelling are held to indicate that it is in the gel state. This condition would, however, apparently include a system similar to that of a weak solution of gelatin which does not set, for Seifriz says that it 'makes little difference what we call protoplasm whether sol or gel so long as we remember that its physical properties are the properties of jellies. If the ability to flow is our criterion of the sol state, then protoplasm is usually, but by no means always, a sol; but there are other indicators of the colloidal state such as elasticity, rigidity, and imbibition, and these are gel characteristics'.

Seifriz therefore pictures protoplasm as forming either an entangled mass of fibres or an orderly arrangement of chain molecules, and it is interesting to note that food substances found in living cells, starches, fats, and proteins are generally regarded as characterized by possessing molecules made up of chains of groupings. Seifriz indeed at one time said that none possess a cyclic structure, but that we must now regard as very doubtful, as it is now generally held that sugars, long supposed to have a chain structure, are really cyclic.

In fixed cells a fibrous structure of the protoplasm has often been observed, but whether this means that the structure of the living protoplasm is also fibrous, is doubtful. Hardy was able to obtain a number of different structures out of the same material by altering the method of fixation so that it is likely that none gave a picture of the living material. To criticism of this kind Seifriz has replied that artefacts arise only from pre-existing conditions which determine them. The fibrous structure seen in fixed cells may be the result of coagulation at death, but only of already existing strands or fibrous units of colloidal or molecular dimensions, the agglutination of which yields linear aggregates of microscopic size.

There are thus in the field at present really three views of the structure of protoplasm which regard it as, respectively, (*a*) a suspensoid sol, (*b*) an emulsoid sol, (*c*) a gel. But the upholders of all these views would readily admit that protoplasm is not a constant material and that it is capable of undergoing wide variations in consistency. We may in any case accept Bayliss's view of protoplasm, that it is 'an extraordinarily complex heterogeneous system of numerous phases and components, continually changing their relations under the influence of electrolytes and other agents'.

Such a view is scarcely in keeping with what was known as the 'biogen' hypothesis, formulated definitely by Verworn in 1903, though traceable as far back as 1867, which supposed protoplasm to be a substance possessed of a very large molecule, or biogen, with a large stable nucleus and side groupings capable of undergoing various reactions which constituted the activities of the cell. While there was never much evidence for such a view, rather have we to-day to think of protoplasm as possessing its characteristic qualities, partly no doubt on account of the actual chemical composition of the constituent phases, but especially on account of its physical structure, which is, of course, partly conditioned by the chemical composition of its constituents.

In this medium we know that a large number of different reactions proceed. We can conceive of a number of different reactions taking place contemporaneously in the same cell by supposing that different reactions take place in the different phases or at their surfaces. It has been noted above that besides the dispersion medium, Lepeschkin differentiates at least three disperse phases, and besides these such bodies as the nucleus and plastids must be regarded as phases of the protoplasm. All the same the separate phases must not be thought of as independent of one another, for in such a polyphase system a disturbance in any one phase, by upsetting the equilibrium of the system, may result in changes throughout the whole protoplasm.

With such a view of the living substance we can understand that it can vary not only from plant to plant, but from cell to cell in the same plant and in the same cell at different times. It is essentially a system capable of continual change and of infinite variety. Indeed, the different structure and behaviour of different species must be attributed to differences in the composition, structure, and organization of the protoplasmic system.

In such heterogeneous systems as the protoplasm, one of the most important characteristics is the great extent of surface, that is, of boundaries between phases, and many of the reactions taking place in the cell are no doubt surface reactions taking place at the junction of two phases. But apart from these internal phase boundaries, the importance of which no one would deny, considerable discussion has centred round the limiting surfaces of the protoplasm itself. At its outer surface the protoplasm is in contact with cell-wall or external medium, at its inner surface, in the case of vacuolate cells, with the cell-sap in the vacuole. At these boundaries it is generally held that there are thin layers possessing properties different from those of the bulk of the protoplasm and known as the plasmatic membranes, or plasma-membranes. A discussion of this question follows in the next section of this chapter.

THE PLASMATIC MEMBRANES

There are a number of reasons for supposing that the outermost layer of the protoplasm differs in constitution from the bulk of it, so that it forms a very thin membrane separating the body of the protoplasm from the bordering medium. Such membranes appear to have been hypothesized by Pfeffer in 1877 who gave to them the name plasmatic or plasma-membranes, the outer one bordering the cell-wall (or external medium in the case of cells devoid of walls) being the outer plasmatic membrane, and the inner bordering the vacuole being the inner plasmatic membrane. De Vries used the terms ectoplast and tonoplast (or vacuole wall) respectively for these membranes. To-day the membranes are most usually called the inner and outer plasma-membranes.

Since these membranes are not morphologically obvious, it is necessary to inquire into the arguments for their existence. These arguments are as follows:

1. In the preceding section of this chapter it has been shown that there are in the field three views of the structure of protoplasm according to which it is regarded respectively as (1) a suspensoid sol with an aqueous dispersion medium, (2) a sol with a protein-lipoid-water dispersion medium immiscible with water, and (3) a gel. It has also been pointed out that there is reason to believe that protoplasm is capable of passing from one to another of these conditions.

With the second and third of these views it is possible to understand that protoplasm can remain intact in contact with water, but under the first view it would appear necessary for the protoplasm to be separated by a membrane of some kind under such a condition, as otherwise it would mix with water and so disappear. Such, indeed, is sometimes the case, when the protoplasm is said to exhibit 'decomposition by diffluence', but more usually drops of protoplasm in water remain separate from it and intact. Such behaviour of protoplasm has been observed many times during the last hundred years, and in very many and various species, although it is more easily observable and noticeable in some species than in others. In *Vaucheria*, for instance, the formation of droplets of protoplasm exuded from the filament is very easily observed. Also there are quite a number of lower forms in which, no cell-wall being present, the protoplasm appears to remain normally in direct contact with the external medium.

This affords one argument for the existence of a plasma-membrane, but it is based on one particular view of the structure of protoplasm and on special cases: isolated droplets of protoplasm and naked cells. There are, however, other reasons for supposing the presence of a membrane surrounding the protoplasm.

2. If protoplasm forms a substance immiscible with water, where it comes into contact with water or with any other medium with which it does not mix, according to physico-chemical laws any substances present in either protoplasm or external medium, which lower the surface tension at the interface, will tend to accumulate at the surface, so that the concentration of such substances in the surface layer will be higher than in the bulk of the protoplasm or external medium, respectively. Now the proteins, and particularly the fatty substances which form so considerable a part of the protoplasm, are substances which lower surface tension considerably at a water-air interface, and it is highly probable that they will behave similarly in the case of the protoplasm in contact with cell-wall, cell-sap, or other medium. Thus there is every reason to suppose that at the surface of the protoplasm there is a thin layer of material differing physically and chemically from the bulk of the protoplasm. Ramsden showed that at the surface of separation of a protein sol with some other phase a thin rigid film can form, possibly by the protein which accumulates at the surface undergoing some change. The formation of a similar membrane at the surface of the protoplasm is, at any rate, a possibility.

3. Observation of some cells under the ordinary microscope reveals the presence of a surface layer differing in appearance from the general mass of the protoplasm. In Myxomycetes there is a clear outer layer to which the name hyaloplasm is given, but this layer is not of general occurrence in plant cells, and it is presumably not to be identified with the plasma-membrane supposed to be present in all cells.

Sometimes in cells that have been killed the protoplasm appears to be bordered by a very fine membrane. Such a membrane is not visible in living cells and it is open to question whether the membrane observed in the killed protoplasm is a former membrane present in the living cell changed and rendered visible by the killing process, or whether it is produced for the first time when the cell is killed.

Observations with the ultra-microscope are conflicting. In some cells, for example those of the leaf of *Elodea canadensis*, Price could discern no indication of a differentiation of the protoplasm into an inner and outer layer, while in other cases, for example the cells of the hairs of the stem and leaf of *Cucurbita*, he thought he could make out a very thin membrane at the protoplasmic surface. In the hairs of the tomato there also appears to be an outer layer of protoplasm differentiated from the rest, although it seems likely that this is to be identified with hyaloplasm rather than with a plasma-membrane.

4. Study of protoplasm by the method of micro-dissection has led Seifriz to conclude that in Myxomycetes and *Amoeba* there is definite

evidence of an exceedingly thin plasma-membrane forming a film outside the hyaloplasm.

5. The behaviour of cells when put in water or aqueous solutions can be explained on the supposition that the outermost part of the protoplasm acts as a membrane allowing easy passage to water but barring the passage of dissolved substances, and in particular, those within the cell. The water is, in fact, absorbed by osmosis through a semi-permeable membrane. The question of the water relations of the cell will be dealt with in detail in the next chapter, but it may be pointed out here that in the case of a vacuolate cell the whole body of the protoplasm can be regarded as forming a membrane separating the cell-sap in the vacuole from the outer medium, so that there is no need to hypothesize the presence of additional membranes limiting the protoplasm on its outside and inside. Further, some writers have urged that the absorption of water by non-vacuolate cells can be explained without invoking a plasma-membrane.

6. Just as the behaviour of the cell towards the absorption of water has been explained by means of a plasma-membrane, so has the behaviour of the cell in regard to the absorption of dissolved substances. Some substances pass readily into cells while others enter slowly or not at all. Such behaviour at once suggests that of non-living membranes of different kinds in regard to the passage of dissolved substances through them. However, other explanations in terms of absorption and chemical combination are possible and equally credible, while a quantitative examination of the problem shows that the passage of dissolved substances into cells is a more complex one than the passage of similar substances through artificial membranes. From a consideration of the data of solute absorption by cells it can be concluded that the phenomena of this solute absorption shed little light at present on the question of the plasma-membrane.

7. Living cells and tissues offer an unexpectedly great opposition to the passage of an electric current; that is, they possess a high electrical resistance. The same cells and tissues when killed may exhibit a resistance of only one-tenth of that of the living material. This phenomenon has been explained on the ground that the plasma-membrane in the living cell offers a considerable resistance to the passage of the ions which carry the current; this resistance more or less disappears when the cell is killed. Here again other explanations are possible. Thus the death of the cell may involve the breaking down of complex compounds with a consequent increase in the concentration of electrolytes. Also, if the resistance of the membrane to an electric current is greatly reduced by the death of the cell, so may be the resistance of the whole of the protoplasm, and if this is so the need for a membrane to explain the observed facts disappears.

Although much of the evidence for the existence of plasma-membranes does not withstand criticism, there is, nevertheless, probably a balance of evidence in favour of the presence of a thin limiting membrane surrounding the cell. It is clear that such a membrane must be very thin while the actual thickness can only be estimated very roughly. Seifriz thinks the thickness of the plasma-membrane is of the order of magnitude of 0·1 μ, or 0·0001 mm. while Collander and Bärlund, from considerations of the rate of penetration into cells of non-electrolytes and their solubility in fats come to the conclusion that this is probably a maximum value. As regards the composition of the plasma-membrane, it would seem likely, from the preceding considerations, to contain materials present in the protoplasm, though in different concentrations and proportions, and perhaps altered. Thus various workers have urged the presence in the membrane of proteins, fats, and the more complex fatty or lipoid substances. If the membrane has a real existence it is likely that it contains all these substances in a complex colloidal arrangement.

In this connexion it is worth while calling attention to what Heilbrunn has called the 'surface precipitation reaction'. Heilbrunn, it will be recalled, regards protoplasm as a suspension with an aqueous disperse phase which, in contact with water, will mix with it unless separated by a membrane. It has already been mentioned that such a mixing of protoplasm with water, the diffusion, for example, of a drop of protoplasm through a mass of water, sometimes takes place. But, quite commonly, the protoplasm retains its individuality, and, according to Heilbrunn, this must be due to the formation of a film round the surface of the protoplasm. Now, although it is true that protoplasm contains substances which lower the surface tension of water at a water-air interface, and it is also true that such substances will be present at the interface in higher concentrations than in the body of the protoplasm (that is, will be adsorbed at the surface), Heilbrunn holds that the films which form round masses of protoplasm cannot be adsorption films for the following reasons: (*a*) between water and the aqueous phase of protoplasm there is no surface of separation at which such films could form, (*b*) low temperature favours adsorption but retards the formation of surface films on the protoplasm, (*c*) in cases of various animal cells, namely the eggs of *Stentor*, *Arbacia*, and *Echinarachnius* in which the film formation has been studied the presence of calcium is necessary although the calcium can be replaced by strontium, but not by magnesium. In these animal eggs pigment granules are present and it has been shown that these latter are also necessary for membrane formation.

The whole process has been compared by Heilbrunn to blood coagulation. This consists of two stages, a first stage requiring calcium which results in the formation of thrombin, and a second stage in which the thrombin produces a clot by combining with fibrinogen and for which calcium is not necessary. Clotted blood contains thrombin, and so the serum (containing thrombin) which exudes from a clot can bring about the clotting of fresh blood in absence of calcium, as only the second stage in the reaction is involved. Similarly it appears that in the case of these animal eggs a preparation of the broken-up eggs contains a substance which can produce a surface precipitation reaction in absence of calcium.

In the case of protoplasm from the sea-urchin eggs a similar pair of reactions is hypothesized, a substance, ovothrombin, being formed which is comparable to thrombin. The stages in the process of membrane formation are therefore supposed to be:

(i) Calcium + pigment granules → ovothrombin.
(ii) Ovothrombin + protein → membrane.

It is suggested by Heilbrunn that if such a reaction is common to all cells it would be better to give the name cytothrombin to the intermediate product. Heilbrunn suggests that the surface precipitation reaction does not occur in the cell interior because of the absence of free calcium ions. Presumably for some reason these become released at the surface of naked protoplasm. Should any action bring about the release of free calcium ions in the interior the surface precipitation reaction should take place inside the cell. Reference is made to this question in the next section of this chapter.

THE VACUOLE

In meristematic cells the protoplasm occupies the whole of the cell interior but as the cell passes out of its meristematic condition a small number of less viscous inclusions appear within the protoplasm. These watery inclusions, the vacuoles, increase in size until they may merge with one another so that in most adult cells the vacuole occupies nearly the whole of the cell interior, the protoplasm being limited to a thin layer lining the cell-wall.

The development of the vacuole obviously results from the absorption of water by the meristematic cell, the usual explanation of this absorption being that the outer layer of the protoplasm, the plasmatic membrane, acts as a semi-permeable membrane enclosing an osmotically active solution, so that from a solution of lower osmotic pressure water will be absorbed into the cell. The mechanism of this water absorption will be discussed in more detail in the next chapter. But however the water is absorbed some explanation is obviously necessary to account for the separation of water as vacuoles and for the maintenance of these as phases distinct from the protoplasm. For we have seen that the latter is a colloidal system with an aqueous dispersion medium, or at any rate, a dispersion medium containing much water, and at first sight it would appear that absorbed water would mix intimately with the dispersion medium.

Two explanations with some credibility have been put forward to account for this separation of the vacuole from the more viscous protoplasm. Lepeschkin's view has been indicated in the previous section of the present chapter. According to him the protoplasm has no limiting membrane, but its dispersion medium is a protein-lipoid compound with water which can only be taken up to a certain limit, that is, to a sort of saturation point. Any water in excess of this must separate out as vacuoles. The weakness of this view is that it affords no explanation of why water should be absorbed beyond the saturation point of the protoplasm.

Quite a different view of vacuole formation is taken by Heilbrunn. It will be recalled that this writer emphasizes the necessity for a membrane surrounding protoplasm in contact with water if the two liquids do not mix, and he puts forward the theory of the surface precipitation reaction to account for the formation of such a membrane. According to Heilbrunn, the same reaction gives rise to vacuolation, or at least, to that type of vacuolation which consists of the appearance of a number of small vacuoles in the protoplasm, and which appears to be identical with what is called cytolysis in the eggs of sea-urchins, and which can be produced in plant cells by a variety of agents, such as distilled water, acids, alkalies, an electric current, mechanical injury, and high temperatures. The sequence of events is (1) the setting free of calcium ions in the cell interior, (2) the reaction of the calcium with (in the sea-urchin egg) a pigment granule or some constituent of it, or (in other cells) some similar substance, presumably to produce ovothrombin, and (3) the reaction of the ovothrombin with a substance in the protoplasm, probably a protein, to cause vacuolation.

While it is clear in this scheme how the material of the membrane might be precipitated in the protoplasm it does not appear clear why the material should be deposited in the form of spherical films surrounding liquid of a more aqueous consistency than the bulk of the protoplasm. Nor is it clear whether Heilbrunn's theory of the formation of small vacuoles in protoplasm by numerous reagents can be extended to the case of the normal formation of the large vacuoles characteristic of the plant cell. The theory is, at any rate, interesting and ingenious.

The vacuole is generally regarded as consisting of an aqueous solution of various crystalloidal substances, and this is no doubt true in general as the osmotic pressure exerted by the cell, and the chemical analysis of expressed cell-sap, both indicate. These dissolved substances are, however, very varied. Organic acids such as malic and oxalic are frequently prominent, while in some Crassulacae, soluble salts of malic acid may compose half the dry weight of the sap. In some plants inorganic salts are present in quantity. Thus in the leaf-stalks of *Gunnera scabra* potassium chloride appears to be the chief solute in the vacuole, while in the pea, bean and maize, the sap contains much potassium nitrate. In yet other plants, sugars are present in quantity, as in the onion bulb, beet root, and many other storage organs.

It also appears that a certain amount of colloidal material may be present in, at any rate, some vacuoles, for particles of ultra-microscopic dimensions have been observed in vacuoles by means of the ultra-microscope. Also, more rarely, solid particles may be present, as in the case of some desmids, the vacuoles in which contain small granules of calcium sulphate.

As is well known, pigments may sometimes be dissolved in the vacuole. These belong to the group of pigments known as anthocyanins which are glucosidal substances, being formed of a sugar grouping combined with some other radicle. Such pigments, as a rule, give a blue, purple, purplish-red, or crimson colour to the cell-sap. They are familiar in the petals of numerous flowers, in the roots of the red beet and of *Tolmiea menzesii*, frequently in young leaves and sometimes in ageing leaves.

THE CELL-WALL

Although some plant cells such as a few unicellular algae and germ cells are unprovided with cell-walls, the vast majority of plant cells possess this envelope, which must therefore be regarded as a normal part of the plant cell, a feature in which plant cells stand generally in contrast to those of animals.

Cell-walls are very varied in appearance owing to differences in structure and composition, which are closely related to the various functions of cells and tissues.

When a cell divides the first membrane to appear between the two daughter nuclei is the so-called 'cell-plate'. The fate of the cell-plate is in doubt. According to one view it divides longitudinally and wall material is deposited between the two halves, while according to the other view the cell-plate itself develops into the wall by the deposition of fresh material on or within it and possibly by a change in its own composition. However this may be, the cell-plate becomes replaced by the middle lamella. The composition of the cell-plate and middle lamella is unknown, but the material composing the latter has been held by different investigators to be pectin (see Chapter VIII), calcium pectate, a simple cellulose or lignin. On this middle lamella subsequent layers are deposited.

The new wall separating the daughter cells after cell division is an apparently homogeneous thin structure, and so long as cells remain meristematic their walls remain thin. As a cell-wall gradually develops into a constituent of a permanent tissue the wall thickens, new material being added by protoplasmic activity. Such material is apparently added in two different ways known as (1) growth by apposition, in which fresh particles are laid down on the inner surface of the existing wall, and (2) growth by intussusception, where the new particles are deposited among the particles of the existing wall.

Walls of adult cells appear to be composed of microscopically distinct layers. It has been usual to differentiate these layers of the mature wall into three groups : (1) middle lamella, (2) primary wall and (3) secondary wall, but different writers have not all used these terms in the same sense. Recently Anderson has urged that for the sake of clarity the significance attached to these terms by Kerr and Bailey should be generally adopted. These authors define these terms as follows :

1. The middle lamella consists of amorphous isotropic [1] material first deposited by the cytoplasm and more usually considered to consist largely of pectic compounds.

2. The primary (or cambial) wall is the first anisotropic layer of the wall to be deposited and consists chiefly of cellulose and pectic substances. It is capable of growth and reversible changes in thickness.

3. The secondary wall comprises layers deposited on the primary wall. With the formation of the secondary wall the cell no longer retains the power of enlargement.

The outer layers of the cell-wall in the case of parenchymatous tissues are composed chiefly of celluloses, which, as is well known, are substances yielding glucose on hydrolysis, and which are therefore condensation products of glucose, that is, they are glucosans. Similarly, condensation products of other sugars such as mannose, galactose, and xylose may be present as well as pectin. Substances intermediate in structure and complexity between the sugars and their condensation products have also been reported as present in so-called cellulose cell-walls. Apart from these carbohydrate constituents, which certainly make up by far the greatest part of the cell-wall, it is possible that fatty substances may be present, while a certain amount of inorganic material, principally calcium oxide, calcium carbonate and silica, is often present. Cell-walls always contain a certain amount of water.

In 1864 Nägeli put forward a theory of the structure of the cell-wall according to which the latter consists of minute units or micellae, so arranged that although they form definite layers, each micella is separated from its neighbours by a layer of water.

Current views on the structure of the cellulose wall have been greatly influenced by the results of recent work in which polarized light and X-radiation has been used for the determination of the constitution of cellulose and the structure of the wall. As a result of this work it appears that the cellulose molecule consists of a chain of anhydrous glucose groups each such group being connected with the next through an oxygen atom. The chain constituting the cellulose molecule, according to Haworth, comprises about 200 glucose groups. By X-ray examination it has been determined that these cellulose chains are aggregated into bundles of about 60. This aggregate of

[1] Isotropic bodies are similar in physical properties in all directions and are contrasted with anisotropic bodies in which the arrangement of the molecules and, consequently, the physical properties, are different in different directions.

cellulose molecules can be regarded as a colloidal particle, or equally as a micella as hypothesized by Nägeli.

Some differences of opinion exist regarding the arrangement of the cellulose micellae in relation to other constituents of the wall. One group of workers including Frey-Wisselingh, Meyer and others are led to a view which differs little from Nägeli's. According to this view the cellulose micellae are not in contact, but are separated from one another by a colloidal system. Thus, if the cell-wall is allowed to swell in water X-ray examination shows that the micellae do not increase in size but the distance between them does, as would be the case if the inter-micellar material were a colloid capable of imbibing water.

On the view of Hess and others also, the cellulose units are separated from one another, but the units are much larger than the submicroscopic micellae hypothesized by Frey and Meyer, while the thin layer of non-cellulose material separating the cellulose units serves as a kind of adhesive holding the cellulose particles together.

Sponsler rather considers the cellulose as forming a continuous framework of long chains of anhydrous glucose groups running parallel with one another and connected by cross-chains so that the whole forms a three-dimensional lattice-like framework.

Some mention should be made also of the views of Hansteen-Cranner, although these were put forward before the application of X-ray analysis to the cell-wall had been attempted. This worker laid emphasis on the activity of the cell-wall itself in vital phenomena and considered it as a living part of the cell. He regarded the protoplasmic colloids as continued into the cell-wall and constituting the material between the cellulose micellae or framework. Such a view of the cell-wall is not antagonistic to those of recent investigators mentioned above.

Some cell-walls pass into a mucilaginous condition. This appears to result from the conversion of glucosans and related substances into pentosans, condensation products of pentose sugars which have considerable capacity for absorbing water. Other cell-walls become changed by the appearance in the walls of substances known as lignin, suberin and cutin. Lignified walls contain xylan (a pentose condensation product), hadromal (an aromatic aldehyde) and the so-called lignic acids. Lignin is presumably a mixture of these and possibly other substances. They appear to be intimately mixed with the cellulose constituents of the wall; the lignin and cellulose, that is, do not form separate layers. The middle lamella is the first part of the wall to become lignified, and it has been suggested that lignification of other layers may involve a transformation of the pectic constituents into lignin. Cells that become cork are said to develop suberized walls. Suberin, like lignin, appears to be a mixture of substances, and a number of fatty acids, the so-called suberogenic acids, have been isolated from suberized walls. While the substances composing suberin have some fatty characteristics they do not appear to be true fats. Van

Wisselingh thought that the middle lamella of suberized walls consists of a thin continuous layer of suberin on which cellulose might be deposited subsequently. Unlike lignified cell-walls, suberized walls are impervious to water.

Cutin is deposited as a continuous layer on the outside of epidermal cells of the herbaceous stems and leaves of vascular plants, and forms the layer termed cuticle. It resembles suberin in being a mixture of substances and in preventing the passage of water, but it appears to consist of different substances from suberin. In some cases, at any rate, the substances appear to be waxy, and those present in the cuticle of a number of plant organs have been investigated by Chibnall and his co-workers. In the cuticle of the apple fruit, which they examined in detail, they found that the waxy substances consist of long chain alcohols and paraffins including the following:

n-nonacosane, $C_{29}H_{60}$
n-heptacosane, $C_{27}H_{56}$
d-*n*-nonacosan-10-ol, $CH_3.(CH_2)_8.CHOH.(CH_2)_{18}.CH_3$
n-hexacosanol, $CH_3.(CH_2)_{24}.CH_2OH$
n-octacosanol, $CH_3(CH_2)_{26}.CH_2OH$
n-triacontanol, $CH_3.(CH_2)_{28}.CH_2OH$

The amount of fatty acid present was found to be very small, and it would seem that in this cuticle only a small proportion, if any, of the alcohol exists combined as wax esters. How far these are the substances which in general give to cuticle its characteristic qualities must, for the present, remain an open question.

Cell-walls are usually elastic; that is, when acted upon by a force bringing about stretching or change of shape in general, they tend to return to their original length or shape after removal of the force. Thus Schwendener and Krabbe, for example, immersed cylinders of young pith in water, as a result of which they increased in length by from 25 to 30 per cent. On plasmolysis they contracted and their final decrease in length was the same as that of similar cylinders plasmolysed directly without previous immersion in water.

When a body is deformed by a force the deforming force acting on unit area of the surface of the body is called the stress and the deformation produced is the strain. The modulus of elasticity in general is defined as the ratio of stress to the strain produced. When stretching alone is considered, that is, change in length, the modulus of (linear) elasticity, or Young's modulus, is the ratio of the force acting on unit cross-section to the change in length of unit length. This quantity varies greatly with different cell-walls, and with the same cell-wall depends on the amount of water present. It appears that as a wall absorbs more and more water it must become more easily stretched, that is the modulus of elasticity decreases. Also as the stress increases, the increase in length for the same increment in the stress becomes relatively less, at any rate, after a certain point is reached; this may sometimes be due to loss of water, but does not appear to be always so.

When the pressure is increased beyond a certain value the limit of elasticity is reached and the wall, instead of returning to its original length on release from the pressure, remains permanently stretched. Here also different cell-walls exhibit very different behaviour. It was shown by Pfeffer that the cell-walls of the filaments of the stamens of some Cynareae can be stretched to double their original length before reaching the limit of their elasticity, whereas with most phloem and sclerenchyma fibres the limit of elasticity is reached when the stretching is no more than from 0·5 to 1·5 per cent. of the length. The majority of cell-walls lie between these two extremes.

Quite commonly the cell-wall ruptures as soon as the limit of elasticity is passed, and in some cases rupture may occur before the elastic limit is reached. However, in the case of collenchyma it has been shown that whereas the elastic limit is reached with a stress of about 10^5 grams per sq. cm., rupture does not occur until the stress is about 8 times this. In the case of growing cells the walls are not stretched up to their elastic limit and are capable of further stretching, which, if continued, may become permanent.

REFERENCES

HEILBRUNN, L. V., *The Colloid Chemistry of Protoplasm.* Berlin, 1928.
LEPESCHKIN, W., *Kolloidchemie des Protoplasmas.* Berlin, 1924.
SEIFRIZ, W., *Protoplasm.* New York and London, 1936.

CHAPTER III

THE WATER RELATIONS OF THE PLANT CELL

WATER ABSORPTION

THE great part played by water in the life of the plant is familiar to every one and requires no emphasis. Water constitutes 80 per cent. or more of the content of active cells, not only forming the greater part of the protoplasm and vacuole, but also forming an important constituent of the cell-wall. It is the medium in which other constituents of the cell are dissolved or dispersed and through which substances are able to diffuse, and its presence is necessary for the carrying out of the normal functions of the growing plant. While some organs as, for example, many seeds, can withstand the withdrawal of a greater part of water from their cells without killing them, there is no obvious activity in such organs. Only when a further supply of water is absorbed by such desiccated organs is the renewal of cell activity possible. It is a necessary characteristic of all living cells that such an absorption of water is possible.

Ever since the work of Pfeffer on osmotic pressure more than half a century ago his view that the absorption of water is an osmotic phenomenon has dominated botanical thought on this question. On this view, the cell in respect of its water relations may be regarded as a solution of crystalloidal substances surrounded by a semi-permeable membrane which is itself surrounded by the cell-wall, the latter being generally completely permeable both to water and dissolved substances. With regard to the vacuolated cell, a very good case can be made out for this view. The vacuole, as we have seen in the last chapter, consists of an aqueous solution of various crystalloidal substances, and, whether colloidal material is also present or not, will therefore exert an osmotic pressure. The protoplasm, forming a layer inside the cell-wall, acts as a membrane separating the solution contained in the vacuole from the cell-wall, and hence from any solution external to the cell, for if the cell-wall is permeable to water and substances dissolved in it, solutions outside the cell will be imbibed by the wall and so come into contact with the protoplasm.

In the case of non-vacuolated cells the applicability of this simple osmotic view is not so clear. To meet such cases it has been customary to suppose that the outermost layer of the protoplasm, the so-called plasma-membrane, constitutes a semi-permeable membrane separating the bulk of the protoplasm from the external medium. In the non-vacuolated cell, crystalloids in solution in the protoplasm would therefore exert an osmotic pressure just as solutes in the vacuole do in vacuolated cells. But if the outermost layer of the protoplasm acts as a semi-permeable membrane distinct from the rest of the protoplasm in the case of non-vacuolated cells, we must suppose that the same must be the case with vacuolated cells. Moreover, if the outermost layer of the protoplasm is differentiated from the bulk of the protoplasm it is at least credible that the innermost layer of the protoplasm, that bordering the vacuole, is similarly differentiated. If, then, the non-vacuolated cell can be regarded as a simple osmotic system, the vacuolated cell must be regarded as a more complex system consisting of the vacuole surrounded by the inner plasmatic membrane, which itself is surrounded by the bulk of the protoplasm, while this again is separated from the outer medium by the outer plasmatic membrane.

But we have already seen in the last chapter that the evidence for the existence of such membranes is not over-strong, though there is probably a balance of evidence in favour of a differentiation of the external layer of the protoplasm. Whether such membranes really possess the semi-permeable properties at one time generally attributed to them must at present be regarded as remaining in doubt. M. H. Fischer, an observer of the behaviour of animal cells, went so far as to assert that 'there are no membranes about cells', while other writers have spoken of them as hypothetical structures. However this may be, there is certainly at present a preponderance of evidence in support of the more generally accepted opinion that the vacuolated plant cell behaves as an osmotic cell: a solution of crystalloids surrounded by a membrane more or less impermeable to some at least of these crystalloids but readily penetrated by water. Whether the non-vacuolated plant cell behaves similarly is more in doubt; further evidence is needed.

If the osmotic view of water absorption is rejected, some other explanation of the process is required. This is found in the colloidal character of the protoplasm and the behaviour of many non-living colloidal systems towards water. Thus, gelatin, agar-agar and a number of other colloidal substances, when put in water absorb it in considerable quantity, and mixtures of these substances with albumin and other proteins, amino-acids or urea, have been prepared which are said closely to resemble plant cells in their behaviour towards water. We may reasonably suppose that all protoplasm will absorb water in this way, and it may be that in non-vacuolated cells all absorption takes place thus, but it will not account for the absorption of water in vacuolated cells, for the protoplasm forms far too small a proportion of the whole cell volume to imbibe the relatively large amount of water that may be absorbed in this way. We may therefore conclude, at least provisionally, that in vacuolated cells water absorption is chiefly an osmotic process, imbibition by the protoplasm playing a relatively minor, and often negligible part. In non-vacuolated cells, on the other hand, the part played by osmosis through a membrane must be regarded as doubtful.

TURGIDITY

Whatever the mechanism of water absorption by the plant cell, the result is that in contact with water the vast majority of cells will absorb water to such an extent that the protoplasm presses against the cell-wall and stretches it. We have seen in the last chapter that the cell-wall is elastic so that on stretching it tends to return to its original size and shape, that is, it exerts a pressure against its extension, and so against increase in volume of the cell. Within the limit of elasticity this pressure increases with extension of the wall, so that a point will be reached at which the pressure sending water into the cell is balanced by the inwardly directed pressure of the stretched cell-wall. So long as the cell-wall is stretched and the cell contents are pressing against it the cell will be rigid in much the same way that an inflated football bladder or inner tube of a bicycle or motor-car tyre is rigid. The cell is then said to be turgid or in a state of turgor or turgescence. Such a condition is very important for the life of a plant, for the rigidity of herbaceous plants and plant organs depends on it, while cell division and many other life processes can only proceed in turgid cells.

QUANTITATIVE WATER RELATIONS OF THE PLANT CELL

In considering the quantitative relations of the plant cell in regard to water we may take first the simplest case, that of an isolated vacuolated cell in which the protoplasm forms a thin layer separating the vacuole and cell-wall, and occupies a very small part of the whole cell volume. We may then regard water absorption through imbibition by the protoplasm as negligible in comparison with that absorbed by osmosis through the semi-permeable protoplasm.

Our system therefore consists of a solution of crystalloids (the cell-sap) surrounded by a semi-permeable membrane (the protoplasm). The latter is itself surrounded by the permeable and elastic cell-wall, which is permeated by the external liquid which is thus in contact with the protoplasm on its outer side.

Now when an aqueous solution is separated from pure water by a semi-permeable membrane, water will pass through the membrane from the water to the solution. While it is impossible to enter here into the theory of this passage of water across the membrane, it must be pointed out that the pressure tending to send the water through the membrane is, within limits, approximately proportional to the absolute temperature and to the number of molecules, or, in the case of electrolytically dissociated compounds, to the number of ions, present in unit volume of the solution. If the solution is separated by the membrane not from pure water but from another solution, the osmotic pressure of this solution will tend to send water through the membrane in the opposite direction.

In the case of the plant cell the following quantities are therefore concerned. Tending to force water into the cell is the osmotic pressure of the solution occupying the vacuole. If the cell is surrounded by a solution, the osmotic pressure of the latter tends to force water out of the cell. In addition to this, water tends to be forced out of the cell if it is in a state of turgor. In an isolated cell the turgor pressure, that is, the hydrostatic pressure of the water against the wall, is equal and opposite (in direction) to the inwardly directed pressure of the stretched cell-wall. If we denote the osmotic pressure of the solution in the vacuole by P, that of the external solution by P′ and the turgor pressure by T, the actual net pressure tending to send water into the cell is therefore $P - P' - T$. For this quantity the term *net suction pressure* may be employed. Denoting this quantity by S, we have therefore the relation

$$S = P - P' \quad T$$

If the external liquid is pure water which possesses no osmotic pressure $P' = 0$ and consequently

$$S = P - T$$

For this special case the term *full suction pressure* may be employed for the pressure sending the water into the cell.

When a cell absorbs water the cell-wall stretches and the turgor pressure of the cell increases. Consequently the suction pressure continually falls until the cell is in a condition which may be described as saturation when the suction pressure is zero. The turgor pressure is then equal to $P - P'$. A condition of equilibrium should then have been reached in which water is neither lost nor gained by the cell. Such a condition may never be actually realized, perhaps because the cell-wall is not perfectly elastic, or the protoplasm not a perfect semi-permeable membrane or because changes take place within the cell, but a condition of approximate equilibrium is certainly obtainable. If the external liquid is pure water, the turgor pressure is then at its maximum and is equal to the osmotic pressure of the cell-sap.

If the cell is immersed in a solution, the osmotic pressure (P′) of which is greater than the osmotic pressure of the cell-sap less the turgor pressure $(P - T)$, then the suction pressure will be negative, that is water will be withdrawn from the cell. This withdrawal of water will result in (*a*) an increase in the osmotic pressure of the cell-sap and (*b*) a decrease in turgor pressure, and both these processes will tend to bring $P - T$ nearer to P′. When $P - T = P'$ the suction pressure is zero, and a position of equilibrium with regard to the passage of water is reached. But when P′ is big the turgor pressure may be reduced to zero and $P - T$ (now equal to P since T is zero) be still less than P′. The cell is now turgorless and the cell-wall incapable of further contraction. The protoplasm now contracts away from the cell and equilibrium will be reached when the contraction of the protoplasm is such that the concentration of the cell-sap has increased to P′. The cell is now in the well-known condition of *plasmolysis*. At the beginning of plasmolysis the turgor pressure reaches its minimum value, zero. If a cell in this condition is immersed in pure water the turgor pressure gradually increases, while the suction pressure decreases, until the cell is in equilibrium with water and the turgor pressure is a maximum and the suction pressure zero.

Should the protoplasm not form a negligible proportion of the total cell volume, it is possible that imbibition by the protoplasm may account for an appreciable part of the water absorbed. In some cells substances such as starch and pentosans may hold by imbibition an appreciable proportion of the water present in the cell. In all such cases, however, the protoplasm, including the starch and other cell inclusions, is in equilibrium with the solution in the vacuole. Consequently the presence of these imbibing constituents of the cell will only affect the absorption of water by the cell in so far as the dilution of the sap by osmotic absorption disturbs the equilibrium between the sap and protoplasm with its inclusions. Except in cases of very great absorption, the dilution of the cell-sap by osmotic absorption is unlikely to be sufficient to bring about any very appreciable disturbance of this equilibrium. Consequently, we may, in general, neglect the imbibition factor in a consideration of water absorption by vacuolated cells.

In many cases substances dissolved in the water surrounding a cell may affect not only the quantities which determine the net suction pressure, but also those which determine the full suction pressure; the pressure, that is, which is a measure of the capacity of the cell to absorb water. Thus, if the protoplasm is not completely impermeable to the dissolved substance and the latter diffuses unaltered into the vacuole, the osmotic pressure of the cell-sap will gradually rise, that is, the value of P in the relation given on p. 32 will increase, and so also will the suction pressure.

Conversely, if osmotically active material diffuses out of the cell into the external liquid, the value of P and, consequently, of the suction pressure, will be lowered. If a substance in the external liquid should affect the constitution of the cell-wall, its modulus of elasticity might become altered, and so the wall pressure would be affected. If this were lowered, the tendency of water to be pressed out of the cell would be less and the suction pressure would be raised. Thus it can be seen that exposure of a cell to a solution of a certain substance, or of certain substances, can alter its capacity for absorbing water.

It will be observed that when a cell is in equilibrium with water its turgor pressure is at a maximum, being equal to the osmotic pressure of the cell, since the suction pressure is then zero. The turgor pressure is at its minimum when the cell is plasmolysed. Between these limits the value of the turgor pressure depends on the quantity of water present in the cells. It is generally assumed that within the limits of elasticity of the cell-wall the wall pressure, and therefore turgor pressure, increases proportionally with the volume of water taken into the cell, although this relation may only be approximate. Assuming this relationship, if V_z is the volume of the cell and P_z its osmotic pressure when plasmolysis is about to commence, and the turgor pressure consequently zero, and if V is the volume and P the osmotic pressure of the cell when the turgor pressure is T,

$$\frac{V - V_z}{V_z} = kT$$

where k is a constant.

The ratio $\frac{V}{V_z}$ has been called by Höfler the degree of turgoı stretching. From the above equation it is seen that it is equal to $1 + kT$.

In tissues the water relations of the individual cells are complicated by the pressure exerted by neighbouring cells. When a piece of tissue is cut out of an intact plant body a change in its shape may often be observed owing to the removal of this pressure. The curvature observed in the strips of tissue obtained by cutting a dandelion or hyacinth scape longitudinally is an example of this. Here the central tissue must have been compressed and the outer tissue in a state of tension, for on isolation of the strip the latter curves owing to the contraction of the outer and extension of the inner tissue. The general effect of the presence of the surrounding cells must be to reduce the amount of water a cell can absorb, for in addition to the inwardly directed pressure of the stretched cell-wall the cell contents will also be subject to the pressure of the neighbouring cells as they become turgid. The turgor pressure of the cell will then balance the sum of the wall pressure of the cell itself and the pressure exerted by neighbouring cells. Denoting this latter pressure by N and the wall pressure by W, we have,

$$S = P - P' - W - N.$$

Owing to this pressure exerted by surrounding cells, a cell in a tissue can absorb less water than is the case with a similar cell when isolated, and with absorption of water the turgor pressure increases more rapidly in the cell within the tissue than in the isolated cell. As pointed out by Höfler, it follows that suction pressure must increase more rapidly with decreasing volume in the case of a cell in a tissue than in a similar cell when isolated, and within a small range of volume there may be considerable variation in the suction pressure, a condition of affairs which may have considerable physiological significance.

THE RANGE OF OSMOTIC PRESSURE IN PLANT CELLS

Many observations have been made on the magnitude of the osmotic pressure in plant cells either by pressing out the cell-sap and calculating its osmotic pressure from determinations of the freezing-point of the liquid, or by finding the concentration of a sucrose or potassium nitrate solution, the osmotic pressure of which is known, which just brings about plasmolysis of the cell. Such a solution should possess the same osmotic pressure as the cell-sap at the moment when plasmolysis is about to begin. The former method gives an average value of the osmotic pressure of all the cells in a tissue, the latter method enables the osmotic pressure of individual cells to be measured.

The osmotic pressure of plant cells exhibits a very wide range. Among higher plants the osmotic pressure rarely falls below 3·5 atmospheres, while at the other end of the range an osmotic pressure as high as 153 atmospheres has been recorded as occurring in the cells of a plant of *Atriplex confertifolia* growing near the Great Salt Lake. However, osmotic pressures of the cells of land plants are usually from about 10 to 20 atmospheres. Some sort of relation also holds between the habit or ecological type of plant and the value of the osmotic pressure, the latter being generally higher in mesophytes than in succulents, higher in the leaves of woody plants than in those of herbs of the same region, and particularly low in epiphytes. The question naturally arises how far differences in osmotic pressure of different ecological types can be traced to environmental differences. Now many cases have been observed which show that a high concentration of osmotically active substances in soil may result in high osmotic pressures in the cells of plants growing in such soil. The cells of plants growing on soils containing much sodium chloride often exhibit very high osmotic pressures. The case of *Atriplex confertifolia* has already been mentioned, and in general the osmotic pressure of root hair cells of halophytes is higher than that of similar cells of mesophytes. It is not clear, however, that the high osmotic pressure in the case of halophytes is due to the accumulation of sodium chloride rather than to increase of a number of other osmotically active constituents of the cell-sap. According to von Faber, the high osmotic pressure of cells in mangroves is attributable to sodium chloride in some species, but not in all.

Analyses of the cell-sap of different species show that the substances responsible for the osmotic pressure cover a considerable range, including inorganic salts, organic acids and sugars. De Vries found that the principal osmotically active substance in the leaf stalk of *Rheum hybridum* is oxalic acid, whereas in the leaf stalk of *Heracleum sphondylium* it is sugar, and in that of *Gunnera scabra* potassium chloride. In a number of plant organs the osmotic pressure appears to be largely due to potassium nitrate. Cases of this are to be found, according to Copeland, in *Fagopyrum*, *Pisum*, *Phaseolus* and *Zea*.

It will be observed that a substance can only serve to maintain a permanent osmotic pressure in the cell if the latter is impermeable to it, and it might be supposed that the cell, while being impermeable to the relatively larger molecules of organic acids and particularly sugars, would be permeable to the simpler inorganic salts. Such, however, can scarcely be the case if we accept the results of Copeland and de Vries with regard to the responsibility of potassium chloride and potassium sulphate for the osmotic pressure in organs of quite a number of different species.

THE SUCTION PRESSURE

The osmotic pressure of a cell, although an important factor in influencing the amount of water the cell can absorb, does not alone determine this quantity. As we have already noted, the wall pressure of the turgid cell and the osmotic pressure of the external medium both oppose the entrance of water into the cell, so that the actual pressure sending water into the cell, the net suction pressure, is generally less than the osmotic pressure, its value being given by the equation

$$S = P - P' - T.$$

The full suction pressure, that which operates when the external medium is water, is therefore $P - T$, since in this case P' is zero. It may be determined quite simply by finding a solution of sucrose in which the cell undergoes no change in volume (or weight). For then the net suction pressure is zero, and consequently we have the relation:

$$0 = P - P' - T$$

or

$$P' = P - T,$$

which is the value of the full section pressure.

The investigation of suction pressure has been carried on for many years by Ursprung and Blum, who have drawn certain conclusions with regard to the magnitude and variation of suction pressure in plant cells. The movement of water through the living cells of a plant tissue is determined by a gradient of suction pressure, and Ursprung and Blum have amassed a vast quantity of data to show the direction of this movement by determining the suction pressures of the different cells in a tissue and hence the gradients of suction pressure. The method used by Ursprung and Blum consists in cutting thin sections of tissue in liquid paraffin, in which medium the sections are mounted, measuring the size of the cells in the tissue, and then transferring them to solutions of an osmotically active substance covering a range of concentrations. The suction pressure of any particular cell is then equal to the osmotic pressure of the solution in which that cell undergoes no change in volume on being transferred to it from paraffin.

It has recently been pointed out by Miss Ernest that this method of measuring suction pressure involves a source of error which renders invalid all determinations of suction pressure made by it. For the gradients of suction pressure that exist in the living plant will immediately tend to disappear if a piece of tissue is severed from the living plant and all supply of water or pull on the water of the cells removed. The cells of the isolated piece of tissue will immediately tend to proceed to a condition of equilibrium so that all have the same suction pressure, and it seems unlikely that a very considerable time will be required for equilibrium to be reached. Since Ursprung and Blum allowed the sections to remain for 20 to 30 minutes in the osmotically active solution in order that the cells of the tissue might attain equilibrium with the external solution, it seems very likely that the suction pressure gradient in the tissue would be markedly altered in that time. Miss Ernest also points out a second source of error in the method of Ursprung and Blum. Thin sections of tissue include a number of cut cells bordering on the intact cells the suction pressures of which are measured. Now after being ruptured, the cut cells will exert a greater suction pressure than before owing to the disappearance of the wall pressure. Consequently they will tend to withdraw water from the intact cells, the suction pressure of which will be raised in consequence, so that the method of Ursprung and Blum will give values which are higher than those actually obtaining in the isolated tissue before the sections are cut. Miss Ernest has produced experimental evidence in support of her criticisms. The lower epidermis of certain leaves can be stripped off together with a certain amount of mesophyll tissue, the cells of which appear to be completely uninjured. In some monocotyledons the veins also can be torn from the leaves along with a number of intact mesophyll cells. Determinations of the suction pressures of cells obtained in this way from *Crocus*, *Iris* and *Saxifraga umbrosa* showed no evidence of any suction pressure gradient in the pieces of detached tissue. We must conclude therefore that, although suction pressure gradients must be present in tissues through which there is a movement of water, the suction pressure gradients observed in sections cut from pieces of tissue are largely illusions of experimentation.

THE EFFECT OF TEMPERATURE ON THE ABSORPTION OF WATER BY PLANT CELLS

Temperature appears to have little effect on the total amount of water a cell is capable of absorbing, but it influences considerably the rate at which water is absorbed or lost by cells. Reliable data on this question have been obtained by two methods respectively applicable to different kinds of tissue. In one method, used by Miss Delf, the rates of water loss at different temperatures were compared. A solution of sucrose at various temperatures was allowed to flow through the hollow interior of a cylindrical organ such as an onion leaf or dandelion scape. The sucrose solutions were chosen of such a concentration that the cells shrank in them, but not to such an extent that plasmolysis resulted. The contraction resulting from exposure of the cells surrounding the hollow space in the middle of the cylinder to the ' subtonic ' solution was then measured by an optical device giving a linear magnification of 350 times. This method, it will be observed, is applicable to organs which have the form of hollow cylinders. In the second method seeds or discs of tissue are immersed in water at various temperatures and weighed after various times This method has been used by a number of investigators for measuring the effect of temperature on the intake of water by seeds, and by Stiles and Jorgensen for determining the influence of temperature on absorption of water by storage tissue.

In general, the rate of water absorption increases with rise in temperature, but above 35° C., as a result of the effect of the high temperature on the cell organization, at any rate in the case of dandelion scapes, onion leaves and storage tissues, the intake of water after a time gives place to water withdrawal. The general course of water absorption by potato tuber at different temperatures is shown in Fig. 1.

A convenient way of describing concisely the effect of temperature on a process is to state the *temperature coefficients* over different temperature ranges.

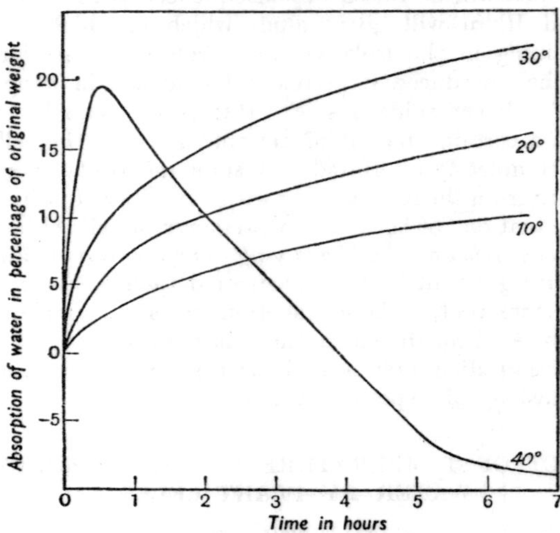

Fig. 1.—Curves illustrating the absorption of water by potato tuber tissue at different temperatures

By the term temperature coefficient is usually understood the ratio of the rate of a process at a certain temperature to the rate of the process at a temperature 10° C. lower. To obtain values for the temperature coefficients of water absorption or excretion it is necessary to compare the rates at which water passes into or out of the tissue at different temperatures when the tissue is in the same degree of swelling. Temperature coefficients determined in this way for water absorption by, or excretion from, various tissues, are summarized in Table V.

Table V

Temperature Coefficients of Water Absorption or Excretion

Temp. range in Centigrade degrees	Onion leaves	Dandelion scapes	Potato tuber	Carrot root	Barley grains	Xanthium seeds Sample— I	II
5–15 . .	1·4	—	—	—			
10–20 . .	1·5	2·3	3·0	1·3			
15–25 . .	2·0	3·3	2·75	1·4	Av. 1·99	Av. 1·55	Av. 1·83
20–30 . .	2·6	3·8	2·7	1·6			
25–35 . .	2·9	3·0	—	—			

The values for onion leaves and dandelion scapes were obtained by Miss Delf, for potato tuber and carrot root by Stiles and Jørgensen, for barley grains by Brown and Worley and for the two samples of seeds of *Xanthium pennsylvaticum* by Shull. The temperature range employed in the case of barley grains was actually 3·8° to 34·6° C.

THE EFFECT OF DISSOLVED SUBSTANCES ON WATER ABSORPTION BY PLANT CELLS

We may consider first the simple case in which a cell is immersed in a solution of a substance to which the protoplasm is impermeable. We have observed that the pressure sending water into a plant cell (the net suction pressure) is given by the expression P – P′ – T where P and

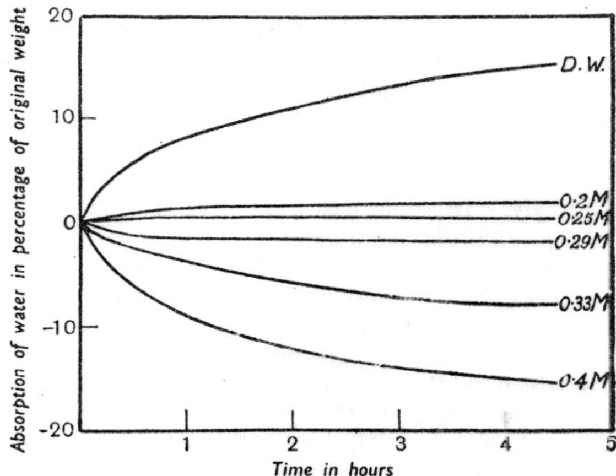

FIG. 2.—Curves illustrating the absorption of water by potato tuber tissue immersed in sucrose solutions of different concentrations

P′ are the osmotic pressures of the cell-sap and the external medium respectively and T the turgor pressure. Consequently, the greater the osmotic pressure of the external medium, the lower the net suction pressure. The higher the concentration of the external medium, the slower therefore will water enter the cell. When P′ > P – T the net suction pressure will be negative and water will be withdrawn from the cell. These relationships are clearly shown by the diagrams in Fig. 2 which show the rate of water intake by disks of potato tuber immersed in solutions of sucrose of various concentrations. It will be observed that not merely is the rate of water entry or exit dependent on the concentration of the external solution, but that the total quantity of water the cells are capable of absorbing is determined by the concentration of the external solution, the more concentrated the external solution, the less the amount of water the cell is capable of absorbing.

This is, of course, only to be expected, since when an osmotically active substance is present in the external liquid this balances to some extent the osmotic pressure of the cell which cannot therefore develop as high a turgor pressure as when the external medium is water and the turgor pressure alone at equilibrium balances the whole of the osmotic pressure.

These considerations only apply when the osmotically active substance present in the external medium is incapable of penetrating the cell. If, on the other hand, the substance should pass through the protoplasm into the cell-sap the matter is not so simple. Should the substance remain unchanged after entering the cell it will bring about an increase in the osmotic pressure of the cell, the water absorbing capacity of which will therefore be increased. If absorption of the substance continues until equality of its concentration within and without the cell results, the cell will be able to absorb about as much water as if the external medium were pure water.

If the substance should not remain unchanged after entering the cell yet other considerations apply. Should the substance react with some cell constituent a change in the osmotic pressure will result and the suction pressure and total amount of water absorbable will be altered accordingly. In some cases the reaction may result in the formation of an addition compound and the consequent complete removal from the entering substance of its osmotically active property. In many cases, the substance may, by its reaction with some constituent of the protoplasm, lead to loss of water from a turgid cell as the latter dies, for as this takes place the protoplasm becomes permeable to the substances dissolved in the cell-sap, and these thereupon diffuse out into the external medium. With a completely permeable membrane no cell can remain turgid. The course of water exchange between cells and their surroundings when these are immersed in a solution of a toxic substance, as, for example, copper sulphate, are shown in Fig. 3. It will be observed that although absorption of water takes place at first from a dilute solution (0·01 N), the absorption is less than from distilled water, while after a time actual water loss sets in.

In some cases a toxic substance may have a strange effect on absorption of water inasmuch as it may bring about an increase in the rate of absorption which obtains when the tissue is surrounded by pure water. Such an effect has been observed when potato tuber tissue is brought into contact with solutions of sulphuric acid, osmic acid, mercuric chloride and aliphatic alcohols. It is not clear how these substances bring about an increase in the suction pressure. Either they must affect the osmotic pressure of the cell-sap by bringing about some change in its composition, or they must affect the modulus of elasticity of the cell-wall so that the latter offers less resistance to stretching. So far this effect has only been observed with toxic substances and sooner or later the absorption of water gives place to the water loss characteristic of the action of all toxic substances (cf. Fig. 4).

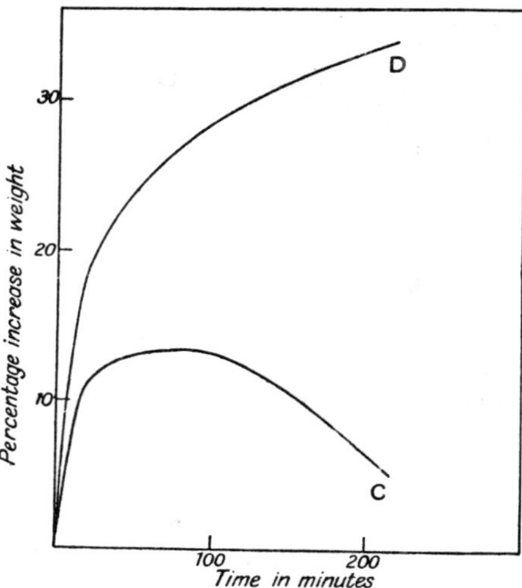

Fig. 3.—Curves illustrating the absorption of water by potato tuber tissue immersed in distilled water (D) and a 0·01 N solution of copper sulphate (C)

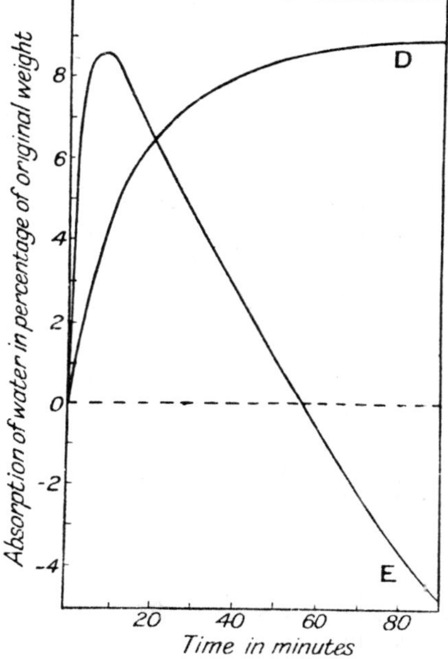

Fig. 4.—Curves illustrating the absorption of water by potato tuber tissue immersed in distilled water (D) and a 2 M solution of ethyl alcohol (E)

ISOTONIC COEFFICIENTS

A solution having the same osmotic pressure as the cell-sap is said to be isosmotic or isotonic with it; a solution, the osmotic pressure of which is lower than that of the cell-sap is said to be hypotonic, while one whose pressure is higher than that of the cell-sap is said to be hypertonic.

If a turgid cell is immersed in a solution of a substance incapable of penetrating the protoplasm and of such concentration that incipient plasmolysis just, and only just results, the solution is isosmotic with the cell-sap. Consequently, if this concentration is found for a number of different substances, the various solutions should have the same osmotic pressure, and by this means the osmotic pressures of various substances may be compared. For comparison of osmotic pressures in this way, certain conditions are necessary for success. These are, (1) the least trace of plasmolysis should be observable, (2) all the cells used for the determinations should just plasmolyse in solutions of the same salt of exactly the same concentration, (3) the substances examined should not penetrate the cells. Such cells are, according to De Vries, provided by the epidermal cells of the outer side of the growing leaf sheath of the dark red form of *Curcuma rubricaulis*, and the violet cells of the lower epidermis at or near the midrib of the leaves of *Rhoeo discolor*. Both these kinds of cells contain coloured sap on account of which they fulfil the first of the three conditions given above.

A number of determinations of this kind were made by De Vries. Actually De Vries determined what he called isotonic coefficients. He used the term isotonic coefficient to indicate the attraction of a molecule of a substance in dilute aqueous solution for water, the attraction of a molecule of potassium nitrate being taken as 3. Thus if a solution of a substance of molecular concentration X just produces plasmolysis whereas the concentration of potassium nitrate required for this is K, the isotonic coefficient is $\frac{3K}{X}$, since a solution of the substance of concentration X has the same osmotic pressure as a solution of potassium nitrate of concentration K, and consequently has the same attraction for water.

A variant of the plasmolytic method just indicated for comparison of osmotic pressures and determination of isotonic coefficients was also devised by De Vries. This is the plasmolytic transport method. The cells employed, after weak plasmolysis in the solution of the substance under investigation, are transferred to potassium nitrate solutions of different concentrations. That solution in which the volume of the plasmolysed protoplast undergoes no further change is isotonic with the solution used to produce plasmolysis.

What is practically a second variant of the method was also described by De Vries and is known as the tissue tension method.

A length of a cylindrical organ, such as a dandelion scape, is cut from a plant and divided longitudinally into a number of strips. On isolation, owing to tensions that were present in the tissue these strips usually curve so that the pith occupies the convex side of the strip. If immersed in water the curvature is intensified owing to absorption of water by the pith cells and the whole strip may roll into a spiral. On the other hand, if immersed in a strong solution, water is withdrawn from the pith cells and the curvature is reduced or even reversed, and for any substance not penetrating the cells there can be found a solution of a particular concentration in which no change of curvature takes place. Such a solution has an osmotic pressure equal to what may be termed the full suction pressure of the strip, using this expression to denote the pressure sending water into the strip from pure water (cf. p. 36). It is thus possible to find the concentrations of solutions of different substances which have no effect on the curvature of different strips from the same length of tissue. Assuming the suction pressure is the same in these different strips, it follows that the solutions have the same osmotic pressure and are therefore isotonic.

It must be regarded as extremely doubtful whether the determination of isotonic coefficients in this way has much practical value, for it is recognized that a great many substances can penetrate plant cells so that the osmotic pressure of cells immersed in a solution of such a substance increases with entry of the substance if the latter remains in solution. Moreover, different substances may enter cells at different rates so that the osmotic pressure, and consequently the suction pressure, of the cells of the strip undergo change with time, the rate of change depending on the nature of the substance in solution.

However, if the term 'isotonic coefficient' is now of little value, the terms hypotonic, isotonic and hypertonic are themselves convenient for indicating solutions of lower, the same, or higher osmotic pressure respectively than that of the cell-sap.

IMPERMEABLE AND SEMI-PERMEABLE CELL-WALLS

It has been noted in the previous chapter that some cell-walls are impervious to the passage of water. The contents of such cells, as, for example, those of cork, must soon die if they are completely enclosed by walls of this character, since exchange of material between them and their surroundings is impossible. In other cases, such as cells of the endodermis of roots where some, but not all, of the walls of each cell are impermeable to water, this is not the case, and the presence of such walls may then have the effect of ensuring the passage of water and dissolved substances in a definite direction through the plant.

There are also cell-walls which are semi-permeable; that is, they permit the ready passage through them of water, but not of dissolved substances. A. J. Brown observed cell-walls of this kind in barley grains, for the latter, when immersed in aqueous solutions of sulphuric acid, copper sulphate, silver nitrate and a number of other salts, absorb water readily but do not absorb the solute. That the semi-permeable property is located in the cell-wall and not in the protoplasm or its limiting layer is clear from the fact that the grains suffer no damage in comparatively concentrated solutions of poisons, including those mentioned above, for if the protoplasm were to come into contact with such substances it would be speedily damaged and rendered permeable.

Semi-permeable cell-walls do not appear to be uncommon in seeds, but apart from these, ready permeability of cell-walls to water accompanied by imperfect permeability to dissolved substances appears to exist in certain cells, such as those of onion bulb scales, onion leaves and beet root. When this is so plasmolysis will take place more slowly than when the cell-wall is readily permeable to the solute, for it will take some time for the solution immediately outside the protoplast to reach a concentration high enough to induce plasmolysis.

REFERENCE

STILES, W., *Permeability.* London, 1924.

CHAPTER IV

THE RELATIONS OF THE PLANT CELL TO DISSOLVED SUBSTANCES

EVERY organism absorbs from the environment not only water but a number of substances dissolved in the water, some of which, but not necessarily all, are necessary for the building up and functioning of the plant body. These substances pass into the plant in solution and are conveyed through the plant for greater or less distances. They may themselves ultimately be stored, or they may take part in reactions which result in the formation of other substances which may still be soluble in water or which may form solid inclusions in the cell. But any transference of such material must be in solution; solid particles are unable to penetrate the cell membranes and pass from cell to cell. Hence we meet in all plants with the problems of the passage of such dissolved substances from the environment into the plant body, and the passage of these substances from cell to cell of the multicellular plant; that is, we have to do with the problem of the permeability of plant cells to dissolved substances.

The term 'permeability' applied to the plant cell is often used somewhat loosely to represent the capacity of the cell to allow the passage into it of substances under consideration without reference to any particular part of the cell influencing this entrance. A more exact way of viewing the question is to regard the outer layers of the vacuolated cell (the cell-wall and protoplasm) as constituting a membrane separating the cell from its surroundings, and the permeability of the cell membranes to a substance as a measure of the rate at which a substance is able to penetrate these membranes under certain defined conditions. Moreover, since it is very frequently, though not always, the case that most dissolved substances readily penetrate through the cell-wall, the measured rate of passage of a dissolved substance through the cell membranes is very often regarded as determined entirely by the plasmatic layer and hence to provide an indication of the permeability of the protoplasm. Whether any retardation of entry of a dissolved substance into the cell is effected uniformly by the whole thickness of the protoplasm or by any particular part of it, such as the limiting layers (plasma-membranes), is a point on which information at present available is insufficient to form any reliable opinion, at any rate for the generality of cases. But it should be remembered that not all cells are vacuolated, and if the entrance of dissolved substance into non-vacuolated cells should be found to follow the same laws as that of the entrance of such substances into vacuolated cells, it would afford a strong argument for the view that the permeability of the cell membranes as a whole is determined by that of the limiting layers. However, we have as yet practically no information regarding the laws governing the entrance of dissolved substances into non-vacuolated plant cells ; practically all our information has been derived from studies of vacuolated cells.

Since we may regard the cell-wall and protoplasmic layer of the vacuolated cell as a membrane separating the cell interior from the surrounding medium, it will be as well if we first consider the laws which have been found to operate in the case of the penetration of dissolved substances through isolated non-living membranes, and further, since in both the cases of such non-living membranes and of the cell, the relationship between the penetration of such substances through the membranes and their diffusion through water or solutions unimpeded by such membranes must arise, the discussion of the permeability of membranes will be preceded by a brief summary of the more important facts of the diffusion of dissolved substances in water.

HYDRO-DIFFUSION

The diffusion of dissolved substances through water was first subjected to systematic investigation by Thomas Graham in 1850, who showed that the rate at which a substance diffused through unit cross-section of a column of water depended on the nature of the substance, its concentration and the temperature. The relationship between these quantities was expressed mathematically by Fick in 1855 by the equation

$$dS = -D\frac{dc}{dx}dt$$

where dS is the quantity of dissolved substance passing through unit cross-section of liquid in a time dt at a point x where $\frac{dc}{dx}$ is the concentration gradient (that is, the change of concentration with distance). The value of D depends on the nature of the substance and varies somewhat with the temperature and concentration. Its value at a defined temperature and concentration is called the diffusion constant, diffusivity, or coefficient of diffusion of the substance at that temperature and concentration. The more rapidly a substance diffuses the larger is the value of D, for inspection of Fick's equation will reveal the fact that the coefficient of diffusion is the quantity of substance diffusing across unit cross-section in unit time when the concentration gradient is unity (that is, when two cross-sections of the liquid at unit distance apart in the direction of the line of diffusion differ by unity in their concentrations). This law has been experimentally verified many times.

The diffusivity of a solute in water thus depends on (1) the nature of the substance, (2) its concentration and (3) the temperature. It should also be noted that the diffusivity will not necessarily be the same in different solvents, and if a substance is diffusing, not through pure water, but through a solution of some other substance, the diffusivity is reduced. We can understand that the presence of additional molecules in the medium results in further resistance to the movement of diffusing solute molecules. As regards the nature of the solute, some values for the coefficients of diffusion of a number of substances at temperatures mostly in the neighbourhood of 10° C. are collected in Table VI. It will be observed that the strong mineral acids and the strong alkalies are the most rapidly diffusing substances while salts of monovalent metals possess as a rule higher coefficients of diffusion than do those of divalent metals. Sulphates diffuse less rapidly than the corresponding chlorides and nitrates. Organic substances diffuse more slowly than the simpler salts and the smaller the molecule the more rapid the diffusion, while colloidal substances such as albumin diffuse very slowly.

The concentration of the diffusing substance has a slight, though definite, effect on its coefficient of diffusion. On the whole the diffusivity of electrolytes decreases slightly with increasing concentration, a state of affairs partly attributable to decrease in the degree of ionization with increase in concentration and greater mobility of the constituent ions as compared with that of the undissociated molecule. The extent of the influence of concentration on diffusivity will become clear from an inspection of Table VII in which are summarized some observations of Öholm on this subject.

Temperature also has a definite effect on the coefficient of diffusion.

Table VI

Coefficients of Diffusion in Water of Various Substances

Substance	Concentration in gm. mols. per litre	Temperature in Centigrade degrees	Coefficient of Diffusion in sq. cm. per sec. × 10^{-5}	Observer
Hydrochloric acid	0·02	18	2·64	Öholm
Nitric acid	0·02	19·5	2·45	Thovert
Sulphuric acid	0·005	18	1·51	,,
Potassium hydroxide	0·02	18	2·19	Öholm
Potassium chloride	0·02	18	1·66	,,
Potassium nitrate	0·02	17·6	1·48	Thovert
Potassium carbonate	—	18	0·76	De Heen
Potassium sulphate	0·95	19·6	0·92	Thovert
Sodium chloride	0·02	18	1·33	Öholm
Magnesium sulphate	—	18	0·42	De Heen
Ethyl alcohol	0·05	11	0·85	Thovert
Glycerol	0·125	10·14	0·41	Heimbrodt
Sucrose	0·97	18·5	0·28	Thovert
Albumin	—	13	0·073	Graham-Stefan

Table VII

Effect of Concentration of a Solute on its Diffusivity

Concentration in gm. mols. per litre	Coefficient of Diffusion at 18° C. in sq. cm. per sec. × 10^{-5}			
	HCl	KOH	KCl	NaCl
0·01 . .	2·69	2·20	1·69	1·35
0·02 . .	2·64	2·19	1·66	1·33
0·05 . .	2·61	2·17	1·63	1·32
0·10 . .	2·58	2·15	1·61	1·29
0·20 . .	2·55	2·13	1·58	1·27
1·00 . .	2·57	2·15	1·54	1·24
2·00 . .	—	2·19	1·53	—
2·8 . .	—	—	—	1·23
3·6 . .	—	—	1·55	—
5·5 . .	—	—	—	1·23

With rise in temperature the diffusivity increases, and temperature coefficients in relation to diffusivity have been determined for a variety of substances by a number of different workers. Calculations, based on these temperature coefficients, of the diffusivity in water of a number of substances at 0° and 40° C. are shown in Table VIII and indicate the extent to which diffusivity is likely to be affected by temperature in most living plants. It will be observed that with electrolytes the effect of a rise of as much as 40° C. results in the diffusivity increasing from 2 to 3 times. With sugars temperature appears to have a somewhat greater effect than with salts.

Table VIII

Diffusivity of Various Substances at 0° and 40°

Substance	Concentration in gm. mol. per litre	Coefficient of Diffusion in sq. cm. per sec. × 10^{-5}	
		0°	40°
Hydrochloric acid	0·02	1·74	3·74
Potassium chloride	0·02	0·98	2·56
Potassium nitrate	—	0·79	2·29
Potassium carbonate	—	0·39	1·22
Sodium chloride	0·02	0·74	2·06
Sodium hydrogen phosphate . .	—	0·48	1·53
Magnesium sulphate	—	0·24	0·65
Sucrose	0·97	0·12	0·48

As mentioned earlier, the diffusivity of a substance is lowered by the presence in the solvent of another dissolved substance, at any rate if this is a non-electrolyte. This was shown by Öholm in the case of potassium chloride diffusing through solutions of glycerol and sucrose of different concentrations. Thus the diffusivity of this salt in a 2 M solution of sucrose was found to be $0{\cdot}255 \times 10^{-5}$ sq. cm. per second in pure water. In a strong solution of glycerol, one of 7·48 M, the diffusivity was only $0{\cdot}201 \times 10^{-5}$ sq. cm. per second.

Diffusion in gels, such as those of gelatin and agar-agar is also slower than in pure water, but the retardation in diffusion brought about by the presence of the gel is, perhaps, not so great as might be expected. The results obtained by Stiles and Adair on the influence of the concentration of gels of agar-agar on the coefficient of diffusion of sodium chloride are shown in Table IX. From the numbers given here it will be observed that the influence of temperature on the rate of diffusion of sodium chloride in a gel is much the same as with diffusion in water.

Table IX

Coefficient of Diffusion of Sodium Chloride in Gels of Agar-agar of Different Concentrations at Various Temperatures

Concentration of agar-agar in per cent.	Coefficient of Diffusion in sq. cm. per sec. $\times 10^{-5}$		
	0·0°	20·1°	40·2°
0·5 . .	0·781	1·386	—
1·0 . .	0·765	1·357	2·143
2·0 . .	0·738	1·328	2·084
4·0 . .	0·711	1·290	2·061

PARTITION COEFFICIENTS

So far we have considered the movement of dissolved substances through a still aqueous medium. If the latter is shaken the mixing of molecules of solute and solvent soon brings about a uniform distribution of solute throughout the solvent and a state of equilibrium results, which is only slowly attained by pure diffusion alone.

Also we have only considered the movement of dissolved substances in a medium which is either homogeneous or in which the heterogeneity is of the colloidal type. But if we have a heterogeneous system consisting of two liquid phases, that is two liquids which do not mix, such as olive oil and water, in both of which a particular substance is soluble, its distribution throughout the whole system when equilibrium is reached will not be uniform, but the concentrations of the substance in the two phases will bear a constant relation to one another. Thus, if the system consists of two liquids with one solute which when equilibrium is reached possesses a concentration of C_1 in one of the liquids and C_2 in the other, then

$$\frac{C_1}{C_2} = K$$

where K is a constant which is called the *partition coefficient*, and which is independent of the presence of other solutes. Should the molecules of a solute undergo aggregation or polymerization in one of the liquids the partition law becomes

$$\frac{C_1}{C_2^{\frac{1}{n}}} = K$$

where n is the number of molecules associated together in each aggregate.

ADSORPTION

The phenomenon of adsorption may play an important part in the relationships of dissolved substances in plant cells. At the surface of separation of two phases a dissolved substance will accumulate if the substance tends to lower the surface tension at the interface, the accumulation being spoken of as adsorption. If a substance in a fine state of division is dispersed through a liquid, molecules of a solute may be adsorbed at the surface of the particles of the dispersed phase which is thus an adsorbent. The adsorption equation, giving the distribution at equilibrium of a solute between a liquid and an adsorbent is generally written in the form

$$\frac{x}{m} = KC^{\frac{1}{n}}$$

where x is the amount of solute adsorbed to a quantity m of adsorbent, C the concentration of the solute in the liquid at equilibrium, while K and n are constants. Now $\frac{x}{m}$ is of the nature of a concentration, and it will be observed that if we write C_1 for $\frac{x}{m}$ and C_2 for C in the adsorption equation, the latter becomes identical with the equation representing the partition law when the solute undergoes aggregation in one of the solvents.

In addition to purely mechanical adsorption attributable to the tendency to lower surface tension, adsorption may also occur on account of electrical conditions. Thus, if a surface possesses a negative charge there will be a tendency towards the deposition on the surface of any particle carrying a positive charge, a deposition of this kind being spoken of as electrical adsorption.

THE PERMEABILITY OF MEMBRANES

By the term membrane, as used in biology, is meant a thin layer of either solid or liquid material. Membranes may be prepared of a large number of non-living materials, and a large number of observations on the permeability of dissolved substances through such membranes have been made in the hope that such observations might throw light on the mechanism of the permeability of living cells. The permeability of many different kinds of membranes has been examined, but two types in particular have formed the subject of such investigations, namely, membranes of collodion and the precipitation membranes of which that of copper ferrocyanide is the best known.

Collodion membranes have proved particularly useful for work on permeability because it is possible to obtain them with widely varying degrees of permeability. The method employed by W. Brown for preparing such membranes consists in immersing air-dried membranes

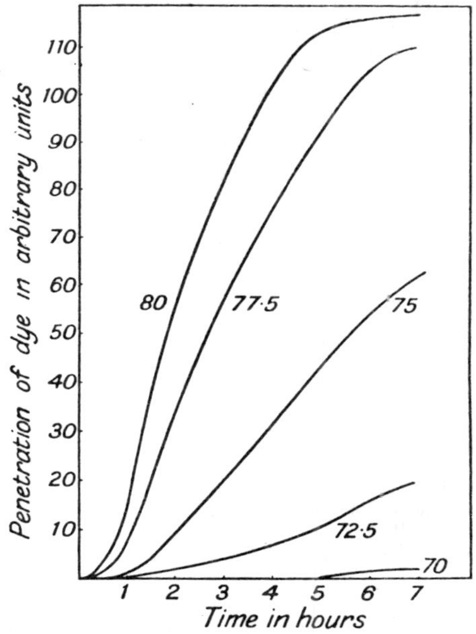

Fig. 5.—Curves showing the rate of passage of methylene blue through membranes of different permeability

(Constructed from the data of W. Brown)

of collodion, which have a very low permeability, in solutions of alcohol in water for a time and then immersing the membranes in water. The permeability of collodion membranes treated in this way increases with increase in the concentration of the alcohol solution employed. Thus in Fig. 5 are shown curves, constructed from data published by Brown, showing the different rates at which methylene blue penetrates through membranes graded in alcohol solutions of five different concentrations (70, 72·5, 75, 77·5 and 80 per cent.). A membrane graded in a solution of ethyl alcohol of lower concentration than 70 per cent. is impermeable to methylene blue. Since usable membranes can be obtained with alcohol solutions ranging in concentration from 0 to 97 per cent. a very wide range of permeabilities can be obtained with such membranes.

Brown obtained an indication of the relative penetrability of various substances through collodion membranes by determining the strength of alcohol in which the collodion had to be graded in order to produce a membrane which permitted the passage through it of each substance examined. He defined the *alcohol index* of a substance as the number representing the concentration of alcohol in per cent. in which the collodion had to be graded in order just to prevent the diffusion of the substance. The higher the alcohol index, the less is the penetrability of the substance through collodion membranes. The alcohol indices of a number of substances are shown in Table X; the higher the substance in this table, the greater its penetrability.

Table X

Alcohol Indices of a Number of Substances

Substance	Alcohol Index
Sodium chloride	0
Potassium permanganate	30 –40
Picric acid	35 –40
Copper sulphate	60 –70
Potassium oxalate	60 –70
Sodium sulphate	60 –70
Bismark Brown	65 –
Methylene Blue	70 –
Neutral Red	72·5–75
Safranin	75 –77·5
Dextrin	85 –87·5
Starch	90
Aniline Blue	92
Litmus (neutral)	93
Congo Red	96
Night Blue	96

More recently the penetration of a number of substances through such collodion membranes has been directly measured by Collander. This worker used membranes of three different degrees of permeability obtained by grading in 0, 68 and 80 per cent. alcohol solutions respectively. By measuring the rate of passage of the various substances through these membranes under known differences of concentration of the substance he obtained values for the permeability of the different membranes to the different substances. When these values are compared with those of the size of the molecule as indicated by the molecular refraction, it appears that in the case of non-electrolytes or slightly dissociated electrolytes the size of the molecule is the dominating factor in determining the penetrability of the substance. An inspection of Collander's results, summarized in Table XI, shows that very generally the permeability to the various substances examined decreases with increase in molecular size. Deviations from the rule are for the most part slight and are confined to a few acids (acetic, glycollic, propionic, monochloracetic and valeric) and to phenol and *m*-nitrophenol, while with some of the acids the deviation is slight enough to be accounted for by experimental error.

Table XI

Permeability of Collodion Membranes to Various Substances

Substance	Molecular refraction (proportional to molecular volume)	Relative permeability of Membrane graded in		
		0% alcohol	68% alcohol	80% alcohol
Ammonia	5·78	25·0	75·0	100·0
Hydrogen peroxide	5·92	10·5	—	—
Methyl alcohol	8·34	8·2	—	—
Formic acid	8·57	7·3	41·4	86·4
Formamide	10·61	6·0	—	—
Ethyl alcohol	12·78	1·6	—	—
Acetic acid	13·02	1·8	20·1	—
Urea	13·67	1·0	—	—
Ethylene glycol	14·40	0·5	—	—
Glycollic acid	14·50	0·6	10·8	37·4
Propionic acid	17·55	1·1	15·4	—
Monochloracetic acid	17·84	1·4	18·8	—
Malonic acid	19·13	0·2	7·3	—
Lactic acid	19·19	0·2	4·3	30·3
Isobutyl alcohol	22·19	c. 0·2	—	—
Pyruvic acid	23·74	0·1	6·2	—
Malic acid	25·27	0·0	1·9	18·2
Isoamyl alcohol	26·74	c. 0·1	—	—
Tartaric acid	26·79	0·0	0·7	9·6
Valeric acid	26·85	c. 0·1	5·2	—
Phenol	27·95	3·2	—	—
Isovaleramide	28·72	0·0	—	—
Methyl-ethyl-malonic acid	32·98	0·0	0·6	11·7
m-Nitrophenol	34·03	0·7	—	—
Citric acid	36·04	0·0	0·2	6·0
Mannitol	39·06	0·0	—	—
Quinic acid	39·96	0·0	0·2	5·6
Antipyrin	55·37	0·0	—	—
Sucrose	70·35	0·0	—	1·6

These results strongly suggest that the collodion membrane can be regarded as a sieve through the meshes of which the molecules of the diffusing substance pass. The meshes may actually be water or an aqueous solution forming pores or tube-like channels in a collodion framework. These pores are likely to vary in size, so that the larger the diffusing molecules the smaller will be the effective area of membrane through which they can pass, for the molecule can only pass through a pore provided the diameter of the latter is greater than that of the molecule of solute. The exceptions to the rule can be accounted for, partly as we have seen, by experimental error, partly by error involved in assuming the volume of a molecule of a solute in water to be directly proportional to the molecular refraction in all cases, but partly also perhaps in some cases by substances diffusing through the collodion phase of the membrane as well as through the pores, which might well be the explanation of the abnormally high permeability of the membranes to phenol and *m*-nitrophenol.

The permeability of collodion membranes to electrolytes appears to be a more complex matter. Michaelis found that the slightly permeable air-dried collodion membrane was permeable for univalent kations, but practically impermeable to all anions and to multivalent kations. In such cases movement of kations through the membrane involves a movement of hydrogen ions in the reverse direction from the water on the far side of the membrane. The order of penetrability of the univalent ions is the same as that of the mobility of the ions when diffusing in water, but the relative differences between the mobilities are much greater in the case of diffusion through the membrane than in that of free diffusion in water. Michaelis concluded that the diffusion of the ions through the membrane is partly determined by the size of the pores in the membrane, but that the movement of the ions is complicated by electrical phenomena. The walls of the pores or tubes in the membrane are supposed to carry a negative charge and this repels the negatively charged anions which are therefore unable to penetrate the membrane. Further, the charge on an ion holds bound to it by electrostatic attraction some surrounding water molecules, with the consequence that the size of the moving particle is increased and a retardation of the movement of this particle must be brought about if it comes near enough to the pore-wall for adhesion to become operative.

Collander's observations agreed with those of Michaelis. His results dealing with the penetration of some strongly dissociated acids are shown in Table XII where the permeability of the membranes to ammonia is also shown for the sake of comparison. The low permeability of the membranes to the acids is attributed to the slow penetration of the anion as explained by Michaelis.

Table XII

Permeability of Collodion Membranes to Acids

Substance	Relative permeability of membrane graded in		
	0% alcohol	68% alcohol	80% alcoho
Ammonia	25·0	75·0	100·0
Nitric acid	0·6	7·2	—
Hydrochloric acid	0·1	2·6	—
Phosphoric acid	0·0	0·5	6·6
Sulphuric acid	0·0	0·2	3·3

Precipitation membranes can be conveniently prepared of a number of ferrocyanides, ferricyanides and cobalticyanides by precipitating them in the walls of a porous pot. Thus the copper ferrocyanide membrane is formed when a solution of copper sulphate is separated from one of potassium ferrocyanide by a porous pot, the solutions meeting in the wall of the pot, where the copper ferrocyanide is deposited in the form of a gel.

The various precipitation membranes vary in their permeability, one of gelatin-tannic acid being the most permeable and the copper ferrocyanide membrane the least permeable, while each membrane exhibits different degrees of permeability to different substances. Thus, in 1867 Traube found the copper ferrocyanide membrane permeable to potassium, sodium and ammonium chlorides, but impermeable to calcium and barium chlorides, potassium and ammonium sulphates and barium nitrate. Although the results of later investigations by various workers exhibit differences in detail, certain conclusions may be drawn regarding the permeability of precipitation membranes to dissolved substances. Thus from the very extensive series of observations published in 1892 by Walden which dealt with experiments made with a variety of membranes and a considerable number of diffusing solutes, it appeared that although the membranes used varied in their permeability, yet the order of permeability of the various diffusing substances was the same in each case. Potassium and ammonium chlorides, bromides and iodides passed through the membranes most easily, and in general, acids and salts containing at least one univalent ion passed through the membranes more readily than those with divalent ions. Thus, molecular size as indicated by the number of atoms in the molecule does not appear to be the only factor determining the penetrability of the substance, for sodium acetate and sodium oxalate have the same number of atoms in the molecule, but precipitation membranes are more permeable to sodium acetate than to sodium oxalate. Similarly, these membranes are more permeable to sodium thiosulphate than to sodium sulphate which contains the same number of atoms as the thiosulphate. Thus Walden concluded that the nature and arrangement of the atoms in the molecule, the chemical constitution, that is, are more important in determining the permeability than the size of the molecule as indicated by the number of atoms in it.

Tammann's observations also published in 1892 on the permeability of three different precipitation membranes to 17 dyes showed certain irregularities in the order of penetration of the different dyes through the various membranes. Thus as regards these dyes he found the gelatin-tannic acid membrane was permeable to eleven, the zinc ferrocyanide membrane to seven and the copper ferrocyanide membrane to five, so that the last-named membrane was the least permeable and the gelatin-tannic acid membrane the most permeable of the three. Yet fuchsin chloride was able to penetrate the copper ferrocyanide membrane but not that of zinc ferrocyanide, while cotton blue was able to penetrate the zinc ferrocyanide membrane but not that of gelatin-tannic acid.

In more recent years the relation between the nature of a substance and its penetrability through the copper ferrocyanide precipitation membrane has been investigated by Collander, his results appearing in the years 1923 and 1924. Using a considerable number of organic substances he determined the quantity of each diffusing through the membrane in a definite time. With non-electrolytes and organic acids Collander found the same general relationship between the molecular volume and penetration through the membrane as with collodion membranes; the smaller the molecule, the more rapidly it diffused through the membrane. Here also there were a few minor deviations, but for the most part the results obtained can be explained on the view that the copper ferrocyanide membrane, like collodion membranes, acts as a molecular sieve or ultra-filter. Indeed, there is here a moderately sharp division between substances which diffuse readily and those which penetrate with difficulty, which would suggest a rather uniform size of pore or capillary in the membrane. Thus, isopropyl alcohol, α-propylene glycol and glycerol with molecular volumes of 82·0, 85·2 and 87·0 respectively in the units employed by Collander, passed through the membrane readily, whereas phenol, isobutyl alcohol and ethyl acetate with molecular volumes of 101·8, 102·0 and 106·0 respectively, passed through very slowly.

The permeability of collodion and precipitation membranes to solutes is thus very largely explicable on the view that the membranes behave like sieves, molecules or molecular aggregates of the solutes passing through the membranes so long as the pores or tubes in the membrane are large enough to allow the passage through them of the diffusing particles. But it is clear that all membranes are not of this kind. Thus a film of oil or paraffin on the surface of water can be regarded as a membrane, and this must be regarded as a homogeneous layer, the only discontinuity being that involved in the arrangement of the molecules. Diffusion of a substance through a membrane of this kind involves the passage of molecules of the solution between the molecules of the material of the membrane, but this simply amounts to solution of the solute in the membrane. A substance might also pass through a homogeneous membrane if its molecules entered into some form of association with those of the membrane, either by chemical combination or adsorption, the association being reversible so that on the far side of the membrane the chemical or adsorption compound breaks down and the solute is released.

Hence in the case of a single-phase membrane such as a film of oil, permeability will be a function of the solubility of the diffusing substance in the membrane or of the combination of the molecules of solute and membrane in some kind of reversible association. When the membrane involves two or more phases, either in the form of a colloid or of a coarser structure, the permeability may be determined by the size of the pores which constitute one of the phases. Even in heterogeneous membranes of this kind a diffusing substance may also be more or less soluble in the major phase of the membrane, in which case it will diffuse in solution through that phase as through a homogeneous membrane. The possibility of transference of the substance through the membrane by the formation of a reversible compound of substance and membrane material is still possible.

There are thus three theories of membrane permeability. The solution and combination theories are particularly applicable to the case of homogeneous single-phase membranes while the third theory, the sieve or ultra-filtration theory applies only to two-phase membranes such as those of collodion and the precipitation membranes. With such membranes, however, a combination of ultra-filtration through the pores or capillaries of the membrane with solution in the more voluminous phase of the membrane, or with combination between diffusing substance and membrane, is always a possibility. As will be seen later, all three theories have been invoked to account for the permeability properties of plant cells.

DONNAN EQUILIBRIUM

When a membrane separates pure water from an aqueous solution of a substance to which the membrane is permeable, the substance will diffuse through the membrane until there is equality of concentration of the substance on the two sides of the membrane. Similarly, if there are in the system two dissolved substances, non-electrolytes, to which the membrane is completely permeable, the distribution of these at equilibrium is such that there is equality of concentration of each substance on the two sides of the membrane. In the case of two electrolytes without a common ion the same is the case; at equilibrium the concentration of each of the constituent ions and of each non-ionized molecule will be respectively the same on the two sides of the membrane. But if the two electrolytes possess a common ion to which the membrane is permeable, while the other ion can penetrate the membrane in the case of one substance, but not of the other, the distribution of the various ions at equilibrium does not correspond to equal concentration of each ion on the two sides of the membrane.

Thus, let us suppose that we have on one side of the membrane unit volume of a solution of sodium chloride to both ions of which the membrane is permeable, and on the other side of the membrane the same volume of a solution of another sodium salt to the anion of which the membrane is impermeable; such might be a dye consisting of a sodium salt of an organic acid yielding a large colloidal anion, or a sodium proteinate of some kind. If we represent the indiffusible anion by the symbol An, and if, to simplify the case, we assume that both electrolytes are completely ionized and monovalent, then before the commencement of diffusion the system can be represented thus:

$$\begin{array}{c|c} C_1(Na^+) & C_2(Na^+) \\ C_1(Cl^-) & C_2(An^-) \end{array}$$

where C_1 represents the concentration of the ions of sodium chloride and C_2 that of the salt with the indiffusible anion. Since the ion An^- cannot pass through the membrane diffusion is limited to the sodium chloride, which will pass through the membrane until equilibrium is reached.

When equilibrium is established a certain amount of sodium chloride will have diffused through the membrane and the system can now be represented thus :

$(C_1 - x)(Na+)$	$(C_2 + x)(Na+)$
$(C_1 - x)(Cl-)$	$x(Cl-)$
	$C_2(An-)$

Now ions can only penetrate the membrane under two conditions. Either there must be an interchange of equal numbers of ions of the same sign between the two sides of the membrane, or an equal number of ions of opposite sign must pass through the membrane together. That means that at equilibrium the number of Na^+ and Cl^- ions which strike the membrane at the same time during unit time on the one side must equal the number of Na^+ and Cl^- ions which strike the membrane together during unit time on the other side. Now the number of pairs of ions which strike the membrane together in unit time is proportional to the products of the concentrations. Hence at equilibrium we shall have the relation

$$(C_1 - x)^2 = x(C_2 - x)$$

where C_1 is the original concentration of sodium chloride and C_2 the original concentration of the electrolyte with the indiffusible anion and x the quantity of sodium chloride which diffuses through the membrane, and which equals numerically the change in concentration since unit volume of solution is assumed to be present on each side of the membrane. It is clear that at equilibrium the concentration of the diffusing salt will be less in the solution on the side of the membrane containing the indiffusible ion than on the side containing the diffusible salt only. For if this were not so and the concentration of sodium chloride were the same on the two sides, the product of the sodium and chloride ions on the side containing the indiffusible anion would be greater than on the side containing only the sodium chloride.

How uneven the distribution of the permeable salt may be, depends on the relative concentrations of the diffusible and indiffusible salts, the weaker the concentration of the former in relation to the latter, the further the conditions at equilibrium are removed from equality of concentration. This is illustrated by the numbers given in the following table, it being assumed that both electrolytes are practically wholly dissociated into their constituent ions. The numbers are, of course, equally applicable to other substances.

Table XIII

Concentrations at Equilibrium of Sodium Chloride on two sides of a Membrane if a Sodium Salt with an indiffusible Anion is present on one side

Ratio of Concentrations $\frac{NaAn}{NaCl}$	Percentage of Sodium Chloride on two sides of the membrane: On side with indiffusible Anion	On side without indiffusible Anion
0·01	49·8	50·2
0·1	47·6	52·4
1·0	33	67
10·0	8·3	91·7
100	1	99

If instead of electrolytes with a common ion we have to do with a membrane separating two electrolytes with no ion in common, as, for example, potassium chloride in which both ions can penetrate the membrane and a sodium salt with an indiffusible anion as in the case just considered, the question of the equilibrium is more complicated. Since the K^+ Na^+ and Cl^- ions can all penetrate the membrane, at equilibrium there will be a certain concentration of each of these on either side of the membrane. Not only does potassium chloride diffuse through the membrane, but the sodium ion will diffuse through in the opposite direction, so that there is a certain exchange of kations between the solutions on the two sides of the membrane. Here also the concentrations of each ion on the two sides of the membrane are not the same, but the products of the concentrations of each pair of oppositely charged diffusible ions (K^+ and Cl^-, and Na^+ and Cl^- in the example taken) are the same on the two sides of the membrane. The actual relative concentrations of the various ions on the two sides of the membrane for this case of two electrolytes without a common ion, separated by a membrane impermeable to one ion of the four but permeable to the other three, are summarized in Table XIV.

Table XIV

Donnan Equilibrium in the case of Two Electrolytes without a common Ion

Ratio of Concentrations $\frac{NaAn}{KCl}$	Percentage of various Ions on the two sides					
	On side with indiffusible Anion			On side without indiffusible Anion		
	K	Na	Cl	K	Na	Cl
0·1 . .	50	50	50	50	50	50
1 . .	67	67	33	33	33	67
10 . .	90	92	10	10	8	90
100 . .	99	99	1	1	1	99

It will also be noted that, since these considerations refer to equilibrium conditions it is immaterial whether at the beginning of diffusion the completely diffusible electrolyte is on the same side of the membrane as the electrolyte with the indiffusible ion, or on the other side; the same position of equilibrium will be reached in both cases.

The important point emerges, therefore, that when a membrane separates two electrolytes, one of which possesses one ion to which the membrane is impermeable, there will generally result an unequal distribution of ions on the two sides of the membrane. Thus, diffusion may cease long before equality of concentration is reached; in other cases, as an inspection of Table XIV shows, the diffusion may proceed very much beyond the position corresponding to equality of concentration, so that there is a movement of the ion up its concentration gradient and an accumulation of the ion on the far side of the membrane.

THE ENTRANCE OF DISSOLVED SUBSTANCES INTO PLANT CELLS

With these considerations of membrane permeability in mind we are in a better position for examining the question of cell permeability. We may note in the first place that the passage of solutes into the cell is independent of the passage of water, entrance of both proceeding towards their own positions of equilibrium. If the solute is continually removed in some way, the position of equilibrium may never be reached, and for this reason the absorption of a solute by a portion of tissue attached to a plant may differ from the observed absorption by the same tissue when isolated from the rest of the plant.

The rate of entrance of a solute into a cell depends on the concentration of the solute outside and inside the cell and on the permeability of the cell membranes to the substance. It may also be influenced by temperature and possibly by other factors of the environment such as light and the presence of other substances. It has already been noted how, in the case of electrolytes, the matter may be complicated by the presence of electrolytes with non-diffusible ions, so that it cannot be assumed that the rate of entry of an ion into a cell is proportional to the difference in its concentration on the two sides of the cell membranes. Also, removal of the solute inside the cell by chemical combination or adsorption would lead to an accumulation of the substance and its entrance, apparently, though not actually, against its own concentration gradient.

It will thus be clear that the rate of absorption of a substance by a cell must not be regarded as a measure of the permeability of the cell membranes. These latter may be equally permeable to two substances, but if one is accumulated by the cell and the other not, both the total amount of substance absorbed and, for the greater part of the time, the rate of its entrance into the cell may be very different in the two cases.

Unless definitely stated otherwise, the term permeability of the cell used here refers to the permeability of all the membranes (cell-wall and protoplasm) which separate the sap in the vacuole from liquid outside the cell. The permeability of the cell to a substance may be quantitatively defined as the amount of the substance passing through unit area of the cell membrane in unit time under unit difference of concentration of the substance on the two sides of the membrane, provided that the complications introduced by the conditions bringing about a Donnan equilibrium are not operative.

THE QUALITATIVE DETERMINATION OF PERMEABILITY TO SOLUTES

The quantitative measurement of cell permeability is frequently fraught with many difficulties, but the mere qualitative determination of permeability is often quite simple. Thus the entry of a dye into the cell may be proved by the staining of the cell contents, while some cells contain indicators which change colour on the entry of an acid or alkali. The pigment, for instance, present in the cells of the root of red beet changes in colour from red to blue when a piece of the root is placed in a solution of ammonia, thus indicating that the cells are permeable to that substance. In many cases again the cells contain a substance in solution which reacts with an entering substance to produce a precipitate or give a particular coloration. Thus Osterhout found that in the root hairs of seedlings of *Dianthus barbatus* crystals of calcium oxalate are formed when the roots are immersed in a solution of a calcium salt, but not otherwise. Again, the entrance of iron salts into cells containing tannin is rendered obvious by the production of a blue colour. The absorption of copper salts by a tree of *Quercus macrocarpa* was observed by MacDougall owing to an accumulation of metallic copper in the cells. The formation of starch which takes place when leaves are floated on solutions of sugar or glycerol in the dark is likewise evidence that the cells must be permeable to these substances.

Even when the penetration of the substance into the cell is not accompanied by any visible change, its entrance may often be demonstrated by the application of a microchemical test. Thus the penetration of potassium nitrate or of one or other of its constituent ions has been demonstrated in many different kinds of plant cells, by testing with diphenylamine, which gives a blue coloration with nitrate, or with platinic chloride which yields under certain conditions a precipitate of potassium chloroplatinate.

Not only may a visible change in the cell interior indicate the entrance of a substance into the cell, but a visible change in the external medium may do so as well. The absorption of a dye or some other substance which yields a coloured solution may be made obvious by the lessening of the depth of colour of such a solution when brought into contact with plant tissue. While this behaviour indicates absorption of the dissolved substance by the tissue it does not prove that the cell membranes are permeable to the substance, for the solute may be retained by the membranes and not reach the interior of the cell at all. The diffusion of a substance out from the cell may sometimes be observed by visible changes produced in the external medium. Thus, the diffusion of a pigment from the cell-sap to the external medium might be made plain in this way, and would indicate the permeability of the cell membranes to the pigment. The diffusion of an acid or alkali into the external medium might be shown by the addition to the latter of an indicator, or a precipitate might be produced by a substance diffusing out from the cell interior if a suitable precipitant is present in the external solution.

Recovery from plasmolysis may also form an indication of permeability of a living cell to a solute. If a cell is immersed in a strong enough solution of a solute it will become plasmolysed, but if the cell membranes are permeable to the solute, as the latter enters the cell the osmotic pressure of the cell-sap will increase, provided the solute does not enter into any association with materials in the cell resulting in the withdrawal of the solute from osmotic activity. As the osmotic pressure of the cell increases water will enter from the external medium, so that there is a gradual recovery from plasmolysis. Hence deplasmolysis can be used as evidence of the permeability of the cell membranes to a solute.

THE QUANTITATIVE MEASUREMENT OF THE INTAKE OF DISSOLVED SUBSTANCES BY PLANT CELLS

The measurement of the amount of dissolved substance absorbed by plant cells, or of the rate of entrance of a dissolved substance into plant cells, has been made by a variety of ways, but all depend either on determining the loss of material from the external solution or the gain in the substance by the cell.

The loss of material from the external solution may be determined by chemical analysis of the solution, a method which has been much used. In the case of coloured solutions the absorption of the substance has frequently been followed and measured quantitatively by colorimetric determinations by means of a colorimeter. The absorption of hydrogen ions or hydroxyl ions may be measured by determining the hydrogen-ion concentration either by the colorimetric indicator method or by the use of the hydrogen electrode. In the case of acids the quinhydrone electrode is now found to be the most generally satisfactory for this purpose. The absorption of electrolytes can be followed by observing the change in electrical conductivity of the external solution, but with this method a complication is introduced by the exosmosis of electrolytes from the cells, as well as by possible changes in the degree of ionic dissociation of the electrolyte with change in its concentration, a factor which may be of significance if the change in concentration of the electrolyte is considerable.

Similar methods to these can be used for determining the gain in the quantity of a substance by the cells, the cell-sap being pressed out and then analysed chemically or colorimetrically, or having its hydrogen-ion concentration or electrical conductivity measured. Where the cells are large enough, determinations can be made on single cells, and this has been done in cases of *Valonia*, *Chara*, and *Nitella*. Many other observations have been made with masses of cells in the form of pieces of tissue.

The disadvantage of methods for measuring solute intake which involve analysing the cell-sap is that only one determination can be made on one cell or sample of tissue. The analysis of the external solution, on the other hand, often allows the course of absorption to be followed with one sample of tissue. This is the case where small samples of the liquid can be removed from time to time, or where the fall in the electrical conductivity is used as the criterion of absorption, since the measurements can be made without removal of any of the liquid external to the cells. It should, however, be noted that this type of method only provides information regarding the absorption of material; it does not tell us whether the substance is absorbed by cell-wall, protoplasm or vacuole or whether it is distributed between all three phases of the cell. The same drawback applies, though to a less degree, to methods depending on the analysis of cell-sap. Here the separation of cell-wall from the cell interior may be complete, but the separation of protoplasm and vacuole, though attempted, is difficult of achievement.

Much use has been made of the phenomenon of deplasmolysis for measuring the intake of substance into the cell. As already pointed out, the recovery of a cell from plasmolysis indicates the entry of a dissolved substance from the external solution into the cells.

A number of attempts have been made to make the method quantitative, of which special mention may be made of those of Fitting and Höfler. Fitting plasmolysed similar cells of *Rhoeo discolor* in solutions of a salt of different concentrations. The stronger the solution, the greater the degree of plasmolysis, so that as the cells in a stronger solution deplasmolyse owing to entry of the salt, a stage will be reached in which the degree of plasmolysis is the same as that produced at first by a weaker solution. During the time that elapses between the commencement of deplasmolysis in the stronger solution and the arrival at this stage corresponding to the degree of plasmolysis in the weaker solution, a quantity of salt must have entered the cells sufficient to raise the concentration of salt in the cells by the difference in concentration between the stronger and weaker solutions. A little consideration will be sufficient to make this clear. If a plasmolysing solution of concentration X reduces the volume of the protoplast to V_x and one of a weaker concentration Y reduces the volume of the protoplast to V_y, and if C represents the quantity of osmotic material in the cell, we have, before the entry of any solute into the cell, equality of osmotic concentration inside and out, whence

$$X = \frac{C}{V_x} \text{ and } Y = \frac{C}{V_y}$$

or

$$C = XV_x = YV_y.$$

When deplasmolysis in the stronger solution reaches the point where the volume of the cell is V_y, then similarly

$$C + m = XV_y.$$

where m represents the quantity of solute which has entered the cell.

Consequently

$$m = V_y(X - Y)$$

or

$$\frac{m}{V_y} = X - Y$$

that is, the increase in concentration of solute in the cell is equal to the difference between the two plasmolysing concentrations.

A difficulty lies in determining exactly the degree of plasmolysis. Fitting made use of the fact that in a piece of tissue the cells do not all plasmolyse in the same concentration, and he took as a measure of the degree of plasmolysis the proportion of the cells in a preparation which are plasmolysed, such as one-third, one-half and two-thirds. Thus, with a concentration of say X, two-thirds of the cells of a preparation are plasmolysed while with a lower concentration Y only one-half of the cells plasmolyse. As the solute enters the cells surrounded by the solution X those plasmolysed gradually deplasmolyse until a stage will be reached when only one-half of the cells of the preparation are plasmolysed. Enough solute must then have entered the cells to raise their concentration by the difference between X and Y.

Höfler's *plasmometric* method makes use of the same principle, but the degree of plasmolysis is determined by actual measurements of the size of the cells, and the rate of deplasmolysis is followed by means of measurements made on individual cells. The amount of solute which enters the cell over any period of time during deplasmolysis is obviously equal to the product of the concentration of the external solution and the increase in volume during that time.

It will be observed that plasmolytic methods of measuring the intake of substances into the living cell differ from the methods previously considered in that they involve the assumption that the absorbed solute remains osmotically active when it reaches the cell interior. The method further assumes that exosmosis of osmotically active material from the cell does not take place to any significant extent. In all probability these assumptions are justified in some cases and not in others.

Where the assumptions just mentioned are justified, it will be observed that the method of deplasmolysis gives data not only with regard to absorption, but also with regard to the actual permeability of the cell membranes. This question will be considered later in this chapter.

THE ABSORPTION OF ELECTROLYTES BY PLANT TISSUE

Observations on the absorption of salts and other electrolytes during the last 30 years has made it clear that very generally the molecules of electrolytes do not enter the cell as such but diffuse into the cell in the form of their constituent ions, and that, moreover, the two ions of an electrolyte may be absorbed to different extents. The entrance of an excess of one ion cannot, on account of electric attraction of oppositely charged ions, take place without the replacement of the excess by a quantity of another ion carrying the same amount of charge of the same sign. There are two obvious ways in which the replacement could be made. Either one of the ions of water might replace the excess of ion in the external solution, or there could be an exosmosis of ion of the same sign into the external solution. Thus, if we suppose that potassium chloride is absorbed by a plant cell, but that more potassium ions are absorbed than chloride ions, the excess potassium ions might be replaced by an equal number of hydrogen ions resulting from the dissociation of water. In this case the hydroxyl ions corresponding to these hydrogen ions accompany into the cell the excess potassium ions which they equal in number. One result of this will be that the external solution becomes acid owing to the presence of free hydrogen ions. The fact observed many years ago that the solutions used for water cultures in some cases become acid or alkaline can be accounted for in this way, though other explanations are possible. On the other hand, the excess potassium absorbed might be replaced by the diffusion out from the cell of some other kation such as magnesium or calcium, and there is evidence that in some cases at any rate such an exchange of ions is involved in the unequal absorption of the constituent ions of an electrolyte.

That the constituent ions of a single salt are absorbed to unequal extents has been shown to be the case for a large number of storage tissues when immersed in solutions of a variety of single salts. In 1909 both Meurer and Ruhland observed that slices of beetroot and root of carrot placed in solutions of potassium chloride, sodium chloride and calcium chloride absorbed more kation than anion, while from potassium nitrate more anion than kation was absorbed by carrot tissue. If the tissue is killed first, the two ions are absorbed in equal quantities so that the unequal absorption is a function of living cells. Some results with a variety of salts and tissues published by Stiles in 1924 are summarized in Table XV.

Table XV

The Unequal Absorption of the Constituent Ions of a Single Salt by Storage Tissue

Salt	Conc. of Salt in Normalities	Tissue	Duration of Experiment in Hours	Absorption in per cent. of Original Amount	
				Kation	Anion
Ammonium chloride . .	0·01	Carrot	48·0	57·2	32·2
Ammonium phosphate . .	0·02	,,	23·5	28·85	6·0
,, ,, . .	,,	,,	27·5	34·1	6·95
,, ,, . .	,,	Parsnip	24·33	34·5	8·2
,, ,, . .	,,	Beetroot	24·5	19·6	5·1
Ammonium sulphate . .	0·1	Carrot	22·0	3·9	4·2
Potassium chloride . . .	,,	,,	22·0	4·6	6·3
Sodium chloride	,,	Artichoke	22·6	5·4	1·9
,, ,, ...	,,	Carrot	47·1	18·5	8·5
,, ,, ...	,,	Bryony	50·2	21·0	8·4
Sodium sulphate	,,	Carrot	22·0	15·0	8·0

In these experiments determinations of the hydrogen-ion concentration of the solutions external to the tissue showed that they remain practically neutral through the period of absorption, so that the excess of kation absorbed is not replaced in the external solution by hydrogen ions. On the other hand, direct evidence was obtained of the diffusion out from the tissue of other kations. Thus, taking a definite case, carrot root tissue was immersed for 47 hours in a solution of sodium chloride of concentration 0·1 N. After removal of the tissue at the end of this time the concentrations of various ions, assuming complete ionic dissociation, were, in normalities: chloride 0·092, sodium 0·075, calcium 0·009, potassium 0·003 and magnesium a trace. The excess of sodium absorbed was thus replaced by calcium, potassium and magnesium. That the molecules do not exactly balance could be attributed to experimental error since this might be fairly large on account of the small quantities of material involved.

The unequal absorption of ions is not limited to storage tissue but has been observed by various workers to occur in the absorption of salts by the roots of a large number of flowering plants growing in water culture as well as by yeast and various algae. The most considerable body of experiments on absorption of ions by whole plants was described by Pantanelli in 1915; a selection of data from his results is presented in Table XVI.

Table XVI

Absorption of Ions from Living Plants

Species	Salt	Concentration in gm. mols. per litre	Duration of Experiment in Hours	Absorption of mg. ions	
				Kation	Anion
Elodea canadensis .	Calcium chloride	0·05	2	3·0	0·41
,, ,, .	Potassium sulphate	0·05	2	3·44	0·29
,, ,, .	Ammonium sulphate	0·05	2	0·29	2·05
Azolla caroliniana .	Calcium chloride	0·05	2	0·26	0·00
,, ,, .	Potassium nitrate	0·05	2	1·89	0·91
,, ,, .	Aluminium nitrate	0·0125	2	0·16	0·23
,, ,, .	Potassium sulphate	0·025	2	2·67	2·5
,, ,, .	Ammonium sulphate	0·025	2	0·17	2·28
,, ,, .	Magnesium sulphate	0·025	2	3·58	2·48
,, ,, .	Ferrous sulphate	0·025	2	0·76	0·61
,, ,, .	Aluminium sulphate	0·0125	2	0·95	0·062
Phaseolus multiflorus	Calcium chloride	0·01	8	1·18	1·50
,, ,,	Barium chloride	0·01	8	0·04	0·77
Cicer arietinum . .	Potassium chloride	0·025	8	0·35	0·23
,, ,, . .	Calcium chloride	0·025	8	0·025	0·14
,, ,, . .	Potassium nitrate	0·025	8	2·74	1·95
,, ,, . .	Aluminium nitrate	0·0125	8	0·68	2·31
Ulva lactuca . . .	Calcium chloride	0·025*	2	3·29	2·79
,, ,, . . .	Potassium sulphate	0·025*	2	1·48	0·18

* 10 c.c. of 0·25 M salt + 90 c.c. sea water.

Here again the excess absorption of one ion is accompanied by excretion of ions of the same sign from the cells. Miss Redfern found this to be the case with plants of *Pisum sativum*, grown with their roots in a solution of calcium chloride, and Stoklasa, Sebor, Týmich and Cwacha found the same to be true of plants of *Carex riparia, Phragmites communis* and *Eriophorum vaginatum* growing in solutions of aluminium and ferric salts.

It may be of significance that the diluter the solution the less is the relative inequality in the absorption of the two ions. This appears to apply to both storage tissues and whole plants. In Table XVII is shown the influence of concentration of the external solution on the absorption of the ions of ammonium chloride and sodium chloride by slices of carrot root tissue recorded by Stiles in 1924, and on the absorption of the ions of calcium chloride by the roots of living plants of *Pisum sativum* recorded by Miss Redfern in 1922.

Table XVII

Influence of Concentration on the Unequal Absorption of the Ions of a Salt by Plant Tissue

Plant Material	Salt	Duration of Experiment in Hours	Initial Concentration of Solution in Normalities	Percentage Absorption	
				Kation	Anion
Carrot root.	Ammonium chloride	48	0·1 0·01 0·001	12·9 57·2 97·1	5·9 32·2 80·7
Carrot root.	Sodium chloride	48	0·1 0·01 0·001	25·2 36·1 38·1	8·3 15·8 22·8
Pisum sativum (living plants)	Calcium chloride	36	0·1 0·01 0·001	17·74 19·61 23·10	3·578 12·47 15·09

It may thus be concluded that the absorption of a salt and possibly of any electrolyte, by plant cells is not a simple process involving the diffusion of the salt through the cell membranes until the salt has the same concentration on each side of the membrane. On the contrary very generally the two ions of the salt are absorbed at different rates and to different extents and the absorption is accompanied by the excretion of ions from the tissue, so that absorption of ions involves an interchange of ions between the external solution and the tissue.

Before dealing with the explanations which have been offered of the unequal absorption of ions, another aspect of salt absorption by plants, namely, the position of equilibrium reached in the intake of salts, must be discussed.

THE POSITION OF EQUILIBRIUM REACHED IN THE ABSORPTION OF ELECTROLYTES BY PLANT CELLS

It has been noted earlier in this chapter that when a membrane separates two electrolytes one of which possesses an ion to which the membrane is not permeable, then the position of equilibrium is generally one in which the ions are not uniformly distributed throughout the system, but in which each ion is in different concentrations on the two sides of the membrane. Since it is highly probable that impermeable ions, such as those of proteins, are present inside living cells, it is not to be regarded as unlikely that each of the ions of a salt, when absorbed by living cells, may come to be in different concentrations on the two sides of the cell membranes. We certainly have no right to assume, without experimental evidence, that when absorption of a salt is complete its concentration, or that of each of its ions, is the same inside the cell as outside.

It is not always easy to determine the actual concentration of ions inside the cell. We can, indeed, by the methods mentioned earlier in this chapter, find the amount of solute absorbed from a solution, but we cannot always decide what happens to the solute after entering the cells. It might remain as such, or it might become adsorbed to some constituent, or it might enter into chemical combination with a substance already present. It may get no further than the cell-wall, it may be retained in the protoplasm or it may diffuse through both these membranes into the vacuole.

If a salt is adsorbed by, or combines chemically with, some constituent of the cell, its actual concentration in solution inside the cell may be very low, although its concentration judged from its absorption, may appear to be high. In such cases the entry of an ion might appear to proceed well beyond the point which would correspond to equality of concentration inside and outside the cell, since, in fact its concentration inside the cell is kept low. But there is now quite a considerable amount of experimental evidence indicating that ions may pass through the cell membranes into the vacuole and accumulate there so that when equilibrium is reached the actual concentration of the ion is greater in the vacuole than in the external solution. On the other hand, equilibrium may be reached at a point where the ion is in less concentration inside the cell than outside. Illustrating this are the analyses that have been carried out during recent years of the expressed vacuolar sap of large-celled algae. It is established that in such cells the concentration of the ions may be quite different from their concentration in an external aqueous medium. Two examples will suffice. In Table XVIII are shown the molecular proportions of various ions found by Osterhout to be present in the cell-sap of *Valonia macrophysa* and in the surrounding sea water. The amounts are expressed in relation to the total halide (chloride + bromide) present. It should be noted that the halide concentration of the sea water is about 0·6 M and that of the sap is 1·08 times as much, so that in an absolute comparison of the concentrations of each ion without and within the cell the numbers for the sap should be raised by 8 per cent.

Table XVIII

Relative Proportions of Ions present in *Valonia*-sap and in surrounding Sea Water

Ion	Sap	Sea Water
Chloride + bromide	100	100
Sodium	15·08	85·87
Potassium	86·24	2·15
Calcium	0·288	2·05
Magnesium	0	9·74
Sulphate	? trace	6·26

Considerable differences between the ionic concentration of cell-sap and the surrounding medium were also found in the case of *Chara ceratophylla* examined by Collander. From the data given in Table XIX it will be seen that potassium is 63 times and nitrate about 80 times as concentrated in the sap as in the surrounding brackish water.

Table XIX

Relative Concentrations of Ions in the Cell-sap of *Chara ceratophylla* and in the surrounding Brackish Water

Ion	Sap	Sea Water
Potassium	88	1·4
Sodium	142	60
Calcium	10·5	3·6
Magnesium	31	13
Chloride	225	73
Sulphate	7·8	5·6
Nitrate	0·4	ca. 0·005
Phosphate	4·1	trace

Where the plant is living in fresh water the divergence may be greater still, as analyses of the cells of the fresh-water alga *Nitella clavata* and its surrounding pond water published by Hoagland, Davis and Martin indicate.

Thus, in the cases quoted above, the ions are generally, though not invariably, present in greater strength in the cell-sap than in the solution surrounding the cells under natural conditions. Many examples are now on record of tissue of various kinds under experimental conditions absorbing the ions of a salt well beyond the position of equality of concentration within the cell, although, for the reasons stated earlier, it is not generally known in such cases whether the ions retain their individuality in the cell interior or whether they become adsorbed or involved in a chemical reaction.

In considering the absorption of a substance by plant tissue it is convenient to have a term to express the relationship between the concentration of a substance or ion inside the tissue to its concentration in the solution outside with which it is in equilibrium. For the ratio of the internal concentration to the external concentration Stiles and Kidd proposed in 1919 the term *absorption ratio*. Many determinations of the absorption ratio of salts or their ions by various plant tissues have been made by different workers. These determinations have most frequently been made by analyses of the external solution, the internal concentration being calculated from the loss of

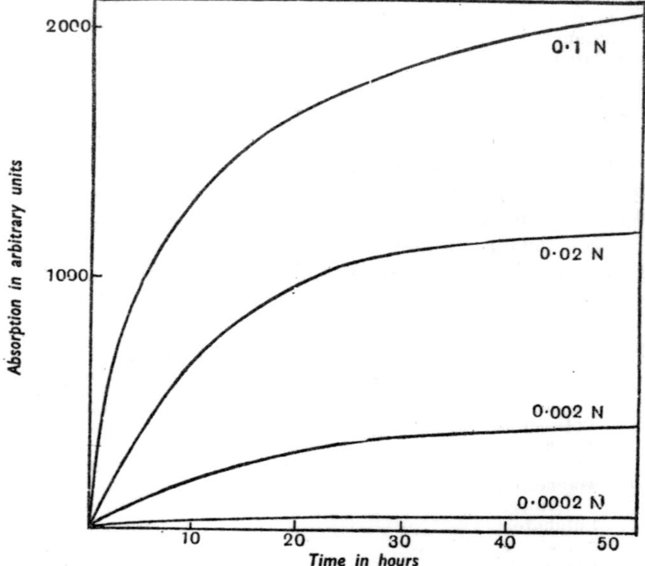

FIG. 6.—Curves showing the absorption of potassium chloride by carrot tissue from solutions of various concentrations

material from the external solution, and estimates of the internal volume throughout which the absorbed material is distributed. It is evident, therefore, that the actual values obtained are likely to be only approximate, while they indicate only the extent to which absorption can take place and not the state or location of the substance in the cell after absorption.

The influence of concentration of the external solution on the extent of absorption of the ions of a salt is very striking.

This question was investigated in 1919 by Stiles and Kidd, who estimated the absorption of various salts by carrot root tissue from determinations of the change in electrical conductivity of the external solution. This method actually gives an estimate of the less absorbed of the two ions of a salt if they are absorbed to different extents, since, as we have already seen, the excess of the one ion absorbed must be replaced by another ion of the same sign. The method is not an ideal one, but Stiles and Kidd were able to show that in the case of potassium, sodium and calcium chlorides absorption proceeded towards an equilibrium position which was greatly dependent on the concentration of the solution, the diluter the solution the greater the proportion of salt or ion absorbed by the tissue. Their results showing the internal and external concentrations of the various salts at the approach of equilibrium are summarized in Table XX. It will be observed that in the diluter solutions the calculated internal concentration of the absorbed material is many times that of the final external concentration. It may be noted that no account is taken of the possible presence of any of the ions concerned inside the tissue at the beginning of the experiments, so that the values given for the absorption ratios are minimum ones.

If graphs are plotted between the logarithms of the final internal concentrations and the logarithms of the final external concentrations they are approximately straight lines. Hence the relation between

Table XX

Absorption Ratios in the Intake of Various Chlorides from Solutions of Different Concentrations

Salt	Initial Concentration in Normalities	Duration of Experiment in Hours	Relative Final Concentration		Absorption Ratio
			Int.	Ext.	
Potassium chloride . . .	0·0002	52	0·06	0·024	25·0
,, ,, . . .	0·002	,,	4·20	0·238	17·6
,, ,, . . .	0·02	,,	11·90	4·882	2·4
,, ,, . . .	0·1	,,	20·50	26·200	0·78
Sodium chloride	0·0002	48	0·56	0·012	46·7
,, ,, ...	0·002	,,	4·00	0·148	27·0
,, ,, ...	0·02	,,	10·80	3·990	3·5
,, ,, (dead tissue)	0·02	,,	4·45	5·060	0·88
,, ,, ...	0·1	,,	17·80	21·550	0·83
Calcium chloride	0·0002	42·5	0·46	0·030	15·3
,, ,, ...	0·002	,,	1·17	0·420	2·8
,, ,, ...	0·02	,,	2·50	4·930	0·51
,, ,, ...	0·1	,,	5·30	22·120	0·24

final internal and final external concentration can be expressed by the equation

$$\log I - m \log E = \log K$$

or

$$I = KE^m$$

where I is the final internal concentration, E the final external concentration and K and m are constants. This equation is that representing the relationship between concentration and amount of adsorption, but, as repeatedly pointed out by the writer, it does not prove that the absorption of the salts is a process of adsorption. What is proved is that the absorption of a salt or of its ions is greatly influenced by concentration and that the weaker the solution the greater is the absorption relative to the concentration.

This relation between absorption and concentration was subsequently confirmed by Stiles by direct chemical analysis of the external solution, the salts used being sodium chloride and ammonium chloride. The concentration of each ion was determined and it was found that the law was obeyed in both cases (cf. Table XXI and Figure 7). More recently, Steward has found that the accumulation of the bromide ion by potato tuber and carrot root follows the same law.

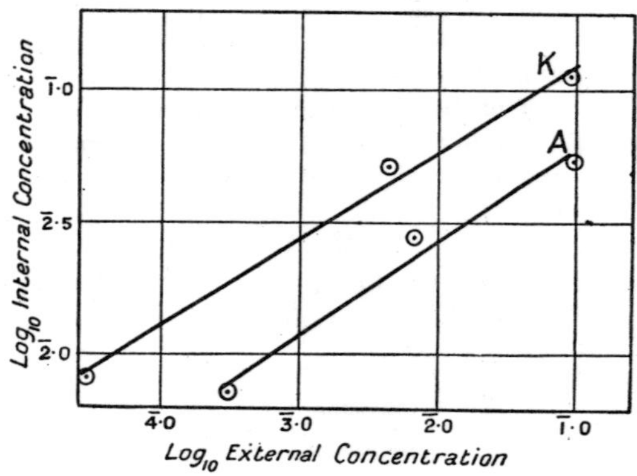

FIG. 7.—Curves showing the relation between the final internal and external concentrations of the kation and anion of ammonium chloride after absorption by carrot root tissue

Moreover, this relationship between accumulation of salt and concentration is not limited to storage tissue. Laine has recently found that detached roots of *Phaseolus multiflorus* placed in solutions of potassium, calcium and manganese chlorides of different concentrations exhibit similar behaviour, the relation between internal and external concentration of each of the kations being given by the adsorption equation. Also Hoagland, Davis and Hibberd have found that *Nitella* cells absorb relatively more bromide ion the diluter the solution of bromide in which the cells are immersed.

The same general relation between absorption and concentration is observed with a number of dyes. This will be evident from an inspection of Table XXII which gives data provided by experiments published by Miss Redfern in 1922. The agreement with the adsorption equation is close with some of the dyes, but deviations from the law are observed with others.

Table XXI

Influence of Concentration on the Position of Equilibrium attained in the Absorption of the Constituent Ions of Sodium Chloride and Ammonium Chloride by Carrot Root

Salt	Initial Concentration in Normalities	Final Concentration of Kation in Normalities		Final Concentration of Kation in Normalities		Absorption Ratio	
		External	Internal	External	Internal	Kation	Anion
Sodium chloride	0·1	0·0748	0·223	0·0917	0·07345	2·98	0·801
	0·01	0·00639	0·03195	0·00842	0·0140	5·00	1·66
	0·001	0·000619	0·00337	0·000772	0·00202	5·45	2·62
Ammonium chloride	0·01	0·0871	0·1141	0·0941	0·0520	1·31	0·553
	0·01	0·00428	0·0506	0·00678	0·0285	11·8	4·20
	0·001	0·0000293	0·00859	0·0001923	0·00715	293·2	37·2

Table XXII

Absorption Ratio at the Approach of Equilibrium in the Absorption of Various Dyes by Disks of Storage Tissue

Dye	Character of Dye	Concentration at Equilibrium		Absorption Ratio
		External	Internal	
Neutral red	Basic ; semi-colloid	0·06	2·0	33·3
		0·03	1·0	33·3
		0·0005	0·475	950
		0·000125	0·244	1952
Methylene blue . . .	Basic ; crystalloid	0·032	3·4	106
		0·01	1·9	190
		0·0008	0·46	575
		0·0003	0·235	783
Methyl violet . . .	Basic ; semi-colloid	0·03	3·5	117
		0·01	2·0	200
		0·0018	0·41	228
		0·001	0·20	200
Aniline blue	Basic ; colloid	0·1	—	—
		0·045	0·25	5·56
		0·008	0·1	12·5
		0·003	0·1	33·3
Eosin	Acid ; crystalloid	0·1	—	—
		0·045	0·25	5·56
		0·0085	0·075	8·82
		0·004	0·05	12·5

Table XXIII

Absorption Ratios of a Number of Chlorides, Sulphates, Nitrates and Potassium Salts presented to Carrot Tissue in a Concentration of 0·02 N

Group	Salt	Duration of Experiment in Hours	Absorption Ratio
I	Potassium chloride . . .	91	3·58
	Sodium chloride . . .		3·49
	Lithium chloride . . .		1·16
	Calcium chloride . . .		1·09
II	Potassium sulphate . .	64·5	0·51
	Sodium sulphate . . .		0·46
	Magnesium sulphate . .		0·097
III	Potassium nitrate . . .	71·5	4·65
	Sodium nitrate		3·30
	Calcium nitrate . . .		1·19
	Aluminium nitrate . . .		0·53
IV	Potassium chloride. . .	42	1·99
	Potassium sulphate . .		0·55
	Potassium nitrate . . .		2·20

With the simpler salts different substances are accumulated to different extents. The absorption ratios obtained by Stiles and Kidd for a number of chlorides, sulphates and nitrates are shown in Table XXIII. These numbers are the results of four sets of experiments. As in each set of experiments a different sample of tissue was used, and as the experiments in the different groups were terminated at different times, only the numbers within each group are strictly comparable.

These numbers indicate that the extent to which a salt is absorbed by carrot root depends on the kation and anion of the salt. The order of absorption of kations is K, Na, Li, (Ca, Mg), Al, while the order of absorption of anions is NO_3, Cl, SO_4. It thus appears that monovalent kations are absorbed to a greater extent than divalent kations, and the trivalent aluminium ion to a less extent than the divalent ions. Similarly, the monovalent anions NO_2 and Cl are absorbed to a much greater extent than the divalent SO_4 ion.

There is a considerable variation in the extent to which different dyes are absorbed; in general basic dyes [1] are absorbed to a greater

[1] A basic dye is a salt in which the colour radicle is a base, whereas in an acid dye the radicle to which the colour is due is an acid. Thus Bismark brown is a basic dye, the base being triamidoazobenzene $NH_2.C_6H_4N:NC_6H_3:(NH_{22}$ and the dye itself triamidoazobenzene hydrochloride. As an acid dye Congo-red may be taken as an example; it is the sodium salt of a complex acid.

extent than acid dyes. Thus Collander found that a number of sulphonic acid dyes (light green, orange G, cyanol, fuchsin S) were absorbed to a very slight extent by a number of tissues such as the parenchyma of the bulb scales of the onion and hyacinth, the cortical cells of the root of *Pisum sativum*, the spongy parenchyma cells of the leaves of *Rhoeo discolor* as well as cells of *Spirogyra*. In very few cases did the absorption ratio in Collander's experiment reach 0·125 with concentrations of the dye of 0·1 or 0·8 per cent.

The characteristics of the absorption of electrolytes by plant cells are thus (1) the unequal absorption of the two ions of an electrolyte accompanied by the ionic exchange between the cell and the surroundings, (2) the absorption of the ions towards an equilibrium which is apparently not that of equality of concentration, but which varies with the concentration of the ion in the same way as its adsorption varies, and (3) as regards the simpler salts different kations and different anions are not absorbed to the same extent, monovalent ions being more absorbed than divalent ones.

In attempting to find an explanation of this behaviour it is obviously of importance to know the fate of the absorbed substance in the cell; whether, that is, it is held in cell-wall, protoplasm or vacuole. We have already noted that in the case of some large-celled algae there is evidence that the ions accumulate as such in the vacuole. Steward also concluded from analysis of expressed sap that potassium bromide absorbed by potato tuber tissue accumulates as such in the sap. The data on the question are scanty, but there is definite evidence that in some cases salts or their ions penetrate into the cell-sap towards a condition of equilibrium which is not that of equality of concentration on the two sides of the cell membranes.

Where salts are absorbed by the roots of living plants it is possible and, indeed, probable that they diffuse from the actual absorbing cells and find their way eventually to parts of the plant where they enter into chemical combination with other substances; at any rate this is to be expected in the case of salts containing elements essential for the growth of the plant, and in this way an actual equilibrium condition will never be reached in the absorbing cells. Where, however, an isolated piece of tissue, or an individual cell, is concerned, an approach to an equilibrium condition in salt intake is more likely to occur. That this equilibrium is not one of equality of concentration inside and outside the tissue might be due to adsorption of the ions by a colloidal cell constituent, or it might be due to chemical combination of the ions with a cell constituent. That other possibilities must be reckoned with, is, however, clear from the observations already mentioned indicating that ions in some cases, at any rate, can accumulate in the cell-sap.

The adsorption theory of absorption of electrolytes by living cells has found some support in the past, notably from the work of Moore and Roaf in 1908 and of Szücs in 1912, among others. It is supposed that some cell constituent, most probably in the plasmatic layer, adsorbs the ions. Subsequently this adsorption compound may break down and the ions be released, but the actual absorption process is regarded as one of adsorption by a colloidal constituent of the cell. The observed relation between the internal and external concentrations of ions as equilibrium is approached certainly supports the theory, but it has not found much favour during recent years, chiefly on account of the evidence to which allusion has already been made, that in some cases it appears fairly certain that the ions accumulate in the vacuole. The same objection can be raised against a theory of chemical combination of the absorbed ion with some cell constituent, which would otherwise account for the apparent position of equilibrium in absorption not being one of equality of concentration within and without the cell.

The theories of salt absorption that have been put forward during recent years have therefore attempted to explain the accumulation of ions in the cell-sap. The distribution of ions at equilibrium resembles that occurring in a Donnan equilibrium, and it is not surprising that attempts have been made to explain the accumulation of ions in the vacuole in terms of a Donnan equilibrium. However, Briggs and Petrie, from data of the concentrations of various ions inside and outside cells determined experimentally, calculated that the product of the apparent internal concentrations of kation and anion of a salt is greater, in the cases examined, than the product of the concentration of the two ions in the external solution, whereas the two products should be equal. This obviously cannot be so where both ions are in higher concentration in the internal than in the external solution, as has been shown to hold in several cases investigated. The absorption can thus proceed farther than would be the case if its amount were controlled by a condition of Donnan equilibrium on the two sides of a membrane separating homogeneous phases. But, as Briggs and Petrie have pointed out, the plant cell is not simply a membrane enclosing a homogeneous solution possessing an indiffusible ion. As we have seen in Chapter II the protoplasm, and perhaps the vacuole as well, constitute heterogeneous systems containing more than one phase, and a Donnan equilibrium might result between any two of them. Yet even so, in such a system where distribution of ions is determined by Donnan equilibria, if the ratio of the apparent internal concentration to the external concentration of the kation of a salt is greater than unity, then the corresponding ratio of the anion should be less than unity. But, as just pointed out, the ratio of internal to external concentration may be greater than unity with both kation and anion. Briggs and Petrie, however, showed that this distribution could result if one internal phase contained indiffusible kations and another internal phase indiffusible anions. Subsequently Briggs has shown that the experimental data obtained regarding the absorption of ions by mature plant cells from weak solutions can be explained on these lines, it being suggested that kations are exchanged between solution and cytoplasm and anions between solution and cell-sap. But with strong solutions kations also pass into the cell-sap.

While, then, it would appear that the accumulation of ions in the interior of the cell can be accounted for in terms of Donnan equilibria, provided more than one phase is present inside the cell, a number of workers during recent years have emphasized the connexion between absorption of salt and metabolic activity, particularly the production of carbon dioxide in respiration, which is a continuous process in living cells. The carbon dioxide dissolves in the water present in the cells with formation of carbonic acid which partially dissociates into H^+ and HCO_3^- ions. Hence if a living cell is surrounded by a solution of potassium chloride we have the cell membranes separating a solution containing K^+ and Cl^- from one containing H^+ and HCO_3^-. There will result an exchange of ions between the cell and external solution, but a variety of views have been expressed regarding how this exchange is brought about.

In 1929 S. C. Brooks put forward a theory of salt absorption based on experiments with *Valonia*. He supposed the cell membrane to be composed of a mosaic of areas permeable only to kations and of areas permeable only to anions. The production of carbon dioxide and hence of carbonic acid in the cell leads to the formation of H^+ and HCO_3^- ions. Entry of kations can then only take place through the kation-permeable areas by exchange with an equivalent quantity of hydrogen ions, and similarly the entry of anions can only take place through the anion-permeable areas by exchange with an equivalent quantity of anions. Equilibrium, it is said, would be attained when the ratio of the concentrations of any kation in the cell and outer medium is the same as the corresponding ratio for the hydrogen ion. Since the concentration of hydrogen ions in the *Valonia* cell is very many times that of sea water, the potassium ions can thus enter the cell in exchange for hydrogen ions until their concentration is much higher than that in the external solution. Similarly, as regards the entrance of anions, these, notably chloride ions, should enter through the anion-permeable areas until the ratio of the concentrations of chloride ions inside and outside the tissue is the same as that of the bicarbonate ions. Actually the ratio of the concentration of chloride ions inside and outside the tissues is more than unity, while according to Osterhout and Dorcas the concentration of bicarbonate ions is greater in the sea water outside the cell than in the cell-sap.

Briggs has pointed out that although the observed accumulation of ions cannot be explained in relation to a membrane consisting of a mosaic of kation-permeable and anion-permeable areas, accumulation of both kations and anions could take place in a respiring cell in which the protoplasmic membrane as a whole undergoes rhythmic changes so that it is alternately permeable to kations and to anions.

There is, however, no definite evidence that such changes in the protoplasm actually occur, and it must be realized that the actual mechanism by which salts are absorbed by, and accumulate in, plant cells is still a matter of conjecture and not of certain knowledge.

THE ABSORPTION OF NON-ELECTROLYTES BY PLANT CELLS

According to work published by Bärlund in 1929 in which the intake of solutes by the epidermal cells of *Rhoeo discolor* was investigated by a plasmolytic method, and to observations of Collander and Bärlund described in 1933 in which absorption of solutes by cells of *Chara ceratophylla* was determined by chemical analysis, the position of equilibrium reached in the intake of a number of non-electrolytes by these cells is one in which there is equality of concentration of the solute within and without the cell. It is therefore reasonable to suppose that with these substances the mechanism of absorption is one of simple diffusion, and determinations of the amount of various solutes which had entered the *Chara* cells after different times of immersion showed that the course of absorption proceeded just as would be expected if the process of solute intake were simple diffusion through cell membrane.[1]

In such cases it is possible to obtain values for the permeability of the cell membrane to different substances.

The permeability of the cell membrane (wall + protoplasm) may be defined as the quantity of solute diffusing through unit area of the membrane in unit time when there is unit difference of concentration of the solute on the two sides of the membrane. Hence if the concentration of the external solution is kept constant at the value C, and if the concentration of the solute in the cell-sap after a time t has reached the value c, the rate of entry of solute at this time is given by the equation

$$\frac{dq}{dt} = \text{PA}(\text{C} - c)$$

where dq is the quantity of solute entering in the time dt, q being the quantity of solute in the cell-sap, A is the area of the membrane surface through which entry of the salt takes place, and P is a value depending on the permeability of the membrane to the solute in question, and is, indeed, a measure of the permeability of the membrane to the solute. It is called by Collander and Bärlund the permeability constant. If the volume of the cell is V, then

$$c = \frac{q}{\text{V}} \text{ and } \frac{dq}{dt} = \text{V}\frac{dc}{dt}$$

Hence

$$\frac{dc}{dt} = \frac{\text{PA}}{\text{V}}(\text{C} - c)$$

[1] It cannot be assumed that the absorption of all non-electrolytes is with all plant cells so simple a matter. Further reliable investigations with other kinds of cells are necessary before a decision can be taken on this point.

which on integration gives

$$P = \frac{V}{At}\log\left(\frac{C}{C - c}\right)$$

Since isolated cell-wall is readily permeable to such substances as sucrose and orange G, and since also non-electrolytes enter dead cells much more rapidly than living ones, it is not unreasonable to suppose that the experimentally determined values of the permeability constant can be taken as measures of the protoplasmic permeability.

By the use of this equation Collander and Bärlund calculated the permeability constants for a number of substances diffusing into cells of *Chara ceratophylla* by measuring the quantity of solute which entered cells of known dimensions in a certain time. The permeability constants found by them for a number of substances are shown in Table XXIV. The values represent the number of mols of the substance passing through 1 sq. cm. of protoplast surface in one hour when the difference in concentration of the solute in the external solution and the cell sap is 1 mol per c.c., and the temperature 19·2° C.

It will be seen from this table that the permeability of the plasmatic layer to different substances varies greatly with the substance. Very few reliable data with regard to the permeability of other cells to a range of substances are available, but Bärlund's earlier observations on the epidermal cells of *Rhoeo discolor* show a general agreement with those on *Chara*, except that amides, in relation to other substances,

Table XXIV

Permeability of *Chara* Cells to various Non-electrolytes

Substance	Permeability Constant
Methyl alcohol	0·99
Ethyl alcohol	0·56
Urethane	0·43
Antipyrin	0·22
Butyramide	0·17
Propionamide	0·13
Propylene glycol	0·087
Formamide	0·077
Acetamide	0·053
Ethylene glycol	0·043
Ethyl urea	0·012
Methyl urea	0·0068
Urea	0·0040
Glycerol	0·00074
Arabinose	<0·000031
Glucose	<0·00003
Sucrose	<0·00003

appear to penetrate the protoplasts of the latter more readily than those of the former. In 1934 Helene Bonte attempted a comparison of the penetrating capacity of 23 organic compounds into cells of the liverwort (*Hookeria lucens*), an alga (*Hydrodictyon utriculatum*) and a fungus (*Basidiobolus ranarum*). While her method of estimating permeability is open to criticism, her results indicate on the whole a similar order of permeability of these cells to different substances as that found by Collander and Bärlund. The bearing of these results on theories of cell permeability will be considered later.

There is no doubt that the permeability of different cells to the same substance may vary greatly For example, Höfler and Stiegler examined by the plasmometric method the permeability to urea of different kinds of cells in plants of *Gentiana sturmiana*. They found that in this species the permeability varies considerably among the different tissues, the most permeable cells being those of the epidermis of the peduncle, while the permeability of the parenchymatous cells of cortex and pith of the peduncle is less, and that of the cells of the petals still less, the permeability of these last being only from 2·5 to 5 per cent. that of the epidermal cells. Similarly, in the root of this plant the cells of the piliferous layer are more permeable to urea than those of the cortex.

THE ABSORPTION OF TOXIC SUBSTANCES

In the previous discussion on absorption of both electrolytes and non-electrolytes by plant cells it has been assumed that the cells are alive and remain so throughout the process of absorption. But many substances, both electrolytes and non-electrolytes, have a toxic effect on plant cells if presented to them in sufficiently high concentration, and bring about their death, wherein the cells lose their turgid condition and become, as a rule, completely permeable both to solutes outside and inside the cell. Among such toxic substances are many salts, such as those of copper, silver and mercury, acids and alkalies among electrolytes, and alcohols, aldehydes, ketones, phenols and many other organic compounds among non-electrolytes.

It is obvious that such substances, to exert a toxic action, must enter the cell, and it is also clear that the substance does not merely diffuse through the protoplasm, but must act upon it in some way. What this action may be is not clear, but there is little doubt that it varies in nature in different cases. Thus there is reason to suppose that the toxic action of acids is a chemical effect, possibly due to a reaction of the acid with protoplasmic proteins. In the first place it was found by Stiles and Jorgensen that the temperature coefficient of absorption of the hydrogen ions of hydrochloric acid by potato tissue is about 2, a value characteristic of many chemical reactions, while Stiles and Rees have found that when potato tissue is immersed in solutions of various organic acids, namely, formic, acetic, propionic and other acids of the aliphatic series, the hydrogen ions are removed from the solutions but are not returned to them on the death of the tissues, thus indicating an irreversible reaction between the protoplasm and the acid.

On the other hand, there is reason for supposing that in the case of alcohols or ketones, the toxic effect may be due to some physical property of the substance. For the various members of a homologous series of alcohols or ketones are not equally toxic; the higher the position of the member in the series, the more poisonous it is. Thus in the series of normal monohydric aliphatic alcohols, ethyl alcohol is more poisonous than methyl alcohol, normal propyl alcohol more poisonous than ethyl alcohol, and so on. Roughly speaking, the toxicity of any member of the series is 3 times that of the member next below it; that is, a molecule of the alcohol $CH_3(CH_2)_nCH_2OH$ is 3 times as toxic as a molecule of the alcohol $CH_3(CH_2)_{n-1}CH_2OH$. There appears to be no chemical property of the alcohols which varies in this way with increasing molecular size, but there is a definite parallelism between the surface activity of the various alcohols and their position in the series. The results of Stiles and Stirk on the comparative toxicities of normal monohydric alcohols to potato tissue and the relative surface activities of these alcohols as deduced from measurements of the lowering of the surface tension of water resulting from the presence of alcohol in the water, are summarized in Table XXV.

Table XXV

Relative Toxicities and Surface Activities of the Molecules of Normal Monhydric Alcohols

Alcohol	Relative Toxicity	Relative Surface Activity
Methyl	1·00	1·00
Ethyl	2·62	2·74
Propyl	6·10	8·77
Butyl	15·34	31·04
Amyl	66·9	100·5
Hexyl	233	442
Heptyl	849	1480
Octyl	2139	4444

Whether the toxicity of alcohols and other substances which lower the surface tension of water considerably (surface active substances) is indeed related to this physical property cannot be definitely stated,

Table XXVI

Relative Toxicities of the Molecules of Normal Fatty Acids

Acid	Relative Toxicity
Formic	1·00
Acetic	0·262
Propionic	0·211
Butyric	0·259
Valeric	0·335
Caproic	0·616
Oenanthic	1·23
Caprylic	2·55
Pelargonic	4·03

but it seems to be a likely factor. A comparison of the numbers in Table XXV with those in the above table where the toxicities of the corresponding series of acids are shown indicates a very definite difference in the relationship between molecular size and toxicity in the two cases. This difference is explained if it is assumed that in the case of the acids the toxicity is partly chemical, due to the action of the hydrogen ion, and partly physical. As the series is ascended the toxicity due to the hydrogen ion tends to decrease, since on the whole the degree of dissociation of the higher acids is less than that of the lower ones, while the toxicity related to molecular size increases. Hence a solution of acetic acid is less toxic than an equimolecular solution of formic acid, because of its lower concentration of hydrogen ions, and for the same reasons propionic is a little less toxic than acetic. But as the series is ascended above propionic the toxicity increases, for although the degree of dissociation of these acids is much the same up to caprylic acid, the toxicity related to molecular size, which may or may not depend on surface activity, increases regularly as in the alcohol or ketone series.

While the actual causes of toxic action are, therefore, in doubt, there seem very definite reasons to suppose that toxicity may be due both to chemical and physical properties of the toxic substance. Chemical action may result in the removal of an essential constituent of the protoplasm, while physical properties may be responsible for altering the state of aggregation of substances composing the cell colloids and so for disturbing essential space relations of protoplasmic constituents. Where such changes are reversible no permanent damage may be done to the cell if the poison is not allowed to act for too long a time, but if the change is irreversible the cell becomes permanently damaged or killed.

ANTAGONISM AND SYNERGY

In the previous sections of this chapter the relations of the cell to dissolved substances have been considered when only one solute has been presented externally to the cell. When more than one substance is present in solution outside a cell the relations are more complex. Very generally the entry of a substance is retarded by the presence of another, a phenomenon to which the name *antagonism* is given. The opposite state of affairs, in which the entry of a substance is accelerated by the presence of another, is much rarer; for this the name *synergy* has been suggested.

A great many observations are on record regarding antagonism between salts. Thus in 1906 Osterhout showed that certain marine plants which can live for some time in distilled water, die considerably sooner in a solution of sodium chloride isotonic with sea water than in distilled water. The presence of a small quantity of calcium chloride along with the sodium chloride enabled the plants to live nearly as long as in distilled water, while the addition of a little potassium chloride as well enabled the plants to live longer than in distilled water, while with the further addition of some magnesium chloride the plants could live practically as long as in sea water. Many similar observations with other mixtures of salts were recorded by Osterhout. The reduction of toxicity of the single salt resulting from the presence of a small quantity of another was attributed to a hindrance in the entrance of the toxic salt into the cells of the plant. The evidence is, of course, indirect, but in 1907 Benecke produced direct evidence that the presence of one salt retards the entry of another into plant cells. The cells of many species of *Spirogyra* contain tannin. The entrance of an iron salt into such cells is rendered evident by the formation of a green or blue compound. Benecke found that if a quantity of a calcium salt is added to a solution of ferrous sulphate, the entrance of the iron into *Spirogyra* cells is much retarded, an observation which was later confirmed and extended by Szücs, who showed that not only calcium salts, but those of other metals, could bring about the same result, the antagonistic action depending on the valency of the kation, the greater the valency, the more effective the antagonism.

Since these early observations antagonism has been demonstrated by a variety of methods both direct and indirect, but mostly depending on the reduction of toxic action or increase in growth rate in mixed, as compared with pure, solutions. All kinds of plant cells have been used in such experiments, including those of freshwater and marine algae, leaves and actively growing roots of higher plants, and storage tissues. A few cases may be quoted as examples. Osterhout found that roots of wheat seedlings grew more rapidly in solutions containing both salts of sodium and potassium (or ammonium, calcium or magnesium) than in pure solutions of the same concentration, there being a definite ratio of the concentrations of the two salts for which the rate of growth of the roots was greatest. Again, Chien in 1917 showed that a peculiar contraction of the chloroplasts of *Spirogyra* produced by chlorides of barium, strontium and cerium in certain concentrations was inhibited by addition of calcium chloride in a certain proportion. Cerium bromide could itself antagonize barium chloride and strontium chloride, but no antagonism was observed between the chlorides of barium and strontium.

As another example may be cited the antagonism between salts of copper and aluminium investigated by Szücs. In 1911 this worker showed that when seedlings of *Cucurbita pepo* are placed with their roots in a pure solution of copper sulphate the copper is rapidly absorbed. With a solution of 0·025 N copper sulphate enough copper is soon absorbed to inhibit the response of the hypocotyl and root to the stimulus of gravity and of the hypocotyl to light (cf. Chapters XXII, XXIII) and to give a definite qualitative chemical reaction for copper. Addition of some aluminium chloride to the copper sulphate solution delays the inhibition of the geotropic and phototropic reactions, while the amount of copper absorbed in a definite time is much less. With copper sulphate in a concentration of 0·025 N the maximum antagonistic effect of aluminium is exhibited with aluminium chloride in a concentration of 0·15 N.

The phenomenon of synergy between simple salts is apparently only rarely met with. It was recorded as occurring between sodium citrate and a number of other sodium salts including the chloride, sulphate, nitrate, iodide and thiocyanate. Mixed solutions of these salts were observed by Raber to be more toxic to *Laminaria* cells than pure solutions of the single salts possessing the same electrical conductivity.

Not only does antagonism occur between simple salts; it has also been recorded as occurring between a simple and a complex salt, such as an alkaloid hydrochloride or sulphate or a dye. Thus Szücs in 1910 showed that potassium, calcium and aluminium nitrates retarded the entrance of methyl violet into cells of *Spirogyra*, and in 1912 he demonstrated a similar antagonism between quinine hydrochloride and these three nitrates. Osterhout in 1919 showed that antagonism occurred between sodium chloride and sodium taurocholate and sodium chloride and cevadine sulphate in regard to their toxic effect on cells of *Laminaria*. Further, a number of observations by different workers have shown that antagonism occurs between electrolytes and non-electrolytes. Weevers, for example, using the exosmosis of the red pigment from cells of the root of red beet as a criterion of toxic action, showed in 1914, that the action of a number of organic substances in effecting this exosmosis could be antagonized by aluminium chloride and in some cases by zinc sulphate, cobalt chloride and manganese acetate. The organic poisons included formaldehyde, ethyl alcohol, amyl alcohol, ether, chloroform and chloral hydrate. Later, in 1921, Collander showed that antagonism could exist between non-electrolytes and sulphonic acid dyes. Thus the absorption of the dyes cyanol and orange G was greatly retarded by a 2 per cent. solution of ether or a 0·5 to 1 per cent. solution of chloral hydrate.

The mutual hindrance to the entry of each of two or more substances into plant cells is thus a phenomenon of wide occurrence. In attempting to find an explanation of it emphasis has been laid on certain relationships discovered by different workers. In the first place the valency of the ions has a very considerable effect on the efficiency of one salt in retarding the entry of another. This was shown by Szücs in 1910 in the case of the absorption of methyl violet by cells of *Spirogyra*, and by Mann in 1924 for the absorption of methylene blue, neutral red and orange G by cells of mangold root, the rule being that the higher the valency of the kation of the salt the less the absorption. It will be observed that all the dyes in question are basic dyes in which the coloured ion is the kation, so that actually the measured absorption of dye is that of the dye kation. We are therefore justified in regarding the results of Szücs and Mann as indicating a relationship between the valency of the ion of a salt and its antagonistic effect on ions of the same sign. Some of Mann's results are summarized in Table XXVII.

Table XXVII

The Absorption of Methylene Blue and Neutral Red by Mangold Tissue at 20° C. from 0·01 per cent. Solutions of the Dyes containing Various Chlorides in 0·01 N Concentration.

Salt	Percentage of dye absorbed in 12 hours	
	Methylene Blue	Neutral Red
Ammonium chloride . . .	55·0	81·0
Magnesium chloride . . .	29·0	70·5
Aluminium chloride . . .	5·0	14·0
Lanthanum chloride . . .	5·5	20·0

As the curves in Fig. 8 indicate it would appear that the presence of the antagonizing salt does not merely retard the rate of entry of the

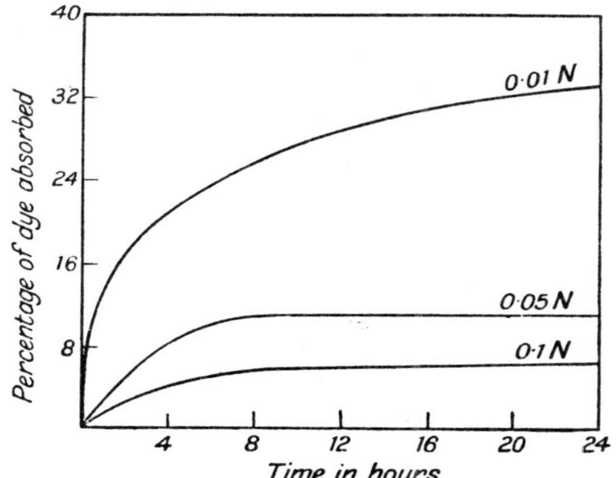

FIG. 8.—Curves showing the absorption of methylene blue by mangold tissue immersed in solutions of this dye containing magnesium chloride in various concentrations (0·01 N, 0·05 N, 0·1 N) as shown on the figure

(Constructed from the data of C. E. T. Mann)

dye, but alters the position of the equilibrium attained in the intake of the dye. In the case shown graphically it appears that when the position of equilibrium was approached in the absorption by mangold tissue of methylene blue from a 0·01 per cent. solution of the dye containing 0·01 N magnesium chloride, five times as much dye had been absorbed as when the concentration of the magnesium chloride was 0·1 N.

In Mann's experiments the molecular concentration of the dye was too low for the latter to exert any considerable retarding effect on the entrance of the metallic kation. In the more recent experiments of Asprey published in 1933 the same effect of the valency of various kations on the intake of ammonium and calcium ions by potato tuber has been clearly demonstrated. This relationship is clearly shown by the curves in Fig. 9, which show the rate of absorption of ammonium ions from a pure 0·02 N solution of ammonium chloride, and from solutions containing this salt in the same concentration with the addition of sodium, calcium and lanthanum chlorides respectively in the same concentration. A similar reduction in intake of calcium ions

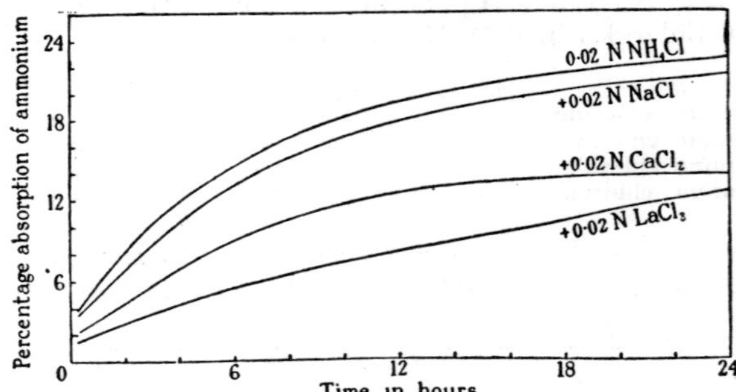

Fig. 9.—Curves showing the rate of absorption of ammonium ions from a pure 0·02 N solution of ammonium chloride and from solutions of 0·02 N ammonium chloride containing in addition 0·02 N sodium chloride, 0·02 N calcium chloride and 0·02 N lanthanum chloride respectively

(From G. F. Asprey)

as the result of the presence of magnesium or lanthanum ions was shown by Asprey (cf. Fig. 10).

An exception was found by Asprey in the case of a solution containing 0·02 N ammonium chloride and 0·02 N lithium chloride. In

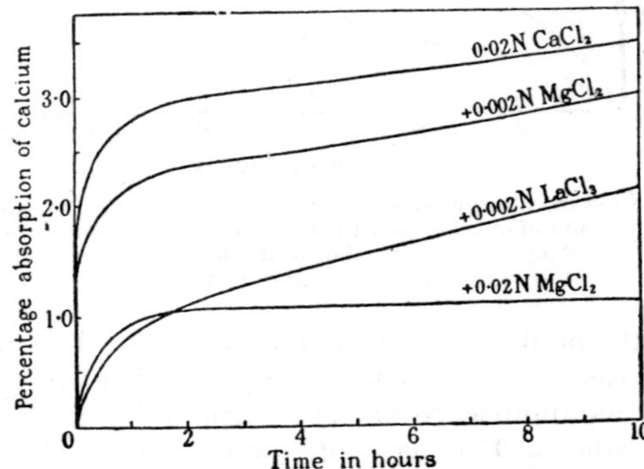

Fig. 10.—Curves showing the rate of absorption of calcium ions from a pure solution of calcium chloride and from mixtures of calcium and magnesium chlorides and calcium and lanthanum chlorides respectively

(From G. F. Asprey)

presence of the latter the absorption of ammonium was greatly increased.

As might be expected, the antagonistic action of one ion on another depends on the concentration of the former if that of the latter is kept constant. Thus Asprey found that after 10 hours' immersion in mixed solutions containing 0·02 N ammonium chloride and various concentrations of calcium chloride, the intake of ammonium is lowered by 14 per cent. by 0·002 N calcium chloride, by 36·5 per cent. by 0·02 N and by 50 per cent. by 0·1 N calcium chloride. Similar results were obtained by Mann in his work on the antagonistic action of metallic salts and dyes.

Another relationship between ions in their antagonistic action was suggested by Kahho in 1921 as a result of observations on the absorption of salts by roots of young yellow lupin plants. In descending order of absorption kations can be arranged in the series K, Na, Li, Mg, (Ba, Ca) when the roots are exposed to solutions of single salts. The order of absorption of anions is (Br, I, NO_3) Cl, tartrate, SO_4, citrate. Now if two salts are present in the external solution Kahho found that the entrance of any kation in this series is retarded by the presence of any other kation to the right of it in the series, the retardation being greater the further the kation is to the right. Thus potassium chloride enters the cells more rapidly than lithium chloride from a solution of the same osmotic pressure, but from a solution possessing this osmotic pressure and containing both salts, salt enters much more slowly than from pure potassium chloride but a little faster than from pure lithium chloride. Kahho's method of measuring salt intake, depending on the rate of recovery of the original volume of the tissue with entry of the salt following an initial contraction, is similar in principle to a deplasmolytic method, and only gives the total salt absorbed from a mixed solution.

The usual way of explaining antagonism is to regard it as due to competition between the solutes in their absorption. Thus if they were adsorbed, the adsorbing surface being limited, the adsorption of a second solute would necessarily reduce the surface on which the first solute is adsorbed. This was at one time the more usual view of antagonism, but since the realization of the possibility of the condition of Donnan equilibrium between a cell and the surrounding medium the view has gained ground that the absorption of electrolytes is largely a question of interchange of ions between the cell and external medium.[1] Since the amount of ionic material in the cell is limited, antagonism between ions is equally understandable on this view. The observed antagonism between salts and non-electrolytes does not appear to be explicable on these grounds, but on the other hand Collander's recent work on the absorption of non-electrolytes by *Chara*, in which it was shown that these substances are absorbed until there is equality of concentration between internal and external solutions, indicates that adsorption plays no significant part in the absorption of non-electrolytes in these experi-

[1] Steward has proposed the term 'primary salt absorption' to describe the absorption of salt by actively metabolizing cells which absorb both kation and anion in the vacuole and gain rapidly in salt content, and the term 'induced absorption' for the readjustment of salt content, without change in salt concentration, of relatively mature cells. While the latter involves *only* an interchange of ions, such an interchange may also be involved in absorption by more active cells.

ments. Hence the question of antagonism still presents difficulties. Actually the experimental data are still insufficient for coming to a decision with any confidence. In particular insufficient attention has been given to differentiating between an antagonism which simply reduces the rate of entry of a solute and one in which the total amount of substance absorbed is altered.

REVERSIBLE CHANGES IN CELL PERMEABILITY

When a cell is acted upon by a toxic substance the usual result is that the organization of the protoplasm is broken down so that the latter becomes very much more permeable to dissolved substances. Not only simple electrolytes but complex substances can then diffuse out from the cell, the unequal absorption of ions no longer takes place, and the position of equilibrium attained in their intake is now generally one of equality of concentration inside and outside the cell. These changes in the permeability of the cell induced by toxic substances are irreversible and are associated with the death of the cell.

But it is of interest to inquire whether the cell membranes undergo reversible changes of permeability. External factors which can reasonably be regarded as possibly bringing about such reversible changes in permeability are temperature, light, and the composition of the medium surrounding the cell. Metabolic changes in the cell might also in themselves lead to changes in permeability by bringing about chemical or physical changes in the protoplasm.

Temperature. It has been noted earlier in this chapter that the rate of diffusion of solutes increases with a rise of temperature. Hence it seems likely that solutes would diffuse more rapidly through the cell membranes the higher the temperature, provided the temperature is not high enough to bring about a disorganization of the protoplasm. There are, however, few data dealing with the effect of temperature on permeability of the cell membranes to dissolved substances. Some scattered observations by various workers on the intake of dyes by plant cells indicate that, within limits, the rate of absorption of these substances is increased by temperature, while Stiles and Jørgensen in 1915 found that temperature had a considerable effect on the rate of absorption of hydrogen ions of hydrochloric acid by potato tissue, the temperature coefficient of the reaction over a range of temperatures from 0° to 30° being about 2. But whether this temperature effect is related to permeability is extremely doubtful, since the acid is toxic to the tissue and it may be that the cells are soon rendered irreversibly very permeable and the measured temperature coefficient is largely related to a chemical action between acid and some cell constituent, possibly protein. All the same, it seems likely that temperature influences the rate of entry of solutes into plant cells, and that such changes in rate of entry are reversible.

Light. A number of observations are on record which are supposed to indicate that light has an effect on cell permeability. Actually most of these observations can be interpreted otherwise, but there are one or two which do indeed suggest that light may be effective in this way. Thus Blackman and Paine in 1918 found that exosmosis of electrolytes took place more rapidly from the cells of an excised pulvinus of *Mimosa pudica* in the light than in the dark.

Dissolved Substances in the External Medium. Here again it has frequently been concluded that various substances influence the permeability of the cell membranes. The irreversible increase in permeability produced by toxic substances has already been mentioned. It is possible that the increase in permeability produced by such toxic substances might be reversible if the poison were in a dilute enough concentration or if the time of its action were sufficiently short. Although definite evidence on this point is lacking there is evidence that reversible changes in permeability may be brought about by non-toxic solutions, although generally more than one interpretation of observed experimental results is possible. Among these alleged cases of change in permeability is one recorded in 1905 by Wächter, who found that the presence of potassium nitrate in the medium external to onion bulb scales inhibited the diffusion of sugars from the cells. This can be interpreted in terms of a reduction in permeability produced by potassium nitrate.

Recently Asprey has shown that previous immersion of potato tuber tissue in various salts greatly affects the absorption of ammonium ions by the tissue. Sodium, potassium and lithium chlorides increase the rate of subsequent absorption of ammonium as compared with that taken up by tissue previously treated with distilled water. Previous treatment with calcium, magnesium or aluminium chloride, on the other hand, brings about a lessening of the rate of absorption of ammonium. Whether these changes in the rate of intake of ammonium ions as a result of pre-treatment of the cells with other salts is simply due to the intake proceeding to a different position of equilibrium, or whether there is actually a change in the permeability of the protoplasm to ions is not certain, but the latter alternative is at least a possibility. These observations by Asprey enabled him to bring the phenomena of synergy into line with those of antagonism. Thus he considers the effect produced by the presence of one ion on the absorption of another to be the result of two factors, the competition for the ions in their absorption and the specific effect of the ion on the absorptive capacity of the tissue, an effect which might very well be a change in the permeability of the protoplasm. Should the permeability be increased sufficiently the synergic effect would result.

THEORIES OF CELL PERMEABILITY

Most workers on the relations of plant cells to dissolved substances have adopted the view that the cell can be regarded as a solution surrounded by a protoplasmic membrane which may comprise the whole thickness of the protoplasm or only a thin layer of it. The cell-wall forms a second membrane surrounding the protoplasm, but since much evidence suggests that this is typically readily permeable to most solutes, it cannot as a rule play much part in determining the permeability of the cell membranes as a whole.

If this view is adopted, then the same possibilities exist with regard to the mechanism of the passage of solutes through the protoplasmic membrane as through artificial membranes. It will be recalled that there are three possibilities. If the membrane is heterogeneous with a porous structure, the molecules of the solute may only be able to pass through the phase constituting the pores and hence the membrane acts as a sieve or 'ultrafilter'. On the other hand, if the membrane is homogeneous a dissolved substance must either diffuse through the membrane in solution or form a reversible adsorption or chemical compound with the substance of the membrane. With a heterogeneous membrane on the other hand, it is obviously a possibility that some substances might pass through the phase constituting the pores, while others pass through in solution in the continuous phase. Thus if the protoplasmic membrane is heterogeneous it might act as an ultrafilter to some solutes, while others pass into the cell in solution in the continuous phase, or, for that matter, by combination with this phase. It has already been noted in Chapter II that although the view has been held that the outermost layer of the protoplasm is a film of fat, it is more usual to regard both the protoplasm as a whole and its limiting layers as possessed of a heterogeneous structure.

The ultrafiltration theory of cell permeability finds its chief support from work by Küster and Ruhland on the staining of cells by dyes. The method of investigation described by Küster in 1911 consists in placing the cut end of a shoot in a solution of a dye which is then carried through the plant in the vascular bundles, and from these passes into the neighbouring living cells if the latter are permeable to it. Using this method, Ruhland examined the capacity of 89 acid dyes to penetrate into cells of young plants of *Vicia faba*, and came to the conclusion that the penetrability of the dyes runs parallel with the diffusivity, which itself depends on the size of the molecule or molecular aggregate, the smaller the latter the greater the diffusivity. A similar conclusion was drawn by Ruhland from experiments on the penetration of 30 basic dyes into cells of *Spirogyra* and epidermal cells of onion bulb scales. Thus, since the smaller the molecule or molecular aggregate the more rapid the absorption of the dye, the plasmatic membrane behaves just as an ultrafilter in regard to the entrance of these dyes. Later observations published by Ruhland and Hoffmann on the penetration of a large number of substances of various kinds, both crystalloidal and colloidal, into the cells of the bacterium *Beggiatoa mirabilis*, completely supported the ultrafiltration theory.

A number of exceptions have been recorded to Ruhland's rule

regarding the entry of dyes into plant cells, but although ultrafiltration may not be the only mechanism by which solutes enter the cell, Ruhland's observations afford strong evidence that to many dyes and other substances the cell membranes act as a sieve or ultrafilter.

The chief solution theory of permeability is that put forward by Overton in 1895 and succeeding years, according to which the capacity of solutes to penetrate the protoplasm is a function of their solubility in lipoid substances, which as we have already seen in Chapter II, are generally regarded as essential constituents of the protoplasm and particularly of its limiting layers.

Overton himself examined the permeability of many different kinds of cells to a wide range of substances, including dyes and a number of non-electrolytes, and found that those substances which readily enter cells are soluble in fatty substances (' lipines '), while those which are insoluble in lipines do not readily enter living cells. As with the ultrafiltration theory, so here, also, a number of exceptions to the rule, particularly among dyes, have been recorded, while it is obvious that the penetration of all substances into living cells cannot depend on their solubility in lipines, since water and inorganic salts which are not soluble in fatty substances readily enter cells.

The recent observations of Collander and Bärlund on the permeability of *Chara* cells to non-electrolytes are rather illuminating. Experimenting with some 35 organic compounds, they found a striking parallelism between the lipoid solubility as measured by the partition coefficient of the substance in the system olive oil/water and the rate of entry into the cell. There were, however, some exceptions to this rule. About 20 per cent. of the substances penetrated the cells more readily than was to be expected if the penetrating power depended on solubility in lipines. All these were substances with comparatively small molecules (methyl alcohol, formamide, acetamide, propionamide, cyanamide, ethylene glycol) and the conclusion can be drawn that if the molecules are small enough they can pass not only through the liquid phase of the plasmatic layer, but through the pores in this phase, the membrane thus acting as an ultrafilter. These results support a lipoid-filter hypothesis of cell permeability according to which entrance of solutes into the cell is mainly by solution through a lipoid plasmatic layer. Between the lipoid particles, however, are pores presumably occupied by another phase, through which molecules with a diameter smaller than that of the pores can pass. According to the estimate of Collander and Bärlund, the solute molecules can pass through the spaces between the lipoid micellae if they are less than 0·4 $\mu\mu$ in diameter, which thus gives a measure of the size of the pores. Electrolytes were not dealt with by Collander and Bärlund, but if the phase occupying the spaces between the lipoid micellae is an aqueous one, we may suppose that the passage of ions, at any rate the smaller ones, might take place by diffusion through those pores.

The mechanisms suggested to explain cell permeability so far considered have involved the passage of the solutes into the cell by simple diffusion through one phase or another of the protoplasmic layer. Other views that have been propounded involve a more complex mechanism. Thus Kahho regarded the permeability of salts as related to their capacity for coagulating certain of the protoplasmic colloids, probably the lipoid constituents. Other writers have laid stress on the possibility, as a result of the action of solutes, of alterations in the space relationships of the different phases of the protoplasm, as, for example, by the transference of water from one phase to another, with consequent alterations in permeability, particularly if the solutes pass through only one phase of the plasma-membrane.

There is little experimental evidence at present to enable us to evaluate these theories, but it will be observed that they are not contrary to either ultrafiltration or solution theories, to which, indeed, they can be regarded as additions or further developments. But there are those who deny the very existence of anything functioning as a membrane in living cells. This view has certainly found support chiefly among animal physiologists, for it is clear that in the vacuolated cell, in which protoplasm lines the cell-wall, the whole body of the protoplasm forms a membrane which with the cell-wall separates the vacuole from the external surroundings. We have already seen that the position in non-vacuolated cells requires investigation, and at present it is impossible to say whether the relations of this type of cell to the intake of solutes are best explained in terms of the presence of limiting plasma-membranes or not. However this may be, the passage of a substance into such a cell, if there is no limiting membrane, must depend on diffusion or, if the substance accumulates in the cell beyond the amount demanded by equality of concentration within and without the cell, either chemical combination or adsorption. Such theories of solute absorption by cells, depending on chemical combination or adsorption, have also been applied to absorption through protoplasmic membranes. Szücs, indeed, thought the intake of dyes and various non-electrolytes might be due, as Overton supposed, to their solubility in lipines, while salts enter through adsorption by, or chemical combination with, the protoplasm. The relationship between concentration of solute and amount of its adsorption might be held to support this view, and so might the influence of valency on the antagonistic effect of different ions. However, dyes appear to behave for the most part like inorganic salts in these respects, and should therefore be included rather with the salts than with non-electrolytes. Further, as we have already seen, the tendency at present is to relate the accumulation of inorganic ions in the cell to the establishment of a Donnan equilibrium. There can, however, be no doubt that the various theories of the mechanism of the absorption of electrolytes by plants must still be regarded as tentative, and it is equally certain that the last word has not been written with regard to the way in which non-electrolytes enter living cells.

CHAPTER V

ENZYME ACTION

A VERY important aspect of the life of a plant is the building up of complex materials from less complex substances and the breaking down of the complex substances into simpler ones. While certain specialized cells of the plant, those cells coloured green owing to the presence of chlorophyll, are alone capable of manufacturing carbohydrate from inorganic material, every cell in the plant body appears to possess certain potentialities in regard to the synthesis of more complex organic compounds from simpler ones, or, at any rate, in regard to the breaking down of complex substances into simpler materials. Actions involving the transformation of organic materials are to be regarded as characteristic of all living cells, and therefore to fall into the province of what we regard as the general physiology of the cell.

It is a feature of these actions that they take place readily in the living cell at ordinary temperatures, whereas outside the plant they will in many cases only take place at very much higher temperature and with the aid of reagents which are certainly not present in the plant. To take a simple example, it has been known for many years that in many plant organs starch is hydrolysed into glucose. Outside the plant the rate of conversion of starch and water alone into glucose is so slow as to be negligible. To bring about the hydrolysis of starch outside the plant at a rate comparable with that taking place in the living cell it is necessary to add to the reactants a quantity of acid rendering the medium very much more acid than the cell contents, and to employ a temperature much higher than that present in the living cell.

Now it is well known that substances exist which increase the rate of reactions, but which are not used up themselves, since at the end of the reaction they remain unchanged. These substances are called catalysts, and it is to the presence and action of catalysts in living cells that the comparatively rapid material transformations in these cells are to be ascribed. Substances can be extracted from living cells and tissues which enable these transformations to take place readily outside the plant at ordinary temperatures, and it is natural to conclude that it is owing to the presence of these substances in the cell that the actions in question proceed so readily. To such a catalyst present in living cells the name *enzyme* or *ferment* is given.

Certain actions which we know are brought about by enzymes have been recognized for a very long time. Among such, alcoholic fermentation, the production of ethyl alcohol from sugar through the agency of the enzymes present in yeast, is the most noteworthy, for the operation of this process has undoubtedly been practised for thousands of years. Other long-recognized processes due to the action of enzymes are digestion and putrefaction. It was not, however, until the nineteenth century that the mechanism of enzyme action came to be in any way understood, and even now, after a hundred years of development of our own knowledge of enzymes, much obscurity remains concerning them and their mode of action.

The history of enzyme action is generally regarded as commencing with the work of Dubrunfaut, who in 1830 prepared from germinating barley (malt) an extract which possessed the power of converting starch into sugar. In 1833 Payen and Persoz prepared from such an extract, by precipitation with alcohol, a substance which would bring about this conversion of starch into sugar. To this substance the name diastase was given.

In the same year de Saussure had prepared a substance from germinating wheat which brought about the same action, and which he recognized as possessing the same properties as Payen and Persoz's diastase. The discovery of other enzymes in plants and animals followed: in plants notably that of emulsin in bitter almonds which was found to break down the glycoside amygdalin present in those seeds.

Meanwhile the process of alcoholic fermentation had attracted the attention of chemists. Lavoisier and Gay-Lussac had shown that it consisted in the breaking down of sugar into alcohol and carbon dioxide, while Berzelius, who was responsible for the conception of catalysis, regarded alcoholic fermentation and analogous processes, such as acetic acid and lactic acid fermentations, as catalytic actions.

Pasteur showed that alcoholic fermentation was brought about by the action of the living yeast cells. It thus appeared that such catalytic actions could be induced either by living cells as in the case of alcohol, acetic acid and lactic acid fermentations, or by non-living substances such as diastase and emulsin. It thus came about that such catalysts or ferments were regarded as forming two classes; organized ferments such as yeast and certain bacteria, which were living cells, and unorganized ferments like diastase and emulsin which acted independently of living cells. The use of the term enzyme for unorganized ferments was proposed by Kühne in 1878.

The distinction between organized and unorganized ferments disappeared as the result of the work of Buchner in 1897. Yeast was ground up with fine quartz sand and the juice pressed out from the broken cells. The liquid so obtained was found to be capable of fermenting glucose and it therefore appeared that the fermenting power of yeast is due to the presence in it of an enzyme which can be extracted and retain its activity outside the organism, just as diastase, after extraction from the plant, can bring about the hydrolysis of starch. The words ferment and enzyme thus became synonymous and although in this country the word enzyme is generally used, on the Continent the word ferment is also used quite frequently.

At the present time a very great number of enzymes have been recognized in living plants, and the leading part they play in material transformations in the plant realized. For the substance acted upon by an enzyme the term *substrate* is now generally used. While much has been learnt about their properties and mode of action during the last thirty years, our knowledge of enzymes must be regarded as still in its infancy, for so long as the chemical constitution of no single enzyme is known, they must be regarded as in some considerable degree mysterious.

GENERAL PROPERTIES OF ENZYMES

Enzymes, being catalysts, possess the general properties of those substances, but at the same time exhibit very definite characteristics in which they differ from inorganic catalysts. Various definitions of enzymes have been proposed by different authorities and it is difficult to find a really satisfactory definition which will include all undoubted enzymes and exclude substances definitely not enzymes. Bayliss defined enzymes simply as ‘ catalysts produced by living organisms ’. E. F. Armstrong's definition was that they are ‘ selective, colloidal catalysts, present in living cells and destroyed by heat ’. J. B. S. Haldane defined an enzyme as ‘ a soluble, colloidal, organic catalyst, produced by a living organism ’. Bayliss objected to their being described as selective, as some, for example, emulsin, act on a great variety of compounds ; moreover, some inorganic catalysts may be described as selective. He also disapproved of the introduction of the word ‘ colloidal ’ on analogous grounds ; it is doubtful whether the active constituent of an enzyme is always colloidal, while some inorganic catalysts are colloidal. Destruction by heat, which is involved in Armstrong's definition, was also not accepted by Bayliss since the actual temperature at which the enzyme is rendered inactive varies with the enzyme, while again, some inorganic catalysts are thermolabile. It appears that Bayliss would definitely have included as enzymes systems that Haldane excludes. In the present inadequate state of our knowledge there is perhaps, as yet, insufficient reason to depart from Bayliss's point of view that it is merely a matter of convenience to separate enzymes from other catalysts. Nevertheless there are substances which would be included in Bayliss's definition which appear to be unlike enzymes in most of their characteristics, and at present it is convenient to limit enzymes at any rate to colloidal substances. The selective action of enzymes varies greatly, and it seems inadvisable to emphasize this as a distinctive quality of enzymes, although it is often of the highest importance. The destruction of the enzyme by heat is also a very general property, but Keilin has demonstrated the presence of thermo-stable substances in yeast and in other organisms, which have enzymatic properties. To exclude certain surfaces which appear to bring about oxidations and reductions in the living cell, it has also been proposed to limit the term enzyme to soluble catalysts of the kind mentioned, but the fat-splitting enzyme of *Ricinus* is insoluble in water. There are thus a few substances or systems which would be included as enzymes by some authorities and not by others.

Certain properties of enzymes are shared by all catalysts. Thus they alter the rate of chemical reactions but are generally considered not to initiate them. Although they change the rate of the reaction they are not themselves used up, that is, they are not bound up in the end-products of the reaction but remain qualitatively and quantitatively unchanged at the end of it.

This statement that catalysts alter the rate of reactions requires some amplification. It has been pointed out by J. B. S. Haldane that in organic chemistry side reactions are usual; thus under a particular set of conditions, and in absence of a catalyst, 99 per cent. of a substance, X, may be converted into another, A, but the remaining 1 per cent. may be converted into a different substance, B. A catalyst may bring about an acceleration of the latter process but not of the former, and so the whole of the reactions in which X is involved are quantitatively altered by the catalyst. The same considerations hold with regard to enzyme actions. One enzyme may, for example, accelerate the production of A from X, another the production of B, and so, according to the enzyme present, X will meet with different fates. Hence enzymes may actually determine the direction of actions in the organism.

Catalytic actions are reversible and it has been shown that a number of enzyme actions also are reversible, the action proceeding to a definite equilibrium point, the direction of the action depending upon the active masses of the reacting substances present. The question of reversibility of enzyme actions will be dealt with in some detail later.

The other characteristics of catalytic actions are also exhibited by enzymes. A very small quantity is often sufficient greatly to alter the rate of the reaction, while, with low concentrations of enzyme the degree of acceleration of the action is proportional to the concentration of catalyst present relative to the concentration of the substance acted upon.

Other characteristics of enzymes are not shared by catalysts in general. In the first place, since no enzyme has been isolated in sufficient purity for its chemical composition to be determined, enzymes can only be identified by their activity. Secondly, although Bayliss objected to their selective action being emphasized in defining them, enzymes are nevertheless specific in their action, but the degree of specificity varies. Thus the enzyme invertase, sucrase or saccharase which catalyses the hydrolysis of cane sugar into glucose and fructose, is not known to act on any other substance; zymase, the enzyme complex responsible for alcoholic fermentation, acts on three hexose sugars, *d*-glucose, *d*-fructose, and *d*-mannose, but not on any other, while one protein-splitting enzyme appears to possess the power of breaking down many different proteins. But whether many similar actions, as for example the hydrolysis of various fats, are brought about by one enzyme, or whether a different enzyme is responsible for each action or a relatively small number of similar actions is really not known; it is not known, that is, whether there are relatively few or many fat-hydrolysing enzymes, and the same ignorance exists in regard to many other enzyme actions.

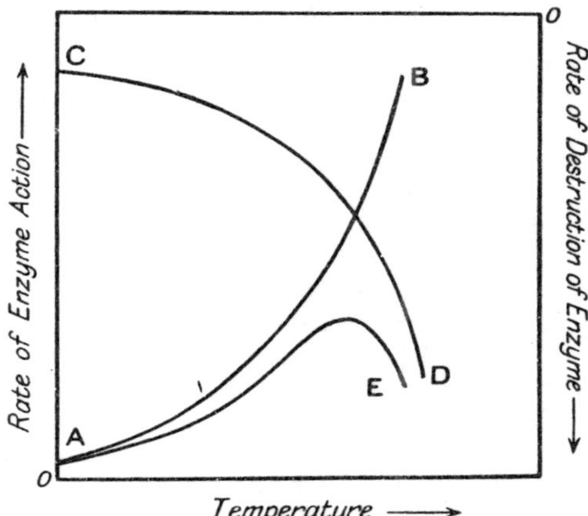

FIG. 11.—Curves showing the relation between temperature and the rate of enzyme action (After Duclaux)

As is generally the case with chemical reactions taking a measurable time, the rate of an enzyme action increases with rise in temperature, the rate being very frequently increased about 2 to 3 times for a rise in temperature of 10° C., although for some enzyme reactions temperature coefficients outside the ordinary range of values have been observed. It should also be noted that the coefficient is not necessarily constant over a wide range of temperatures, and in the case of enzyme actions, owing to the destruction of the enzyme by heat, the range is often very small. Above 50° C. the instability of most enzymes rapidly increases with the temperature. The observed effect of temperature on enzyme actions is thus the combined result of a direct effect of temperature on the rate of the reaction and the influence of temperature on bringing about destruction of the enzyme. A reference to Fig. 11 will make it clear that there will thus be apparently an optimum temperature for the enzyme action. Here the curve AB represents the temperature effect of the action alone and CD the relation between temperature and rate of destruction of the enzyme.

The observed relation between rate of enzyme action and temperature will then be represented by the curve AE. But, as pointed out by Kanitz in 1915, the optimum is not a fixed point but will depend on the quantity of the enzyme present. With a larger quantity of enzyme the action will proceed more rapidly than with a smaller; consequently, in the former case at high temperature, more of the substrate will have been converted after a certain time than in the latter, so that at the same relative stages of substrate conversion in the two cases a greater proportion of the original quantity of enzyme will remain active in the former than in the latter case. The high temperature will thus have less effect in reducing the rate of enzyme action with the higher concentrations of enzyme, and so the optimum temperature will appear to be higher. Again, since the quantity of enzyme inactivated will increase with the time of action, the optimum observed will depend on the time for which the experiment has proceeded when the rate of the action is measured; the longer the period during which the enzyme has been exposed to the high temperature, the lower will the optimum temperature appear to be.

Co-enzymes. In some cases it is possible to separate an active enzyme preparation into two parts, one of which is colloidal and thermo-labile and which is therefore to be regarded as containing the enzyme proper, while the other is crystalloidal and thermo-stable and which has therefore not the general characteristics of enzymes, but which nevertheless must necessarily be present for the enzyme to function. Such a substance is termed a co-enzyme. Thus an active preparation of zymase from yeast can be separated by dialysis into a colloidal and a crystalloidal portion. Both fractions are inactive, but when brought together again the enzymatic activity is renewed. The colloids in the non-dialysable portion are regarded as including the enzyme proper, zymase, which is not only colloidal but also destroyed in solution by a temperature of 50° C.; the dialysable fraction which can withstand a temperature of 100° C., is said to contain the co-enzyme. This fraction contains much soluble phosphate, and it was at one time thought that the co-enzyme might be phosphate. However, addition of soluble phosphate to the enzyme fraction does not restore its activity, whence it is concluded that some other substance is the co-enzyme required for the functioning of the zymase. This substance, according to the researches of Euler and his collaborators, appears to be a complex organic substance involving a base adenine and pentose sugar groupings. But, in addition to this organic substance, phosphate is necessary for the functioning of zymase. Bayliss therefore considered zymase required two co-enzymes for its functioning. J. B. S. Haldane would limit co-enzymes to organic substances of a high degree of specificity, which would exclude phosphate from the category of co-enzyme. In the present state of our knowledge it is perhaps not of the highest importance whether we so limit our definition of co-enzymes or not.

Accelerators. Frequently the rate of an enzyme action is increased by the presence of some other substance. The acceleration of laccase action by a low concentration of a manganese salt may be cited as an example. Such substances, a number of which have been recorded as accelerating the action of different enzymes, have been termed accelerators or activators.[1]

Inhibitors. Just as some substances accelerate enzyme actions so others retard them or stop them altogether. Among such substances are salts of heavy metals, alcohols and aldehydes. In some cases the enzyme suffers no permanent damage, recovering its activity on removal of the harmful substance; in other cases the reduction of enzyme activity is irreversible. Such substances are termed inhibitors, inactivators, paralysers or poisons. Different inhibitors act differently on different enzymes.

Reaction of the Medium. The hydrogen-ion concentration of the medium in which the enzyme action takes place has a very considerable influence on the activity of the enzyme. An optimum *p*H can usually be determined for each enzyme action, but the optimum may vary somewhat according to the substrate and accompanying substances. The majority of enzymes exhibit an optimum between *p*H 4 and *p*H 8, so that they tend on the whole to be more active in weakly acid media. Practically all enzymes are only active over a certain limited range of *p*H, and exposure to a too strongly acid or alkaline medium may result in destruction of the enzyme.

Preparation and Purification of Enzymes. The general principle involved in the preparation of most enzymes is to render the cell membranes permeable to the enzyme, extract the latter with a solvent and then precipitate the enzyme with a suitable reagent. In this way a crude preparation of the enzyme is obtained. For purifying this preparation various methods are employed, such as re-solution followed by re-precipitation, dialysis, and adsorption followed by elution.

For rendering the cell membranes permeable a number of different methods have been used. Sometimes the cells are ground up with fine sand; this results in disintegration of the cells, and their liquid contents can then be pressed out of them. Another method consists

[1] The term co-enzyme was originally introduced by Bertrand in 1897 for a substance which brings about a great increase in enzymatic activity, the particular case involved being that of manganese salts which in very dilute concentration greatly increase the activity of the oxidizing enzyme laccase. Bayliss distinguished between a co-enzyme, which is necessary for an action to proceed at all, and an accelerator, which increases the rate of an action which does proceed without it. J. B. S. Haldane employs a different criterion, that of specificity, and regards a co-enzyme as a thermo-stable crystalloidal organic substance of fairly high specificity, and an activator as a non-specific substance which 'permits or increases' enzyme activity. Waksman and Davison speak of 'specific accelerators or co-enzymes' and include inorganic salts among these. It should be noted, therefore, that the conceptions of co-enzyme and activator (or accelerator) of different authorities differ.

in treating the cells with alcohol or acetone or some other reagent for a short time. The soluble contents of the cell are then capable of diffusing out of it. Yet another method consists in drying the tissue, generally at a moderately low temperature (20° C. to 30° C.) since too high a temperature, as we have already noted, brings about destruction of the enzyme. In exceptional cases, however, higher temperatures can be employed for a short time, especially after the cells have lost most of their water, since enzymes are more resistant to high temperature when dry. In yet other cases the cells are rendered permeable to certain enzymes by *autolysis*. The cells are allowed to stand for some time in water in presence of toluene or some other antiseptic to prevent bacterial action. Some of the enzymes in the cell bring about destruction of the protoplasm which thus becomes permeable to the cell enzymes.

Most enzymes are soluble in water and glycerol, and these are the solvents most generally used to extract the enzymes. The water or glycerol extract contains a very great deal of dissolved material besides enzyme. Precipitation of the latter is usually effected with alcohol, and a considerable amount of the accompanying substances remains in solution.

A certain amount of purification can often be effected by re-solution and re-precipitation. Crystalloidal substances can be separated from the colloidal enzymatic and other material by dialysis, using membranes impermeable to the enzyme but permeable to crystalloids. In one case, that of invertase from yeast, the separation of coloured impurities has been effected by using membranes permeable to the enzyme but impermeable to the coloured material. It is often possible to separate the enzyme from impurities by adsorbing it. A large number of adsorbents have been used for this purpose, the chief being charcoal, kaolin and aluminium hydroxide. The enzyme is then freed from the adsorbent by adding a substance which is preferentially adsorbed, or by altering the reaction of the medium.

It should be noted that very few enzymes have been obtained in a state of even approximate purity.

CLASSIFICATION OF PLANT ENZYMES

Plants have now been shown to contain a very large number of enzymes and it is not merely convenient, but necessary, to classify them. If their chemical composition were known chemical constitution would obviously form a sound basis for their classification, but as the composition of no single enzyme is known the only satisfactory method of classifying them must depend on the reactions they catalyse. Thus many of the best-known enzymes catalyse hydrolytic actions such as the hydrolysis of starch to maltose, effected by diastase (amylase), the hydrolysis of sucrose to glucose and fructose by invertase (sucrase or saccharase) and the hydrolysis of fat into fatty acid and glycerol by lipase. Such enzymes have been termed hydrolases. Other enzymes break down the C—C linkage. Such is carboxylase, which breaks down α-ketonic acids to aldehyde and carbon dioxide. Enzymes of this type have been called desmolases. A third group of plant enzymes of great importance are concerned in oxidations and reductions in the plant. For these the name oxido-reductases has been suggested, but, as will be seen later, in the present state of knowledge it is better to use the term oxidizing enzymes, for the term oxido-reductase is conveniently restricted to what may be called one particular class of oxidizing enzyme. Many so-called enzymes it is impossible to classify in one of these groups as they are undoubtedly mixtures of enzymes; among such, zymase is a notable example. Other enzymes do not fall into any of them, as, for example, catalase, which catalyses the splitting of hydrogen peroxide into water and molecular oxygen.

Of the hydrolytic enzymes a number act on complex carbohydrates, breaking them down to simple carbohydrates, generally sugars; these can be grouped together as carbohydrases. Other hydrolytic enzymes, the glycosidases, break down glycosides into the constituent sugar and the other component or components of the glycoside. Yet other hydrolases break down fats and other esters into their constituent acid and alcohol, while others, the proteases, bring about the hydrolysis of proteins.

Zymase has been regarded as the most definite example of a desmolase, splitting up hexose into alcohol and carbon dioxide, but it is now recognized that zymase is a mixture of a number of enzymes which fall into different groups, one certainly being a desmolase, but others belonging respectively to the categories of hydrolases and oxidizing enzymes.

The oxidizing enzymes are perhaps the most difficult to classify, as indicated by the different systems of classification adopted by different authorities. The system adopted here is based on the source of the oxygen utilized for oxidation, enzymes catalysing oxidation by molecular oxygen being termed oxidases, those in which oxidation is effected by hydrogen peroxide being called peroxidases, while those in which the oxygen is provided by water being termed dehydrases or oxido-reductases.

Such a classification is only to be regarded as temporary, and no doubt as more is discovered concerning enzymes and their mode of action, a generally acceptable system of classification will be evolved.

Hydrolytic Enzymes

In Table XXVIII are recorded the more important hydrolytic enzymes found in plants, with the substrate on which they act, and the products of the respective actions. It must be recognized that often the so-called enzyme is actually a mixture of enzymes or represents a group of related enzymes.

Table XXVIII

Plant Hydrolytic Enzymes

Enzyme	Substrate	Products
Diastase (Amylase)	Starch	Maltose
Maltase	Maltose	Glucose
Invertase (Sucrase)	Sucrose	Glucose + Fructose
Cellulase	Cellulose	Cellobiose
Cellobiase	Cellobiose	Glucose
Cytase (Hemicellulase)	Hemicelluloses, Mannans, Galactans, Pentosans	Glucose and other monosaccharides
Inulase	Inulin	Fructose
Pectinase	Pectin	Reducing sugars
α-glycosidase (Maltase)	α-glycosides	Glucose + other products
β-glycosidase (Emulsin) (Amygdalase + Prunase)	β-glycosides	Glucose + other products
Sinigrinase (Myrosin)	Sinigrin	Glucose + $KHSO_4$ + Allyl thiocyanate
Lipase (Esterase)	Esters	Acid + Glycerol
Tannase	Tannin	Glucose + Gallic acid
Phosphatase	Fructose diphosphate	Fructose + Phosphate
Proteases, including		
Pepsin	Proteins	Albumoses + Peptones
Trypsin	Proteins	Peptides + Amino-acids
Erepsin	Peptones, peptides	Amino-acids
Urease	Urea	Ammonia + Carbon dioxide
Chlorophyllase	Chlorophyll	Chlorophillide + phytol

Diastatic Enzymes

As mentioned earlier, a principle that would convert starch into sugar was obtained from germinating barley by Dubrunfaut and from germinating wheat by de Saussure, both in 1833. The name diastase was given to the principle by the former worker and for long was retained as the name for the enzyme, or rather group of enzymes, bringing about the conversion of starch into glucose. Subsequently diastase has been found in a wide range of plant organs, notably in seeds containing starch and in starch-forming leaves, but also in other plant material, including various Algae, Fungi and Bacteria. The complete breaking down of starch into sugar is now recognized as taking place in a number of stages for which at least two distinct enzymes are responsible. Amylase, or diastase in a restricted sense, brings about the degradation of the polysaccharide starch, $(C_6H_{10}O_5)_n$ to the disaccharide maltose, $C_{12}H_{22}O_{11}$, while maltase effects the hydrolysis of the maltose to glucose, $C_6H_{12}O_6$.

There is also evidence that amylase is a mixture of at least two enzymes, amylase proper which brings about the conversion of starch to dextrin, and dextrinase which breaks down the dextrin to maltose. Samples of diastase obtained from different plants differ in their properties, as for example, in regard to the apparent optimum temperature of action and in other ways. In this respect particular mention may be made of the diastase obtained from the fungus ***Aspergillus oryzae***, the enzyme generally known as taka-diastase, in comparison with that obtained from higher plants such as malt diastase. Pure preparations of taka-diastase have a much greater power of liquefying starch than presumably equally pure preparations of malt diastase, while the latter have a greater power of producing maltose. From this difference in behaviour it might be concluded that fungus diastase contains a relatively higher proportion of amylase and maltase and lower proportion of dextrinase than diastase prepared from malt. The different behaviour of different diastase preparations may also be due to differences in the impurities accompanying the enzyme. Thus Brown and Morris found that tannin inhibits the action of diastase, while the hydrogen-ion concentration of the medium also affects the activity of the enzyme.

Maltase is probably widely distributed through the plant kingdom, but its investigation has not been easy as it is not readily extracted by water from plant tissues, while it is sensitive both to temperature and certain antiseptics. The optimum temperature for its action is about 40° C., while it is destroyed at temperatures a little above 50° C. It is also destroyed by chloroform and alcohol. Its presence has, however, been demonstrated in barley, various leaves including those of *Tropaeolum*, *Dahlia*, sunflower and mangold, and certain Fungi and Bacteria, and it is reasonable to suppose that it is present in all starch-containing plants.

Invertase

This enzyme brings about the hydrolysis of sucrose into glucose and fructose according to the equation :

$$\underset{\text{sucrose}}{C_{12}H_{22}O_{11}} + H_2O = \underset{\text{glucose}}{C_6H_{12}O_6} + \underset{\text{fructose}}{C_6H_{12}O_6}$$

This is the change known as inversion of cane sugar, whence the older name invertase given to this enzyme by its discoverer Berthelot in 1860. It is now frequently termed sucrase or saccharase to emphasize the nature of the substrate on which it acts. It is more resistant to heat than maltase, the optimal temperature for its action being in the neighbourhood of 52° C., while it is not affected by alcohol at temperatures below 30° C.

The presence of invertase has been demonstrated in many plant tissues, including leaves, stems, and other organs of higher plants, and various fungi, including yeasts and species of *Aspergillus* and *Penicillium*.

Cellulose-destroying Enzymes

The enzymes bringing about the decomposition of cellulose are less known than the diastatic enzymes and invertase. They are more generally considered as falling into two groups : (*a*) cellulases and cellobiases which between them break down true cellulose to glucose, and (*b*) cytases, which hydrolyse hemicelluloses such as mannans, galactans, pentosans and related substances.

True cellulases have so far been found in very few organisms, these, among plants, being limited to only a few fungi and bacteria. Among the latter are thermophilic bacteria which contain both cellulase and cellobiase. The former breaks down the polysaccharide cellulose to the disaccharide cellobiose, while the cellobiase completes the breakdown to glucose. These actions may be represented by the equations :

$$\underset{\text{cellulose}}{(C_6H_{10}O_5)_n} + nH_2O = \underset{\text{cellobiose}}{nC_{12}H_{22}O_{11}}$$

$$C_{12}H_{22}O_{11} + H_2O = \underset{\text{glucose}}{2C_2H_{12}O_6}$$

Cytases appear to be very much more widely distributed. They occur in many fungi and in the seeds of various plants such as the date palm and lupin in which so-called cellulose forms a food reserve. It seems very probable that the cytases comprise a number of distinct enzymes.

Inulase

An enzyme bringing about the hydrolysis of the polysaccharide inulin into fructose was first described in 1888 by Reynolds Green . sprouting tubers of the Jerusalem artichoke (*Helianthus tuberosus*), in which considerable reserves of inulin are stored. It has since been found to occur in many monocotyledons as well as in yeasts and a number of other fungi, and in bacteria. The active enzyme does not appear to be present in resting tubers of the Jerusalem artichoke but develops when these sprout ; it is therefore said to be present in the resting tuber as a zymogen, that is, in a form capable of changing into, or giving rise to, the active enzyme under certain conditions.

Enzymes Acting on Pectic Substances

Pectic substances, which form an essential constituent of plant cell-walls, are complex derivatives of sugars in which both pentose and hexose units are included. They appear to be built up on a basis of pectic acid, the molecule of which consists, according to Nanji, Paton and Ling, of a closed ring of one arabinose unit, one galactose unit and four galacturonic acid units with four carboxyl groups forming side chains. Pectic acid is insoluble in water and so are the alkaline earth pectates ; the salts of pectic acid with the alkali metals are, however, soluble in water. On addition of acid to solutions of alkali pectates the free pectic acid separates as a gel.

The greater part of the pectic substances of plants, including that in the cell-wall and cell-sap, are probably methyl esters of pectic acid. They may be distinguished as *soluble pectins* or simply *pectins*. Other derivatives of pectic acid are insoluble ; such *insoluble pectins* or *pectoses* are probably compounds of pectic acid partly with methyl and partly with cellulose groupings.

While the chemistry of pectic substances cannot be regarded as by any means settled, we may thus provisionally consider pectic substances in plants to fall into the following categories : (1) free pectic acid, (2) salts of pectic acid (pectates), (3) methyl esters of pectic acid (soluble pectin), (4) compounds of pectic esters and cellulose (insoluble pectin or pectose).

There is evidence that at least three different enzymes are concerned in the breaking down of pectic substances. These are (1) pectosinase, which acts on pectose, breaking it up presumably into soluble pectin and cellulose, (2) pectase, which hydrolyses soluble pectin to pectic acid and methyl alcohol, and (3) pectinase, which hydrolyses pectic substances to their constituent hexose and pentose sugars. It has been usual to regard pectase as a 'clotting' enzyme, because it brings about the production of the pectic acid in the form of a gel or gelatinous precipitate, but if the view of its action given here is correct, it would appear actually to be a hydrolytic enzyme. In the breaking down of cell-walls in the plant, either in the normal course of the life of the individual, or as a result of fungal attacks. it seems probable that pectin-destroying enzymes play a much more important part than enzymes attacking cellulose. Pectinase is also present in a number of bacteria, as for instance *Bacillus carotovorus*, the organism responsible for the soft rot of carrots. The presence of pectin-destroying enzymes in various micro-organisms is made use of in the process known as *retting* of flax, in which the fibres of the flax are disintegrated by removal of the pectin through the agency of the pectin-destroying enzymes in various bacteria and fungi.

The pectic enzymes are widely distributed, not only in fungi and bacteria, but also in higher plants. Among tissues in which pectase has been found are carrot root, potato tubers, clover, lucerne and maize, while pectinase can be prepared, not only from various fruits and storage organs, but from the fungus *Rhizopus tritici*.

Glycosidases

The glycosides are compounds of glucose with other substances including alcohols, aldehydes, phenols and many other substances, the condensation of the glucose and other molecule taking place through a hydroxyl group. Thus α-glucose with methyl alcohol yields α-methyl glycoside :

```
 H      OH                              H      OCH3
   \   /                                  \   /
     C                                      C
   /   \                                  /   \
HCOH     O                             HCOH     O
 |       |       + CH3OH  =             |       |        + H2O
HOCH   HC.CH2OH                        HOCH   HC.CH2OH
   \   /                                  \   /
     C                                      C
   /   \                                  /   \
 H      OH                              H      OH
  α-glucose                          α-methyl glucoside
```

Similarly, β-glucose yields β-methyl glycoside. Other sugars may form similar combinations; thus we have, for example, galactosides, pentosides and so on.

Where the sugar molecule condenses with another sugar molecule, either of the same, or of different kind, a disaccharide results. Thus the disaccharides maltose, cellobiose and gentiobiose are all glycosides of glucose, while sucrose is a fructose glycoside.

The glycosidases are the enzymes which break up the glycosides into their constituent sugar and other substance. The best known are emulsin and myrosin. The former breaks down amygdalin, a glycoside occurring in almonds, plum-stone kernels and elsewhere, into its constituents, glucose, benzaldehyde and hydrocyanic acid. It is undoubtedly a mixture involving certainly two, and possibly three, distinct enzymes. The first stage in the degradation of amygdalin is brought about by the enzyme amygdalase which splits off a molecule of glucose leaving as a residue the glycoside prunasin:

$$\underbrace{C_6H_5.CH(CN).O.C_6H_{10}O_4.O.C_6H_{11}O_5}_{\text{amygdalin}} + H_2O = \underbrace{C_6H_5.CH(CN)O.C_6H_{11}O_5}_{\text{prunasin}} + C_6H_{12}O_6$$

The prunasin is now hydrolysed by means of prunase into glucose and mandelonitrite:

$$C_6H_5.CH(CN)O.C_6H_{11}O_5 + H_2O = \underbrace{C_6H_5.CHOH.CN}_{\text{mandelonitrite}} + C_6H_{12}O_6$$

The mandelonitrite is next converted into benzaldehyde and hydrocyanic acid, probably by the action of a third enzyme, oxynitrilase:

$$C_6H_5.CHOH.CN = C_6H_5.CHO + HCN$$

The last enzyme, it will be observed, is not actually a glycosidase, although it may be regarded as forming part of the emulsin complex.

Myrosin, myrosinase or sinigrinase occurs in mustard and many Cruciferae, especially in the seeds, as well as in some plants belonging to other orders. It brings about the hydrolysis of the glycoside sinigrin to glucose, potassium hydrogen sulphate and allyl thiocyanate (oil of mustard):

$$C_{10}H_{16}O_9NS_2K + H_2O = C_6H_{12}O_6 + C_3H_5NCS + KHSO_4$$

The glycosidases can be grouped into α-glycosidases and β-glycosidases according as the glycosides they hydrolyse are derived from α-glucose or β-glucose. Under this classification emulsin is a mixture of β-glycosidases, while maltase can be regarded as an α-glycosidase.

Lipases

The lipases constitute in all probability a group of enzymes which hydrolyse esters into their constituent alcohol and acid. They have been differentiated into lipases proper which act upon true fats, converting them into fatty acids and glycerol, and butyrases, which break down esters of lower fatty acids, as, for example, ethyl butyrate. Preparations of lipase from most sources hydrolyse not only fats but also lower esters; it appears, from the work of Falk, that such preparations contain two enzymes, a true lipase and an esterase, which can be separated by their differential solubility in water and 1·5 N sodium chloride, the esterase being more soluble in the former and the lipase in the latter. Much work, however, remains to be done to establish definitely the specificity of the lipases and esterases.

Since the seeds of approximately 80 per cent. of the species of flowering plants contain reserve food in the form of fat, it is to be expected that lipases are widely distributed in seeds. Lipase was, in fact, first reported from germinating seeds of *Ricinus communis* by Reynolds Green in 1890. He found, however, that the active enzyme is not present in the resting seed, but that addition of acetic acid to the latter resulted in the production of the active enzyme, whence it is concluded that lipase is present in resting *Ricinus* seeds in the form of a zymogen which is activated in some way by the acid. In some resting seeds the active enzyme has been demonstrated. This is the case with hemp, flax and rape, and a number of other seeds. Lipase has also been obtained from *Penicillium* and *Aspergillus*. The reversibility of lipase, or rather esterase, action is one of the easiest cases of reversibility of enzyme action to demonstrate. Further reference to this will be made later.

Proteases

Enzymes concerned in the breaking down of proteins fall into three groups: pepsin, trypsin and erepsin. Pepsin brings about the hydrolysis of the more complex proteins to the relatively simple peptones and albuminoses (see Chapter X), while erepsin hydrolyses these simpler substances of protein character into amino-acids. Trypsin brings about the hydrolysis of the complex proteins to amino-acids and thus is equivalent in its action to the combined actions of pepsin and erepsin.

Although first recognized in the animal body, proteolytic enzymes are widely distributed through the plant kingdom. Their presence in seeds containing reserves of protein was established in 1874 by Gorup-Besanez, who demonstrated the presence in seeds of vetch, hemp, flax and barley, of an enzyme capable of converting fibrin, a complex protein, to peptone and therefore to be regarded as a pepsin. Subsequent investigators have found proteolytic enzymes, or, at any rate zymogens, present in many other seeds.

In 1874 the presence of a proteolytic enzyme in the liquor found in the pitchers of *Nepenthes* was demonstrated by Hooker, and this enzyme was later shown by Gorup-Besanez to resemble a pepsin.

In 1880 Wurtz described a proteolytic enzyme found in the fruit and other parts of the papaw (*Carica papaya*), and subsequent investigation has shown that this enzyme is capable of breaking up natural proteins to amino-acids. It was therefore regarded as a 'vegetable trypsin' and the name *papain* given to it. Later an enzyme of similar character was demonstrated in the fruit of the pineapple; to this enzyme was given the name *bromelin*. In the latex of the fig a third trypsin, *cradein*, was discovered.

In the early years of the present century a series of investigations on the proteases of plants was made by Vines. He found that an erepsin is present in every plant, and practically every part of the plant, he examined, including such varied material as the pileus of *Agaricus*, the root of *Dahlia*, and germinating barley.

Vines also investigated the vegetable trypsins and came to the conclusion that they are not single enzymes, but mixtures of pepsin and erepsin. Thus he was able to separate the pepsin constituent of papain from the erepsin and obtained a preparation of the pepsin which digested fibrin, but which was incapable of decomposing peptone. He also succeeded in effecting a separation of the pepsin and erepsin present in hemp seeds.

As far as plants are concerned, there is therefore at present no evidence of the existence of trypsins as distinct enzymes. In general it may be concluded that these are mixtures of pepsin and erepsin. The latter appear to be universally distributed in plants, but pepsins have not, so far, been shown to be of such wide occurrence.

Urease

An enzyme which hydrolyses urea has been known for a long time, but it is only during the present century that its wide distribution through the plant kingdom has been recognized. It occurs in many bacteria and fungi, and its presence has been demonstrated in various organs of a number of higher plants, notably in the seeds of the soya bean. Its action is to hydrolyse urea into carbon dioxide and ammonia:

$$(NH_2)_2CO + H_2O = 2NH_3 + CO_2$$

Zymase

The isolation from yeast of the enzyme capable of breaking down glucose into ethyl alcohol and carbon dioxide was first effected by Buchner in 1897. About 70 per cent. of the water was first pressed out of the yeast and the residue then ground up with a mixture of quartz sand and kieselguhr. Such treatment breaks up the yeast cells and from the resulting mixture a juice containing the enzyme complex zymase can be pressed out. This juice contains, however, a number of other enzymes as well, including invertase, maltase and protease.

In the tissues of higher plants zymase was first shown to be present by Stoklasa and Czerny in 1903. It has subsequently been shown to occur in various moulds and bacteria as well as in yeasts. It may be universally distributed throughout the plant kingdom.

Zymase acts upon three sugars only, namely *d*-glucose, *d*-mannose and *d*-fructose. Yeast and yeast juice will also break down sucrose as this is first hydrolysed into glucose and fructose by invertase contained in the yeast. Similarly, maltose is hydrolysed by the maltase present in the yeast into glucose which is then broken down by the action of the zymase.

Although for a time after its isolation zymase was regarded as a single enzyme it is now recognized as a mixture of enzymes, each one of which appears to be responsible for one particular stage in the breaking down of sugar into alcohol and carbon dioxide. The enzymes forming part of, or associated with, the zymase complex are hexosephosphatase (or phosphatese), phosphatase, glycolase, carboxylase and ketonaldehydemutase.

Hexosephosphatase and Phosphatase. Reference has already been made to the fact that an active preparation of zymase can be separated into three parts: (1) the enzyme itself which is organic and colloidal, (2) co-zymase, which is organic and crystalloidal, and (3) phosphate. The last named is necessary for the formation of compounds with the hexose, the first stage in fermentation being the formation of hexosephosphates. There are probably both hexose monophosphates and hexose diphosphate formed, it being suggested that the formation of the monophosphates precedes that of the diphosphate. It is stated that whichever of the three fermentable sugars is the substrate for this action, the diphosphate produced is always fructose diphosphate. The enzyme responsible for this synthesis is called hexosephosphatase or phosphatese. The fructose diphosphate is broken down into fructose and phosphate by a different enzyme, phosphatase, which has the power of breaking down organic phosphates generally. The result of the action of these two enzymes appears to be the production of fructose, probably in an active form.

Glycolase. An enzyme which breaks down hexose to methyl glyoxal was shown to be present in yeast by Neuberg. The same or a similar enzyme has also been demonstrated in tobacco leaves by Kobel and Scheuer. To this enzyme the name glycolase has been given. It is not to be confused with the enzyme present in animal tissues which breaks down sugar to lactic acid. To this latter enzyme the name glycolase was also given, but it is better to follow J. B. S. Haldane's nomenclature and refer to this enzyme as myozymase. It is probably a mixture of enzymes possibly including the true glycolase.

In zymase action glycolase probably acts on the fructose produced by the hydrolysis of the fructose diphosphate by phosphatase.

Ketonaldehydemutase. An enzyme bringing about the production of lactic acid from methyl glyoxal was found to be present in yeast by Neuberg in 1918. Such an enzyme, called glyoxalase, has been known in animal tissues since 1913. It may also be, perhaps, a constituent of myozymase. It has been shown by Neuberg to occur not only in yeast but also in a number of higher plants.

Carboxylase. This is an enzyme of the zymase complex which deserves special notice, for it effects the breaking up of a chain of carbon atoms with the evolution of carbon dioxide. It acts on certain α-ketonic acids, that is, acids with the general formula R.CO.COOH, splitting off carbon dioxide and leaving an aldehyde with fewer carbon atoms in the molecule. The best-known action catalysed by carboxylase is the breaking down of pyruvic acid, $CH_3.CO.COOH$, to form carbon dioxide and acetaldehyde :

$$CH_3.CO.COOH = CH_3.CHO + CO_2$$

In the plant its function appears to be the catalysis of this particular reaction, but it can also effect the breaking down of a number of other α-ketonic acids, as, for example, α-keto-butyric acid, $C_2H_5.CO.COOH$, and oxalacetic acid, $COOH.CH_2.CO.COOH$:

$$C_2H_5.CO.COOH = \underset{\text{propionaldehyde}}{C_2H_5.CHO} + CO_2$$

$$COOH.CH_2.CO.COOH = \underset{\text{acetaldehyde}}{CH_3.CHO} + 2CO_2$$

There is little doubt that carboxylase is widely distributed throughout the plant kingdom. Its presence has been demonstrated by Zaleski and Marx in seeds of *Pisum sativum*, *Vicia faba* and *Lupinus luteus*, by Bodnár in potatoes and beetroot, and by Dox and Neidig in fungi.

Aldehydemutase. An enzyme bringing about the transmutation of aldehydes into the corresponding acid and alcohol or the ester resulting from the combination of these two, has been reported as present in yeast. Such mutase action would bring about the formation of acetic acid and ethyl alcohol (or ethyl acetate) from acetaldehyde :

$$\begin{matrix} CH_3.CHO & & H_2 & & CH_3.CH_2OH \\ + & + & | & = & + \\ CH_3.CHO & & O & & CH_3.COOH \end{matrix}$$

Knowledge of this enzyme appears to be somewhat indefinite. It should perhaps be regarded as a specific oxidizing enzyme of the oxido-reductase type.

Oxidizing Enzymes

The presence of a number of different oxidizing and reducing systems has been demonstrated in living cells. Not all of these systems involve enzymes as defined on p. 95, but many of them are definitely enzymatic. Although a very great deal of research has been carried out in recent years with the object of elucidating the mechanism of oxidations and reductions in living cells, knowledge of oxidizing enzymes is still in a very confused state. Opinions differ radically as to the mechanism of vital oxidations, and no two authorities classify the oxidizing systems of living cells in the same way.

Systematic work on oxidizing systems in plant cells was practically commenced by Bach and Chodat in the early years of the present century, and continued by them and their collaborators for many years.

Bach himself thought that during oxidation in living cells the oxygen molecule is, in current terminology, 'activated' so that the molecule may be represented as —O—O— instead of O=O. This activation leads to the ready combination of the oxygen with substances to form peroxides, as, for example, hydrogen peroxide,

$$\begin{array}{c} H-O \\ | \\ H-O \end{array} \quad \text{or} \quad \begin{array}{c} H \\ H \end{array}\!\!>\!O=O$$

in which part of the oxygen is readily given off to other substances. One essential feature of Bach's view of oxidation is thus an activation of molecular oxygen.

The more recent views of Warburg, put forward in 1921, also regard oxidation as involving an activation of oxygen. The activation is supposed to take place at a surface, and iron is supposed to be necessary. Oxidation is thus a case of catalysis at a surface. These views are largely based on the results of experiments with a ' charcoal model ', certain oxidations which take place in the living cell being effected by absorbing the substrate on charcoal in presence of air, provided that a trace, at least, of iron is present.

The theory of oxidation put forward by Wieland in 1912 regards oxidation as consisting essentially in a removal of hydrogen from the molecule of the substance oxidized. Where oxidation actually involves the addition of oxygen to the molecule this is supposed to be brought about by the production of a hydrate by addition of water, hydrogen then being removed from the hydrate. Examples of oxidation on these lines are afforded by the oxidation of ethyl alcohol into acetaldehyde, and the latter into acetic acid. The first oxidation involves simply the removal of hydrogen from the molecule

$$CH_3.CH_2OH = CH_3CHO + 2H$$

the second oxidation, involving the addition of oxygen, may be regarded as taking place through the intermediate hydrate stage thus :

$$CH_3.C\left\langle\begin{array}{l} H \\ O \end{array}\right. + H_2O = CH_3.C\left\langle\begin{array}{l} H \\ OH \\ OH \end{array}\right.$$

$$CH_3.C\left\langle\begin{array}{l} H \\ OH \\ OH \end{array}\right. = CH_3.COOH + 2H$$

According to Wieland, oxidization in living cells is due to the ' activation ' of hydrogen in the molecule oxidized. This activated hydrogen is readily removed or ' accepted ' by certain substances which are thus at the same time reduced. The substance oxidized is thus a ' hydrogen donator ', giving hydrogen to the substance thereby reduced, which is thus a hydrogen acceptor.

On this view an oxidizing enzyme functions by bringing about an activation of hydrogen in the molecule of the substrate. For this reason Wieland called such enzymes dehydrases. There seems no doubt that enzymes capable of bringing about oxidation in this manner, that is, by effecting withdrawal of hydrogen, are present in many cells. Thus Thunberg showed that frog's muscle contains an enzyme which will bring about the oxidation by dehydrogenation of a number of substances if methylene blue is present. The latter acts as an acceptor of the activated hydrogen in the substrate molecule which is, owing to the removal of the hydrogen in this way, thereby oxidized, while the methylene blue is reduced to a leuco-compound. Enzymes apparently capable of acting in a similar way have been demonstrated in a variety of plant and animal cells. They have also been called dehydrogenases, and, because both an oxidation and reduction process is involved in their action, oxido-reductases.

Some supporters of this view of the mechanism of oxidation would hold that all oxidations are brought about by the activation of hydrogen, and that all oxidizing enzymes are therefore dehydrases. There does not at present seem sufficient evidence to justify this view, and oxidizing enzymes at present known appear to fall into three well-defined groups, oxidases, peroxidases and dehydrases. Of these, the oxidases bring about oxidation of a substrate by means of molecular oxygen, the peroxidases catalyse oxidation by means of hydrogen peroxide, while the dehydrases, as already mentioned, bring about oxidation by removal of hydrogen from the substrate molecule and require neither free oxygen nor hydrogen peroxide for the oxidizing action, although water, as we have seen, may sometimes be necessary.

Oxidases

The term oxidase was first applied by Bach and Chodat to the enzymatic system present in many plants, both higher and lower, which brings about the production of a blue colour in a tincture of guaiacum gum. This colour is due to the formation of an oxidation product of guaiaconic acid, one of the constituents of the gum. Bach and Chodat regarded an oxidase as comprising two parts, (1) an oxygenase, which is capable of forming a peroxide with molecular oxygen, and (2) a peroxidase, which, as we have seen, brings about oxidation by means of the oxygen of the peroxide. In the guaiacum reaction the guaiaconic acid acts as an acceptor of this oxygen, but in the plant cell other substances would act as oxygen acceptors and so become oxidized.

In 1911 Miss Wheldale (later Mrs. Onslow) showed that all plants giving an oxidase action contain a substance or substances possessing the ortho-dihydroxy grouping

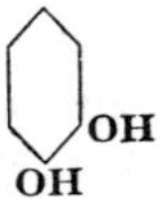

OH
OH

of catechol. No species has so far been found which gives an oxidase action but which does not contain a catechol compound, while there are few, perhaps no, species which contain a catechol compound and no oxidase.

Miss Wheldale also showed that catechol in solution undergoes autoxidation in air with formation of a peroxide. It thus appeared that if Bach's idea of the oxidase system were accepted, his oxygenase could be identified as a compound possessed of a catechol grouping. Subsequently, however, Mrs. Onslow showed that the oxidation of the catechol compounds is catalysed by an enzyme, which she called oxygenase, and that the substances resulting from this oxidation would bring about the blueing of a tincture of guaiacum.

The enzyme oxygenase therefore catalyses the oxidation of compounds with the catechol grouping by means of molecular oxygen. It may therefore be appropriately called 'catechol oxidase'. The actual reaction involved is, however, in doubt. On one view a catechol peroxide is first formed by addition of oxygen, the peroxide being then decomposed by water to give hydrogen peroxide and some other oxidation product of catechol:

$$\begin{aligned} X + O_2 &= XO_2 \\ XO_2 + H_2O &= XO + H_2O_2 \end{aligned}$$

The alternative view supposes that an orthoquinone is formed by removal of hydrogen to form hydrogen peroxide with atmospheric oxygen:

OH, OH (catechol ring) $+ O_2 =$ O, O (orthoquinone ring) $+ H_2O_2$

There is evidence that both hydrogen peroxide and orthoquinone are produced by the catechol oxidase action, but it is not certain whether the orthoquinone arises directly from the catechol as suggested by the second of the schemes given above, or whether it arises by the action on catechol of the enzyme peroxidase, from which oxygenase preparations are never completely free, in presence of the hydrogen peroxide produced by oxygenase action as suggested by the first of the above schemes.

Mrs. Onslow has pointed out that products resulting from the oxidation of catechol, probably orthoquinone, can bring about secondary oxidations, such as the oxidation of monophenols to diphenols, and of other substances with monophenolic groupings to similar substances with diphenolic groupings. As examples we may cite the oxidation of *p*-cresol to homocatechol and of tyrosine to dihydroxyphenylalanine:

CH_3 / ring / OH (*p*-cresol) → CH_3 / ring / OH, OH (homocatechol)

$CH_2.CH(NH_2).COOH$ / ring / OH (tyrosine) → $CH_2.CH(NH_2).COOH$ / ring / OH, OH (dihydroxyphenylalanine)

The monophenolic compounds so oxidized have the hydroxy group in the para or meta position:

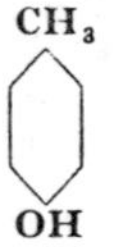

CH_3 / OH — *p*-cresol

CH_3 / OH — *m*-cresol

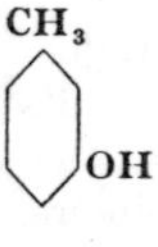

the result of the oxidation in either case is to produce a diphenol with two adjacent hydroxy groups:

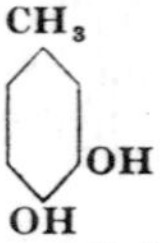

CH_3 / OH / OH — homocatechol

that is, with the catechol grouping. These can then be acted upon by the oxygenase itself.

According to Mrs. Onslow, during extraction of the catechol oxidase, some orthoquinone is formed which is absorbed by colloidal constituents present in the enzyme preparation, so that such preparations not only bring about the oxidation of compounds with the catechol groupings, but also effect the oxidation of monophenolic substances to the corresponding diphenolic ones.

In yeast there is present an oxidase which appears to be distinct from catechol oxidase of the higher plants. Unlike catechol oxidase it does not blue guaiacum, but oxidizes the so-called indophenol reagent, which consists of a mixture of *p*-phenylenediamine and α-naphthol giving rise to indophenol blue.

Other enzymatic systems which have been regarded as oxidases are laccase and tyrosinase. Laccase occurs in the latex of the lac tree (***Rhus vernicifera***). The fresh latex is white, but on exposure to air rapidly darkens, drying to the black substance with a shiny surface known as Chinese or Japanese lacquer. That this change is due to an enzyme was shown as long ago as 1883 by Yoshida. A preparation of the enzyme, or enzyme system, was subsequently obtained by Bertrand, who gave the name laccase to it. Laccase can bring about the oxidation of hydroquinone and pyrogallol to quinone and purpurogallol respectively, and also acts similarly on other phenols. For this reason it is sometimes called phenolase. Enzymes with a similar action are present in other plants, including apple, pear, carrot, turnip and clover. Others have been isolated from the higher fungi ***Russula*** and ***Lactarius***.

Like the catechol oxidase, laccase can blue guaiacum, and it has been regarded as identical with the former enzyme by some authorities. Mrs. Onslow thinks that laccase of the higher plants is the catechol oxidase with more or less adsorbed orthoquinone.

Tyrosinase is the name given to an enzyme also first isolated by Bertrand from ***Russula nigricans***. It oxidizes the monophenolic amino-acid tyrosine (see p. 113) and a number of other phenolic substances related to it. The course of oxidation is characterized by definite colour changes: a pink colour is first produced which slowly darkens to red and then to black, insoluble black compounds being the final oxidation products.

As well as in *Russula* and some higher fungi, tyrosinase has been found in a large number of flowering plants. Mrs. Onslow gives a list of 21 higher plants which contain tyrosinase; these all also give the reaction for catechol oxidase. Comparatively few plants which give the catechol oxidase reaction do not give the reaction for tyrosinase.

It has been usual to regard tyrosinase as a definite oxidizing enzyme, but Mrs. Onslow regards the tyrosinase reaction as intimately connected with catechol oxidase, and suggests that the oxidation of tyrosine is brought about by ortho-benzoquinone which is formed by the action of catechol oxidase as already described.

Peroxidase

Peroxidase brings about the oxidation of a number of phenolic compounds in presence of hydrogen peroxide and possibly of certain other peroxides. Among the substances so oxidized are catechol, pyrogallol, α-cresol and guaiacum. The peroxidase can be regarded as catalysing the splitting of hydrogen peroxide into water and active oxygen, the latter then effecting the oxidation of the phenolic compound.

Peroxidase is a very widely distributed enzyme in plants and may be universally present. Out of a large number of flowering plants examined by Mrs. Onslow only about 5 per cent. failed to give a peroxidase reaction.

Pure preparations of peroxidase have been prepared by Willstätter and his co-workers, but it has not so far been possible to discover the chemical constitution of the enzyme. The purified enzyme was free from protein and sugar, and although the final product contained some iron, it is to be observed that during the various stages in the preparation of the purified enzyme the percentage of iron decreased, so that it is doubtful whether, as had been supposed, iron forms part of the enzyme molecule or is essential for peroxidase action. It is, however, possible that there may be a number of different peroxidases, for both thermo-labile and thermo-stable preparations have been obtained with the characteristics of peroxidase. Thus Keilin described a substance, or group of substances, cytochrome, present in yeast and other organisms, as having the properties of peroxidase, but these substances are thermo-stable and possibly not colloidal, and so are not true enzymes if we limit such substances to colloidal, thermo-labile catalysts.

Dehydrase

The essential characteristic of a dehydrase is that it brings about oxidation by removal of hydrogen from the substance oxidized. This is usually expressed by stating that the enzyme activates the hydrogen in the substrate molecule, or sometimes that it activates the substrate molecule itself. An acceptor is required for the hydrogen removed, the acceptor being thereby reduced; hence a reduction as well as an oxidation is involved, so that the enzymes concerned are sometimes called oxido-reductases.

The best-known enzyme of this type is the Schardinger enzyme of milk. This enzyme catalyses the oxidation of acetaldehyde in presence of methylene blue, the acetaldehyde being oxidized to acetic acid and the methylene blue being reduced to the leuco compound:

$$CH_3.CHO + H_2O + \text{(MB)} = CH_3.COOH + \text{(MB)}H_2$$

The best-known plant dehydrase is the nitrate-reducing enzyme of potato and other plants. This enzyme reduces nitrate to nitrite in presence of aldehyde, the latter being oxidized at the same time to the corresponding acid.

$$NaNO_3 + CH_3.CHO = NaNO_2 + CH_3.COOH$$

In this particular action the above equation would suggest that no oxygen has been supplied by any substance other than the nitrate. In many cases, however, to explain oxidations as due to the action of a dehydrase it is necessary to suppose that a hydrate is first produced by the formation of an addition compound with water, the dehydrase then effecting oxidation by removal of hydrogen from the hydrate.

Thus in the case of the enzyme mentioned above, the aldehyde is supposed to form a hydrate with water:

$$CH_3.C\begin{matrix}\diagup H \\ \diagdown\!\!= O\end{matrix} + H_2O = CH_3.C\begin{matrix}\diagup H \\ - OH \\ \diagdown OH\end{matrix}$$

This hydrate is then oxidized by dehydrogenation, the hydrogen so removed forming water with an atom of oxygen from the sodium nitrate:

$$NaNO_3 + CH_3.C\begin{matrix}\diagup H \\ - OH \\ \diagdown OH\end{matrix} = NaNO_2 + H_2O + CH_3.C\begin{matrix}= O \\ \diagdown OH\end{matrix}$$

Enzymes catalysing the so-called 'Cannizzaro reactions' are probably to be regarded as belonging to the category of dehydrases. In a Cannizzaro reaction two molecules of an aldehyde react together, one being oxidized and the other reduced with the result that the corresponding acid and alcohol are produced simultaneously. Thus in alcoholic fermentation by yeast, actions of this type are supposed to take place by means of enzyme action. Reference to these will be made in the next chapter.

Catalase

This enzyme catalyses the decomposition of hydrogen peroxide to water and molecular oxygen. It is frequently classed as an oxidizing enzyme, but as the oxygen is not produced in an active state the justification for this classification appears to be wanting.

Catalase is widely distributed throughout the plant kingdom and is generally supposed to be universally present in all living tissues. Its function is obscure, but since it acts only on hydrogen peroxide it is inferred that it has a protective function, destroying hydrogen peroxide formed in the course of metabolic processes.

THE FUNCTION OF ENZYMES IN THE PLANT

There can be no doubt of the supreme importance of enzymes in the life of the plant. Their universal presence in itself is evidence of this. In some cases their function is understood at any rate in part, whereas in others it can only be guessed. Among enzymes, the functions of which appear to be adequately comprehended, are the large group of hydrolysing enzymes. Wherever more complex reserve products are mobilized into simpler mobile or active substances, hydrolysing enzymes such as amylase, lipase or protease are the agents catalysing the hydrolysis. Thus in starch-containing leaves, seeds and perennating organs, hydrolysing enzymes appropriate for the breaking down of the stored reserve product become active when certain conditions are fulfilled.)

Since catalytic actions are reversible, and enzymes have the general characters of catalysts, it has been very generally assumed that enzyme actions are reversible and that, in consequence, they may be not only agents furthering hydrolysis but also agents of synthesis. Although it has not been found easy to demonstrate the reversibility of enzyme actions, this reversibility has now been established in a number of cases, and it seems likely that it only requires improved technique to demonstrate the reversibility of any enzyme action.

The actions in the plant catalysed by enzymes are balanced actions, that is, the hydrolysis, for example, is not complete but reaches a position of equilibrium in which the hydrolytic and synthetic products bear a certain quantitative relation to one another. In the action, for example, of maltase on maltose, when equilibrium is reached most of the maltose will have been converted into glucose, but there will be a small proportion of maltose left. In the case of many enzyme actions the position of equilibrium is so near that of complete hydrolysis that some observers have thought the reactions complete. However, careful investigation has shown that this is not so. For example, it was thought that the glycoside salicin was completely hydrolysed by emulsin (β-glycosidase), but it was shown by Visser and also by Bourquelot and Bridel that at equilibrium there is present a small proportion of salicin.

Nevertheless, in general, under ordinary conditions, the position of equilibrium is seldom far removed from that of complete hydrolysis. In spite of this, the fact that synthesis can take place at all is of fundamental importance. Thus, to take a specific example, if we start with an aqueous solution of glucose and add maltase, an equilibrium point will be reached when a small quantity of maltose is formed.

If the maltose is removed from the sphere of action, by, for example, combination with some other body, the equilibrium is only of momentary duration and a further small amount of maltose will be produced. Thus, in general, provided the product of the synthesis is removed as it is formed, the synthetic action of the enzyme will continue since the equilibrium is not maintained. It is to be noted that the concentration of the enzyme does not alter the position of the equilibrium, since both the hydrolytic and synthetic actions are furthered to about the same degree whatever the concentration.

We may now examine some of the cases in which the reversibility of enzyme action has been demonstrated. Synthesis by means of an enzyme was first recorded by A. Croft Hill in 1898. By adding maltase to a strong aqueous solution of glucose, this worker obtained a quantity of disaccharide, which he concluded was maltose. It subsequently appeared that the synthesized disaccharide was a mixture of maltose and an optical isomer of it, isomaltose. Isomaltose itself is not hydrolysed by maltase, but by emulsin (β-glycosidase). The explanation of this result given by Bayliss appears to be the true one, namely, that the maltase preparation used by Croft Hill contained some emulsin and that both enzymes acted as synthetic agents, the maltase bringing about the synthesis of maltose from glucose, the β-glycosidase that of isomaltose from glucose.

The synthetic action of emulsin has been indubitably shown by Bourquelot and his associates, who have synthesized a number of glycosides by its means and isolated them in crystalline form. The synthesis of urea by means of urease has also been shown by Barendrecht and others.

The most easily demonstrated case of reversible enzyme action is that of esterase (butyrase). If this enzyme is allowed to act on ethyl butyrate, hydrolysis takes place with formation of ethyl alcohol and butyric acid. Conversely, when allowed to act on a mixture of ethyl alcohol and butyric acid a certain amount of synthesis takes place with formation of ethyl butyrate. Butyric acid and ethyl butyrate possess characteristic and different odours and the hydrolytic and synthetic actions can both be recognized by the definite odours of the respective products. It is easy, and perhaps more satisfactory, to determine the products by chemical analysis.

Enzymes are not only agents of hydrolysis and synthesis. As pointed out earlier (p. 96) they may also determine the direction of the reaction where a substance can be broken down in more than one way, and the respective reactions concerned are catalysed by different enzymes.

The oxidizing enzymes and catalase are generally supposed to be concerned in respiration, since the latter usually involves an oxidation of material. The question of the relation of these enzymes to respiration will be dealt with in the next chapter.

THE NATURE OF ENZYMES

Only a very few enzymes have been obtained in even an approximately pure state; these include sucrase, lipase, urease and peroxidase. Willstätter and Pollinger obtained from horseradish a preparation of the last-named enzyme which was 20,000 times as active as the original material, while the preparation of lipase obtained by Willstätter and Waldschmidt-Leitz from *Ricinus* was only 100 times as active as the original material.

Even in these cases where purification has been carried out to a very considerable extent, there is doubt as to the chemical nature of the enzymes. At one time it was thought that enzymes were proteins, but it seems likely that the proteins present in enzyme preparations were frequently impurities. Nevertheless, the purest preparations of urease, obtained by Sumner and Hand, consisted of a crystalline protein, and these workers regarded pure urease as having this constitution. It is still possible, however, that the actual enzyme only formed a small proportion of the whole preparation, though undoubted evidence to this effect is wanting. In other cases, such as those of lipase and sucrase, purification results in a great decrease in the protein content, and the purest preparations of these substances appear to contain little or no protein. There is some evidence that peroxidase is related to the porphyrins, substances themselves related to chlorophyll and haemoglobin.

Although the enzymes, or at least some of them, thus do not appear to be proteins, they appear to be associated with proteins in some very definite way, for the removal of the last traces of protein leads to a considerable reduction in stability of the enzyme.

Owing, probably, to the difficulty of identifying enzymes chemically, it has even been doubted whether enzymes are definite substances. Willstätter regards sucrase, and presumably enzymes in general, as consisting of a chemically active group attached in some way to a large colloidal nucleus or carrier, the whole forming a 'symplex'. De Jager, and also Arthus, believed that enzymes were not substances with any definite chemical composition, but that various substances might develop the catalytic property characteristic of enzymes. A more definite view of this kind was put forward by Barendrecht who considered that the peculiar properties of enzymes are due to radiations emitted by them; that, in fact, enzymes are radioactive substances. These views have, however, received little support.

THE MECHANISM OF ENZYME ACTION

We may, then, accept the general view and regard enzymes as definite substances, and inquire how it is they accelerate reactions. It is almost universally assumed that this is brought about by the enzyme entering into some kind of combination with the substrate from which subsequently the enzyme is released, but opinions differ as to whether this union takes the form of a chemical combination or of an adsorption compound. Michaelis and others, for example, regard an enzyme action as involving the formation of a chemical compound between enzyme and substrate molecules, followed by the breaking down of this compound into the free enzyme and the products of the reaction. Bayliss was responsible for the idea of the formation of an adsorption compound between substrate and enzyme, but he regarded the adsorption as a preliminary stage to the true chemical reaction in which the substrate was acted upon. The adsorption results in an increase in the active mass of the substrate at the seat of the reaction and will therefore increase its rate, but Bayliss pointed out that it might not be possible to explain all the chemical activities of enzymes as due to their adsorptive properties. He also held that true chemical compounds might be formed by a union of enzyme and substrate.

The kinetics of a number of enzyme actions have now been examined, with the result that current opinion leans to the view that chemical compounds are formed between enzyme and substrate, and that, owing to the relative sizes of the molecules of enzyme and substrate, in many cases only relatively few substrate molecules can be adsorbed to one enzyme molecule. Hence enzyme action cannot in general be accounted for by an increase of the active mass of the reactant as a result of adsorption.

THE FORMATION OF ENZYMES IN PLANTS

It seems reasonably clear that enzymes are to be found in every living cell, and it may be that they form an integral part of the protoplasm. At any rate they are to be regarded as an essential part of the vital mechanism of the cell. It appears likely that they are produced in the cells in which they are found, for their colloidal character renders them indiffusible. While some cells are particularly rich in one particular enzyme, a number of enzymes may be active in one and the same cell. This is, of course, obvious in the case of unicellular and filamentous organisms. Yeast, in particular, has formed a source of extraction of a large number of enzymes, and a dozen or more enzymes have been recognized as active in the yeast cell.

In some cases the enzyme appears to be present in a non-active form which can be extracted and rendered active by treatment with some reagent. Thus lipase in a non-active form can be extracted from seeds of *Ricinus* and rendered active by treatment with acetic acid. Such a non-active form of an enzyme is termed a pro-enzyme or zymogen. It may be that the so-called pro-enzyme or zymogen consists of an enzyme united with some other substance through its active grouping, but actually the relationship of zymogen to enzyme is not understood.

Nothing definite is known of the way in which enzymes are produced.

It has been held that they are formed in the nucleus and then extruded into the protoplasm, while on another view it is supposed that mitrochrondria are the seat of enzyme formation. Neither view has received any general support or confirmation.

REFERENCES

BAYLISS, W. M., *The Nature of Enzyme Action.* London, 3rd Edit., 1914.

EULER, H., *General Chemistry of the Enzymes.* English Translation by T. H. Pope. New York and London, 1912.

FALK, K. G., *The Chemistry of Enzyme Actions.* New York, 1921.

GREEN, J. R., *The Soluble Ferments and Fermentation.* Cambridge, 1899.

HALDANE, J. B. S., *Enzymes.* London, 1930.

VERNON, H. M., *Intracellular Enzymes.* London, 1908.

WAKSMAN, S. A., and DAVISON, W. C., *Enzymes.* London, 1926.

CHAPTER VI

RESPIRATION

A CHARACTERISTIC of all living matter is the continuous release of energy resulting, usually, from the breaking down of complex substances present in living cells. The most usual manifestation of this function is the absorption of oxygen and evolution of carbon dioxide, and it was to this exchange of gases between organism and environment that the term respiration was first applied. However, it came to be realized that the essential feature of the function in which this gaseous exchange is involved is the release of energy, and that such energy release is sometimes effected by reactions which do not involve the usual gaseous exchange. This being so, the term respiration is now generally understood to refer to the whole of the processes, of whatever kind, involving a release of energy from substances present in the cells, and not merely to the familiar exchange of oxygen and carbon dioxide. To avoid ambiguity the term *energesis* was at one time suggested to refer to the whole process of energy release, but has scarcely survived.

KINDS OF RESPIRATION

While in the vast majority of plants growing under natural conditions the respiratory process involves the absorption of oxygen and evolution of carbon dioxide, there are types of respiration in which such an exchange of gases does not take place. These unusual methods of respiration are of very limited incidence, occurring either in absence of oxygen, or in certain groups of bacteria. The various types of respiration met with in plants are summarized below.

Normal Aerobic Respiration

(In all plants and all cells of plants, with only a few exceptions, respiration, in presence of air, consists essentially in the oxidation of carbohydrate or fat, and perhaps in some cases of other organic material. Generally, the oxidation is complete so that the end-products are carbon dioxide and water. Where the substance oxidized, or *respiratory substrate,* is hexose sugar, as is generally the case, for example, in leaves, the whole of the respiratory process can be summed up in the equation

$$C_6H_{12}O_6 + 6O_2 = 6CO_2 + 6H_2O$$

This oxidation of sugar involves the release of energy, the breaking down of one gram-molecule of sugar releasing 674 calories, It may be taken as certain, for reasons which will appear later, that the sugar is not broken down to carbon dioxide and water in one stage, but that a whole chain of reactions is involved between the beginning and end of the whole process.

In seeds of 80 per cent. of the species of flowering plants the principal storage materials are fats, and there is no doubt that these substances are utilized in respiration, although it is generally supposed, and quite probable, that hexose sugar is first formed from the fat by processes at present quite obscure. At any rate, the end-products of respiration when fat is utilized are the same as when carbohydrates are the respiratory substrate. It will, however, be observed that owing to the small amount of oxygen relative to the carbon in a fat, as compared with the amount in a carbohydrate, more external oxygen is required to oxidize the carbon of a fat completely to carbon dioxide than to oxidise that of a carbohydrate. Thus the equation representing the oxidation of the fat triolein, $C_{57}H_{104}O_6$, is :

$$C_{57}H_{104}O_6 + 80O_2 = 57CO_2 + 52H_2O$$

Another variant of the aerobic respiratory process occurs in succulent plants where, possibly owing to the slow diffusion of gases through their tissues, the oxidation of carbohydrate, at least in absence of light, tends to be incomplete and leads to the formation of organic acids, particularly malic acid, thus :

$$2C_6H_{12}O_6 + 3O_2 = 3C_4H_6O_5 + 3H_2O.$$

Some fungi appear to be able to utilize a variety of materials for respiratory purposes. Thus 30 years ago Kostychev showed that *Aspergillus niger* could utilize not only tartaric acid and quinic acid, but also peptone. The equations for the complete respiration of the first two substances are respectively :

$$2C_4H_6O_6 + 5O_2 = 8CO_2 + 6H_2O$$

and

$$C_7H_{12}O_6 + 7O_2 = 7CO_2 + 6H_2O$$

Other substances which can provide material for the respiration of *Aspergillus niger* are lactic acid, glycerol and mannite.

The fate of peptone when this forms the respiratory substrate is however, not clear, although apparently amino-acids are first produced, which substances can themselves serve as respiratory substrate. To what extent proteins can be utilized in the respiration of higher plants must at the present time be regarded as very doubtful.

Anaerobic Respiration

Plants which in presence of oxygen respire in the normal manner continue to respire when deprived of oxygen, as, for example, when they are placed in a vacuum or in an atmosphere of nitrogen or hydrogen. The only known exception to this rule is *Elodea*. When deprived of oxygen the substrate is only partially broken down, the end-products as a rule being carbon dioxide and ethyl alcohol, so that where the substrate is hexose the whole process may be represented by the equation :

$$C_6H_{12}O_6 = 2C_2H_5OH + 2CO_2$$

The energy released in this process is very much less than that released in the complete oxidation of hexose to carbon dioxide and water, being only about 26 calories for every gram-molecule of hexose broken down.

This type of respiration was originally termed intramolecular respiration, but the expression is not a suitable one and the term 'anaerobic respiration', proposed by Kostychev in 1902, is now generally employed for this process.

Certain bacteria live only in the absence of oxygen or in a very low concentration of this gas, and such normally respire anaerobically. Among such 'obligate anaerobes' are *Bacillus denitrificans*, and certain butyric and lactic bacteria, and it may be that the anaerobic respiration of these forms is in part the same or a similar process to that which obtains in higher plants respiring anaerobically. In these lower forms the sugar or other carbohydrate is supposed to be broken down into carbon dioxide, alcohol, lactic acid and butyric acid. In *Bacillus denitrificans* nitrate is absorbed and reduced to ammonia or nitrogen, with a consequent release of oxygen, which it is supposed is utilized in the further oxidation of carbohydrate, so that in spite of the organism being an obligate anaerobe, part of the respiratory process may, in fact, be the same as that in normal aerobic respiration.

Nitrifying Bacteria

The energy-releasing process in the nitrifying bacteria (*Nitrosomonas*) consists in the oxidation of ammonium salts into nitrous acid, the equation representing the process being usually given as :

$$2NH_3 + 3O_2 = 2HNO_2 + 2H_2O$$

Some controversy has resulted over the question whether the source of carbon for these plants consists of the carbonates of the soil or the carbon dioxide of the air. It seems possible that under certain conditions either source may be used, but in neither case does the carbon compound form a respiratory substrate, this being provided by ammonium compounds, the oxidation of one gram-molecule of ammonia releasing 79 calories.

Nitrating Bacteria

Similarly, the nitrating bacteria of the soil, *Nitrobacter*, obtain energy by the oxidation of nitrous acid (or nitrites) to nitric acid (or nitrates), according to the equation :

$$2HNO_2 + O_2 = 2HNO_3$$

the oxidation of one gram-molecule of nitrous acid releasing 18 calories.

Sulphur Bacteria

The sulphur bacteria, *Beggiatoa*, *Thiothrix* and *Hillhousia*, respire by oxidizing hydrogen sulphide. The oxidation appears to take place in two stages, sulphur being first formed and appearing in the form of particles in the cell protoplasm. The elemental sulphur is then further oxidized to sulphuric acid which, reacting with bases present in the cell, produces sulphates, the usual one being calcium sulphate. The respiratory process in these bacteria may be summed up in the equations :

$$2H_2S + O_2 = S_2 + 2H_2O$$
$$S_2 + 2H_2O + 3O_2 = 2H_2SO_4$$

Thiosulphate Bacteria

Apparently distinct from the sulphur bacteria are the thiosulphate bacteria, *Thiobacillus*, which obtain their energy by the oxidation of thiosulphates to sulphates, the equation representing the process being :

$$6K_2S_2O_3 + 5O_2 = 4K_2SO_4 + 2K_2S_4O_6$$

Iron Bacteria

The iron bacteria, of which the best known is *Spirophyllum ferrugineum*, have been thought to respire by oxidizing ferrous iron to the ferric state, and it has been suggested that the process might be represented by the following equation :

$$2Fe(HCO_3)_2 + O + H_2O = Fe_2(OH)_6 - 4CO_2$$

However, considerable doubt exists regarding the mode of life of these organisms, and Molisch and others have asserted that they can grow in absence of iron salts.

Hydrogen Bacteria

A number of bacteria, including *Bacillus hydrogenes*, *B. pantotrophus* and *Hydrogenomonas* spp., obtain their energy by oxidizing hydrogen to water.

It will be observed that all the unusual forms of respiration, with the exception of anaerobic respiration of normally aerobic plants, are confined to bacteria, and are thus of relatively rare occurrence, and found only in specialized groups of plants. They are, however, of general interest, as they serve to emphasize the fact that respiration consists essentially in a release of energy which, although generally effected in one particular way, can nevertheless be brought about in a variety of different ways involving the utilization of very diverse materials.

THE MATERIALS USED IN NORMAL RESPIRATION

The aerobic respiration of the vast majority of plants proceeds at all times and in all organs. Every living cell respires, and continuous respiration is to be regarded as an essential function of living matter. As already stated, the normal respiration of aerobic plants involves the breaking down of organic material, nearly always carbohydrate or fat but perhaps sometimes more complicated substances involving such substances in their molecules. Where more complex carbohydrates form the respiratory substrate there is a probability that hexose sugar is first formed from them by the action of hydrolytic enzymes, while it is generally held, although direct evidence is not available, that where fat is the respiratory substrate sugar is first formed. Thus it may be that whatever the substance used for respiration, the actual aerobic respiratory process is the same and consists of an oxidation of hexose sugar, this being provided in some cases by hydrolysis or oxidation of other substances present in the plant as reserves.

Where sugar is the substance utilized it will be observed from the equation on p. 122 that the volume of carbon dioxide produced is equal to that of oxygen absorbed. The ratio of carbon dioxide evolved to oxygen taken in is called the *respiratory quotient,* and where sugar is the substrate its value will thus be unity. The same is also the case where starch, cellulose and other complex carbohydrates provide the respiratory material. But where fat is employed relatively more oxygen is required for the complete combustion of the carbon, as fats contain relatively little oxygen, and in the case of the respiration of triolein quoted above, an inspection of the equation given on page 123 will show that the respiratory quotient in this case will be about 0·71.

Determinations of the carbon dioxide evolved and oxygen absorbed by tissue give an indication of the materials utilized in respiration. Some recent observations by Stiles and Leach on the respiratory quotients of seeds will serve to illustrate this point. In Fig. 12 are shown the values of the respiratory quotient of a germinating grain of maize and a germinating seed of castor oil from the beginning of germination. The grain of maize contains a large quantity of carbohydrate, roughly 60 per cent., but only a very moderate quantity of fat, namely 7 or 8 per cent. It will be observed that the respiratory quotient in the early stage of germination is very nearly unity, and then slowly falls to a minimum of about 0·74, after which it slowly rises towards unity. These results can be interpreted on the view that at the beginning of germination carbohydrate, probably sugar, is utilized for respiratory purposes, but that as time goes on the fat is mobilized and utilized, thereby bringing about a lowering of the quotient. As time goes on the fat reserve tends to become exhausted and carbohydrate again becomes the dominant constituent of the respiratory material, with the result that the quotient rises towards unity. In *Ricinus* seeds, on the other hand, the content of carbohydrate may be as low as 2 per cent., whereas the fat content is high, often comprising more than half of the dry weight. Here the respiratory quotient at a very early stage in germination is well below unity, being only about 0·8 after 7 hours in the example quoted, and falling to as low as 0·5 after about 120 hours. Here the results may be interpreted as indicating at first a utilization of sugar and fat together. The sugar is soon exhausted and fat alone provides the respiratory material. But if the first stage in the utilization of fat is the formation of sugar, and if the latter should be formed faster than it is respired, and so accumulate, the measured respiratory quotient will fall below that characteristic of the oxidation of fat

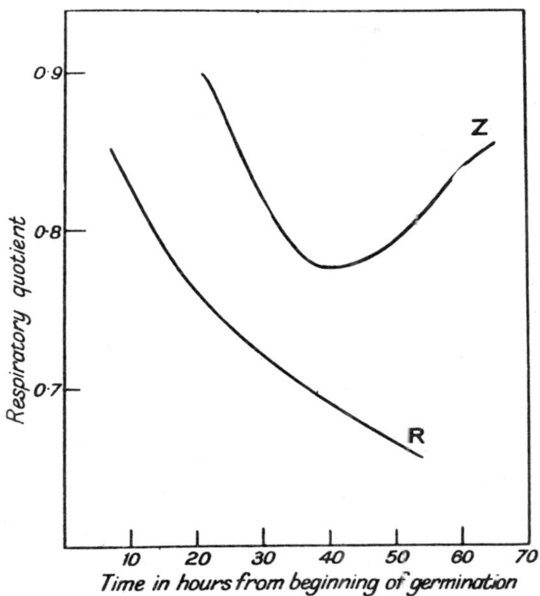

Fig. 12.—Curves showing changes in the respiratory quotient of a grain of *Zea mais* and a seed of *Ricinus communis* during germination
(Constructed from the data of W. Stiles and W. Leach)

which is about 0·7. Thus, the principal fat in the seeds of ***Ricinus*** is probably the triglyceride of ricinoleic acid with a probable formula of $C_{57}H_{104}O_9$. If this fat is wholly oxidized to carbon dioxide and water the equation summing up the whole of the reactions involved would be :

$$2C_{57}H_{104}O_9 + 157O_2 = 114CO_2 + 104H_2O$$

and the respiratory quotient would thus be :

$$\frac{114}{157} \text{ or } 0{\cdot}73$$

If, however, the fat should be entirely converted to sugar the equation summing up the changes involved would be

$$2C_{57}H_{104}O_9 + 10H_2O + 43O_2 = 19C_6H_{12}O_6$$

Hence if, during any time, half the fat which disappears is oxidized into carbon dioxide and water and half converted into sugar, the apparent respiratory quotient is

$$\frac{114}{157 + 43} \text{ or } 0{\cdot}57$$

while if only one-third is oxidized to carbon dioxide and water and two-thirds to sugar, the apparent quotient will be only 0·47.

The observed changes in the value of the respiratory quotient with time of germinating seeds of *Ricinus* thus indicate not only a utilization of fat for respiratory purposes, but afford indirect evidence of a conversion of fat to sugar. Confirmatory evidence of this is derived from direct analyses of seeds and seedlings. Thus in Table XXIX are shown, for example, analyses of *Ricinus* seeds and seedlings, made by Leclerc du Sablon.

Table XXIX

Changes in Fat and Sugar Content of *Ricinus* Seeds during Germination

Time of Germination in days	Length of Radicle in cm.	Fat in per cent. of dry weight	Glucose in per cent. of dry weight	Sucrose in per cent. of dry weight
0	0·0	71·4	0·0	1·1
2	0·7	63·9		
5	2·0	48·8	3·1	9·0
8	3·5	25·0	8·4	12·4
13	9·0	14·1	13·7	12·0
16	10·0	4·9	17·4	5·5

As the fat becomes transformed to sugar we should expect a rise in the value of the respiratory quotient, as sugar must become the principal, and, in time, probably the only substrate. Such a rise was observed by Bonnier and Mangin to take place in seedlings of flax (*Linum usitatissimum*) 4 days after the commencement of germination and to

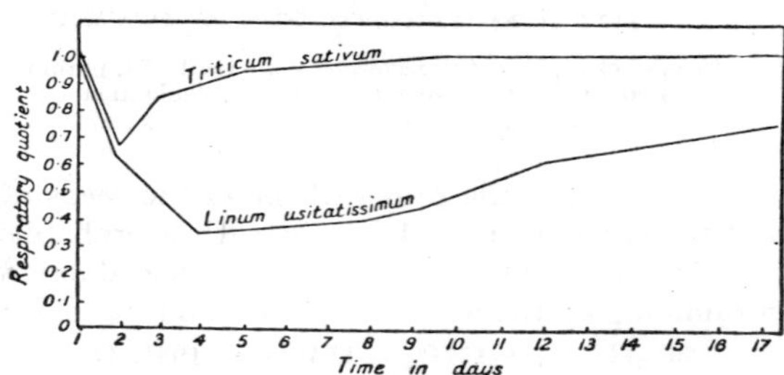

Fig. 13.—Curves showing changes in the respiratory quotient of wheat and flax seedlings over a period of 17 days
(From W. Stiles and W. Leach, after G. Bonnier and L. Mangin)

continue more or less regularly for 17 days, when measurements were discontinued (cf. Fig. 13).

Whether other substances besides carbohydrates and fats can in general provide material for respiration is not very certain. It is at any rate possible that anthocyanins may serve such a purpose, and this would not be very surprising as they are glycosidic in nature, containing sugar in combination in their molecules.

Reference has already been made to the fact that various organic substances other than carbohydrates and fats, and notably tartaric acid, quinic acid and peptone, can serve as respiratory substrate for *Aspergillus niger*. A glance at the equation given on p. 123 for the respiration of tartaric acid will show that the theoretical respiratory quotient when this substance is the sole source of respired material is 1·6. An actual determination of the quotient made by Kostychev was 1·54, a result in sufficient agreement with the theoretical value to indicate the complete oxidation of the tartaric acid. With quinic acid an even more satisfactory agreement between the theoretical and observed quotients was obtained, the values being respectively 1·00 and 1·01.

The observations of Kostychev and Butkewitsch leave no doubt that the simple protein peptone can be respired by *Aspergillus niger*. In this case Kostychev obtained values of 0·45 and 0·52 for the respiratory quotient.

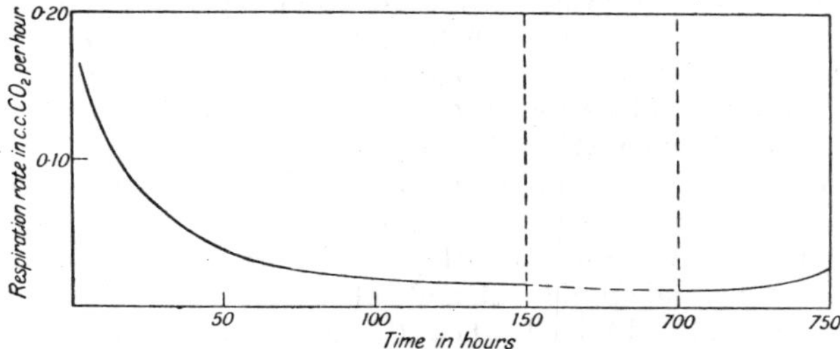

Fig. 14.—Curve showing the course of respiration in a starved leaf of privet (*Ligustrum vulgare*)

Whether protein can be respired by higher plants, as was at one time thought, is doubtful. Many leguminous seeds contain much protein, but there is no indication that in these seeds any of this is utilized for respiration. The behaviour of leaves starved by maintenance in the dark requires mention in this connexion. The respiration rate of leaves subjected to this treatment falls regularly until it reaches a low constant rate which is maintained for a long time. An example of this is shown in Fig. 14, which illustrates the respiration of a privet leaf kept in the dark for 5 weeks. It will be observed that here the minimum constant rate is only about 7 per cent. of the initial rate of respiration.

From observations on starved leaves of *Prunus laurocerasus* by Godwin and Bishop, it seems that in that species, at any rate, the course of respiratory activity varies with the age of the leaf, and that although there is always a rapid fall in rate, a constant low level rate is not always attained. Where it does occur it appears in some cases, at any rate, to be about 25 per cent. of the initial rate. In starved leaves of barley Yemm also obtained variable results; but where an approximately level rate of respiration was observed it appeared to be about 60 per cent. of the initial rate.

This minimum rate was called by F. F. Blackman 'protoplasmic respiration' as distinct from the 'floating respiration' in excess of the former. It has been suggested that in floating respiration reserve carbohydrates and fats are utilized, whereas in protoplasmic respiration protein is utilized for respiration after the reserve materials have been exhausted.

Three pieces of evidence have been advanced in favour of this supposition. In the first place, Yemm's observations indicate a rapid fall in sugar content of starved barley-leaves. Secondly, according to Deleano, there is rapid loss of nitrogen in starved vine-leaves after the disappearance of carbohydrate. Thirdly, the respiratory quotient of starved leaves of barley was found by Yemm to be of the order of about 0·82 during the period of the constant minimum respiratory rate, while a value of the quotient of the same order has been obtained by the writer for starved privet-leaves. The complete oxidation of protein would yield a quotient of about 0·8, so that contemporaneous respiration of both carbohydrate and protein would yield a quotient between unity and 0·8. Other explanations are, however, possible of the fall in the respiration rate under starved conditions.

From what has been said above it will be clear that the value of the respiratory quotient is closely related to the nature of the substrate, the quotient having a value of 1·6 in the case of tartaric acid, 1·0 in the case of carbohydrates and quinic acid, and about 0·7 with fats. These values, however, are only obtained when the oxidation of the substrate is complete. If the concentration of oxygen surrounding the respiring material is reduced below a certain value the quotient is increased, presumably on account of imperfect oxidation, which may take the form of a certain amount of anaerobic respiration in which carbon dioxide is produced but no oxygen absorbed. Some data obtained by Stich on this question are summarized in Table XXX.

Table XXX

Effect of Concentration of Oxygen on the Respiratory Quotient

Plant Material	Oxygen Concentration	Respiratory Quotient
Triticum vulgare seedlings	20·8	0·98
	9·0	0·94
	5·0	0·93
	3·0	3·34
Zea maïs seedlings	20·8	0·89
	9·0	0·96
	5·0	1·35
	3·6	1·37
Pisum sativum seedlings	20·8	0·83
	9·3	0·86
	3·5	2·31
Narcissus poeticus bulb	20·8	0·96
	10·2	1·04
	7·5	2·36

In succulent plants kept in the dark, even in normal atmosphere, the oxidation of the respired carbohydrates is incomplete, and instead of a complete breakdown of the substrate to carbon dioxide and water, organic acids such as malic and oxalic accumulate, and in extreme cases no carbon dioxide is evolved at all. Usually, however, a certain amount of this gas is formed, but in quantity much less than that of oxygen taken in, with the result that the quotient is much below unity. Some values actually obtained by Aubert in his classical investigation of this question are summarized in Table XXXI.

Table XXXI

Respiratory Quotient of Succulent Plants in the Dark

Plant Material	Respiratory Quotient
Sedum reflexum, young stems	0·83
S. Telephium ,, ,,	0·60
Crassula arborescens, leafy branch	0·24
,, ,, ,, ,,	0·52
Phyllocactus grandiflorus, adult branch	0·33
Opuntia tomentosa	0·00
,, ,,	0·047

The values of the respiratory quotients of these plants are by no means characteristic; they vary considerably according to the period of exposure to light or darkness. Thus Bennet-Clark found that when leaves of the succulent plant *Sedum praealtum* are maintained in the dark the rate of carbon dioxide output first falls, reaching a minimum value in about 6 hours, after which the rate rises to a value which remains approximately constant for many days. The rate of oxygen intake also changes with time, but not always in the same direction as the rate of carbon dioxide output. The rate of oxygen absorption first rises to a maximum, next falls to a minimum, and then rises again until it exceeds the rate of carbon dioxide evolution. The resulting respiratory quotients for a succession of two-hour periods are shown in Table XXXII.

Table XXXII

Respiratory Quotient of Leaves of *Sedum praealtum* during Maintenance in Darkness

Average time in Hours from beginning	Respiratory Quotient
1	1·46
3	0·55
5	0·17
7	0·30
9	0·78
11	0·86
13	1·40
15	1·1

With regard to the accumulation of malic acid, Bennet-Clark found the acid content of the leaves of *Sedum praealtum* reaches a maximum about 5 hours after the leaves are first darkened. A fall in acid content was found to follow for the next 8 to 12 hours, after which there is a slight rise followed by a slow fall. It would thus seem that after darkening carbohydrate is converted to malic acid, as a result of which process the respiratory coefficient falls much below unity. During the following period, when the acid disappears, the respiratory quotient rises. While the utilization of malic acid as respiratory substrate might in part account for this, the whole of the rise in the quotient cannot always be so explained, for if respiration were wholly due to the utilization of malic acid the quotient would only be 1·33 whereas in some experiments higher values were observed by Bennet-Clark, in one instance a quotient of 2·05 being obtained. Bennet-Clark therefore suggested that the breakdown of carbohydrate is not directly through malic acid, but through some other compound requiring less oxygen for its oxidation, the malic acid being formed as a side product from the main line of katabolic reactions. This view can thus be represented by the following scheme :

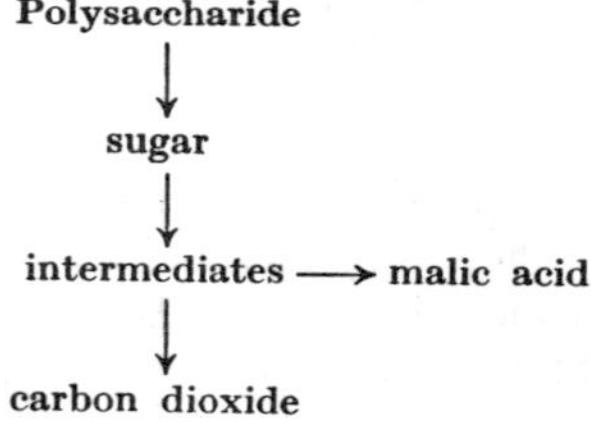

According to Nicolas, there is also an accumulation of organic acids in leaves containing anthocyanin, and here also quotients considerably below unity have been observed.

It will thus be seen that the value of the respiratory quotient may give indications of the nature of the respiratory substrate and the processes involved in its breakdown. It must, however, be remembered that more than one substrate may be utilized at the same time, and that there may be more than one chain of reactions involved in the breaking down of the substrate with a diversity of final products, on the nature of which, as well as on the substrate itself, the value of the quotient depends. Where one substrate or kind of substrate only is present, or greatly predominates, as, for example, carbohydrates in leaves, and the oxidation is complete, the quotient will be characteristic of that substrate and the chain of reactions involved in its breakdown. But where there are more than one of these chains of reactions as, for example, where there are several substrates, or where the same substrate has more than one fate, the drawing of conclusions regarding the nature of the respiratory process from the observed values of the quotient requires great care. Two simple examples may be given. Firstly, substrate consisting of a mixture of tartaric acid (with a quotient of 1·6) and fat (with a quotient of 0·7) in the right proportions, if both are completely oxidized. will give a quotient of unity, that characteristic of carbohydrate. Secondly, in an atmosphere poor in oxygen, a substrate of pure sugar will, owing to the presence of a certain amount of anaerobic respiration, give a quotient above unity, and there will be one concentration of oxygen in which the observed quotient will be that characteristic of tartaric acid.

THE INTENSITY OF RESPIRATION

The intensity at which respiration proceeds, and the way in which this intensity is modified by altered conditions, is a matter of interest in itself and also because it may shed light on the mechanism of the respiratory process. There is, however, an initial difficulty met with in dealing with this question, as it is not always a simple matter to find a mode of expressing the intensity of respiration. In the first place the rate of respiration of a plant or organ might be expressed as the quantity of carbon dioxide evolved by the material in unit time, or as the quantity of oxygen absorbed, or as the amount of substrate decomposed. If the substrate is always the same, and if the chain of reactions involved remains the same, then any one of these three quantities will give a measure of the rate of respiration of the material as a whole. But if the substrate and reaction chain alter with time, then the quantity of substrate disappearing, the oxygen absorbed and carbon dioxide evolved, will not bear a constant relation to one another, and it may not always be clear which of the three values will give the best criterion of respiration. Also, if some other process involving carbon dioxide or oxygen is active in the cell the question of a criterion of respiration is further complicated. The change in the value of the respiratory quotient during the germination of seeds has already been mentioned and affords an illustration of this point.

Having these difficulties in mind, the most satisfactory measure of respiration rate in most cases is generally regarded as given by the rate of carbon dioxide evolution, although clearly in some cases, notably that of succulent plants, the carbon dioxide evolution over a period of time may give no measure of the breakdown of substrate. Sometimes oxygen absorption has been used as a measure of respiration, but the loss in substrate, which might be a better criterion of respiratory activity than either carbon dioxide evolution or oxygen absorption, is not as a rule easily measurable. Still less has it been found possible to measure the rate at which energy is set free, which would give a truer measure of respiration.

It is frequently desirable to express the actual measurements of respiration as respiratory intensity by referring the quantities measured to the dry weight of the tissues, the fresh weight of the tissue, the area of leaf surface, or the amount of nuclein nitrogen in the tissues. If this could be done satisfactorily there would then be a satisfactory basis for comparing the respiratory activities of different material.

However, a moment's reflection will show that a simple mode of expressing respiration intensity which is universally applicable is not at present possible. Fresh weight, depending so much on water content liable to fluctuations, is obviously unsatisfactory, while dry weight may include variable amounts of cell-wall and storage products which have no direct relation to respiratory activity. Such is very obviously the case with seeds and storage organs generally. Area of surface may be a moderately satisfactory basis when the leaves of one species are considered, but even here variations in thickness of leaves may be a disturbing factor, and this will of course operate even more with leaves of different species. For these reasons Palladin attempted to measure respiratory intensity as carbon dioxide evolved in unit time per unit of protoplasm, a value for this last being given by the amount of nuclein nitrogen as measured by the amount of protein not digested by gastric juice. How far a value for the amount of protoplasm present in different tissues can be obtained in this way is, however, extremely doubtful.

The data that have been obtained for respiratory activity of various plants and plant organs must be accepted having in view the limitations imposed by necessity on the significance of the determinations. In spite of these limitations it is clear that the intensity of respiration varies widely not only among different plants but in the same plant or organ at different stages of its life, and under different external conditions. Some representative examples are given below.

RESPIRATORY INTENSITY OF PLANTS OF DIFFERENT SPECIES

Considerable variation in respiration intensity exists among plants of different species. Some data regarding higher plants, obtained by Aubert, in which the intensity of respiration is given in terms of the volume of oxygen in cubic millimetres absorbed per hour per gram of fresh weight, are recorded in Table XXXIII.

Table XXXIII

Intensity of Respiration of Plants of Different Species

Species	Temperature in degrees C.	Respiration Intensity in c.mm. Oxygen per hour per gm. fresh weight
Cereus macrogonus	12	3·00
Opuntia cylindrica	13	6·80
Opuntia tomentosa	13	11·40
Crassula arborescens	13	16·60
Picea excelsa	15	44·10
Galanthus nivalis	13	77·60
Vicia faba	12	96·60
Mirabilis jalapa	15	120·0
Triticum sativum	13	291·0

Much more intense respiratory activity has been observed in fungi and bacteria. Thus, if we assume the dry weight of wheat to be about one-quarter of the fresh weight, the highest respiration intensity observed by Aubert is about 1164 c.mm. or 1·164 c.c. of oxygen per hour per gm. dry weight. Now Kostychev found that a culture of *Aspergillus niger* grown on a quinic acid medium for 2 days liberated from 72·5 to 78 c.c. of carbon dioxide per hour per gm. dry weight.

Not only do different plants thus respire at very different rates, but the same plant at different times of its life, and different organs of the same plant at the same time, may exhibit very different intensities of respiration. As regards the variation in respiratory activity of the same plant at different stages of development, a number of observations have been made on the respiration of seedlings during development. Some observations on this point with a number of different species (*Fagopyrum esculentum*, *Lathyrus odoratus* and *Helianthus annuus*) are summarized graphically in Fig. 15. In these experiments the covering of the seed (testa or pericarp) was removed and under this condition it appears that the respiration intensity expressed as carbon dioxide evolved in unit time per unit of dry weight increases to a maximum after which the intensity falls. If the protective coverings of the seeds are not removed there may be a delay in reaching this maximum respiration rate, as described in a later chapter. In the case of seeds the respiration intensities referred to dry weight of the seed are of little use for comparison of the activities of different species, as so much of the dry matter of the seed is frequently non-living reserve material. Bearing this in mind, it is clear that germinating seeds or young seedlings appear to exhibit very considerable differences in respiratory activity. Thus a seedling of *Helianthus annuus* 3 days old was observed by Stiles and Leach to respire at

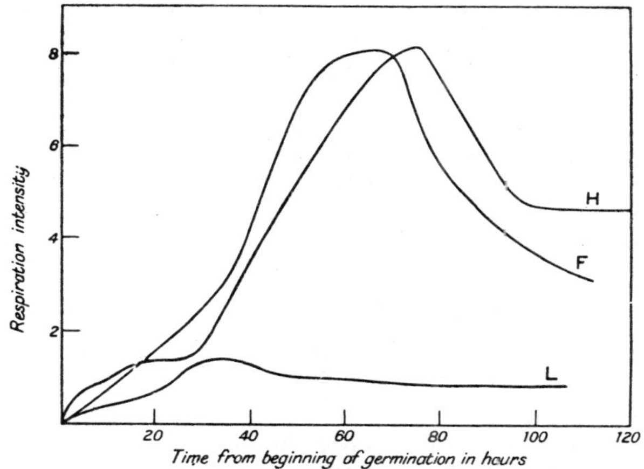

FIG. 15.—Curves showing the course of respiration in germinating seeds of *Fagopyrum esculentum* (F), *Lathyrus odoratus* (L), and *Helianthus annuus* (H)
(Constructed from data obtained by W. Stiles and W. Leach)

the rate of 7·9 mg. of carbon dioxide (or about 4 c.c.) per hour per gram of air-dry seed. Rates of respiration as high as this were also observed in young seedlings of *Fagopyrum esculentum*, but in most cases the maximum rates reached were not so high. In young seedlings of *Lathyrus odoratus*, for example, the maximum rate of respiration was rarely as high as 1·5 mg. of carbon dioxide per gram of air-dry seed.

The most complete series of observations on the change in respiratory intensity during subsequent development of the plant has been made by Kidd, West and Briggs, also with *Helianthus annuus*. Their results, as regards the whole plant, are summarized in Table XXXIV.

Table XXXIV

Change in Intensity of Respiration of Plants of *Helianthus annuus* during Development

Age of Plant in days	Dry weight of a Single Plant	Respiration Intensity in mg. carbon dioxide per hour per gm. dry weight
1	0·0225	2·90
2	0·0223	3·00
4	0·0242	3·00
13	0·1009	2·80
22	0·630	3·00
29	4·065	2·30
36	12·85	1·21
43	22·05	1·03
50	45·15	0·94
59	93·20	0·66
64	98·30	0·71
89	294·7	0·48
99	377·4	0·37
112	818·3	0·26
136	419·5	0·39

It will be observed that the respiration intensity, after reaching the early maximum observed by Stiles and Leach, falls off regularly as the plant grows older, with a slight rise during flowering late in the life of the individual. Similar results, unpublished at the time of writing this, have been obtained by Mehta for *Phaseolus vulgaris*.

THE INTENSITY OF RESPIRATION OF DIFFERENT ORGANS

Kidd, West and Briggs also determined the respective intensities of respiration of stem, leaf, stem apex and flowers of *Helianthus annuus* during stages of development. Their results are summarized in Table XXXV. From these results it will be observed that the respiration intensity of the stems and leaves falls very considerably with advancing age and that when flowers are produced the respiratory intensity of the reproductive parts of the flower is considerably greater than that of the vegetative parts at the same time. The same observation has been made by Maige with a number of other species, the stamens and gynaecium of *Papaver rhoeas*, for example, respiring at respective rates of 1·041 and 0·690 c.c. carbon dioxide per hour per gram of fresh weight, whereas the rates at which sepals, petals and ordinary foliage leaves respired were respectively only 0·390, 0·367 and 0·332 c.c. carbon dioxide per hour per gram of fresh weight.

Table XXXV

Intensities of Respiration of Separate Organs of *Helianthus annuus*

Age of Plant in days	Respiration Intensity in mg. CO_2 per hour per gm. dry weight				
	Stem	Leaf	Stem Apex	Whole Inflorescence	Flowers on Lateral Shoots
36	0·81	1·56			
43	0·69	1·38			
50	0·46	1·52	2·56		
59	0·33	1·32	1·78		
64	0·34	1·24			
89	0·31	0·90	1·13	1·13	
99	0·25	0·45	0·89	1·04	1·1
112	0·098	0·375	0·75	0·85	0·95
136	0·081	0·44	0·96	0·965	0·97

PERIODICITY IN RESPIRATION INTENSITY

There is evidence that perennial plants, even during their active periods, exhibit a seasonal periodicity in respiratory activity. In the case of such plants growing in temperate regions Bonnier and Mangin found the respiration intensity of the whole plant was greatest in spring, decreased somewhat in summer and fell to a minimum in winter. Expansion of leaves in spring and opening of flowers in winter might give rise to subsidiary maxima. The course of respiration of such a perennial plant during one year thus resembles on the whole that of an annual plant like *Helianthus annuus*. In the tropical plant *Artocarpus integrifolia* Inamdar and Singh also observed seasonal variations in respiratory activity, the minimum value here being reached at the beginning of summer and no rise occurring until autumn when the respiration rate increases to the maximum which continues through winter and spring. These periodic changes appear to be in part, but not entirely, related to the amount of substrate present, but some depressing factor operating during the summer appears to result in the falling off of respiratory activity during that season.

THE INFLUENCE OF EXTERNAL CONDITIONS ON THE INTENSITY OF RESPIRATION

While the variations in respiratory activity noticed above must be attributed to changes, for the most part unknown, occurring within the plant, external conditions also influence the intensity of respiration. Temperature, concentration of oxygen and concentration of carbon dioxide may all affect the rate of respiration, while various chemical substances have been observed to have an effect. Light appears to have no direct effect on respiration although it may influence it indirectly in chlorophyll-containing tissues by bringing about an increase in the amount of respiratory substrate through photosynthesis. It may also affect gaseous exchange between the tissue and the surrounding medium in the case of succulent plants by bringing about the decomposition of organic acids, and so affecting the apparent respiratory quotient.

Temperature. The general effect of temperature is to increase the rate of respiration of a tissue. Between 0° and 30° C. an increase in temperature of 10° C. generally results in the rate of respiration being increased from 2·1 to 2·5 times. Such a temperature coefficient of respiration has frequently been observed, as an inspection of Table XXXVI will show.

Table XXXVI

Temperature Coefficient of Respiration

Plant Material	Temperature Range in degrees Centigrade	Temperature Coefficient	Observer
Triticum sativum seedlings . .	0–20	2·5	Clausen, 1890
Pisum sativum seedlings. . .	0–30	2·5	Fernandes, 1923
Prunus laurocerasus leaves . .	16–30	2·1	Matthaei, 1904
Salix glauca leaves	0–10	3·3	Müller, 1928
,, ,, ,,	10–20	1·8	,, ,,
Chamaenerium latifolium leaves	0–10	2·0	,, ,,
,, ,, ,,	10–20	2·5	,, ,,
Syringa vulgaris flowers . . .	0–20	2·5	Clausen, 1890
Fragaria vesca fruits . . .	5–25	2·5	Gerhart, 1930

The coefficient is, however, not constant, falling in value with increase in temperature, at any rate in the case of the roots of *Pisum sativum*. Thus for the temperature intervals 0°–10°, 10°–20° and 25°–30° the coefficients calculated from the data of Fernandes, are respectively 2·69, 2·53 and 2·37. Above 30° C. what has been called a ‘time factor’ becomes operative. Whereas below 30° C. the respiration remains reasonably constant, above this temperature the rate of respiration falls off rapidly with time, behaviour which is probably related to the destruction or inactivation of some part of the protoplasmic system, possibly an enzyme, at the higher temperatures. The higher the temperature, the more rapid is the falling off in the respiratory activity. The operation of a time factor, as observed by Fernandes in the case of *Pisum sativum*, is illustrated in Fig. 16.

Oxygen Concentration. It was at one time thought that oxygen concentration had little effect on respiratory activity over a wide range of concentrations. Normally, of course, plants growing in air are subjected to a partial pressure of oxygen equal to one-fifth of an atmosphere. Not until this pressure is raised to 2 or 3 atmospheres has a notable increase in respiration rate been observed, and this appears to prelude the death of the tissues. It has generally been accepted that a reduction in the concentration of oxygen in the air has no effect on respiration until it is reduced to 2 per cent. when the respiration rate decreases with decrease in oxygen concentration.

It has, however, been fairly recently stated by F. F. Blackman that the respiration intensity of apple fruits is high in an atmosphere of nitrogen devoid of oxygen, and that with increase in oxygen concentration the respiration rate falls until it reaches a minimum between

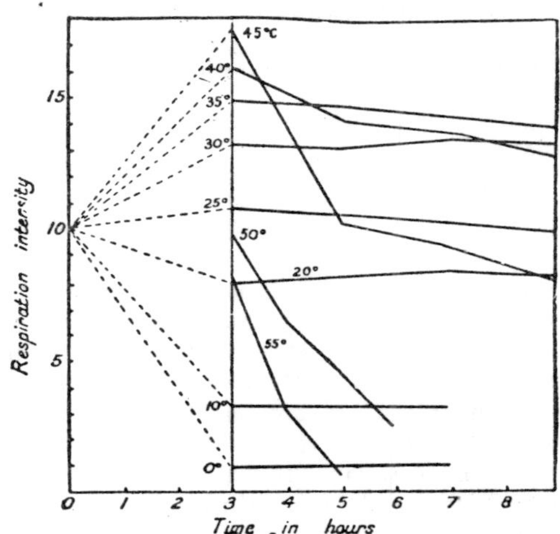

Fig. 16.—Curves showing the influence of temperature on the respiratory activity of seedlings of *Pisum sativum*

(From W. Stiles and W. Leach, after Fernandes)

5 and 9 per cent. With further increase in oxygen concentration respiration intensity rises steadily right up to a concentration of oxygen of 100 per cent.

Mack also found the relation between respiratory activity of wheat seedlings and oxygen concentration was not simple. At any definite temperature the intensity of respiration was found to increase with increase in oxygen concentration until at a certain value of the latter, varying with the temperature between about 8 and 13 per cent., the respiration intensity reached a maximum. With further increase in oxygen concentration the respiratory activity fell to a minimum and then, with still further increase in oxygen concentration rose to a second maximum in the neighbourhood of 90 or 95 per cent. oxygen concentration, above which there was a further fall in respiration rate.

The meaning of this complex relationship between oxygen concentration and respiration rate is not at all clear.

Carbon Dioxide Concentration. With increase in carbon dioxide concentration of the medium surrounding respiring tissue the respiration rate decreases. The falling off in respiratory activity, as measured both by oxygen absorption and carbon dioxide evolution, is well shown by observations on the respiration of germinating seeds of *Sinapis alba* made by Kidd and summarized in Table XXXVII. In these experiments 20 per cent. oxygen was originally present in each case.

Table XXXVII

The Influence of Carbon Dioxide Concentration on the Respiration of Germinating Seeds of *Sinapis alba*

Original concentration of Carbon Dioxide in per cent	Oxygen absorbed in 14 hours	Carbon Dioxide evolved in 14 hours
0 . . .	71	58
10 . . .	57	48
20 . . .	49	38
30 . . .	45	33
40 . . .	38	26
80 . . .	32	17

Sugar Concentration. As sugar forms the respiratory substrate it is to be expected that in cases where sugar can be absorbed by the tissues from the external medium, the concentration of this substance should affect respiratory activity. That this is the case with various fungi has been demonstrated by, among others, Maige and Nicolas, who found that respiratory activity increased, within limits, with increasing concentration of sugar. It was further shown by Palladin that floating starved leaves of *Vicia faba* on sucrose solution greatly increased their respiratory activity, the rate of respiration of the starved leaves averaging 89·6 mg. carbon dioxide per hour per 100 gm. of leaves, while after floating on sucrose solution for 2 days in the dark the average rate of respiration was 147·8 mg. carbon dioxide per hour per 100 gm. of leaves.

The relation between concentration of sugar and respiratory activity of potato tubers was examined by Hanes and Barker. In the course of an investigation on the effect of hydrocyanic acid on the respiration of this tissue they found that variations in respiration rate brought about by exposure to an atmosphere containing 0·14 c.c. to 0·3 c.c. of hydrocyanic acid per litre ran parallel with changes in sugar content of the tissue, and it was thought that the respiration rate was directly related to the concentration of sugar. In a higher concentration of cyanide of about 1·5 c.c. per litre the ratio of respiration rate to concentration of sugar soon fell, a result which is very reasonably attributed to inactivation of enzymes concerned in respiration as a result of the cyanide action.

In dealing with the respiration of potato tubers without cyanide Barker found that the investigation of the relationship between sugar concentration and respiration was complicated by a temperature effect introduced by exposing the tubers to low temperature. Such a treatment appeared to have a depressing effect on respiratory activity, an effect which lasted for several weeks even after the tubers were transferred to a higher temperature. When this depressing factor was not operating, the respiration rate appeared to be definitely affected by the sugar concentration in the way indicated by the work of Hanes and Barker.

Salts, Acids, Alkalies and Organic Substances. A considerable number of observations have been made on the effect of various inorganic salts, acids, alkalies and organic poisons in various concentrations on the respiratory activity of different plants. In general it may be said that small concentrations of these substances bring about an increase in respiration intensity, while higher concentrations have the reverse effect.

The Effect of Wounding on Respiration. One of the best-known results of wounding a plant tissue is a rise in the respiration rate. This was shown by Böhm in 1887 to be the case when potatoes are cut. Since then the same behaviour has been observed in many plant species and organs. In storage tissues such as potatoes and carrots the increased carbon dioxide output on wounding is no doubt to be attributed in part to the release of gas previously held in the intercellular spaces of the tissues, but apart from this, there is a definite increase in respiratory activity as a result of wounding, an increase which may continue for some days. This increased respiration rate is no doubt connected with the renewal of growth which frequently results from wounding, but the mechanism of the processes involved is not understood. The wounding of potato tubers by cutting results in an increase in the concentration of sugar in the cells in the neighbourhood of the wound, and this increase in the concentration of the respiratory substrate might account for the increase in respiration rate, although it has been argued that this is not the case.

HEAT EVOLVED IN RESPIRATION

The respiratory process, as has been emphasized, is essentially one involving a release of energy, some of which is necessary for the carrying out of the vital activities of the organism. But a proportion of the energy so released is dissipated as heat, and it might therefore be supposed that, as every cell in the plant continually respires, the temperature of the plant would be maintained above that of its surroundings. Any difference in temperature, however, between the plant and the medium in which it lives is generally very slight, and it has been stated that the temperature of growing shoots is not as a rule more than 0·3° C. above that of the surrounding atmosphere.

However, during seed germination and the opening of flowers, a considerable development of heat may take place and raise the temperature of the plant or organ considerably above that of the surrounding air. This was observed to be the case by Lamarck in the developing inflorescence of *Arum italicum* in 1778, a finding confirmed by de Saussure in 1822 by direct measurement of the temperature. Later Kraus observed the temperature developed in the spadix of this species to be 27·7° C. above the surrounding air.

The evolution of heat by rapidly respiring tissue is now generally demonstrated by means of vacuum flasks, that is, double-walled vessels with a vacuum between the two walls, through which the transference of heat is very slow. When germinating seeds or opening flowers are placed in such vessels a considerable rise of temperature is readily observable in many cases. By this means Pierce found that germinating seeds of *Pisum sativum* over a period of 7 days evolved on the average 4·93 calories per day per gram of peas. This value is actually very much smaller than that previously observed by Bonnier, namely 59 calories per minute per kilogram of peas, which is equivalent to a rate of 85 calories per day per gram of peas. Bonnier's observations were, however, carried out over short periods of time, for example, 36 minutes, whereas Pierce's observation period was a week. From the amount of carbon dioxide actually evolved during germination there can be calculated the quantity of energy released by the breaking down of that amount of carbohydrate necessary to produce that quantity of carbon dioxide. Results that have been obtained by calculations of this kind are contradictory. In experiments with barley cited by Palladin it was found that the heat evolution actually measured during the early stages of germination exceeds that which can be accounted for by the oxidation of carbohydrate. It would thus appear that the whole of the heat evolved is not to be attributed to the respiratory process alone, and that other actions such as those involved in the absorption of water by the seeds, contribute also to the evolution of heat. The excess of heat evolved over the energy released in respiration is no longer observable after the germination period is past, and in growing stems the reverse is the case.

THE FATE OF THE ENERGY RELEASED IN RESPIRATION

The energy released by the respiratory process appears, as we have seen, partly as heat. The part played by this in raising the temperature of the plant is, for the most part, negligible, and no doubt most of it is dissipated by radiation and conduction. The significance of respiration must be looked for in the fate of the released energy which is not lost as heat. Yet, here, where it is most needed, precise information is wanting. The manufacture of many essential substances in the plant, such as proteins, fats and the more complex lipoid substances, require energy, and it seems probable that the reactions leading to the manufacture of these substances may in the first place be linked in some way with the actual respiratory processes, and that the actual transference of energy may take place through these linked reactions. The movement of protoplasm, and, in some cases of whole organisms, and the passage of substances from cell to cell, which often appears to take place against a diffusion gradient, requires energy, some of which, at least, has originally been provided by respiration. Yet, when all is said, one is left wondering whether these activities really afford a complete explanation of the universality of the respiratory process. Respiration is an essential property of living matter and is thus met with in every living cell, and it would seem that a continuance of respiration is necessary for, or is involved in, the maintenance of the protoplasm. It may be that, as protoplasm is a system in a state of dynamic equilibrium, it contains substances which automatically break down and have to be continually replaced by fresh material, for the formation of which a supply of energy is necessary.

ANAEROBIC RESPIRATION

When a plant which normally respires aerobically is deprived of oxygen, the evolution of carbon dioxide continues, although the rate of evolution of this gas may be changed. This evolution of carbon dioxide by plant tissue in absence of oxygen has been known since 1797 when it was observed with germinating barley by William Cruickshank, although little attention was paid to the phenomenon for about 75 years. It was described as 'intramolecular respiration' by Pflüger in 1875, and as 'anaerobic respiration' by Kostychev in 1902. While both terms are still in use, the latter is much more descriptive and unambiguous and is to be preferred.

Anaerobic respiration has been observed in all plants in which it has been sought, and which normally respire aerobically, with the possible exception of *Elodea canadensis*. It appears to be immaterial whether the tissue is contained in a vacuum or in a gas inert as far as the tissues are concerned. Thus seedlings of *Vicia faba* were observed by Wortmann to give out carbon dioxide as rapidly in a Torricellian vacuum as in air, while they have been found to respire as rapidly in nitrogen and in hydrogen. But from many cases examined it appears that anaerobic respiration will not continue indefinitely, though no general statement can be made as to how long a plant can survive in absence of oxygen. Chudiakow observed that seedlings of *Zea mais* die in 24 hours at 18° C. in absence of oxygen, while it is known that some fruits, such as apples and pears, remain alive for months in an atmosphere of nitrogen.

Along with carbon dioxide, ethyl alcohol is a very common, though not universal, product of anaerobic respiration. This suggests at once a comparison of anaerobic respiration with yeast fermentation of sugar, and the two reactions can thus be roughly expressed by the same equation, already given on p. 124. According to this, equal molecular quantities of carbon dioxide and ethyl alcohol are produced, and as the molecular weights of these substances are respectively 44 and 46, the respective quantities of the two products should be almost equal in weight (actually $C_2H_5OH/CO_2 = 1{\cdot}05$). Actually the quantity of alcohol produced nearly always falls short of this theoretical amount, the divergence sometimes being negligible but in other cases being very considerable, as some data obtained by Boysen Jensen, and summarized in Table XXXVIII, show. In blue grapes, for example, the ratio in one case was found to be nearly unity, whereas in potato tuber only minute quantities of ethyl alcohol, or even none at all, were found. It is possible that the anaerobic respiration process takes a different course in different tissues, or that ethyl alcohol is actually formed in all cases, but that in some tissues it is immediately utilized in some secondary reaction. In either case we should expect other substances besides ethyl alcohol to be formed, and actually a variety of substances, including a number of aldehydes and organic acids, have been identified in anaerobically respiring tissues, and have been presumed to result from the respiratory process.

Table XXXVIII

Ratio of Alcohol to Carbon Dioxide produced in Anaerobic Respiration

Plant Material	C_2H_5OH/CO_2
Pisum sativum cotyledons	0·81
,, ,, ,, . .	0·65
Daucus carota root	0·91
,, ,, ,, ...	0·86
,, ,, ,, ...	0·72
Sinapis sp. seedlings . . .	0·60
,, ,, . . .	0·32
Vitis vinifera blue grapes . . .	0·95
,, ,, ,, ,, . . .	0·88
,, ,, ,, ,, . . .	0·74
Solanum tuberosum tuber . .	0·07
,, ,, ,, . . .	0·02
,, ,, ,, . . .	0·00

The capacity of a plant to survive under anaerobic conditions is ascribed to the continuance of its respiration in absence of oxygen. The energy released in the breaking down of glucose to carbon dioxide and ethyl alcohol is, however, small in comparison with that produced by the complete oxidation of the same amount of hexose to carbon dioxide and water, the calories released from a gram-molecule of glucose being respectively 26 and 674 in the two cases. This means that for every gram-molecule of carbon dioxide produced anaerobically 13 calories are formed as compared with 112 for every gram-molecule of carbon dioxide produced aerobically. Since, moreover, the rate of anaerobic respiration, as measured by carbon dioxide output, is rarely as high as that of aerobic respiration (cf. Table XXXIX) it is clear that anaerobic respiration is a very poor substitute for the normal process as far as release of energy is concerned. Thus the common idea that anaerobic respiration is of the nature of an adaptation of advantage to the plant when deprived of oxygen, is scarcely supported by a consideration of these facts. Moreover, under natural conditions a shortage of oxygen is only likely to be met with in the cases of bulky organs and of roots growing in poorly aerated soils. As regards the former, Devaux showed in 1891 that the internal atmosphere of large fruits such as those of *Cucurbita maxima* and *C. melanosperma* contains nearly as high a concentration of oxygen as does atmospheric air. Again, the harmful effect of poor soil aeration on the vitality of most roots, suggests that anaerobic respiration is not very effective as a substitute for aerobic respiration in these organs, although there are species the roots of which grow satisfactorily under poor conditions of aeration, and here, it may be perhaps, that anaerobic respiration is an effective substitute for the normal process. But from a consideration of all the facts we must conclude that anaerobic respiration has no teleological significance in the life of the plant and it is to be regarded purely as the physical and chemical consequences of a disturbance of the normal life conditions of the plant.

Reference has already been made to the fact that transference of a plant to anaerobic conditions generally results in a lowering of the rate of carbon dioxide output. That this is very generally the case will appear from the numbers obtained some 50 years ago in Pfeffer's laboratory at Tübingen (see Table XXXIX). Of 17 subjects of investigation, only one, germinating seeds of *Vicia faba*, respired at the same rate, as measured by carbon dioxide output, in hydrogen as in air.

Transference of aerobically respiring material to an atmosphere of nitrogen does, however, sometimes result in an increase of carbon dioxide output. This has been observed in apples by F. F. Blackman and Parija, while an output of carbon dioxide as high as, or higher than, that occurring in air, was observed by G. R. Hill in the anaerobic respiration of a number of other fruits, namely, grapes, cherries and blackberries, and by Gustafson in the anaerobic respiration of tomato fruits. How far these high rates of anaerobic respiration are confined to senescent fruits cannot at present be said, but if the sugar utilized in aerobic respiration is oxidized to carbon dioxide and water, and the whole of that in anaerobic respiration is broken down to carbon dioxide and alcohol, it would appear that a more rapid destruction of sugar under anaerobic conditions than under aerobic conditions occurs in other tissues, for it will be observed that to produce a certain amount of carbon dioxide anaerobically three times as much sugar is required as to produce it aerobically. Consequently, wherever the ratio of anaerobic to aerobic respiration exceeds 0·33 the rate of sugar destruction is greater in absence of oxygen. A glance at Table XXXIX shows that this is so in quite a number of cases.

Table XXXIX

Ratio of Respiration Rates in Hydrogen and in Air

Species	Organ	$\frac{R_{hydrogen}}{R_{air}}$
Vicia faba	germinating seed	1·03
Triticum vulgare	seedling	0·49
Cucurbita pepo	,,	0·35
Sinapis alba	,,	0·18
Brassica napus	,,	0·25
Cannabis sativa	,,	0·34
Helianthus annuus	,,	0·33
Lupinus luteus	,,	0·24
Heracleum giganteum	unripe fruit	0·42
Abies excelsa	young leafy shoot	0·077
Orobanche ramosa	flowering stem	0·32
Arum maculatum	,, ,,	0·615
Ligustrum vulgare	leafy shoot	0·816
Lactarius piperatus	fruit body	0·32
Hydnum repandum	,, ,,	0·256
Cantharellus cibatus	,, ,,	0·666
Beer Yeast (on lactose)		0·310

The values in this table probably have little exact significance, as in a number of cases examined the rate of anaerobic respiration changes

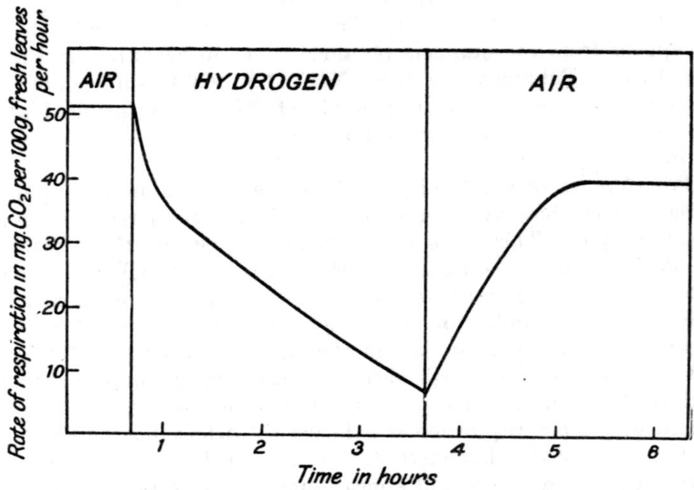

Fig. 17.—Curve showing the course of respiration of leaves of *Tropaeolum majus* when transferred from air to hydrogen and then back to air
(Constructed from the data of P. Boysen Jensen)

with time. Thus Boysen Jensen found that the respiration of both leaves of *Tropaeolum majus* and seedlings of *Sinapis alba* in hydrogen fell rapidly with time, the rate rising again on re-transference to air (cf. Fig. 17). The same has been shown by Leach and Dent to be the case with young seedlings of *Ricinus communis*, *Helianthus annuus*

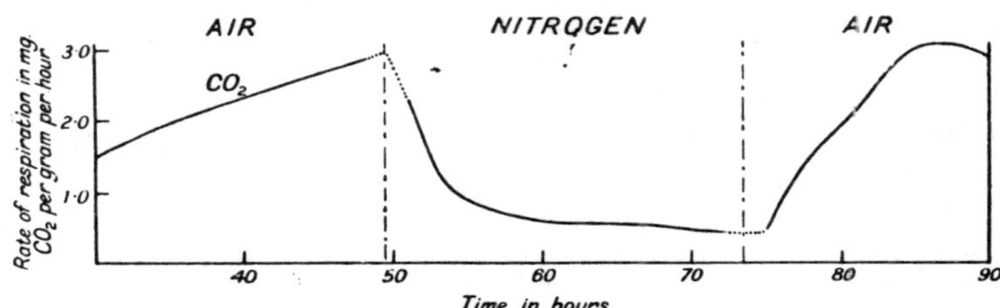

FIG. 18.—Curves showing the course of respiration of young seedlings of *Helianthus annuus* when transferred from air to nitrogen and then back to air
(From W. Leach and K. W. Dent)

and *Cucurbita pepo* (cf. Fig. 18). In the case of apples examined by Blackman and Parija there was also a falling off in the rate of anaerobic respiration after the initial rise. The actual course varies with the physiological type of apple, the fall being more rapid and greater with

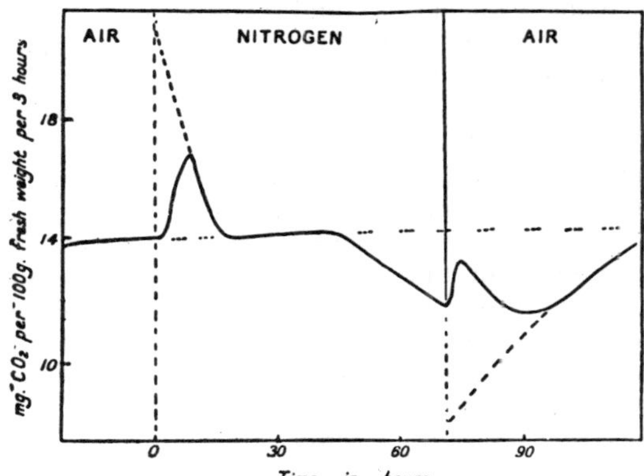

FIG. 19.—Curve showing the course of respiration of an apple when transferred from air to nitrogen and then back to air
(From W. Stiles and W. Leach, after F. F. Blackman)

apples which ripen more slowly. The course of respiration of an apple of a more rapidly ripening type is shown in Fig. 19. The values of the respiration rate in presence and absence of oxygen may, as will appear later, shed considerable light upon the mechanism of respiration.

THE CONNEXION BETWEEN ANAEROBIC RESPIRATION AND ALCOHOLIC FERMENTATION

It has already been pointed out that the most frequent product of anaerobic respiration, apart from carbon dioxide, is ethyl alcohol, and that the substrate and end-products are thus the same as in the fermentation of sugar by yeast. Now if the two processes of anaerobic respiration and fermentation by yeast should be the same, the fact would be of considerable importance to the student of respiration, for although little direct information is available as regards the mechanism of anaerobic respiration, a considerable amount of information has been obtained regarding the stages in yeast fermentation.

Because the substrate and end-products of fermentation and anaerobic respiration are the same, it does not necessarily follow that the breaking down of sugar in the two processes takes the same course. There is, however, evidence suggesting that this is the case. Thus the enzyme complex known as zymase which breaks down hexose to alcohol and carbon dioxide in yeast fermentation, was shown in 1903 by Stoklasa and Czerny to be present in the cells of higher plants where it would presumably effect the same decomposition as in yeast. It has already been pointed out in the preceding chapter that zymase is to be regarded as a mixture of a number of enzymes, including glucolase, ketonaldehydemutase and carboxylase. These individual enzymes, which play their various parts in the decomposition of sugar by yeast, have also been identified in the tissues of higher plants.

Again, Neuberg showed that acetaldehyde is probably formed as an intermediate product in fermentation, for, by addition of sulphite to the fermenting solution the aldehyde can be fixed as aldehyde sulphite and so accumulates. Now Kostychev and his collaborators were able to demonstrate the formation of acetaldehyde during the anaerobic respiration of flowers of *Populus*, while Neuberg and his co-workers have succeeded in fixing acetaldehyde during the anaerobic respiration of certain fungi and of germinating pea seeds. There is thus evidence that fermentation and anaerobic respiration do actually involve the same series of changes.

THE RELATION BETWEEN AEROBIC AND ANAEROBIC RESPIRATION

The earlier authorities, Nägeli and Sachs, were of opinion that the evolution of carbon dioxide in absence of oxygen resulted from injury to the tissues and was thus to be regarded as a pathological phenomenon unconnected with normal respiration. Pfeffer, however, suggested that normal aerobic respiration involved two stages, or series of stages, the first consisting of the breaking down of sugar to ethyl alcohol and carbon dioxide, the second the oxidation of the alcohol, or some precursor of the alcohol formed in the first stage, into carbon dioxide and water. For the first stage oxygen is unnecessary, so that under anaerobic conditions the reaction proceeds as far as the production of ethyl alcohol, but there stops as oxygen is required for the production of carbon dioxide and water. This theory was made a little more precise by Kostychev, whose view of the matter may be made clear by the following scheme:

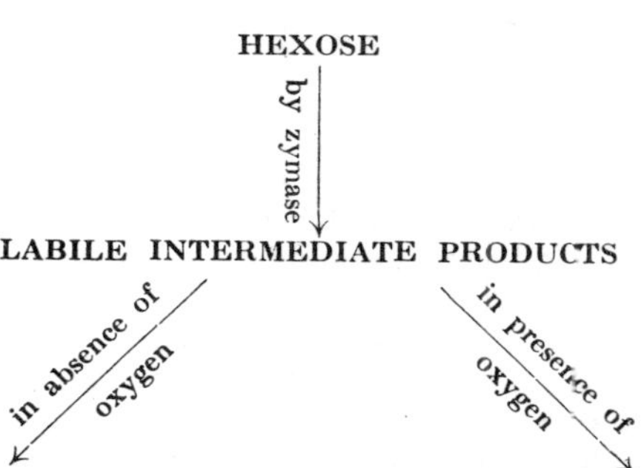

The arguments in favour of such a connexion between aerobic and anaerobic respiration are these. In the first place all plants which normally respire aerobically also respire anaerobically, as far as they have been examined, with the one possible exception of *Elodea*. Secondly, the enzyme zymase which effects the breaking down of hexose to ethyl alcohol and carbon dioxide, appears to be present in all plant tissues, so that its function would appear to be a normal one in all plants. Next, it sometimes happens that after a period of anaerobiosis, transference of tissue to oxygen results in an increase of the respiration rate to a value above the normal. This would be expected if, during anaerobic respiration, a substance accumulated which formed an intermediate product of normal respiration and which would thus be oxidized to carbon dioxide and water on exposure to oxygen. It must, however, be admitted that this argument is not very cogent as this increase in respiration after a period of anaerobiosis is not always observed. It has been suggested that where the increase does not occur toxic substances formed during anaerobic respiration are responsible for reducing respiratory activity. Lastly, acetaldehyde, which it has already been pointed out is probably an intermediate product in anaerobic respiration, has also been found during normal aerobic respiration.

Certain arguments have been urged against the view that anaerobic and aerobic respiration are connected in the way suggested. Thus Boysen Jensen pointed out in 1923 that in some tissues the ratio of the rates of anaerobic and aerobic respiration falls below 1 : 3 without there being any evidence of toxic action in the anaerobically respiring material, from which it can be urged that in such cases the sugar split up by the anaerobic process is insufficient to account for the whole of the aerobic respiration. Again, an enzyme, to which the name glucose oxidase has been given, has been prepared by D. Müller from *Aspergillus niger*, which will oxidize glucose to gluconic acid in absence of zymase, thus suggesting that the first stage in the oxidation of hexose to carbon dioxide and water is not necessarily the splitting of the sugar by the action of zymase or one of its constituent enzymes. Yet again, it has been shown by Lundsgaard that the action of zymase is stopped by monoiodoacetic acid; nevertheless bakers' yeast in presence of this substance can oxidize glucose to carbon dioxide and water, although its fermenting power is feeble, from which it can be concluded that here also the oxidation of hexose to carbon dioxide and water is an entirely separate process from fermentation.

It thus appears that in certain cases glucose can be oxidized in the plant without the intervention of zymase, but at present there is no evidence of the extent to which such an oxidation actually takes place throughout the plant kingdom, and it may be an occasional rather than a general phenomenon. In the present state of our knowledge it is generally regarded as much more probable that anaerobic and aerobic respiration are connected in the manner hypothesized by Pfeffer and Kostychev, and the evidence in favour of this view is fairly convincing.

THE STAGES IN THE RESPIRATORY PROCESS

In devising a scheme of reactions to account for the breaking down of the respiratory substrate to carbon dioxide and water, it is necessary in the first place to decide on what is to be regarded as the actual substrate. We have noticed that a number of different substances may provide material for respiration, such as sugars, starch, cellulose, fats and perhaps other substances. It is generally supposed that, whatever this material, it is first transformed to hexose sugar, and that this should be regarded as the general respiratory substrate. The evidence for this view as regards the more complex carbohydrates such as sucrose, starch and cellulose, is that along with them hydrolytic enzymes are present which can convert these various substances into glucose, while chemical analyses show that such hydrolysis does indeed take place. Where fat is present as a reserve, it is not clear how a conversion of this to sugar can take place, but chemical analyses of, for example, fat-containing seeds during germination, leave no doubt that there is a conversion of fat to sugar in such cases.

Assuming then that hexose sugar is the respiratory substrate, it is supposed that the first stage of respiration is the conversion of the ordinary stable sugar into an unstable active form. That this is so seems likely from a consideration of the modern view, due to Haworth, of the constitution of sugars. According to this view, the molecule of ordinary glucose consists of a ring of five carbon atoms and one oxygen atom, the sixth carbon atom forming part of a $-CH_2OH$ group which is attached to one of the carbon atoms of the ring. There are two stereoisomeric forms of glucose of which the formula of the α form may be shown thus:

```
           OH        OH
           |         |
          /C---------C*\
   H     / |         |  \
   |    /  H         H   \
   C   <                  > O
   |    \  OH        H   /
   OH    \ |         |  /
          \C---------C /
           |         |
           H        CH2OH
```

In this formula the side groups or atoms are to be regarded as lying above or below the plane of the ring according as they are shown in the formula above or below the carbon atom to which they are attached. The β form is similar except that the position of the hydrogen atom and hydroxyl group attached to the carbon atom marked with an asterisk are reversed.

These forms of glucose are stable and are known as pyranose sugars. The first stage of respiration is supposed to involve their transformation to unstable forms termed furanose, labile or γ-sugars. In these sugars there is a five-atom ring, the formula ascribed to γ-glucose being:

```
          CHOH
         /    \
     HCOH      O
       |       |
     HCOH -----C-------CHOH
               |       |
               H-------CH2OH
```

This sugar does not appear to exist in the free state, but it is in this form that it enters into certain derivatives.

Similar considerations hold with regard to other sugars, including fructose, which, like glucose, is of wide distribution in the plant kingdom. Here again, although the active γ-sugar is not found free, it is in this form that it enters into combination with such important products as cane sugar and inulin.

The actual course of production of the active sugar from the stable form is suggested by certain facts regarding alcoholic fermentation. It is known that fermentation by dried yeast or yeast juice is accelerated by phosphate, and that what takes place on addition of phosphate is the formation of compounds of phosphoric acid and hexose, namely, hexose diphosphate and hexose monophosphate, the former of which, at any rate, is then split into hexose and phosphate. But whatever hexose sugar is used, the hexose diphosphate is always *fructose* diphosphate, in which the fructose is present in the active γ-form. The accelerating effect of phosphate would thus appear to be due to the conversion of stable hexose into γ-fructose. Whether activation of the hexose is brought about in the same way in respiration cannot at present be said, but Lyon found that phosphate accelerates the respiration of *Elodea canadensis* and wheat seedlings.

If we accept the view of Pfeffer and Kostychev with regard to the connexion of anaerobic and aerobic respiration, and also agree that anaerobic respiration and alcoholic fermentation are the same process, then we must suppose that the breaking down of the active hexose, or glycolysis, is effected by enzymes of the zymase complex. There are now two main views regarding the stages in this breakdown, the earlier one of Neuberg and the recent one of Embden and Meyerhof.

According to Neuberg, the production of ethyl alcohol from hexose takes place in several stages, and it would appear from what has been said earlier that acetaldehyde is an intermediate product. The stages hypothesized by Neuberg are these:

1. The breaking up of the sugar molecule into two molecules of methyl glyoxal with elimination of water:

$$C_6H_{12}O_6 = 2CH_3.CO.CHO + 2H_2O$$

It has been mentioned in the last chapter (see p. 109) that an enzyme glycolase effecting this glycolysis has been found in yeast and in tobacco leaves.

2. The methyl glyoxal then, according to Neuberg's scheme, takes part in a Cannizzaro reaction, that is, a reaction between two molecules of aldehyde in presence of water, in which one molecule becomes oxidized to an acid and the other reduced to an alcohol. In this particular case it is supposed that one molecule of methyl glyoxal is oxidized to pyruvic acid and the other reduced to glycerol:

$$\begin{array}{c} CH_3.CO.CHO \\ + \\ CH_3.CO.CHO \end{array} + \begin{array}{c} H_2 \\ \| \\ O \end{array} + H_2O = \begin{array}{c} CH_2OH.CHOH.CH_2OH \\ + \\ CH_3.CO.COOH \end{array}$$

Actually, glycerol has been detected as an intermediate product of alcoholic fermentation, although the same cannot be said of the pyruvic acid, which, it is assumed, is immediately decomposed.

3. This decomposition results in the splitting off of carbon dioxide with formation of acetaldehyde, thus:

$$CH_3.CO.COOH = CH_3.CHO + CO_2$$

An enzyme bringing about this decomposition, carboxylase, is known to be of wide distribution in the plant kingdom (cf. p. 109). Moreover, as has already been pointed out, the formation of acetaldehyde during fermentation is a well-established fact.

4. The nature of the next stage in respiration is probably determined by the conditions: whether oxygen is present or absent. In absence of oxygen, according to Neuberg's scheme, a second Cannizzaro reaction now takes place between a molecule of acetaldehyde and one of methyl glyoxal (pyruvic aldehyde) produced in the earlier stage of the process. This reaction leads to the formation of a molecule of pyruvic acid and one of ethyl alcohol:

$$\begin{matrix} CH_3.CO.CHO \\ + \\ CH_3.CHO \end{matrix} + \begin{matrix} O \\ \| \\ H_2 \end{matrix} = \begin{matrix} CH_3.CO.COOH \\ + \\ CH_3.CH_2OH \end{matrix}$$

The ethyl alcohol constitutes an end-product of the process under anaerobic conditions, while the pyruvic acid is immediately broken down by carboxylase to give more acetaldehyde.

Where, in anaerobic respiration, ethyl alcohol is not produced in quantity equivalent to the carbon dioxide evolved, it must be supposed that the course of events hypothesized by Neuberg for alcoholic fermentation is modified in some way. It is significant that modifications in the course of yeast fermentation have been effected by Neuberg in various ways. Thus addition of sodium sulphite fixes the acetaldehyde as it is formed, and the reactions therefore stop at the production of this substance, and ethyl alcohol is therefore not formed.

The theory with which the names of Embden and Meyerhof are associated arose from the discovery made by the former and his co-workers of the presence of phosphoglyceric acid among the products of carbohydrate breakdown in muscle and also in alcoholic fermentation. According to the Embden-Meyerhof scheme, the stages in sugar breakdown are as follows:

1. Hexose diphosphate reacts with glucose and phosphoric acid to produce phosphoglyceraldehyde:

$$C_6H_{10}O_4(PO_4H_2)_2 + C_6H_{12}O_6 + 2H_3PO_4 \\ = 4CH_2(PO_4H_2).CHOH.CHO + 2H_2O$$

2. A Cannizzaro reaction then takes place between pairs of phosphoglyceraldehyde molecules to produce the corresponding acid and alcohol, phosphoglyceric acid and phosphoglycerol (glycerophosphoric acid) respectively:

$$\begin{matrix} CH_2(PO_4H_2).CHOH.CHO \\ + \\ CH_2(PO_4H_2).CHOH.CHO \end{matrix} + \begin{matrix} O \\ \| \\ H_2 \end{matrix} = \begin{matrix} CH_2(PO_4H_2).CHOH.COOH \\ + \\ CH_2(PO_4H_2).CHOH.CH_2OH \end{matrix}$$

3. The phosphoglyceric acid now breaks down to pyruvic acid and phosphoric acid:

$$CH_2(PO_4H_2).CHOH.COOH = CH_3.CO.COOH + H_3PO_4$$

4. The pyruvic acid then breaks down to acetaldehyde and carbon dioxide as postulated in Neuberg's scheme:

$$CH_3.CO.COOH = CH_3.CHO + CO_2$$

5. In Neuberg's scheme under anaerobic conditions the acetaldehyde is supposed to react with methyl glyoxal (pyruvic acid) in a

Cannizzaro reaction. In the Embden-Meyerhof scheme a similar reaction is supposed to take place between acetaldehyde and phosphoglyceric aldehyde produced in stage 1 :

$$\begin{matrix} CH_3.CHO \\ + \\ CH_2(PO_4H_2).CHOH.CHO \end{matrix} + \begin{matrix} H_2 \\ \| \\ O \end{matrix} = \begin{matrix} CH_3.CH_2OH \\ + \\ CH_2(PO_4H_2).CHOH.COOH \end{matrix}$$

Thus ethyl alcohol is produced, while the phosphoglyceric acid is immediately broken down to pyruvic acid and phosphoric acid according to the equation of stage 3.

The theory of Embden and Meyerhof is thus similar in general lines to that of Neuberg, but a more important part is attributed to phosphate in the more recent theory. In the latter also, methyl glyoxal is not supposed to be an intermediate product, its place in the scheme being taken by phosphoglyceraldehyde.

Under aerobic conditions it must be supposed that an intermediate product, perhaps acetaldehyde, but quite possibly some other substance, is oxidized, but the course of this oxidation is obscure. Various oxidizing mechanisms in cells have been described and examined, but it is not evident that any of these can lead to the evolution of carbon dioxide. Considerations of the relative intensities of anaerobic and aerobic respiration indicate the extreme improbability of the carbon dioxide split off by the action of carboxylase forming the whole of this substance produced in aerobic respiration.

However, according to the theory of respiration proposed by Palladin in 1912, the whole of the carbon dioxide in aerobic respiration is evolved during the earlier anaerobic stages. According to this theory, during the first and anaerobic phase water dissociates and the respiratory substrate is oxidized by the oxygen from the water, yielding ultimately carbon dioxide. The hydrogen of the water, along with the hydrogen in the substrate, unites with a substance termed a hydrogen acceptor. This stage of the respiratory process will then be represented thus :

$$C_6H_{12}O_6 + 6H_2O + 12R = 6CO_2 + 12RH_2$$

Certain pigments found in plants do accept hydrogen and are thereby reduced to 'chromogens' which, in presence of oxidase and atmospheric oxygen, are oxidized back to the pigment. According to Palladin the process summed up in the above equation is brought about owing to the presence of these pigments. Under anaerobic conditions it can, however, only proceed until the pigments have taken up their maximum possible quantity of hydrogen, that is, until the whole of the pigments are reduced to chromogens ; after this has taken place the hydrogen then reduces some intermediate product of the action to ethyl alcohol. Under aerobic conditions, however, this state of affairs is not reached because the chromogens are oxidized by oxidase action and the respiration pigments re-formed, and so the removal of the hydrogen by the respiration pigments is able to continue indefinitely. The aerobic stage is thus simply concerned with the regeneration of the respiration pigment or hydrogen acceptor and is not directly responsible for the evolution of carbon dioxide but only for the elimination of water. It may be summed up by the equation :

$$12RH_2 + 6O_2 = 12R + 12H_2O$$

It is difficult to bring Palladin's theory into line with the more recent views based on Neuberg's theory of alcohol fermentation. All the same it is deserving of attention.

Lastly, reference should be made to the views of F. F. Blackman on the respiratory process in plants, based on an ingenious analysis of data obtained with apples. Blackman's views of the first stages in respiration, namely hydrolysis of reserves to hexose, the activation of this, and the glycolysis of the activated sugar by enzymes of the zymase complex, agree essentially with the scheme described earlier. The most important part of Blackman's theory deals with the fate of these products of glycolysis : methyl glyoxal, pyruvic acid, acetaldehyde and possible other substances. This fate is different according as oxygen is present or not. In absence of oxygen the products, as we know, are alcohol and carbon dioxide. In presence of oxygen, on the other hand, part of the products of glycolysis are completely oxidized to carbon dioxide and water, but Blackman shows that in the apple the carbon dioxide so produced accounts for only a part of the carbon involved in glycolysis. The remainder of the carbon broken down in glycolysis must therefore undergo some other change. Since no other substance appears to accumulate the conclusion is drawn that some of the products of glycolysis are built back either to hexose or to some intermediate product. The evidence for this is derived from experiments with apples, which, after a period of normal respiration in air, were transferred to an atmosphere of nitrogen. It was found that the initial rate of respiration in nitrogen, that is, at the moment of transference to this gas, was 1·33 or 1·5 times that in air immediately before transference, the actual value being found by extrapolation (cf. Fig. 19).

Now in nitrogen one-third only of the carbon involved in glycolysis appears in carbon dioxide, the other two-thirds is present in the ethyl alcohol. Consequently, if the amount of carbon involved in glycolysis is represented by Gl and that in the carbon dioxide initially respired in nitrogen as NR, Gl = 3NR. Since this value NR is actually 1·33 or 1·5 times the carbon present in the carbon dioxide respired aerobically, the latter is only a quarter (or even less) of the carbon in the material subjected to glycolysis. If we represent the carbon evolved as carbon dioxide in aerobic respiration as OR, then NR = 1·33 (or 1·5)OR and Gl = 4 (or 4·5)OR, that is, for every 4 or 4·5 atoms of carbon subjected to glycolysis only one is given off as carbon dioxide in aerobic respiration, and the remainder of the carbon is built back into the system. This phenomenon was called by Blackman 'oxidative anabolism'. It may be mentioned that Meyerhof and others have shown that a comparable building back of material may take place in muscle, while Genevois has pointed out that in young seedlings of *Lathyrus odoratus* such an anabolism must take place for reasons similar to those advanced by Blackman in the case of apples.

It seems doubtful, however, whether evidence for oxidative anabolism will be forthcoming from all plants. At present only very few cases have been examined so that it must remain in doubt whether oxidative anabolism is a general feature of aerobic respiration or a phenomenon limited to a relatively restricted number of cases.

REFERENCES

KOSTYCHEV, S., *Plant Respiration.* Authorized edition in English with Editorial Notes, Translated and Edited by Charles J. Lyon. Philadelphia, 1927.

STILES, W. and LEACH, W., *Respiration in Plants.* London, 1932.

BOOK II

PLANT METABOLISM

CHAPTER VII

PLANT NUTRITION

THE most significant characteristic of the green plant is that it absorbs materials of comparatively simple composition from its surroundings and builds up from them the manifold complex substances which make up, and are present in, its body. These substances absorbed by the green plant are not merely non-living, but are inorganic, inasmuch as they form no part of the dead bodies of plants and animals. In this respect the green plant stands in marked contrast to animals and to non-green plants in which the absorbed materials are organic and either form part of the bodies of living or dead plants or are substances immediately derived from these bodies. The green plant thus forms the starting-point for practically all other living things, for it alone, with the exception of a few species of bacteria, is capable of transforming inorganic material into living matter. The bodies of green plants so produced form the food for animals and non-green plants, either directly, or through the medium of other animals or non-green plants which derive their food from the material provided by the body of the green plant. Traced back to its source the food of practically all living things is provided by the activity of the green plant.

Plants thus fall into two groups with regard to their nutrition. Green plants, which obtain their food from the inorganic materials of air, water or soil, are classed as *autotrophic*, while those which, like the animal, obtain their food more or less ready made from the bodies of other organisms, are described as *heterotrophic*. The completely heterotrophic plant is devoid of the green substance chlorophyll to which autotrophic plants owe their colour and which is essential for their power of utilizing inorganic materials from the environment, but there are a number of species which are not completely heterotrophic. Thus, in addition to the completely heterotrophic plants, which are described as parasites and saprophytes according as they derive their food from living or dead organic material, there are some green heterotrophic plants which manufacture part of their food like the normal green plant, but which obtain some of it parasitically from other plants. Such are the well-known hemi-parasites mistletoe (*Viscum album*), and members of the Rhinantheae group of the Scrophulariaceae as, for example, rattle (*Rhinanthus crista-galli*) and cow-wheat (*Melampyrum pratense*).

By chemical analysis it was shown that certain elements are always present in the compounds which make up the body of the plant. These are carbon, hydrogen, oxygen, nitrogen, sulphur, phosphorus, potassium, calcium, magnesium and iron, while sodium, silicon and a number of other elements are usually present as well. From the results of growing plants in water cultures, that is, by supplying the roots only with a solution of salts of known composition, the conclusion was drawn by Sachs and Knop that only the first ten elements mentioned above are necessary for the nutrition and growth of plants. Recent work has, however, indicated that other elements, notably boron, are necessary for the full development of many plants, albeit in only very small quantity. The water culture method also helped to prove, beyond a shadow of doubt, that not only are water and the sulphur, phosphorus and metallic constituents of the plant absorbed through the roots, but that the nitrogen is also, and that the carbon alone is absorbed from the air by the aerial parts of the plants.

In our inquiry into the metabolism of the plant it is our aim to determine the laws governing the intake of material from the outside world and the fate of this material after it enters the plant. The study of plant metabolism must therefore be fundamentally a physical and chemical study, for we have to determine the physical and chemical processes involved in the absorption of material by the plant and in the production of the various substances which make up the plant body. In the manufacture of these substances it is clear that we have to do in the autotrophic plant with processes of synthesis in which more complex substances such as carbohydrates, fats, proteins and even more complex compounds are built up from the comparatively simple substances absorbed from the environment. Such processes are described as *anabolic*, and the general phenomenon of this building up of substances is called *anabolism*. This contrasts with *katabolism* in which the more complex substances are broken down. The principal katabolic process of plants is respiration, and perhaps all katabolic processes are actually to be regarded as processes of respiration inasmuch as they involve a release of energy. The respiratory process, constituting, as it does, a constant feature of the living cell, has been treated here as an aspect of cell physiology, and it is, of course, impossible to draw a hard-and-fast line between what shall be regarded as cell physiology and metabolism respectively. On the whole, however, the processes of metabolism, such as the formation of carbohydrates, the transference of these and other substances through the plant, the absorption of water from the soil and the transport of water through the plant, largely take place in strictly limited regions of the plant; in organs, or tissues, that is, specialized in a manner particularly suited for the particular process.

Since the autotrophic plant absorbs from the outside world comparatively simple substances and builds them up into complex compounds of higher energy content, it is obvious that it must also obtain a supply of energy for this purpose from outside. This energy is provided by the sun which thus plays an essential part in rendering plant metabolism possible. For the heterotrophic plant, on the other hand, this supply of energy is not necessary, as it is provided in the chemical energy of the organic substances absorbed by the parasite or saprophyte from the substrate on which it lives.

The problems of plant metabolism thus include the absorption of carbon dioxide by the aerial parts of plants and of water and salts by the roots, and the manufacture from these simple substances of carbohydrates and more complex substances. For a proper understanding of how this is brought about it will be necessary to examine the laws governing the transport of water and salts through the plant. We must also inquire into the fate of the substances produced and examine how they are conveyed to their temporary or final destinations in the plant. An examination of these questions forms the subject of the following chapters in this book.

CHAPTER VIII

THE ASSIMILATION OF CARBON BY AUTOTROPHIC PLANTS

THE absorption of carbon dioxide by plants and the formation from it and water of carbohydrates may be claimed as the most important metabolic process of all living things. For it is in this synthesis of carbohydrate that energy is absorbed from the light of the sun and transformed to chemical energy, the substances produced being of higher energy content than the simpler substances out of which they are formed. Thus the production of a gram-molecule of hexose sugar from carbon dioxide and water necessitates the absorption of 674 calories. It is the energy so stored that is released in respiration for use in other plant processes such as the production of more complex substances, and it is again this same energy which is acquired by the animal when it feeds on the materials of the plant body.

In the assimilation of carbon dioxide and its utilization in the formation of carbohydrate, there is thus not only an absorption of material, but also of light energy, and hence the process is therefore a synthesis which is effected by means of light energy, whence it is now usually denoted by the name of *photosynthesis*.

The first information relating to this most fundamental of all plant syntheses concerned the interchange of gases between the plant and its environment which is the outward sign of the process. In 1771 Joseph Priestley showed that air which had become impure through animal respiration was purified by plant activity, thus indicating the evolution of oxygen by plants. By 1779 Ingen-Housz had shown that only the green parts of plants bring about this purification of impure air and that it only takes place in light. By the end of the eighteenth century work by Ingen-Housz and Senebier had made it clear that the evolution of oxygen was accompanied by an absorption of carbon dioxide.

In 1804 de Saussure showed by chemical analysis that the green parts of plants absorb carbon dioxide and give out oxygen, and though it has often been stated that he found the volume of oxygen evolved was equal to the carbon dioxide absorbed, this was not so, and it was actually Boussingault who many years later established this relation. The ratio of the number of molecules of oxygen evolved to those of carbon dioxide absorbed is termed the assimilatory quotient or assimilatory coefficient. More recent determinations of it, particularly by Maquenne and Demoussy and by Willstätter and Stoll, have shown that this ratio is unity in a wide range of plants, although in succulents the assimilatory quotient like the respiratory quotient may show wide deviations from unity, a fact related to the peculiar metabolism of these plants.

Hence in photosynthesis only the carbon of the absorbed carbon dioxide is retained. The fate of this carbon was for long in doubt, and, indeed, for many years after de Saussure had established the absorption of carbon dioxide by leaves, it was not realized that the whole of the carbon of the plant was obtained in this way. That this was indeed the case was strongly advocated by Liebig, while the water culture experiments of Sachs and Knop showed without a doubt that plants could be grown throughout their whole life-cycle without any carbon being absorbed through the roots.

Although it was suspected by von Mohl that the starch observed by him to be present in green leaves arose as an assimilatory product, it was Sachs who actually obtained proof that the carbon of this starch was provided by the carbon dioxide absorbed by the leaf. Sachs showed that starch disappears from leaves kept in the dark for several days and reappears when the leaves are again exposed to light. He referred to starch as the first visible product of assimilation, but actually quite a number of plants, including many monocotyledons and a large proportion of the Umbelliferae, Gentianaceae and Compositae among the dicotyledons, do not form starch. In such plants it has been shown by chemical analysis that sugar is produced when the leaves are exposed to light. Moreover, it seems highly probable that even in leaves which form starch this is preceded by sugar, and that the starch is to be regarded as a temporary reserve substance, accumulating during rapid assimilation when the channels of conduction of the assimilatory products of the leaf are insufficient for conveying away these products as rapidly as they are formed. Thus in 1924 Weevers found that when plants of *Pelargonium zonale*, which have been kept in the dark until the leaves are deprived of carbohydrate, are exposed to light, hexose sugars are first formed, then sucrose, and finally starch appears.

If we assume that the first product of carbon assimilation is hexose sugar, we may represent the process by the equation

$$6CO_2 + 6H_2O = C_6H_{12}O_6 + 6O_2$$

This equation, it will be observed, is the reverse of that generally taken to represent normal aerobic respiration. As the latter process involves the release of energy, so therefore, the assimilatory process involves an intake of the same amount of energy, namely, 674 calories for every gram-molecule of hexose sugar produced. While there has been considerable argument as to whether one of the hexoses or sucrose, which are known to be formed in the assimilating leaf, is the first sugar formed, the evidence is rather on the side of the hexoses. It will, however, be observed that the assimilatory quotient is unity whatever the substance produced so long as it is a carbohydrate.

It has been pointed out earlier that every living cell respires, and green leaves and green plant cells generally, when in darkness, absorb oxygen and give out carbon dioxide. It is generally assumed that the green parts of plants in light are no exception to the rule that respiration is a constant accompaniment of life, so that actually in light two processes precisely opposed to one another are taking place in green cells, one, photosynthesis, consisting of a building up of sugar from carbon dioxide and water with absorption of carbon dioxide and evolution of oxygen, the other a breaking down of sugar into carbon dioxide and water with evolution of this carbon dioxide and absorption of oxygen. Although in general assimilation in light exceeds respiration, it may happen under certain conditions of weak illumination or low temperature that respiration exceeds assimilation. When assimilation exceeds respiration the latter is completely masked, for any carbon dioxide produced in respiration is at once utilized for photosynthesis and does not leave the assimilating organ, while oxygen evolved in photosynthesis can provide all of this gas necessary for respiration. Thus we must distinguish between the real and the apparent assimilation, the latter being actually determined from the carbon dioxide absorbed by the plant from its environment, or the oxygen evolved into it, or from increase in the carbohydrate present. This apparent assimilation will be less than the real assimilation by the amount of the respiration, which utilizes part of the carbohydrate in the organ and of the oxygen evolved, and which provides part of the carbon dioxide utilized for photosynthesis.

Another point to be noticed in this connexion is that there may be conditions in which the respiration and assimilation just equal one another; there will then be no gaseous exchange between the assimilating tissue and its surroundings and no net production or loss of carbohydrate apart from any conveyed between the assimilating cells and other parts of the plant.

The chief methods of measuring photosynthesis embody the same principles as the methods used for measuring respiration. Either the amount of carbon dioxide absorbed or of oxygen evolved is taken as a measure of apparent assimilation. To obtain the real assimilation it is usual to correct the value obtained for the apparent assimilation by adding to it an amount equal to the respiration which takes place in the same time when the plant is kept in the dark. This assumes that the respiration rate is the same during a period of illumination as in darkness, but while the evidence indicates that respiration is not affected directly by light in organs which have no power of photosynthesis, there is at least the possibility that photosynthesis, by increasing the quantity of carbohydrate in the leaf, may indirectly lead to increased respiration rate by increasing the concentration of the respiratory substrate.

Besides the more usual ways of estimating photosynthesis by measuring the rate of absorption of carbon dioxide or of evolution of oxygen, attempts have been made from time to time to measure photosynthetic activity by determining the gain in dry weight of the assimilating tissue.

The first measurements of this kind were made by Sachs in 1884 by what is known as the half-leaf method. At the beginning of an experiment one half of a bisymmetrical leaf is cut off, leaving the other half attached to the plant by the midrib. Pieces of the detached half-leaf having a known area are then cut out from it, so that the larger veins are avoided, and the dry weight of these pieces obtained. The attached half-leaf is similarly treated after exposure to light for a definite period. In this way the gain in dry matter by the leaf over a definite period of time can be calculated. The values so obtained have to be corrected not only for respiration, but also for the removal of material which may be translocated from the attached half-leaf during the period of assimilation. Later workers, notably Thoday, have done much towards improving the practice of the method, but it does not, in general, approach in accuracy methods depending on the measurement of the absorbed or evolved gases. There may, however, be occasions on which it may be a more suitable method to use. Other variants of the method, in which the gain in carbon, or of total carbohydrate, is determined, have been tried occasionally, but have not come into general use.

THE PATH OF ENTRANCE OF CARBON DIOXIDE

In lower plants devoid of stomata, and in water plants, carbon dioxide must penetrate the assimilating organs by entering into solution in the outer walls of the surface cells, and then diffusing in solution to the seat of the assimilatory process. In vascular land plants, the intercellular space system of the leaves is connected with the outer air by the stomata, so that the carbon dioxide can diffuse from the outer air through the stomata up to the surface of the assimilating cells of the mesophyll, with the result that the path of diffusion of the carbon dioxide in solution is very short. While the very presence of the stomata suggests that they constitute the chief path by which carbon dioxide enters a leaf, the fact remains that carbon dioxide enters plants without stomata so that always there is at least a possibility of the entrance of some carbon dioxide through the cuticle of the epidermis. Experiments made by Stahl, Meissner and by F. F. Blackman showed, however, that in leaves with relatively thick cuticle and the stomata confined to the under surface of the leaf, blocking the stomata of this surface with wax, cocoa butter or vaseline prevents the formation of starch in the leaves, thus suggesting that no appreciable amount of carbon dioxide diffuses through the cuticle of the upper surface.

More definite information on this point was obtained by F. F. Blackman and by Horace Brown and Escombe, by measuring the rates of carbon dioxide absorption by the two surfaces of leaves with different distributions of stomata on the two surfaces. Thus Blackman found that where the stomata are limited to the lower surface, as in *Ampelopsis hederacea*, *Platanus occidentalis* and *Polygonum sachalinense* carbon dioxide only enters the leaf through the lower surface; none enters through the upper surface. But where, as in *Alisma plantago-aquatica*, the distribution of stomata on the upper surface is more frequent than on the lower, there is a greater absorption of carbon dioxide through the upper surface. Some of Brown and Escombe's results on the same subject are shown in Table XL.

Table XL

Absorption of Carbon Dioxide through the Upper and Lower Surfaces of Leaves

Species	Ratio of CO_2 absorbed.		Ratio of Stomatic Frequencies.	
	Upper surface	Lower surface	Upper surface	Lower surface
Colchicum speciosum	100	72	100	119
Rumex alpinus	100	137	100	269
Nuphar advenum	100	0	100	0
Catalpa bignonioides	0	100	0	100

These numbers confirm the conclusion that where the stomata are confined to one surface practically all the carbon dioxide absorbed by the leaf enters through that surface. Where, however, stomata are present on both surfaces there is relatively more carbon dioxide absorbed by the upper surface per unit of stomatal aperture than by the lower surface. This behaviour was not related by Brown and Escombe to any absorption of carbon dioxide through the cuticle, on which point, indeed, the results from these plants give no information, but to the influence of the higher light intensity to which the upper surface was subjected in comparison with the lower, in inducing firstly a greater degree of stomatal opening, and, secondly, a more rapid assimilation in the palisade cells as compared with that in the spongy mesophyll. This would lead to a steeper carbon dioxide diffusion gradient over the upper surface, and therefore a more rapid entry of the gas into the leaf. From what will be stated later it would seem that this latter explanation of Brown and Escombe's results is the more likely.

As regards the question of the path of entrance of carbon dioxide into the leaf it thus appears that where the cuticle is well developed it is practically impervious to carbon dioxide, but where the outer walls of the epidermal cells are thin and cuticle is only poorly or moderately well developed it must be held as likely that some carbon dioxide enters the plant through these walls as well as through the stomata.

Although there can, then, be no doubt that in plants with a well-developed cuticle the path by which carbon dioxide enters the leaf is limited to the stomata, it is at first sight difficult to understand how carbon dioxide can enter the leaf by this restricted path, at the rate observed, from a medium with such a low carbon dioxide content as the atmosphere. To take a specific case, according to observations of Sachs and Thoday, a leaf of *Helianthus annuus* can absorb from the normal atmosphere about 0·14 c.c. of carbon dioxide per sq. cm. of leaf surface per hour. The stomata occupy about 3 per cent. of the area of the leaf surface, so that, if all the carbon dioxide is absorbed through the stomata, the rate of diffusion amounts to 4·67 c.c. per sq. cm. of stomatal aperture. Similar calculations made by Horace Brown and Escombe show that in leaves of *Catalpa bignonioides* the rate of diffusion through the stomata may be as high as 7·77 c.c. per sq. cm. of aperture per hour. Now these last-named workers found that a normal solution of sodium hydroxide under similar conditions only absorbs carbon dioxide from still air at the rate of 0·12 c.c. per sq. cm. of absorbing surface and at the rate of about 0·18 c.c. if the air is in rapid movement.

The explanation of this anomaly is revealed by a consideration of the laws governing the rate of diffusion through small apertures. The question was investigated by Brown and Escombe, who determined the rate of absorption of carbon dioxide by a normal solution of sodium hydroxide contained at the bottom of a flask, the neck of which was closed except for a small circular hole of known dimensions. The rate of absorption of carbon dioxide was taken as a measure of the rate of diffusion through the hole. The results obtained show that over the range of hole areas employed, the rate of diffusion is proportional, not to the area, but to the diameter, of the aperture. Some of the results are given in Table XLI. The third column shows how, with reduction in area of the pore, the rate of diffusion per unit area of pore increases, while the last two columns show clearly the proportionality between the diameter of the pore and the rate of diffusion through it.

Table XLI

Rate of Diffusion of Carbon Dioxide through Small Apertures

Diameter of Aperture in mm.	CO_2 diffused per hour in c.c.	CO_2 diffused per sq. cm. of Aperture per hour	Relative Areas of Apertures	Relative Diameters of Apertures	Relative Rates of Diffusion
22·7	0·2380	0·0588	1·00	1·00	1·00
12·06	0·0928	0·0812	0·28	0·53	0·39
6·03	0·06252	0·2186	0·07	0·26	0·26
3·233	0·03988	0·4855	0·023	0·14	0·16
2·117	0·02608	0·8253	0·008	0·093	0·10

It is easy to understand how it is that with small apertures the 'diameter law' of diffusion should obtain, if we compare the lines of flow of molecules of the gas in the respective cases of open diffusion and of diffusion through a small pore. In open diffusion, or diffusion towards a large absorbing plane surface, it is clear that over all the middle part of the surface the concentration of the carbon dioxide will be the same at equal distances from the surface so that the so-called 'shells' of equal carbon dioxide concentration are plane surfaces parallel to the absorbing surface and the gas will flow vertically towards the surface. The rate of diffusion will here obviously be proportional to the area. But at the edges of the absorbing area gas will flow in from the air at the sides of the disk as well as from that vertically above it. Hence at the margins we may expect the rate of diffusion or absorption to be greater than over the middle part of the area. But since in the case of a large disk the marginal region will only form a small proportion of the total area, the rate of diffusion or absorption will be practically proportional to the area. With decreasing area of an absorbing surface or of an opening, the marginal region forms an increasing proportion of the total area, and in the case of a small enough pore the marginal region will occupy the whole of the area of the pore, and so the rate of diffusion will be proportional to the length of the margin, that is, to the linear dimensions of the pore, or, in the case of a circular opening, to the diameter. Thus, with decreasing area of the aperture, there is an increase in the rate of diffusion per unit area of aperture. For example, a reduction in the area of a small pore to one-quarter results in a reduc-

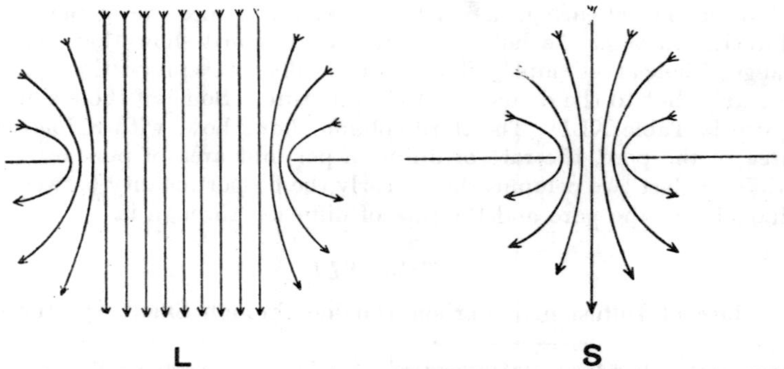

FIG. 20.—Diagrams to illustrate the lines of flow of the molecules of a gas diffusing through a large (L) and a small (S) aperture in a septum

tion in the rate of diffusion by only one-half, so that the rate of diffusion per unit area is doubled. The diagrams in Fig. 20 will help to make the matter clear.

In applying this result to the case of the stomata there are some further points to be noted. Firstly, a stoma cannot be regarded as a simple perforation in which the depth is negligible in comparison with the area, but rather as a tube with a definite length. In this tube a diffusion gradient of carbon dioxide will be set up, and this will have the effect of lowering the rate of diffusion through the stoma. The longer the tube the greater will be the retardation, and Brown and Escombe calculated that the rate of diffusion in such a case is given by the formula $\frac{k\rho\pi r^2}{l + \frac{1}{2}\pi r}$, where k is the coefficient of diffusion of the gas, ρ the density of the gas in the air just outside the opening, r the radius of the tube and l its depth. Hence the deeper the stomatal opening, the slower the rate of diffusion through it.

The second point to be noticed in regard to diffusion through the stomata is the possibility of mutual interference in the diffusion through neighbouring stomata. That there is such a possibility will be obvious from the following considerations. Let us suppose that the rate of diffusion of a gas through an opening 1 mm. in diameter has a value x. Then by the diameter law the rate of diffusion under the same conditions through an opening 0·01 mm. in diameter is 0·01 x. If, then, a septum the size of the larger pore were perforated by 100 holes 0·01 mm. in diameter, the total diffusion, provided there was no interference, would equal that through a hole of the area of the septum. But actually the holes only occupy an area of 0·01 of the whole septum, so that the latter could be perforated by many more holes. It is, however, manifestly impossible for more gas to pass through a perforated septum than through an open pore the size of the septum, hence we must conclude that in the case of a multiperforate septum where the holes are less than a certain distance apart there will be mutual interference in diffusion through the pores, and the diffusion through a single pore in such a multiperforate septum will be less than that through a solitary pore of the same dimensions. Or, expressed in a slightly different way, the interference in diffusion through neighbouring pores in a multiperforate septum becomes less the further removed the pores are from one another, until when a certain distance apart is reached the interference is negligible. This minimum distance for non-interference was found experimentally by Brown and Escombe, in experiments where the pores were 0·38 mm. in diameter, to be 10 times the diameter of the pore. The numbers in Table XLII show the results of their experiments in which diffusion of carbon dioxide took place through multiperforate septa with apertures 0·38 mm. in diameter, but varying in frequency. The septa covered tubes 1 cm. long at the bottom of which was a solution of sodium hydroxide.

Table XLII

Diffusion through Multiperforate Septa

Number of Holes per sq. cm.	Percentage Area of Holes referred to Area of Septum	Distance of Holes apart in Diameters	Diffusion of CO_2 through Septum per hour in c.c.	Percentage of Septum Diffusion referred to Open Tube Diffusion	Ratio of Diffusion to Total Area of Holes
100	11·34	2·63	0·433	56·1	0·038
25	2·83	5·26	0·401	51·7	0·14
11·11	1·25	7·8	0·312	40·6	0·25
6·25	0·70	10·52	0·241	31·4	0·34
4·0	0·45	13·1	0·156	20·9	0·35
2·77	0·31	15·7	0·106	14·0	0·34

An examination of these numbers shows that the ratio of diffusion to total area of holes increases as the distance between the holes increases until the distance apart is about 10 times the diameter of the opening. With further increase in distance no further increase in this ratio occurs.

It will be observed also that although, up to a point, the efficiency of the individual holes increases as their distance apart increases, the total diffusion decreases within the limits employed in Brown and Escombe's experiments.

It is interesting to consider the effect of maintaining this relation between the diameter of the apertures and the distance between them, while reducing the actual size of the apertures. Then as the size of

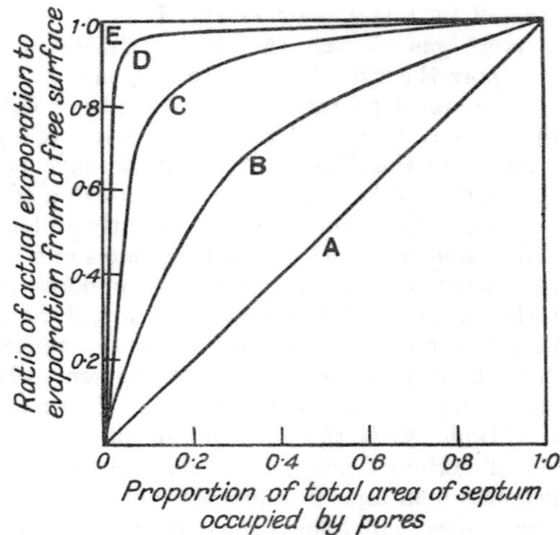

Fig. 21.—Curves to illustrate the rate of diffusion or evaporation through multiperforate septa

A. Theoretical curve for infinitely large pores.
B. Pores of 1 sq. cm. area.
C. „ „ 1 sq. mm. „
D. „ „ 2000 μ^2 „
E. Theoretical curve for infinitely small pores.

(After B. Huber)

the individual aperture is reduced the actual total pore area remains the same for the same total area of septum, but owing to the number of pores in unit area varying inversely as the square of the pore diameter, while the rate of diffusion through the individual pore is proportional to the diameter itself, it follows that with constant pore area the rate of diffusion increases with diminution in the size of the individual pore. Experimental verification of this has been obtained by Huber, whose results are summarized in Fig. 21. The curves in this figure show the relation between the rate of diffusion through multiperforate septa (in terms of free diffusion through an area equal to the whole septum) and the total pore area (in terms of that of the whole septum), for septa with different sizes and numbers of pores. The area of the pores is indicated with each curve. It will be observed that for a given pore area the rate of total diffusion increases with diminution in the size of the pores. Where the area of the individual pore is only 0·002 sq. mm. the diffusion amounts to about 95 per cent. of open diffusion, when the total area of the pores only amounts to 5 per cent. of the total area of the septum.

Now according to Brown and Escombe, the area of the stoma of *Helianthus annuus* is little more than 0·0001 sq. mm., while the total area of the pores amounts to about 3 per cent. of the total leaf area. Taking these data in relation to the findings of Huber we can see that diffusion through the stomata of *Helianthus annuus* will be practically as rapid as free diffusion through an aperture of the size of the leaf. The data given by Lloyd for the stomata of *Verbena ciliata* (see p. 234) are very similar to those obtained for *Helianthus annuus*.

From mathematical considerations Jeffreys concluded that there must be mutual interference in diffusion through a multiperforate septum if $\frac{nrA}{2l}$ exceeds (roughly) unity, where n is the number of perforations per unit area, r the radius of the perforation, A the area of the septum and l of the order of the linear dimensions (e.g. diameter) of the septum. Applying these results to the case of the leaf of *Helianthus annuus*, the number of stomata per unit area is about 33,000, the radius of the stoma may be taken as 0·00053 cm., and $\frac{A}{2l}$ about 3 cm., giving a value of $\frac{nrA}{2l}$ of about 50. This means that the stomata of this plant would have to close until their diameters were only $\frac{1}{50}$ that of the fully open stomata before mutual interference in diffusion ceased. The difference between this conclusion and that of Brown and Escombe is due to the difference in the dimensions of the pores in the latter's experimental septa and in those of the stomata. As a matter of fact, taking into account the approximations made by Jeffreys in obtaining his formula, Brown and Escombe's experimental results fall fairly into line with Jeffreys' conclusion.

Jeffreys' conclusion is of great importance, and further references will be made to it in dealing with transpiration in the next chapter. Here it may be pointed out that Brown and Escombe themselves observed that stomata can close to $\frac{1}{15}$ of their diameter without assimilation of carbon dioxide being affected.

Another point to be noted in regard to diffusion through small perforations is the effect of a wind outside the perforations.

Jeffreys came to the conclusion that in moving air the absorption by the stomata will be much the same as if the whole leaf were absorbing. In this case the rate of absorption will be proportional to $l^{1 \cdot 5}$, where l is a measure of the linear dimensions of the leaf, if the distribution and size of the stomata is similar to that of *Helianthus annuus*.

But if the total area of stomatal aperture per unit area should be much less, or air velocity very high, each stoma may act independently, in which case the diameter law would hold.

From the work summarized above we may conclude that in plants like *Helianthus annuus*, which are well provided with stomata, absorption of carbon dioxide can take place through the latter practically as rapidly as if the whole leaf were an absorbing surface, and that closure of stomata can take place to a very considerable extent without thereby bringing about any appreciable reduction in the rate of entrance of carbon dioxide into the leaf. Where the stomata are smaller or less frequent (cf. Table LII, p. 226), as, for example, in *Oxalis acetosella*, the closure of the stomata will reduce the rate of diffusion into the leaf much more readily.

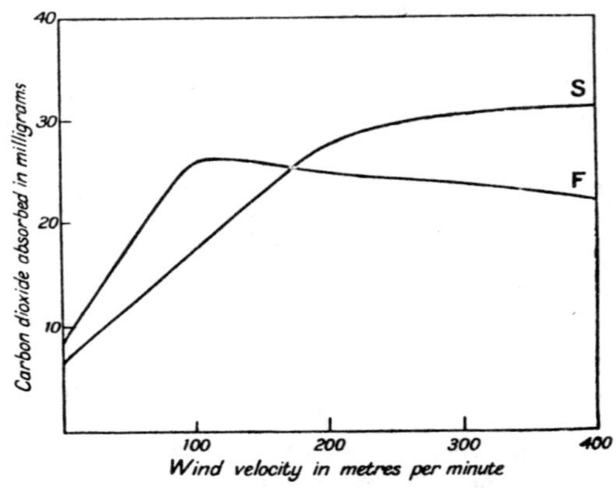

Fig. 22.—Curves showing the effect of wind velocity on the diffusion of carbon dioxide through a multiperforate septum

(After H. Deneke)

There is no doubt that the rate of diffusion of carbon dioxide through a multiperforate septum is increased if the gas containing the carbon dioxide is in movement over the septum. Deneke examined this question with artificially prepared multiperforate septa of paraffined cardboard and zinc foil, but also with what he called 'natural' multiperforate septa. These consisted of pieces of epidermis of *Ficus elastica*, which were fixed with the stomata wide open. In all the septa used, wind brought about an increase in rate of diffusion, the latter being measured, as in Brown and Escombe's experiments, by absorbing the diffused carbon dioxide with potassium hydroxide.

In Fig. 22 (Curve S) are shown Deneke's results on the influence of wind velocity on diffusion of carbon dioxide through a septum having an area of 24·5 sq. mm. with pores 0·4 mm. in diameter set at 10 diameters apart. It will be observed that the rate of diffusion increases at first rapidly with increasing wind velocity, but that when a velocity of 300 metres per minute is reached further increase in wind velocity has only a negligible effect. In the same figure is shown by curve F the effect of wind on absorption of carbon dioxide by a free surface; that is, by the potassium hydroxide solution when the septum is removed. Here the result is on the whole similar, although not so marked, and the maximum value of diffusion is reached with a lower wind velocity. Deneke's results thus agree on the whole with Jeffreys' calculations previously mentioned.

THE ASSIMILATORY PIGMENTS

Already before the end of the eighteenth century it was recognized that the evolution of oxygen in light only takes place from the green parts of plants. Microscopical examination shows that the green pigment is not diffused throughout the cell, but is localized in the small bodies known as chloroplasts. In higher plants these are ellipsoidal, disk-shaped or biconvex lens-shaped, and there are several or many of them in each cell. In the brown algae the coloured plastids are similar in size, having a diameter of the order of 5 μ, but are brown in colour owing to the presence of a brown substance in addition to the green. In many green algae, on the other hand, the pigment is contained in comparatively large chloroplasts of which there are a few, or even only one, in each cell. The chloroplasts are surrounded by protoplasm and are to be regarded as part of it, that is, as part of the living mechanism of the cell rather than as non-living inclusions in it.

The pigments to which the green colour is due had actually been extracted by Nehemiah Grew in 1689, but the name chlorophyll was first given to the green substance by Pelletier and Caventou in 1818.

Grew had already shown that both a green and a yellow substance could be extracted from leaves, and numerous researches since 1860 have confirmed the presence of both green and yellow pigments in the assimilatory cells. The researches on the nature of these pigments may be regarded as culminating in those of Willstätter and Stoll and their collaborators carried out during the years 1906 to 1913. These investigations not only confirmed the presence of two green and two yellow pigments in the material known broadly as chlorophyll, but established the constitution and chemical properties of these substances.

The two green and two yellow pigments of the chloroplasts are constant throughout the plant kingdom, although their relative quantities may vary in different groups of plants. The mixture of the two pigments is usually spoken of simply as chlorophyll, the two components of which are chlorophyll *a*, with a formula $C_{55}H_{72}O_5N_4Mg$, and chlorophyll *b*, with a formula $C_{55}H_{70}O_6N_4Mg$; these are both microcrystalline substances, the former blue-black in the solid state and green-blue in solution, the latter a green-black solid giving a pure green solution. They both dissolve in various organic solvents, although they exhibit a certain differential solubility in different solvents, a property of which use is made in their separation. These solutions exhibit a characteristic red fluorescence. Both chlorophylls have characteristic absorption spectra. That of chlorophyll *a* exhibits a strong band in the red, four bands in the yellow and green decreasing progressively in strength with shortening of the wave-length, a somewhat stronger band in the blue, a stronger band still in the indigo blue and a very strong absorption at the violet end of the spectrum. The absorption spectrum of chlorophyll *b* is very similar to that of chlorophyll *a*, but there are two absorption bands in the red, five in the yellow and green, a strong absorption band in the blue and here also strong absorption at the violet end of the spectrum.

A colloidal solution of chlorophyll in water can be obtained. Such a sol is pure green in colour and does not fluoresce. Some divergence of opinion exists as to whether chlorophyll in the living leaf is present in true or in colloidal solution. Stern's view, for example, was that the chlorophyll is present in true solution in the fatty phase of the chloroplasts, the chlorophyll-fatty phase being dispersed either as an emulsion or emulsoid through an aqueous-protein phase.

Recently an investigation on this question has been carried out by a group of workers, a brief account of which has been given by Baas Becking. From an examination of the absorption spectra of chlorophyll in the plastids and in a variety of solvents, and from a study of its colloidal properties, it is concluded that chlorophyll cannot be present in the plastid either in true or colloidal solution or in the dry state. A study of the properties of chlorophyll indicates that it can form adsorption compounds with both hydrophilic and lipophilic compounds, such as proteins and fatty compounds respectively. Further, microscopic observations on plastids in red light indicate that the chlorophyll is discontinuously distributed in the outer part of the plastid. It is therefore suggested that chlorophyll is held in the peripheral region of the plastid in the form of small particles in adsorptive association with both protein and fatty material.

Stoll thinks the chlorophyll exists in combination with a colloidal carrier in the form of a ' symplex ', chloroplastinsymplex, comparable to those which, on Willstätter's hypothesis (cf. p. 119) constitute enzymes. Just as enzymes owe their activity to this type of construction, so does chlorophyll, which can therefore be regarded as a specific assimilatory enzyme.

The chlorophylls are neutral substances acted upon by acids and alkalies to give a series of definite decomposition products. The phaeophytins (*a* and *b*) are produced by the action of dilute acids on chlorophylls. They are olive-green substances containing no magnesium, this being replaced by hydrogen. By suitable treatment this hydrogen can in turn be replaced by other metals, so that compounds resembling chlorophyll, but in which the place of magnesium is taken by some other metal, are produced. The bright green copper chlorophyll is the easiest of these compounds to prepare. When the phaeophytins are treated with alkalies there are produced nitrogen-containing acids and a nitrogen-free alcohol, phytol, with a formula $C_{20}H_{39}OH$.

When a mixture of the chlorophylls themselves is heated with alkali the green colour changes first to brown and then back to green. This reaction constitutes a test for chlorophyll known as the brown phase test. The brown colour is actually produced by the mixture of the yellow produced by chlorophyll *a* with the red produced by chlorophyll *b*.

In many plants there appears to be present an enzyme, chlorophyllase, which can act in alcoholic media and split off the phytyl group $C_{20}H_{39}$ from the chlorophyll molecule, the residue of which combines with the ethyl group of the alcohol. Thus in the case of chlorophyll *a* we can express the action of chlorophyllase by the equation

$$(MgN_4C_{34}H_{33}O_3)(COOC_{20}H_{39}) + C_2H_5OH = \\ (MgN_4C_{34}H_{33}O_3)(COOC_2H_5) + C_{20}H_{39}OH$$

A similar reaction takes place with chlorophyll *b*. These new ethyl compounds form green crystals, and the mixture of the two had been frequently mistaken for chlorophyll in the past and was known as crystalline chlorophyll. The names given to these compounds by Willstätter are ethyl chlorophyllide *a* and ethyl chlorophyllide *b*. Similar compounds are formed with methyl alcohol. These substances are more readily obtained from some plants than from others; among those giving crystalline chlorophyll easily, and which are therefore presumed to be rich in chlorophyllase, are *Heracleum sphondylium*, *Stachys sylvatica* and *Galeobdolon tetrahit*.

The yellow pigments which accompany the chlorophylls are known collectively as carotinoids. The two universally present in the chloroplasts are xanthophyll and carotin (carotene). The latter is a hydrocarbon with the formula $C_{40}H_{56}$, while to xanthophyll is ascribed the formula $C_{40}H_{56}O_2$. Carotin is the yellow pigment of the carrot root, from which it derives its name. It is soluble in a number of organic solvents and is rapidly oxidized in air, losing its colour in the process. Xanthophyll is similar to carotin in its general properties, and the two substances have rather similar spectra, both possessing two absorption bands in the blue with absorption also at the violet end of the spectrum. There is some difference in the relative positions of the bands, but the actual position in each case depends on the solvent. Both pigments are present in etiolated leaves from which the chlorophylls are absent. There is some evidence that xanthophyll may not be a single substance, but a group of closely related isomers. Thus, when an extract of the leaf pigments is allowed to percolate through a column of powdered chalk or some other adsorbent, the different pigments are adsorbed in succession, so that the respective pigments are held in different layers of the column. This method of separating the leaf pigments is due to Tswett, and was called by him the chromatographic method, and the stratified column of adsorbent with adsorbed pigments a chromatogram. In this chromatogram Tswett distinguished six layers, two green, due to the two chlorophylls, and four yellow, which he regarded as due to four different xanthophylls, for the carotin was not adsorbed at all. Evidence for the existence of four xanthophylls in the perianth leaves of a variety of *Narcissus* was found by Miss Coward.

The brown colour of the brown algae is due to the presence in the chromatophores of a third carotinoid in addition to carotin and xanthophyll. The presence of this pigment in brown algae was demonstrated spectrographically by Sorby in 1873, and the name fucoxanthin given to it. It was separated chromatographically by Tswett in 1906, and in 1914 isolated and examined by Willstätter and Page, who ascribed to it the formula $C_{40}H_{54}O_6$.

In the red algae the usual green and yellow pigments are present, and in addition a red pigment phycoerythrin, and sometimes a blue pigment as well, phycocyanin. The composition of these pigments is in doubt, but the evidence points to their being related to proteins.

The absolute and relative quantities of the pigments vary in different groups of plants. The results of some of the analyses of Willstätter and Stoll are given in Table XLIII.

Table XLIII

Quantities of Pigments present in Different Assimilatory Organs in parts per thousand of Fresh Weight

Pigment	Green Leaves (average)	*Ulva lactuca*	*Fucus*
Chlorophyll *a*	2·0	0·165	} 0·528 (at least 97% is chlorophyll *a*)
Chlorophyll *b*	0·75	0·117	
Carotin	0·17	0·024	0·089
Xanthophyll	0·33	0·064	0·087
Fucoxanthin	—	—	0·169

It will be observed that the green alga *Ulva* only contains about one-tenth of the chlorophyll, estimated on fresh weight, of that in a green leaf.

THE CONDITIONS INFLUENCING PHOTOSYNTHETIC ACTIVITY

By the time of de Saussure it was clear that a supply of carbon dioxide, light, and the green pigment of the leaf, were all necessary conditions for photosynthesis. To these we may at once add a suitable temperature, and some protoplasmic factor apart from chlorophyll, since carbon dioxide, water and chlorophyll exposed outside the living cell to light at ordinary temperatures do not give rise to any carbohydrate. It is not known what this factor is, but it might quite possibly be an enzyme. The essential factors for photosynthesis thus include three of the environment, or external factors, carbon dioxide supply, light and temperature, and two internal, chlorophyll and an unknown protoplasmic factor. There is also the factor of water supply, for water is a necessary material for the production of carbohydrate from carbon dioxide. Although the supply of water to the roots is to be regarded as an external factor, in reality the factor concerned in the actual photosynthetic process is the water content of the leaves, and this is undoubtedly an internal factor.

Since these factors are all necessary for the incidence of photosynthesis, it is likely that on their magnitude will depend the rate of the process. Earlier work directed to determining the relationship between any of the external factors and photosynthetic activity nearly always revealed with increasing value of the factor a corresponding increase in the rate of photosynthesis up to a maximum, above which the rate of photosynthesis fell with further increase in the value of the factor until the rate of photosynthesis was zero. Thus arose the conception that for every factor there were minimum, optimum and maximum values, the optimum value being that at which the rate of photosynthesis was highest, the minimum and maximum values marking the limits within which photosynthesis could proceed. It had, however, to be admitted that the optimum was not a fixed point, that it varied, not only from one species to another, but also in the same plant according to the other conditions. Moreover, as we shall see, the falling off in rate above the optimum results from some secondary effect, generally of the factor on the protoplasm, and as the extent of this falling off increases with the length of time the factor acts, the observed rates of photosynthesis will depend on the length of time over which measurements are made. It may thus be difficult in practice to fix a value for the optimum even with a defined set of conditions.

The inadequacy of this older view of the relation between factor and rate of photosynthesis was emphasized by F. F. Blackman in 1905. Blackman pointed out the necessity of taking into account the other factors of photosynthesis when the influence of one particular factor is being examined. Let us suppose, for example, that we are examining the influence of carbon dioxide concentration on the rate of photosynthesis. We should expect that as the concentration of carbon dioxide in the atmosphere is increased from zero there would be an increase in the rate of photosynthesis, and we may for the sake of simplicity suppose that the rate of photosynthesis is proportional to the concentration of carbon dioxide. But for the production of carbohydrate it is necessary that light energy should be absorbed, and if the leaf is assimilating in weak light it may be that as the concentration of carbon dioxide is increased a point will be reached at which the energy supplied in unit time is just sufficient to utilize the whole of the carbon dioxide entering the leaf in that time. With further increase in carbon dioxide concentration no more carbon dioxide can be decomposed in unit time because the energy supply is insufficient to decompose any more; consequently, in spite of the increased diffusion gradient of carbon dioxide there can be no increase in its rate of absorption and utilization. The intensity of illumination is thus limiting the rate of the whole process, and increase in carbon dioxide concentration can only bring about an increase in rate of photosynthesis if the light intensity is also increased. This interaction of two factors is represented graphically in Blackman's scheme shown in Fig. 23. Here the curve ABC shows the relation between carbon dioxide concentration and rate of photosynthesis in one unit of light, carbon dioxide concentration limiting the rate over the part AB and light intensity over the part BC. With double light intensity the rate of photosynthesis will be able to attain a value twice that possible with one unit of light intensity, and so under these conditions the curve takes the form ADE and with three units of light intensity AFG. Thus in order to determine the effect of one factor on the rate of photosynthesis it is necessary not merely to keep the others constant, but to ensure that they are in excess, otherwise the rate of photosynthesis will be determined by one of these others. But

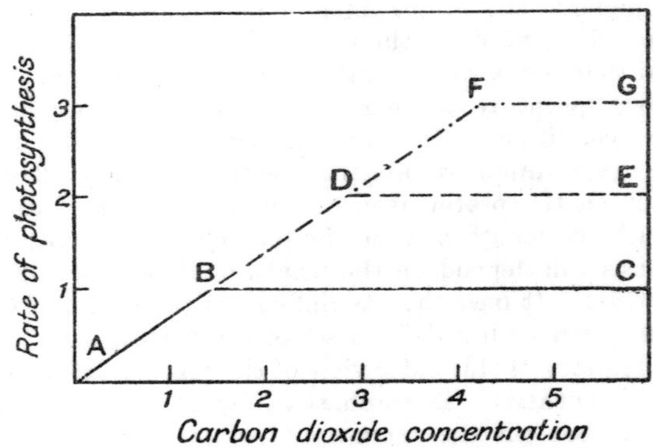

Fig. 23.—Curves to illustrate Blackman's conception of the operation of a limiting factor (After F. F. Blackman)

while, in practice, the value of the external factors of light, carbon dioxide concentration and temperature can be controlled, it is not possible to alter at will the internal factors of chlorophyll content and the protoplasmic factor.

While the introduction of the conception of limiting factors did great service in calling attention to the interaction of the various factors, it is now clear that their mode of interaction is not so simple as that suggested by the scheme expressed graphically in Fig. 23. Actually the curves connecting the value of a factor which is varied with the rate of photosynthesis with other factors constant rarely exhibit a sharp angle between the portions where two different factors are limiting.

There is generally a transitional region, where two factors influence the rate of photosynthesis, between the initial part of the curve where the varied factor chiefly determines (or limits) the rate, and the final region where another factor is mainly determining.

The mode of interaction of the factors may also be considered in regard to the reactions involved in photosynthesis.

As will be shown later, there is evidence that the photosynthetic action takes place in a series of stages, and as Briggs pointed out in 1920, the stage which proceeds at the slowest rate will determine the rate of the whole process. Let us suppose that there is a photochemical stage in which, for example, carbon dioxide and light are involved, and that this is followed by a chemical stage affected by temperature and in which the protoplasmic factor plays a part. The rate at which the photochemical stage proceeds will then depend on both the concentration of carbon dioxide and light intensity. As far as this stage alone is concerned, its rate will be increased both by increase in carbon dioxide concentration or light intensity. But if the rate of the first process is slow, increase in temperature or of the protoplasmic factor, the factors which are operative in the next stage, will have little effect on the rate of the whole photosynthetic process, since the rate at which this stage will proceed will be limited by the rate of production of material in the first stage. In such circumstances either increase in carbon dioxide concentration or of light intensity would bring about an increase in the rate of photosynthesis. On the other hand, if the factors of the second stage have such a low value that the product of the first stage is utilized more slowly than it is formed, although its accumulation may speed up the second process, this accumulation will retard the rate of the first process, and so the general effect is that the second stage will actually limit the rate of the whole photosynthetic process. Increase in temperature and of the protoplasmic factor will then bring about an increase of photosynthetic activity, whereas the effect of increase of light intensity or of carbon dioxide concentration may be negligible.

It is with these considerations in mind that the effect of the various factors on photosynthesis must be examined.

Carbon Dioxide Concentration

Since the first experiments by Godlewski in 1873 on the effect of carbon dioxide concentration on the rate of photosynthesis, this question has been repeatedly examined and observations made on a wide range of material. Thus, Brown and Escombe used leaves of *Helianthus annuus*, Blackman and Smith the water plants *Elodea* and *Fontinalis*, Harder the water plants *Cladophora* and *Fontinalis*, Lundegårdh the leaves of various higher plants, Stålfelt the leaves of *Pinus* and Warburg the green alga *Chlorella*.

In the case of water plants we may consider the results obtained by various workers with the aquatic moss *Fontinalis*. In providing carbon dioxide for the assimilation of water plants either this gas may be dissolved directly in the water or a solution of a bicarbonate may be used. Blackman and Smith used solutions of carbon dioxide, whereas Harder used a solution of potassium bicarbonate. A solution of potassium bicarbonate contains ions of K and HCO_3 from the bicarbonate and H and OH from water, and since both H and HCO_3 ions are present they must be in equilibrium with undissociated H_2CO_3 and this with CO_2. Consequently for a definite concentration of potassium

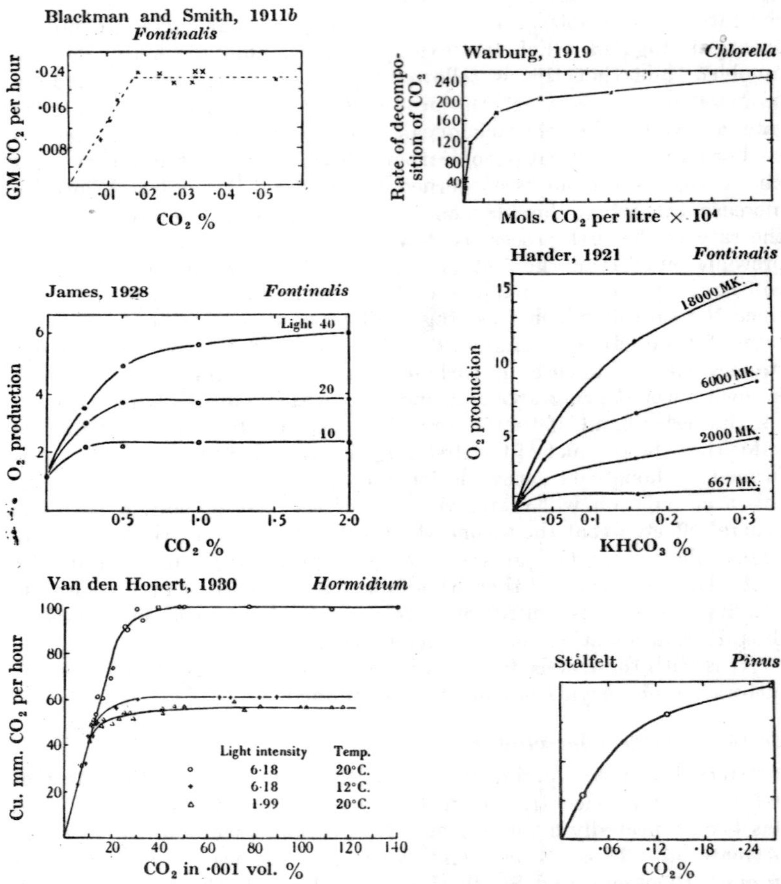

FIG. 24.—Curves to illustrate the relation between rate of photosynthesis and carbon dioxide concentration as determined by various workers

(From W. O. James, after the various authors indicated)

bicarbonate there will be a definite concentration of dissolved carbon dioxide, and although, owing to the change in the degree of dissociation with concentration, the relation between total bicarbonate and carbon dioxide will vary, within a limited range it may be assumed that the carbon dioxide concentration of the solution is proportional to the concentration of bicarbonate.

Blackman and Smith found that with increasing concentration of carbon dioxide the rate of photosynthesis rose in proportion to the carbon dioxide concentration until a point was reached when further increase in carbon dioxide concentration brought about no further increase in rate of photosynthesis. Under the conditions of their experiments the change from limiting carbon dioxide to limiting light intensity was quite sudden. Harder's results showed an approximation to this condition only with the lowest light intensity he used. With higher light intensities in low carbon dioxide concentrations the rate of photosynthesis is approximately proportional to the concentration of carbon dioxide, but although above a certain value as the latter increases the rate of photosynthesis increases less and less, no point was reached where, with further increase in carbon dioxide, there was no further increase in photosynthetic activity (cf. Fig. 24).

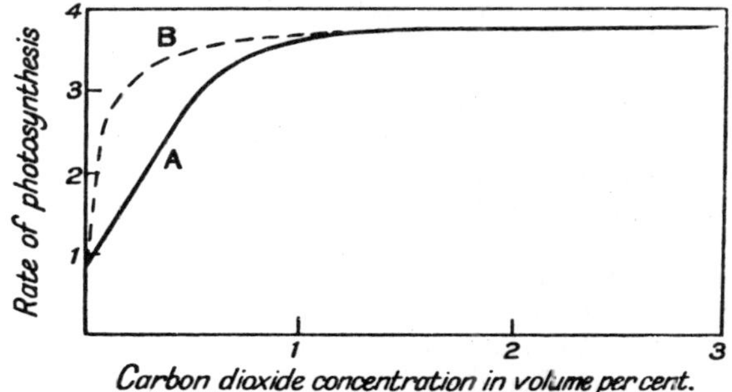

Fig. 25.—Curves showing the relation between rate of photosynthesis by *Fontinalis* and concentration of carbon dioxide when an aqueous solution of the latter flows continuously over the moss

A, rate of flow 400 c.c. per hour; B, rate of flow 600 c.c. per hour

(After W. O. James)

The difference between the shape of the curves obtained by Blackman and Smith and by Harder are probably to be related partly, but not entirely, to the absolute light intensity. The lowest light used by Harder, namely, 667 metre candles, is probably considerably higher than the light intensity used by Blackman and Smith. It will be noted from an inspection of Harder's curves that in a light intensity lower than any he used, the relation between carbon dioxide concentration and photosynthetic activity is such that a rather sharp transition from the carbon dioxide limiting to the light-limiting part of the curve is to be expected.

In a re-investigation of the assimilation of *Fontinalis*, James used both dissolved carbon dioxide and sodium bicarbonate as sources of carbon dioxide for assimilation. He found that if the plant assimilates in a moving current of water containing carbon dioxide, the rate of photosynthesis increases with increasing rate of flow (cf. Fig. 25) of the water up to a maximum which is that obtaining when the same concentration of carbon dioxide is provided by bicarbonate, in which case the rate of photosynthesis is independent of the rate of flow. This is explained by supposing that when a solution of carbon dioxide is used a diffusion gradient of this substance is set up in the solution, so that the concentration of carbon dioxide at the surface of the plant is actually less than the average concentration of the whole volume of solution. With increased rate of flow the concentration at the surface will approach nearer and nearer the average. Now when bicarbonate is used, the concentration of the carbon dioxide is maintained constant at the surface of the plant, as disturbance in the equilibrium relations between the various constituents of the system (ions and undissociated molecules) will be immediately rectified.

The concentration of carbon dioxide at the seat of the photosynthetic process, presumably the surface of the chloroplast, is not that of the external medium, a value which is measurable, but something less than this depending on the diffusion gradient which itself depends on the length of the path and the resistance to diffusion. The path can be regarded as consisting of two parts, one up to the surface of the plant, the other from this to the assimilatory cell. The use of a bicarbonate solution as source of carbon dioxide has the effect of practically eliminating the first part of the diffusion path, while movement of the liquid over the surface of the plant tends in the same direction. With a bicarbonate solution the carbon dioxide concentration of the external solution will thus be more nearly equal to the concentration at the chloroplast surface than is the case when a pure carbon dioxide solution is used, at any rate when this latter is still or in only slow movement. Now James found that the more this external diffusion phase is reduced the more the curves connecting carbon dioxide concentration and rate of photosynthesis tend to be curves of a hyperbolic type throughout their course, and he therefore came to the conclusion that the linear relation between low carbon dioxide concentrations and photosynthesis is determined by the diffusion phase.

Another point to be noted is that even in low concentrations of carbon dioxide an increase in light intensity brings about an increase in rate of photosynthesis. This will be clear from an inspection of Harder's curves and also appears in those published by James.

The relations between carbon dioxide concentration and photosynthetic activity in land plants are in general similar to those found for water plants. Thus Lundegårdh found that in leaves of *Oxalis acetosella*, *Melandrium rubrum*, *Nasturtium palustre*, *Stellaria nemorum* and *Viola tricolor*, as carbon dioxide concentration increases the rate of photosynthesis increases in proportion up to a point, but beyond this a further rise in carbon dioxide concentration produces no further increase in the rate of photosynthesis, the limit of increasing photosynthetic activity being determined by the light intensity. In some cases, however, direct proportionality between carbon dioxide concentration and rate of photosynthesis was not observed, as the latter increased relatively more slowly than the carbon dioxide concentration. Also, even with low light intensities an increased rate of photosynthesis was observed with increase in carbon dioxide concentration even when this was relatively high, thus showing that here, as with water plants, when carbon dioxide concentration is limiting the rate of photosynthesis light intensity also affects the process. Maskell suggests that the influence of light in bringing about this increased rate of photosynthesis in low concentrations of carbon dioxide is due to the increased opening of the stomata resulting from illumination which would have the effect of increasing the concentration of carbon dioxide at the surface of the chloroplast by reducing the resistance to diffusion. In absence of data regarding the influence of light on stomatal opening under the conditions of his experiments in the plants examined by Lundegårdh, it is doubtful how far this offers a complete explanation of the observed behaviour, for Jeffreys' considerations on diffusion through stomata suggest that some considerable degree of closure must take place before any significant reduction in the rate of diffusion can result. Lundegårdh himself considered that the limit of increasing photosynthetic activity might often be determined by stomatal closure, and he cites observations on *Oxalis* in support of this. It is therefore interesting to observe that of the plants examined by Lundegårdh, *Oxalis* has relatively few stomata (see Table LII), and the effect of their closure on diffusion would probably be noticed sooner in this species than in others.

There does not, however, appear to be any need to hypothesize this action of stomatal closure in this connexion, for we have seen that in the aquatic moss *Fontinalis*, where there are no stomata, when carbon dioxide concentration is limiting the rate of photosynthesis, in the sense that with other factors constant, a change in carbon dioxide concentration produces a change in the rate of photosynthesis, change in light intensity will bring about a change in the rate of photosynthesis with constant carbon dioxide concentration.

There can, however, be no doubt that changes in stomatal aperture can influence the rate of photosynthesis, especially when the stomatal aperture is small, by altering the resistance to diffusion of carbon dioxide into the leaf, as Maskell showed from observations of photosynthetic activity and stomatal aperture in leaves of *Prunus laurocerasus*. It may also be taken as obvious that the concentration of carbon dioxide at the assimilating surface of the chloroplast is not likely often to approximate to that of the air outside the leaf. The relation between rate of apparent assimilation and external carbon dioxide concentration can, according to Maskell, be expressed by the equation

$$A = k \frac{C}{r_1 + r_2 + r_3 + r_4 + r_{LT}}$$

where A is the apparent assimilation rate, k a constant, C the concentration of carbon dioxide and r_1 the resistance to diffusion up to the leaf surface, r_2 that due to the stomata, r_3 that of the intercellular space system, r_4 that of the diffusion in the cells up to the chloroplasts, while r_{LT} is a value depending on the rate of the assimilatory processes at the chloroplast surface, these reactions being functions of light and temperature. Thus the extent to which light will influence the rate of photosynthesis when carbon dioxide concentration is limiting in the sense mentioned above, will depend on the relative values of r_{LT} and $r_1 + r_2 + r_3 + r_4$.

If the carbon dioxide concentration reaches a comparatively high value a falling off in the rate of photosynthesis may result. This must be regarded as a secondary effect due to the high concentration of carbon dioxide exerting a toxic effect on the protoplasm and so reducing the value of the protoplasmic factor. It has also been observed that high concentrations of carbon dioxide may induce closure of the stomata, and as this, when it reaches a certain degree, will reduce the effective path of diffusion, the concentration of carbon dioxide in the assimilating cells is lowered, and the rate of assimilation therefore reduced.

Light Intensity

In considering the effect of light intensity on photosynthesis it must be kept in mind that ordinary white light comprises radiations with a wide range of wave-lengths, and that equal intensities of radiation of different wave-lengths have, as might be expected, and as we shall see later, different effects on the photosynthetic process; or, put in other words, different coloured lights are not equally effective in their photosynthetic action. In investigating the influence of light intensity on photosynthesis it is therefore necessary that the light of different intensities should have the same relative composition.

The very numerous observations on the effect of light intensity on photosynthesis show that the relation of this factor to the rate of assimilation is very similar to that of carbon dioxide concentration. With low light intensities the rate of photosynthesis has been found approximately proportional to the light intensity in a considerable number of cases. Such is the conclusion to be drawn from observations made by Weis in 1903 on a sun plant *Oenothera* and a shade plant *Polypodium*, by Boysen Jensen in 1918 on *Sinapis alba*, by Warburg in 1919 on the unicellular alga *Chlorella*, by Harder in 1921 on the aquatic moss *Fontinalis*, by Lundegårdh in 1921 on both sun plants (*Nasturtium palustre*, *Atriplex latifolium*) and shade plants (*Oxalis acetosella*, *Melandrium rubrum* and *Circaea alpina*), to mention a few. With higher light intensities the ratio of photosynthesis to light intensity falls off, owing to the limiting action of some other factor.

We may again consider first the simpler case of water plants, where the complication introduced by change of stomatal aperture is not present. In the case of *Fontinalis* Harder found the relation between photosynthetic activity and light intensity very similar to that between photosynthesis and carbon dioxide concentration. With constant carbon dioxide concentration the initial part of the curve where rate of photosynthesis is approximately proportional to light intensity passes over smoothly, with increase of the latter, into a region where progressive increase in light intensity brings about less and less acceleration of photosynthesis. Also, even in low light intensities, increase in carbon dioxide concentration brings about an increase in the rate of photosynthesis. Somewhat similar results were obtained by Warburg with *Chlorella*, but here the transitional region of the curve between the part in which rate of photosynthesis is limited by light intensity and the part where increase in light intensity produces no further increase in rate of photosynthesis, is relatively short.

In land plants Lundegårdh obtained rather different results with sun and shade leaves. In the shade plants, *Oxalis*, *Melandrium* and *Stellaria*, he found that the rate of photosynthesis increased approximately proportionately to the increase in light intensity until this intensity was about 0·1 that of a clear day in July, above which the rate of photosynthesis remained constant with increase in light intensity, the concentration of carbon dioxide being that of the atmosphere. With the sun plants, *Nasturtium palustre* and *Atriplex liastatum*, on the other hand, the rate of photosynthesis continued to rise until the light intensity was 0·7 that of full sunlight or thereabouts. Similar results to this were obtained by Lundegårdh with spinach, potato and *Atriplex latifolium*, by Boysen Jensen with *Sinapis* and by Stålfelt with *Pinus*, in which plant the assimilation rate had not reached its maximum in full sunlight.

A falling off in photosynthetic activity in high light intensities, for example when this is several times that of full sunlight, has been occasionally noted and may be ascribed to the injurious effect of such high light intensities on the chlorophyll. The depressing effect is thus a secondary effect acting through the chlorophyll factor. Prolonged exposure to light may result in a retardation or inhibition of photosynthesis, which was noted by Ursprung in 1917 in leaves of *Phaseolus multiflorus*. When these are exposed to sunlight for a period of 5 hours, for example, the leaves contain much starch, but during longer exposure the starch disappears and after another 4 hours little may be left. This phenomenon has been called *solarization*.

Reference was made earlier to the fact that under certain conditions the respiration and assimilation will equal one another so that no gaseous exchange is observable. For any particular temperature the light intensity at which gaseous exchange is thus zero is called the *compensation point*. It varies considerably with different species, and also apparently in the same species at different times. For instance, Plaetzer found it in *Elodea canadensis* at about 20° C. to be 2 lux in summer and 18 lux in winter, the lux being the intensity of illumination at 1 metre distance from 1 Hefner candle. In *Cinclidotus aquaticus* in March she found it to be 400 lux at the same temperature.

A question of some interest is whether the leaf or other green organ, on transference to light after a period in the dark, immediately commences to assimilate at a rate which remains constant, or whether there is an induction period during which the photosynthetic activity increases. Osterhout and Haas with *Ulva*, Warburg with *Chlorella* and Briggs with *Mnium undulatum*, have claimed to have demonstrated such an induction phase. In *Chlorella* the induction period was very short, lasting only a few minutes, in other cases it was much longer, lasting at least 2 hours. Briggs has reviewed the experimental data obtained in these different cases, and has come to the conclusion that they harmonize with a scheme according to which a complex of some substance, perhaps chlorophyll, and carbon dioxide is converted through absorption of light energy to another substance which is broken down by a catalyst to give carbohydrate and oxygen, or, alternatively, might decompose an inhibitor which combines with the catalyst.

Temperature

The interpretation of observations on the influence of temperature on photosynthesis is rendered difficult for two reasons. In the first place, only apparent assimilation can be measured, and a correction must therefore be made for respiration. It has already been noted that respiration rate is much influenced by temperature, so that at higher temperatures the correction to be made for respiration may be considerable. Now, when assimilation is rapid and carbohydrate accumulates in the assimilating organ, an increased concentration of respiratory substrate may lead to an increased respiration rate, so that the usual method of correction by adding to the observed assimilation rate the measured rate of respiration in the dark may give different results according to when the measurement of respiration is made; whether, that is, the measurement is made before or after the observation of photosynthesis. Obviously the safest way of dealing with the difficulty would be to make respiration measurements before and after the measurement of photosynthesis and use the mean of the respiration measurements for the correction. The second difficulty arises at comparatively high temperatures where it is found that the rate of photosynthesis does not remain constant, but falls off with time. This *time factor*, as it was called by F. F. Blackman, results from the effect of temperature on some internal factor of photosynthesis, presumably the protoplasmic factor. Consequently the measured rate of photosynthesis at these temperatures depends on the length of time over which the measurement is made, the longer the period the lower will be the average rate of photosynthesis. In the leaves of *Prunus laurocerasus* examined by Matthaei the time factor enters at temperatures above 25° C.

The experiments of Matthaei with leaves of *Prunus laurocerasus* and *Helianthus tuberosus* conducted more than 30 years ago may still be considered as the standard work on the subject of the influence of temperature on rate of assimilation, although a considerable quantity of data with other plants has been obtained since. Miss Matthaei found that from about − 6° C. up to 25° C. the rate of photosynthesis increased regularly with rise of temperature provided the carbon dioxide concentration and light intensity were great enough for neither to act as a limiting factor. She found that over this range of temperature the Van't Hoff rule was followed, photosynthesis increasing about 2·1 times in *Prunus laurocerasus* and about 2·5 times in *Helianthus tuberosus* for a rise of 10° C. Weak light, however, could limit and determine the rate of the process, so that under such a condition temperature would be without effect on the rate of assimilation, and the temperature coefficient would approximate to unity.

For *Elodea*, Blackman and Smith found the temperature coefficient (Q_{10}) between 7° and 13° to be 2·05 ; Osterhout and Haas found that in *Ulva rigida* between 17° and 27° it was 1·81, while Lundegårdh's results with sugar beet in strong light and 1·22 per cent. carbon dioxide indicated a temperature coefficient of photosynthesis of about 2.

With *Chlorella* Warburg found that in higher light intensities the temperature coefficient between 15° and 25° C. was about 2, although with decreasing temperature it gradually increased. In low light intensities temperature was found to be practically without effect on photosynthetic activity over the range of 15° to 25°, a result agreeing with that of Miss Matthaei on the operation of light as a limiting factor. Van den Honert's results with the filamentous terrestrial alga *Hormidium* were similar. With high light intensity the rate of photosynthesis increased continuously from 12° to 20° C., increasing 1·87 times for a rise of 10° C., but with low light intensity temperature changes within the same range had practically no effect.

It is thus clear that when neither light nor carbon dioxide concentration is a limiting factor the rate of photosynthesis in the green cells of plants of a wide variety of species is determined by a reaction, the rate of which is affected by temperature in the manner shown by Van't Hoff to be characteristic of many chemical reactions. When, however, light intensity is low, temperature has little effect, a characteristic of photochemical reactions. It can thus be argued that when light intensity is low a photochemical reaction determines the rate of photosynthesis whereas when light intensity and carbon dioxide are both high a 'dark' chemical reaction, the rate of which is determined by temperature, limits the whole process. It has, of course, been clear for many years that a photochemical reaction is involved in photosynthesis, since light supplies the energy for the process. The existence of the 'dark' reaction was established by work in F. F. Blackman's laboratory and it is now frequently referred to as the 'Blackman' reaction.

The Wave-length of Light

From the time of Senebier in the latter years of the eighteenth century attempts have been made to discover the effect of different-coloured lights on photosynthesis. Much of this work is of little value because no account was taken of the intensity of the light. Obviously, a comparison of the photosynthetic intensity in lights of different wave-lengths may shed little information regarding the influence of the wave-length unless the intensity of the light is known, and undoubtedly effects have been attributed to differences in wave-length that are actually due to differences in light intensity. Observations agree, however, in showing that photosynthetic activity is only affected by radiation with wave-lengths within a limited range. This range corresponds largely with the visible spectrum, but Ursprung observed a little formation of starch in leaves of *Phaseolus vulgaris* exposed for 40 hours to infra-red radiation, while starch formation was also observed under the influence of the ultra-violet.

Within the photosynthetically active range of wave-lengths the rate of photosynthesis is not equal in light of different wave-lengths but of the same intensity. Since the assimilating cells absorb light of different wave-lengths to different extents we should expect those rays which are most absorbed to be the most effective in photosynthesis. The greatest absorption of energy by chlorophyll occurs in the red and to a lesser extent in the blue-violet and therefore it is not surprising that it has been found that photosynthesis is most active in light of these colours. A noteworthy attempt to measure the photosynthetic efficiency of different wave-lengths of light was made by Warburg and Negelein. They measured the energy absorbed by *Chlorella* in different-coloured lights and from measurements of photosynthesis calculated the proportion of the absorbed energy utilized. They found that the ratio of the energy utilized in photosynthesis to the energy absorbed increases with decrease of light intensity. For purposes of comparison they therefore obtained by extrapolation the value of this ratio for the limiting case where light intensity is zero. Their results, summarized in Table XLIV, show that with shortening of the wave-length a smaller proportion of the absorbed light is utilized in photosynthesis. Warburg and Negelein failed to find appreciable photosynthetic activity either in the infra-red or ultra-violet under their experimental conditions.

Table XLIV

Utilization of Light of Different Wave-lengths in Photosynthesis by *Chlorella*

Wave-length in $\mu\mu$	Colour	$\frac{\text{Energy utilized}}{\text{Energy absorbed}}$
660	red	0·59
578	yellow	0·535
546	green	0·444
436	blue	0·338

The values given in the last column of the above table indicate that a large proportion of absorbed energy is utilized in photosynthesis. It is to be noted, however, that these are theoretical values for the limiting case when incident energy is zero; actually the proportion will always be less, and smaller the more intense the illumination. Further, Warburg and Negelein's experiments were arranged so that all the incident energy was absorbed by a thick suspension of *Chlorella*. Hence it is not surprising that in foliage leaves, where some incident energy is reflected, some transmitted through the leaf and often much used in transpiration, a much smaller proportion of the incident energy should be utilized in photosynthesis. Calculations by Brown and Escombe indicated that only from 0·42 to 1·66 per cent. of the energy incident on the leaf of *Polygonum Weyrichii* was utilized in photosynthesis, while Puriewitch's calculations in respect of leaves of a number of species suggested that from 0·6 to 7·7 per cent. of the energy falling on the leaf was utilized in photosynthesis.

Internal Factors in Photosynthesis

It has already been pointed out that in addition to the assimilatory pigments some other component or property of the protoplasm is essential for photosynthesis, so that we have to regard the chlorophyll factor and protoplasmic factor as two separate internal conditions for photosynthesis. There may be others; thus it has been suggested that the water content of the assimilating cells may influence photosynthetic activity.

The investigation of the relation between chlorophyll content and rate of photosynthesis is not so simple a matter as the investigation of the influence of external factors, as chlorophyll content cannot be altered at will. The problem was, however, attacked by Willstätter and Stoll by the use of (*a*) leaves of different species, (*b*) leaves in different stages of development, (*c*) green and yellow varieties of the same species, (*d*) etiolated leaves developing the green pigment and (*e*) chlorotic leaves. The determination of the chlorophyll content was made by a colorimetric method; some subsequent workers have estimated the chlorophyll spectrographically.

The results obtained by Willstätter and Stoll were expressed in terms of what they called the 'assimilation number', which is the rate of photosynthesis per unit of chlorophyll, the rate of photosynthesis being measured in grams per hour and the chlorophyll content in grams. These measurements were made with a high concentration of carbon dioxide (5 per cent.), a high light intensity (48,000 lux or higher) and a temperature of 25° or 30° C. With these external conditions neither carbon dioxide concentration nor light intensity was limiting photosynthesis, for no increase in photosynthetic activity was brought about by increasing either of these factors.

The assimilation numbers of normal leaves of different species, but of apparently a similar stage of development, are of the same order, but by no means identical. Those determined at 25° C. mostly varied between 5·2 (*Acer pseudoplatanus*) and 7·7 (*Acer negundo*) but values as high as 10·9, 14·0 and 16·7 were obtained with *Helianthus annuus*. With hot-house plants, the assimilation numbers determined at 30° C. varied between 6·5 (*Hydrangea opuloides*) and 14·5 (*Pelargonium peltatum*). It would thus appear that the rate of assimilation in different species is not determined solely by external factors and chlorophyll content.

Willstätter and Stoll found that as leaves grow older the assimilation number falls. Actually both the chlorophyll content and the rate of assimilation increase, but the rate of assimilation per unit of chlorophyll decreases. Thus in *Sambucus nigra* the assimilation number of leaves collected in May was 12·2, while that of leaves collected on 14 July was 6·2. As leaves become senescent much divergence in the relation between chlorophyll content and rate of assimilation is observed. In general these results with leaves in different stages of development suggest that some internal factor other than chlorophyll may affect photosynthetic activity.

The same conclusion is to be drawn from observations on the behaviour of leaves from yellow varieties as compared with that of normal green leaves of the same species. Yellow varieties are not devoid of chlorophyll but have a low content of the green pigment so that its presence is masked by yellow pigments. Willstätter and Stoll found that leaves of a yellow variety of *Ulmus* contained only about 0·073 times the chlorophyll of the leaves of the normal green type, but that the rate of assimilation by unit area of the two kinds of leaves was approximately the same even under conditions of high carbon dioxide concentration and light intensity and at 25° C.

More definite information with regard to the action of a second internal factor is derived from observations on the photosynthesis of etiolated leaves exposed to light and becoming green. The chlorophyll content of such leaves gradually increases. Completely etiolated tissues have no power of photosynthesis. In seedlings, with increasing chlorophyll content there may or may not be a corresponding increase in the rate of photosynthesis, what actually takes place depending, according to Briggs, on the age of the seedling. Thus Briggs found that the rate of photosynthesis of etiolated seedlings becoming green was determined by the age of the plants rather than by the chlorophyll content, the older the seedling the greater the rate of photosynthesis resulting on exposure to light. In experiments by Willstätter and Stoll in which plants of *Phaseolus vulgaris* about a fortnight old were used, the rate of photosynthesis increased with the increase in chlorophyll content. These findings suggest that some internal factor other than chlorophyll develops in the seedling whether this is in light or in darkness. Hence, before a seedling has reached a certain age this other factor may act as a limiting factor and change in chlorophyll content will be without effect on photosynthetic activity, while in older etiolated seedlings becoming green this second factor may be fully developed and chlorophyll content may act as a limiting factor, so that with increase in the quantity of chlorophyll there will occur an increase in the power of photosynthesis.

In addition to chlorophyll and the unknown protoplasmic factor which develops independently of it, it is also possible that water content of the assimilating cells may influence the rate of photosynthesis. Various investigators, and particularly Thoday, have found that the photosynthetic activity of leaves decreases along with decreasing turgidity, a result considered by Thoday to be largely ascribable to closure of the stomata, so that the apparent effect of water content is actually due to a lowering of the concentration of carbon dioxide at the chloroplasts. In plants without stomata, such as mosses, reduction of water content appears to have much less effect on photosynthetic activity, although a falling off in the rate of photosynthesis with decreasing water content has been observed in water plants, and in these stomatal movements cannot enter into the matter. Moreover, Dastur has produced evidence to show that differences in water content of various leaves run parallel with differences in photosynthetic activity. In *Abutilon asiaticum*, *Ricinus communis* and *Helianthus annuus*, Dastur and Desai also concluded that the rate of photosynthesis is more closely connected with water content than with chlorophyll content. Dastur had previously shown that in senescent leaves of *Abutilon asiaticum*, *Ricinus communis* and other species, photosynthetic activity is first lost in those cells which are most remote from the water supply provided by the vascular bundles, the cells, that is, of the leaf margin and intravascular regions.

Effect of Various Substances on Photosynthesis

A considerable number of observations are on record with regard to the action of various substances on the rate of photosynthesis. These substances include chloroform, ether, antipyrin, ammonium salts, sulphites, glycerol, quinine, strychnine and other toxic compounds. Generally speaking, any of these has a depressing effect on photosynthesis, and probably this effect is usually due to a harmful action on the protoplasmic factor. These observations, for the most part, have so far shed little light on the photosynthetic process.

There is evidence that photosynthesis will not commence in absence of oxygen. Willstätter and Stoll found that leaves of *Pelargonium*, after exposure to an oxygen-free atmosphere for 2 hours in the dark, did not assimilate on exposure to light and were still unable to do so when they were subsequently brought into contact with oxygen. There is, however, considerable variation among different species in regard to the influence of oxygen on photosynthesis. Thus, in the case of *Cyclamen europeum*, although continued deprivation of oxygen in the dark does result in a lowering of photosynthetic potentiality, this has not quite disappeared even after 15 hours' exposure to an oxygen-free atmosphere in the dark.

A number of explanations of this effect of absence of oxygen are possible. Thus, there may be some at present unknown connexion between respiration and photosynthesis, or anaerobic respiration may result in the production of substances which affect adversely the protoplasmic factor.

THE PRODUCTS OF PHOTOSYNTHESIS

Sachs, we have noticed, regarded starch as the first visible product of assimilation, while Meyer found that many plants which do not form starch produce sugar. While more than one sugar has been identified as arising in green leaves as a result of photosynthesis, there is little evidence that any substances other than carbohydrates or closely related compounds are produced in photosynthesis, at any rate by higher plants. Not only is this the conclusion to be drawn from analysis of leaves, but the same is indicated by determinations of the assimilatory quotient, that is, the molecular ratio of oxygen evolved to carbon dioxide absorbed. This is very generally unity, which should be the case if carbohydrate is formed, whereas if fats, organic acids or other substances were produced the quotient would nearly always be different. In succulent plants, it is true, the assimilatory quotient at times appears to depart widely from unity, but the divergence is here related to the unusual respiration of these plants. In all cases the *apparent* assimilatory quotient depends to a certain extent on the respiratory quotient. Thus if the rates of oxygen absorption and carbon dioxide evolution in respiration are respectively O_r and C_r and the rates of carbon dioxide absorption and oxygen evolution in photosynthesis are respectively C_p and O_p, the true assimilatory quotient is $\frac{O_p}{C_p}$, but the apparent assimilatory quotient actually observed is $\frac{O_p - O_r}{C_p - C_r}$. In most assimilatory organs, the substance used in respiration is carbohydrate, so that the respiratory quotient is unity and hence with a true assimilatory quotient of unity, the apparent assimilatory quotient is also unity. But in the succulents, it will be recalled, imperfect oxidation in respiration results in the formation of organic acids. These accumulate in the dark, but in the light are broken down to carbon dioxide and water. Hence in addition to photosynthesis and contemporaneous respiration gaseous exchange in the light is affected by the production of carbon dioxide from accumulated organic acid. This carbon dioxide is utilized in photosynthesis, so that less will be absorbed from the atmosphere and consequently the apparent assimilatory quotient will be higher than the true quotient. Aubert, indeed, found values up to 7·59 for the apparent assimilatory quotient of succulents. With continuance of illumination the apparent assimilatory quotient of these plants falls and may ultimately approach unity, as the organic acid is removed.

The products of assimilation are thus in general starch and sugars. The most careful analyses indicate that the latter comprise the hexoses, *d*-glucose and *d*-fructose, and the disaccharide sucrose. In addition pentose sugars appear to be present in leaves, according to the researches of Davis and Sawyer, but it is doubtful whether maltose is present in leaves. Other carbohydrates recorded in assimilatory organs are inulin in leaves of *Cichorium intybus* and of species of *Marcgravia*, pentosans in leaves of mangold and potato, dextrin in the latter, and trehalose in certain red algae. The presence of the carbohydrate alcohol mannitol has been reported in leaves of Oleaceae and in the thalli of some Phaeophyceae. Some of these, at any rate, appear to come in the category of assimilatory products. The oil drops of *Vaucheria* have also been regarded as assimilatory products, but their composition is doubtful and they may not be composed of true fat.

Much has been written concerning what substance is to be regarded as the first product of photosynthesis, and the status of the various carbohydrates present in assimilatory organs. Starch is no doubt correctly regarded as a temporary reserve. The various sugars may be either *upgrade*, that is, formed before the production of starch, or *downgrade*, resulting from the breaking down of starch or some other reserve substance.

Various workers have sought to obtain information with regard to the first sugar formed in photosynthesis by making analyses of leaves at different times and so following the changes in carbohydrate content over a 24-hour period and during a season. Observations of this kind leave no doubt that considerable and definite daily and seasonal changes occur in the content of the different carbohydrates in leaves, but it is extremely doubtful whether such data by themselves can give information regarding the first sugar of photosynthesis. As an example of such observations may be taken those of Parkin on the snowdrop (*Galanthus nivalis*), a relatively simple case since starch is not formed and the only carbohydrates present are hexoses (glucose and fructose) and sucrose. Here the hexose concentration in the leaf remains roughly constant over a 24-hour period, while the sucrose content increases during the day and decreases during the night. Similar relations have been observed in other plants and have generally been regarded by the observers themselves as indicating that sucrose is first formed, accumulates as a temporary reserve, and is then subsequently hydrolysed to hexose in which form the sugar is translocated. But that hexose does not accumulate, whereas sucrose does, is equally to be expected if hexose is first formed and is transformed to sucrose when a certain concentration is reached.

Indeed, the observations of Weevers suggest that hexoses precede sucrose in assimilating leaves. In variegated leaves of a number of species, including *Ilex aquilifolium, Hedera helix, Humulus lupulus* and *Pelargonium zonale*, both hexoses and sucrose were identified by this worker in the green portions of the leaves, whereas in the yellow patches, although sucrose was present, hexoses were either entirely absent or present only in small quantity. Further, in leaves of *Pelargonium zonale* exposed to light after prolonged darkening, Weevers found that hexoses first appear, then sucrose, and lastly starch. We have, however, no reliable information as to whether only one monosaccharide is first formed and the other or others formed from it, or whether two or more arise together.

THE MECHANISM OF PHOTOSYNTHESIS

It must be regarded as certain that the production of carbohydrate from carbon dioxide and water in the assimilatory organs takes place in a series of stages. We have already seen that when light is a limiting factor the rate of photosynthesis depends on the light intensity but is practically independent of the temperature, thus suggesting that the reaction is a photochemical one, whereas in high light intensity photosynthesis is influenced by temperature in the same way as many chemical reactions. This behaviour indicates that at least one photochemical reaction and one ' dark ' chemical reaction are involved, one reaction or the other limiting the whole process according to the conditions. What these reactions are we do not know. Many attempts have been made to obtain information with regard to the nature of the intermediate products formed during photosynthesis, but the evidence adduced in favour of the production of any particular substance has never been sufficiently free from objection to induce general acceptance of it. Among substances hypothesized as intermediate products of photosynthesis are various organic acids and aldehydes, the most popular of these being formaldehyde. In spite of the large measure of support which the formaldehyde hypothesis has received in the past, and the many attempts that have been made to obtain experimental evidence in its favour, it does not to-day receive general acceptance. The production of formaldehyde in systems containing water, carbon dioxide and chlorophyll was shown by Jorgensen and Kidd to result from the oxidation of chlorophyll, no formaldehyde being produced in systems containing only water, carbon dioxide and pure chlorophyll in absence of oxygen. Although the presence of formaldehyde had earlier been reported in assimilating leaves, more recent investigators have concluded that there is no definite evidence of this, although there can be no doubt that in the plant there are many substances which will yield aldehydes, though not necessarily formaldehyde, under certain conditions. A number of workers have claimed that leaves can absorb formaldehyde and transform it to sugar. In many cases there can be little doubt that before its absorption the formaldehyde had been oxidized to formic acid and in any case it has been shown that other substances such as glycerol can be absorbed by leaves in this way.

A suggestion that has received some measure of support is that glycollic aldehyde, $CH_2OH.CHO$ is an intermediate product of photosynthesis, either arising subsequently from formaldehyde or directly from carbon dioxide. There appears to be more evidence for the presence of glycollic aldehyde in assimilating leaves than for the presence of formaldehyde, and Rouge in 1921 obtained from assimilating leaves of potato a yield of 0·006 gram of glycollic aldehyde per kilogram, while practically none was present in similar leaves at night. This fact in itself, unsupported by other evidence, cannot be regarded as conclusive proof of the formation of glycollic aldehyde as an intermediate product of photosynthesis.

In formulating a scheme of the photosynthetic process in green plants, theorizers are not assisted by information regarding intermediate products. Most of those who have attempted to hypothesize such a scheme have, however, assumed formaldehyde as an intermediate substance, but if they are correct in this, it would appear that the formaldehyde must be immediately utilized, for, from what has been stated above, it appears clear that it does not accumulate.

The first of such theories is the well-known one of Baeyer suggested in 1870. According to this, the absorbed carbon dioxide is reduced to formaldehyde, which is then transformed through the influence of certain cell contents into sugar, a synthesis which can be effected outside the cell by alkalies. Several subsequent theories of photosynthesis are an elaboration of this theme.

The general acceptance of the view that photosynthesis includes both a photochemical phase and a purely chemical or enzymic phase has led to the recognition that no theory of the mechanism of photosynthesis can be adequate unless these two phases find a place in it. This is the case with the theory which Willstätter and Stoll advanced in 1918. According to this theory, the carbon dioxide on reaching the chloroplast, having already produced carbonic acid by combination with water, now forms an additional compound with chlorophyll, the suggested reaction being represented by the following equation, where the formula of chlorophyll is contracted to R : Mg.

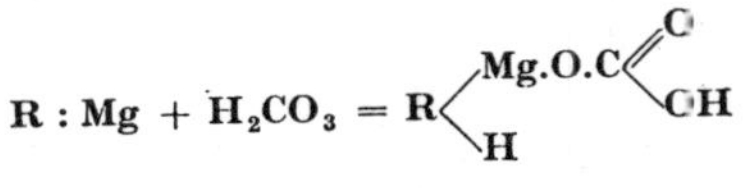

$$R : Mg + H_2CO_3 = R\begin{cases} Mg.O.C\begin{cases} O \\ OH \end{cases} \\ H \end{cases}$$

This purely chemical action is followed by a photochemical reaction in which the chlorophyll-carbonic acid compound undergoes isomerization to a compound with higher energy content, and having a peroxide structure :

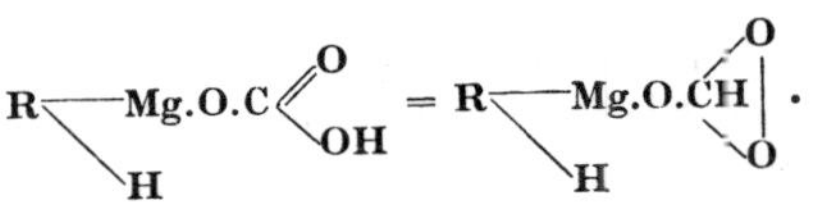

$$R\begin{cases} Mg.O.C\begin{cases} O \\ OH \end{cases} \\ H \end{cases} = R\begin{cases} Mg.O.CH\begin{cases} O \\ | \\ O \end{cases} \\ H \end{cases}.$$

It is thus in this reaction that the actual absorption of light energy takes place.

This peroxide is then supposed to decompose under the influence of an enzyme into chlorophyll, formaldehyde and oxygen, possibly in two stages. The enzyme was regarded by Willstätter and Stoll as the protoplasmic factor of photosynthesis, and might be regarded as a peroxidase, although it must be considered as extremely doubtful whether the protoplasmic factor can be defined in terms of a single enzyme. However this may be, the enzymatic phase in Willstätter and Stoll's scheme of photosynthesis can be represented by the equation :

$$R{<}^{\displaystyle Mg.O.CH{<}^{O}_{O}|}_{\displaystyle H} = R : Mg + H.CHO + O_2$$

The formaldehyde so produced is then assumed to polymerize to hexose, but the mechanism of this is not explained.

The theory of Willstätter and Stoll thus hypothesizes at least five stages in the whole process of photosynthesis, namely, (1) a diffusion phase, (2) a chemical action, (3) a photochemical action, (4) an enzyme action, (5) a chemical action, possibly enzymic.

While this theory thus involves both photochemical and 'dark' chemical phases, it does not explain the presence in assimilating organs of more than one chlorophyll, or of the accompanying yellow pigments. Indeed, no explanation of the function of the yellow pigments has so far been forthcoming which has obtained general acceptance. The final conclusion of Willstätter and Stoll on this matter was that the carotinoids might form part of a system preventing the photo-oxidation of chlorophyll, for these yellow pigments rapidly absorb oxygen.

While various explanations of the presence of two chlorophylls in assimilating cells have been put forward ascribing different functions to the two green pigments, the great similarity in molecular structure and chemical behaviour of these two substances renders it unlikely that they should have different functions, and the final view of Willstätter and Stoll on this point appears credible. They consider that the value to the plant of having both compounds lies in the fact that a mixture of the two chlorophylls will absorb more light than the same quantity of one of the chlorophylls alone, as the absorption bands of the two substances are not in the same position in the spectrum, the chief absorption bands of chlorophyll *b* lying between those of chlorophyll *a*. How far the yellow pigments function in photosynthesis by absorbing light energy remains in doubt.

The theory of photosynthesis of Willstätter and Stoll is only one of many that have been put forward from time to time, but although it is largely speculative it has the merit of not running counter to observed facts. One point of interest in it may be pointed out, and that is that it ascribes to chlorophyll a double rôle ; the pigment plays first a purely chemical part in combining with carbonic acid, and secondly, in the form of chlorophyll-carbonic acid it absorbs light energy and so forms the material for the photochemical reaction. As the absorber of the energy to which practically all living things ultimately owe their activity and being, Shipley's description of chlorophyll as 'the most wonderful substance in the world' is not altogether unjustified.

THE FORMATION OF STARCH

Although in a great number of plants starch is produced in the assimilating organs after exposure to light, we are justified in regarding the photosynthetic process as complete on the formation of sugar. For in the first place many plants do not form starch at all in the assimilating cells, as in the case of a number of monocotyledons, including the bluebell (*Scilla non-scripta*) and snowdrop (*Galanthus nivalis*),[1] and, in the second place, starch undoubtedly arises in non-photosynthesizing organs and in the dark from sugar which has migrated there.

We must therefore regard the starch produced in leaves as a temporary storage product which forms when the concentration of sugar in the leaf reaches a certain value, although we must suppose that not only sugar concentration, but other factors, such as temperature, hydrogen-ion concentration and concentration of some enzyme or enzymes, influence the rate of the change. As regards the question of the enzyme involved, it seems reasonable to suppose that the diastatic system is here operative, for as starch on hydrolysis by means of diastase yields glucose, the same system, catalysing the action in the reverse direction, may be supposed to provide the mechanism for starch synthesis.

The critical concentration of sugar which must be reached before starch is produced we might expect to vary from one species to another, even under the same conditions of temperature, since we may with reason expect the internal conditions of the cells to be different in different species, and it may be that herein is to be sought the explanation of the different amounts of starch produced in the leaves of different species. Certainly in some leaves, for instance those of *Muscari*, which normally produce no starch, this substance is produced when the concentration of sugar is raised considerably above the amounts normally present.

The sugars produced in the assimilating cells are, as we have seen, chiefly glucose, fructose and sucrose, whereas starch is a condensation product of glucose only. The transference of glucose to starch is, as we have seen, thus readily explicable as a synthetic action of diastase, but the transference of fructose and sucrose into starch is not so simply explicable. Whether the fructose and fructose unit of sucrose are converted directly into starch, or whether the fructose is first converted into glucose, or whether only the glucose in the assimilating cells forms a source of starch cannot at present be decided, but as there is not conspicuous increase in the amount of fructose in comparison with glucose during starch formation in leaves, it seems possible that both fructose and sucrose are utilized for starch formation either directly or indirectly. But analyses are difficult and for the most part unreliable, and it is difficult to form a judgement on the matter.

[1] In such cases, however, starch may be present in the guard cells of the stomata and in the sheath surrounding the vascular bundles.

Haworth's recent researches indicate that the molecule of starch comprises about 25 glucose anhydride groups forming a straight chain linked up thus:

```
              CH₂OH                         CH₂OH                         CH₂OH
               |                             |                             |
               C——O                          C——O                          C——O
       H     / H    \       H       H      / H    \      H       H      / H    \      H
       C   <          >   C         C   <          >   C         C   <          >   C
  —O—  |     OH   H /     |   —O—   |     OH   H /     |   —O—   |     OH   H /     |  —O—
             \C——C                        \C——C                        \C——C
              H   OH                        H   OH                        H   OH
```

The differences observed in the starches from different plants are probably due to differences in the physical mode of aggregation of the same starch molecule.

Starch is insoluble in water, and, as is well known, is produced in the plant in the form of granules. It is generally agreed that starch grains are always formed in plastids. Thus in the assimilating organs the starch grains are formed in the chloroplasts, and in places of more permanent storage in the leucoplasts. The starch grains formed as a temporary reserve in the chloroplasts are usually small and several of them may be formed in a single plastid, whereas in storage organs there is usually only one starch grain formed in one leucoplast, and it may be a relatively large structure with a form characteristic of the species.

In those species which do not form starch as a temporary reserve after photosynthesis, sugars, and particularly sucrose, appear to be the substances temporarily stored. Very rarely, as in the leaves of species of *Marcgravia*, inulin is produced and serves as a temporary storage product. It occurs much more frequently as a more permanent storage product in the underground organs of various perennials, but further reference will be made to this question in a later chapter. Here it will be sufficient to point out that the synthesis of sucrose and inulin may be effected by the enzymes sucrase and inulase respectively catalysing in the direction of synthesis.

THE FORMATION OF CELLULOSE AND OTHER COMPLEX CARBOHYDRATES

A number of complex carbohydrates, apart from starch, are common constituents of plants. Of these the most important is cellulose, which is a constant, and the most significant, constituent of cell-walls throughout the plant kingdom with the exception of fungi, bacteria and certain algae. Other complex carbohydrates are the so-called hemicelluloses, which are also found in cell-walls, where they occur sometimes in considerable quantity as reserve substances, often called reserve cellulose. Still other complex carbohydrates are the various pectins, gums, resins and mucilages. The formation of these substances is not limited to the assimilating organs and, in the case of cellulose, occurs in all the growing regions following nuclear division. They must therefore be produced from carbohydrate material transported from the assimilating cells to the seat of their formation.

Cellulose, like starch, on hydrolysis yields glucose, and is therefore to be regarded as a condensation product of this substance, and much of the sugar transported from the leaf is utilized in the production of cellulose in cell-walls. Its empirical formula, like that of starch, is $(C_6H_{10}O_5)_n$, but the value of n appears to be larger than in the case of starch; that is, the molecule is larger. Various views have been held with regard to the structure of the cellulose molecule, but that which appears to find most favour at the present time regards the cellulose molecule as consisting of a straight chain of glucose anhydride groups linked up through an oxygen atom thus:

```
                                  CH2OH
         H    OH                    |                          H    OH
         C----C                     C-----O                    C----C
 --O--| / OH  H \   H      H  /  H       \    |--O--|  /  OH  H \   H
      C          C         C                C      C             C
      H \  H    /   |--O--|   \  OH   H  /  H      H  \  H      /   |--O--
         C----O                  C-----C                 C----O
         |                       H     OH                |
       CH2OH                                           CH2OH
```

The number of glucose residues linked together in this way in the cellulose molecule is, according to Haworth, about 200; other estimates are both higher and lower than this. It also seems likely from the researches of Haworth and his collaborators that the cellulose molecule forms a chain with two ends; that is, the valencies shown in the formula above at each end of the molecule are satisfied by two univalent atoms or groups, such as, for example, —H or —OH, and are not joined up so that the whole molecule forms a large ring.[1]

As regards the mode of formation of cellulose in the plant, it appears always to be secreted by the cytoplasm, either as cellulose itself, or as some simpler substance which polymerizes into cellulose. Possibly the synthetic action of a cellulase or cellobiase is involved in the formation of cellulose, though the fact that such enzymes have not been isolated either from higher or most lower plants renders such a hypothesis unsatisfying. It has been suggested that a starch-like substance may be an intermediate product in the formation of cellulose, but although both the starch and cellulose molecules are condensation products of glucose, it will be observed that the difference between them is not simply that the cellulose molecule contains a larger number of such units than starch;

[1] The molecule of cellulose thus differs from that of starch not only in the length of the chain of glucose residues, but in the space relations of alternate glucose groups. The starch molecule, in fact, consists of a chain of maltose groupings, while that of cellulose consists of a chain of residues of another disaccharide, namely, cellobiose.

the arrangement of the glucose groupings in the two products is different, so that the transformation of starch to cellulose probably involves the breaking down of the former to glucose before the cellulose can be synthesized.

The so-called hemicelluloses, including the reserve cellulose found in the thick cell-walls of certain seeds such as those of *Lupinus luteus*, *Tropaeolum majus*, *Plantago* spp., and notably the date, *Phoenix dactylifera*, are condensation products of various sugars. Thus, the reserve celluloses of the seeds of *Lupinus luteus* are complex substances of this kind built up of various pentose and hexose sugars, the latter including glucose, fructose and galactose. In the hemicelluloses present in the seeds of the date palm and coffee, mannose, as well as galactose, groups are present. It also appears probable that uronic acids (oxidation products of sugars in which the alcoholic $-CH_2OH$ group is oxidized to the acid group $-COOH$) also enter into the molecules of hemicelluloses.

The gums and mucilages met with in some plants appear to have a composition rather similar to that of the hemicelluloses but to involve pentose sugars in their composition. Thus gum produced by cherry and plum trees appears to be a condensation product of the pentose sugar arabinose, while gum arabic, formed in various species of *Acacia*, appears to contain groupings of galactose, galacturonic acid and arabinose. The mucilages which occur in many succulent plants and occasionally in other plants, as, for example, on the surface of the seed coat of species of *Linum*, are also condensation products of various sugars, apparently both pentose and hexose. Both gums and mucilages are produced as cell-wall material, but, like the hemicelluloses, their mode of formation from sugars can only be guessed.

The pectic substances are also general constituents of cell-walls, and much work has been done with a view to determining their composition, but no general agreement with regard to the details of this has been reached. The current opinion appears to be that from plants four types of pectic substances can be isolated; these are referred to as pectic acid, soluble pectin, insoluble pectin or pectose, and pectate. Pectic acid is a compound of galacturonic acid, arabinose, galactose and methyl pentose. It is doubtful if it occurs free in plants. Soluble pectin occurs in both the vacuole and wall of the cells in many plants. It is a derivative of pectic acid in which some of the acidic radicles of the galacturonic acid are combined with methyl alcohol, so that some of the $-COOH$ groups of pectic acids are replaced by $-COOCH_3$ groups. Insoluble pectin is possibly soluble pectin combined with cellulose, although its composition is at present a matter of considerable controversy. As to the fourth type of pectic substance, the view that the middle lamella of the cell-wall is composed of calcium pectate is very generally held, but recently doubt has been cast on the very presence of a middle lamella in adult cells while, as mentioned in an earlier chapter, non-pectic substances have also been suggested as composing the middle lamella in young cells.

From what has been written above it will be seen that many and varied condensation products of the simple sugars are synthesized in plants. The two chief functions of these complex carbohydrates is to provide material for cell-walls, and to serve as reserves. Little, however, is known regarding their mode of origin.

THE FORMATION OF FATS AND LIPOID SUBSTANCES IN PLANTS

Carbohydrates are not the only compounds of carbon, hydrogen and oxygen which enter into the make-up of the plant. In every plant, and indeed in every living cell, fats or substances allied to fats are present, and are often to be observed as globules known as 'oil drops'. These fatty substances, or lipins, include not only true fats, but also more complex compounds in which fat groupings enter into combination with others containing other elements, notably nitrogen and phosphorus.

The true fats are glyceryl esters of the fatty acids of which the triglycerides of caproic acid and of oleic acid, $C_3H_5(OOC.C_5H_{11})_3$ and $C_3H_5(OOC.C_{18}H_{34})_3$ respectively, may be taken as representative of a considerable number of such substances met with in plants. On hydrolysis the fat yields glycerol and the corresponding fatty acid; triolein, the triglyceride of oleic acid, for example, gives glycerol and oleic acid:

$$C_3H_5(OOC.C_{18}H_{34})_3 + 3H_2O = C_3H_5(OH)_3 + 3C_{18}H_{34}.COOH$$

In addition to these true fats similar substances occur, though in much smaller amounts, in which the place of glycerol is taken by a monohydric alcohol. Such substances have been defined as waxes, though the tendency now seems to be to include the alcohols themselves in the category of waxes, and to distinguish their products of combination with fatty acids as wax esters.

A number of such alcohols have been isolated from plants, some of them, those obtained from leaves, possessing molecules with a straight chain of carbon atoms, such as *n*-octacosanol $CH_3.(CH_2)_{26}.CH_2OH$. Others occurring in seeds appear to involve rings of carbon atoms in the molecule and to be related to the animal product cholesterol; such are generally termed phytosterols, of which sitosterol $C_{27}H_{45}OH$ and stigmasterol $C_{30}H_{47}OH$ are the best known. How far these substances occur in the free state as well as combined with fatty acids is not clear. In leaves, at any rate, the constituent waxes have been shown to contain, or to be accompanied by, other substances such as paraffins and sometimes ketones.

As regards the more complex fatty substances in plants, the most important of these are phosphorus-containing compounds of nitrogen bases with fatty acids. They appear to be constant constituents of protoplasm and are known as phosphatides, phospholipins, lecithides and phospholipoids. The best known of these are the lecithins, compounds of glycerol, fatty acid, phosphoric acid and the nitrogen-containing base choline (see p. 256). To dipalmityl lecithin, for example, the constitutional formula given below is ascribed :

$$
\begin{array}{l}
CH_2.COOC_{15}H_{31} \\
| \\
CH.COOC_{15}H_{31} \\
| \qquad\qquad \diagup OH \\
CH_2.O.P \\
\qquad\quad \| \quad \diagdown O.(CH_2)_2.N(CH_3)_3 \\
\qquad\quad \| \qquad\qquad\qquad | \\
\qquad\quad O \qquad\qquad\quad OH
\end{array}
$$

Here the fatty acid involved is palmitic acid, $C_{15}H_{31}COOH$, but it appears that various lecithins occur with other fatty acids in place of palmitic acid.

The true fats in plants are for the most part reserve substances and play the same part in the life of the plant as do reserve carbohydrates. Thus they occur chiefly in storage organs, particularly seeds. While one particular fat may predominate in the fruits and seeds of one species, generally there is present a mixture of fats, while phytosterols are as a rule also present in small quantity. Whether these latter are storage substances is doubtful, for they occur in green parts of plants in greater quantity than in seeds, and in the latter they may be present in greatest amount in the embryo.

Little that is definite is known with regard to the origin of the reserve fats of fruits and seeds. They appear to arise from sugars, and various observations are on record of the sugar content of fruits and seeds falling while the fat content rises. The numbers given in Table XLV are taken from a paper published by Leclerc du Sablon in 1896, and illustrate this point.

Table XLV

Changes in Sugar and Fat Content of Almond Fruits during Ripening

The quantities are given as percentages of dry weight

Date of Collection	Fat	Glucose	Sucrose	Amylose (starch, dextrine, etc.)
9 June	2	6·0	6·7	21·6
4 July	10	4·2	4·9	14·1
1 August	37	0	2·8	6·2
1 September	44	0	2·6	5·4
4 October	46	0	2·5	5·3

How the change from carbohydrate to fat is brought about is not known. Ivanov, in 1911, from investigations on the chemical changes taking place during the ripening of various fat-containing seeds, namely, those of flax, hemp, rape and sunflower, came to the conclusion that the carbohydrate first gives rise to glycerol and saturated fatty acids. The latter are then transformed into unsaturated fatty acids, which, reacting with glycerol by means of the synthetic action of lipase (cf. p.

118), give rise to the fats. Since fats contain considerably less oxygen relative to carbon than do carbohydrates, the formation of fats from carbohydrates will involve the evolution of oxygen which, being available for respiration, will thus bring about a lessening of the volume of that gas absorbed from the atmosphere, so that the measured respiratory quotient will be raised above unity. Such was observed to be the case by Gerber in a number of ripening seeds and fruits. Thus in flax he found the average respiratory quotient during ripening of the seed was 1·22.

Other suggestions regarding the production of fat have, however, been made. Among these may be mentioned that of M'Clenahan, who considered that the fat in walnut seeds may be derived from tannins. Little evidence in favour of such a view is, however, forthcoming at present.

The quantity of wax present in seeds is small, varying in seeds so far examined from 0·12 per cent. in the case of *Vitis vinifera* to 1·58 per cent. in *Ribes rubrum*. Whether these waxes form part of the food reserve or are essential constituents of the protoplasm is not clear.

From the researches of Chibnall and his collaborators the waxes present in leaves appear to be constant constituents of the protoplasm. Sahai and Chibnall found that the wax content of leaves of the Brussels sprout increases with the age of the plant, as the numbers given in Table XLVI show. There is no evidence that the wax changes in composition throughout the life of the plant. The waxes in the cabbage and Brussels sprout, lucerne (*Medicago sativa*), cocksfoot (*Dactylis glomerata*), perennial rye grass (*Lolium perenne*) and wheat have been isolated and analysed by Chibnall and his co-workers, and have been found to involve long-chain primary alcohols and mixed fatty acids. The waxes from all these plants also contain a paraffin, and that from the *Brassica* varieties, a ketone as well. The principal alcohol in the wax of these latter plants is one with 29 carbon atoms, and the paraffin and ketone present are those corresponding to this alcohol. In lucerne the alcohol is *n*-triacontanol, that is $CH_3.(CH_2)_{31}CH_2OH$, and in wheat

Table XLVI

Wax in Leaves of Brussels Sprout of Various Ages

Age of Seedling in days above ground	Stage of Development of Plant	Amount of Wax in percentage of dry weight
—	Ungerminated seed	0
10	Seedlings 2 cm. high	0·42
22	Seedlings 3–5 cm. high	0·85
43	Young plants 8·10 cm. high	1·1
54	Growing plant, laminae 8–10 cm. long	1·2
126	Mature plant, laminae 15–25 cm. long	1·2

n-octacosanol $CH_3(CH_2)_{26}CH_2OH$. The fatty acids present in the cabbage leaf consist for the greater part of the unsaturated linolenic and linolic acids, but the saturated palmitic and stearic acids are also present. It has already been noted in an earlier chapter that the same or similar long-chain alcohols occur in the cuticle of apple fruits.

The more complex lipoid substances, the phospholipins and related substances, appear to be universally present as constituents of the protoplasm. They have been recorded for a number of seeds where the amount of them appears to vary from about 0·25 to 2 per cent., the higher values occurring, according to Stoklasa, in seeds with a high protein content. They have also been examined in various leaves, but in those of the cabbage Chibnall and Channon found that the so-called phospholipin actually contained no nitrogen, all the phosphorus being combined with calcium, glycerol and fatty acids, the chief substance of this type present being the calcium salt of a diglyceride phosphoric acid, to which has been given the name phosphatide acid. To the salt the formula

$$\begin{array}{l} CH_2OOCR_1 \\ | \\ CH\ OOCR_2 \\ | \\ CH_2.O.P=O \\ \qquad\quad \| \\ \qquad\quad Ca \end{array}$$

is ascribed where R_1 and R_2 are fatty acid radicles. Although lecithins are thus absent, a comparison of this formula with that of lecithin (p. 202) shows the near relationship between it and phosphatide acid. Both lecithin and kephalin were, however, found by Smith and Chibnall to be present in the leaves of *Dactylis glomerata*.

REFERENCES

SPOEHR, H. A., *Photosynthesis*. New York, 1926.
STILES, W., *Photosynthesis*. London, 1925.

CHAPTER IX

THE WATER RELATIONS OF THE PLANT

THE importance of an adequate supply of water to a plant cannot be overestimated. Water comprises often 75 per cent. or more of the plant body, and it enters into the composition of practically every part of it. As previously pointed out, it is one of the chief constituents of protoplasm, while the growth of cells and tissues only takes place if these contain so much water that they are in a turgid condition. Many, certainly the vast majority, of the chemical processes taking place in the plant require an aqueous medium for their incidence and continuance, while for the all-important process of photosynthesis water is one of the reacting substances.

In the lower plants, including most algae and the simpler fungi, water is absorbed over the whole external surface of the plant, and the water relations of the plant are those of the water relations of the cell already discussed in an earlier chapter. Such is the case not only with unicellular forms such as *Chlamydomonas*, but with filamentous forms such as *Spirogyra* and *Pythium*, flat thalloid forms such as *Ulva* and more complex algae which are completely submerged. With development of life on land, however, came the differentiation of specialized organs for the absorption of water and substances dissolved in the water, organs which in the Pteridophyta and higher plants may be of great complexity. In some relatively few cases alternative, or supplementary, organs for water absorption, such as water-absorbing hairs on the aerial parts of the plant, have been developed. In terrestrial plants the excretion of water, either in liquid or gaseous form, appears to be as universal a phenomenon as the absorption of water. There is thus a more or less continuous stream of water through the plant, in the typical flowering plant water being absorbed by the roots and passing thence through the stem and branches to the leaves, where the greater part of it evaporates into the air. The water relations of the higher plant are therefore obviously complex, and we have not only a root system developed for absorbing water, but also a conducting system specialized for conveyance of the water to the place of its utilization or excretion. While it is necessary for a proper understanding of the water relations of the plant to regard them as involving a series of interlinked processes, we can conveniently analyse them into three constituent processes: absorption, conduction and excretion.

ROOT ABSORPTION

Although it is usual to speak of the root as the absorbing organ, the actual absorbing region is very much localized. The absorption is generally regarded as carried out by the root hairs which are only formed on a limited region of the root. The root apex itself is free from hairs; these latter appear on that part just behind the elongating region of the root (cf. Chapter XVI), and occupy, as a rule, only a few centimetres of the root. They soon wither and die, and are replaced by other hairs developing nearer the root apex. There is thus a slow, but continuous, march of the root hair region downwards, following the growth of the root and the consequent movement downwards of the root apex.

The root hairs are outgrowths of surface cells. Not only do they serve to increase very greatly the absorbing surface, but by their means a much greater volume of the surrounding medium is explored. The extent to which they increase the absorbing surface depends, however, on conditions. Generally speaking, there is little development of root hairs in an aqueous medium, while, on the other hand, there may be complete inhibition of root hair development in a very dry soil. Root hairs attain their maximum development in damp air, where they may increase the surface of the root from 5 to 19 times.

The extent of the root hair development may be gathered from the observations of Schwarz. The longest root hairs measured by Schwarz were those of *Trianea bogotensis,* roots of which in water were observed to bear some hairs as long as 8 mm. Root hairs 5 mm. and 4 mm. long were observed respectively in *Potamogeton* growing in water and *Elodea canadensis* growing in mud.

As already mentioned, damp air is generally a much more favourable medium for the development of the root hair than is soil, and Schwarz observed root hairs measuring 0·8, 2·5, 3·2 and 2·5 mm. in length in *Vicia faba, Pisum sativum, Tradescantia erecta* and *Avena sativa* respectively, when growing under favourable conditions in damp air. These values are very much higher than the mean lengths of root hairs of the same species growing in either dry or damp soil, as reference to the data in Table XLVIII will make clear.

The whole absorbing area of the root system of most plants must be very great. The root system of an annual plant such as sunflower or maize may have a total length of several miles, and while only a small part of this will bear functioning root hairs, the total area of actively absorbing surface must be very considerable.

The root hair itself consists, as mentioned above, of an outgrowing superficial cell of the root. It has a cellulose wall within which is the living protoplasmic lining surrounding the vacuole, and it is generally assumed that water is absorbed osmotically according to the ordinary laws of osmosis.

Table XLVII

Development of Root Hairs

Species	Condition of Root	Number of Root Hairs on 1 mm. of Root	Number of Hairs per sq. mm.	Length of Hair in mm.	Ratio of Area of Root with Hairs to that of Root without Hairs
Zea mais . . .	In damp air	1925	425	0·31	5·5
Pisum sativum . .	In damp air	1094	232	1·2	12·4
Scindapsus pinnatus	Air roots	4386	313	1·2	18·7
Trianea bogotensis .	In water	12·3	10·9	3·25	6·63

Table XLVIII

Influence of the Medium on the Development of Root Hairs

Species	Mean length of Root Hair in mm. in		
	very damp soil	less damp soil	dry soil
Vicia faba	0·24	0·28	0·52
Pisum sativum	0·33	0·50	0·68
Tradescantia erecta . . .	1·9–2·4	1·1–1·9	0·8–1·3
Avena sativa	1·1–1·9	0·8–1·3	0·5–1·1

Part of the water in the soil is held in some sort of combination with colloidal substances present in the soil, part of it is adsorbed to the surface of soil particles, while more of it may be held in the capillary spaces between the particles of soil. Thus different parts of the water in the soil are removed from it with varying degrees of difficulty, and at one time a distinction was drawn between available and non-available water, the latter being the water left in the soil when a plant growing in it commences to wilt. The available water contains mineral and sometimes organic substances dissolved in it, and forms what is known as the soil solution. While the concentration of this may depend to some extent on conditions, it is usually very dilute, the total concentration of dissolved substances, according to observations made on agricultural soils by a number of different observers being of the order of 0·05 to 0·2 per cent.

The condition for the absorption of water from the soil solution by the root hair is that the suction pressure of the latter should exceed the osmotic pressure of the soil solution. The latter being so dilute it is clear that the osmotic pressure of the soil solution is very much less than that of the root hair, for while the osmotic pressure of the former is usually less than an atmosphere, that of the root hair appears to be frequently of the order of 5 atmospheres or more. While, owing to the pressure of the cell-wall, the suction pressure of the root hair will be less than its osmotic pressure, we must suppose that actually a condition of dynamic equilibrium is maintained in which the suction pressure of the root hair is greater than that of the soil solution, with the result that water is continuously absorbed by the root hair.

EFFECT OF EXTERNAL CONDITIONS ON ROOT ABSORPTION

Various external conditions influence the absorption of water by the root. These conditions include the concentration of the external medium, the nature of the substances present in the medium, the temperature and the aeration of the medium, the last named greatly influencing the oxygen and carbon dioxide concentration.

Concentration of the External Medium. Since the absorption of water by the root depends on the difference between the suction pressure of the root hair and the osmotic pressure of the soil solution it is obvious that an increase in the concentration of the external medium, involving as it does an increase in its osmotic pressure, must reduce the rate at which water is absorbed. Such a reduction in the rate of absorption of water can be observed with plants growing in water culture if, say, 1 per cent. of sodium chloride or of potassium nitrate is added to the normal culture solution. It has been observed, at any rate in some cases, that the reduction in the rate of absorption is only temporary, and it would appear that plants may possess a power of adaptation which, by bringing about an increase in the osmotic concentration of the cell-sap, restores to the root its original capacity for absorption. Such a reduction in the rate of absorption, followed by increased absorption, was observed, for example, by Renner in water cultures of *Vicia faba*, as a result of adding potassium nitrate. It is not clear whether the subsequent increase in suction pressure of the cells is due to absorption of dissolved substances from the surrounding medium, or to the production inside the cell of osmotically active substances from solid or only slightly osmotically active substances.

The adaptation to increased concentration of the medium is limited, and on soils containing a solution of relatively high osmotic pressure, such as saline soils, many species are unable to grow. On soils with a high content of sodium chloride it is well known that a special flora develops, the members of which possess definite anatomical features, and which are capable of absorbing water from solutions of high concentrations. As is to be expected, the cells of such plants may possess high osmotic pressures, and in what appears to be an extreme case, Harris, Gortner, Hofman and Valentine in 1921 found the sap of plants of *Atriplex confertifolia* growing on salt flats possessed osmotic pressures of 74·2, 118·5 and 153·1 atmospheres. Halophytes in general appear to possess a power of rapid adjustment to changes in concentration of the external medium.

Many fungi also appear to be able to absorb water from a medium of high concentration, owing to the development of a high osmotic pressure within the plant. Thus Eschenhagen in 1889 found that *Aspergillus niger* and *Penicillium glaucum* when grown on a nutrient solution of high concentration could develop an osmotic pressure of 157 atmospheres, which appears to be the highest recorded for any plant cell or tissue.

Substances present in the External Medium. It has been held for many years that certain substances present in some soils may affect the absorption of water, on account of their chemical nature and quite apart from the osmotic pressure to which they may give rise. Thus bog and moorland soils were regarded by Schimper as 'physiologically dry' because, although they might contain abundance of water, the so-called humous acids present in such soils retarded water absorption. Livingston and others showed that these soils do contain toxic substances and Dachnowski regarded the soil toxins as adversely affecting water absorption. However, Montfort carried out a series of experiments which show that the water of bog soils exercises no retarding influence on the absorption of water, either by bog plants or non-bog plants. Montfort's method of experimentation consisted in keeping the aerial parts of the plants in a saturated atmosphere by covering them with a bell-jar, and measuring the rate of guttation, that is, the rate at which liquid water was exuded from the leaves. Under these conditions it may be assumed that the plant soon comes to hold its maximum quantity of water and that the rate of guttation is equal to the rate of water absorption. In this way Montfort showed that maize, when supplied with bog water, showed no reduction in rate of water absorption for many days, though after a time a reduction in rate of water absorption took place ascribable to toxic action on the roots by substances present in the bog water. Indeed, at first the rate of absorption of water actually increased. Similarly cotton grass, *Eriophorum vaginatum*, exhibited no decrease in rate of water absorption over a period of weeks.

It has also been suggested that the hydrogen-ion concentration of the soil influences the rate of water absorption, some workers holding that in an acid medium the rate of water absorption is increased, while others hold that acid has a reverse effect. Since acids are toxic to plant cells in extremely weak solutions it is, however, doubtful how these observations are to be interpreted.

Temperature. That the rate of water absorption by the root is influenced by temperature was recognized by Sachs in 1860, and has been shown by other workers since. Possibly the effect of temperature on absorption by the root is of the order of that found for water absorption and excretion by other types of plant cell (cf. Chapter III).

Aeration of the Soil. It is well known that poor aeration of the soil, involving a deficiency of oxygen as this is used up in respiration and an accumulation of carbon dioxide from this same process, will retard growth, and a feebly developed root system will result. But also poor aeration actually results in a reduced rate of water absorption. This has been shown by Kozarov by replacing the atmosphere round the roots by carbon dioxide or hydrogen, and by Livingston and Free in a similar way with the use of nitrogen. This effect may be due in part to the toxic effect of a high concentration of carbon dioxide, and in part to the reduction in respiration rate with increased concentration of carbon dioxide and decreased concentration of oxygen and perhaps also to the incidence of anaerobic respiration under such conditions. In any case it demonstrates that water absorption by the root is linked in some way with other plant processes, since factors which affect respiration thus influence it. But the problem of the connexion between root absorption and respiration still awaits analysis.

INFLUENCE OF INTERNAL FACTORS ON ROOT ABSORPTION

Internal conditions in the plant may also affect the rate of water absorption by the root. Thus, from what has been written above, it is clear that a rise in the osmotic pressure of the root cells will result in an increase in the rate of water absorption, owing to the increased suction pressure. But root absorption may also be affected by changes farther afield. Thus increased rates of utilization or excretion of water may also lead to increased rate of water absorption, while alterations in the channels by which water is conducted from the roots to the places of utilization and excretion may also affect the rate of water absorption

ROOT PRESSURE

If we imagine the cells of the root momentarily in a condition of equilibrium in which the cells all have the same suction pressure and therefore have no tendency either to absorb water from, or lose water to, neighbouring cells, it is obvious that absorption of water from the soil by the root hair will disturb this equilibrium, for the suction pressure of the root hair being lowered, the neighbouring cells will absorb water from them and thereby will suffer a lowering of suction pressure, with the result that water will pass farther into the root. So long as water is absorbed we should therefore imagine the cells of the root to exhibit normally a gradient of suction pressure, the suction pressure being lowest in the root hairs. But unless the water passes out of the root altogether this gradient of suction pressure would ultimately disappear, and the suction pressure of all the cells would finally equalize and come into equilibrium with the osmotic pressure of the soil solution. Now it can be shown in various ways, which will be mentioned later, that water actually passes into the xylem of the vascular cylinder, and if a plant growing, for example, in well-watered soil is decapitated just above the soil, water will exude from the cut surface of the stump, and it is generally assumed that the liquid exudes from the cut ends of the xylem tracheids and vessels. Not merely, then, is water absorbed into the root cells to equalize suction pressures of living cells, but actually a pressure is developed which tends to force or draw water into the non-living xylem elements, and such a pressure occurs even when the aerial parts of the plant are removed, and it is thus present when there is no suction on the root resulting from removal of water in aerial parts. The pressure under which water passes from the living cells of the root into the xylem is termed root pressure or exudation pressure. It can be demonstrated and measured by attaching a glass tube to the cut end of the stump of the decapitated plant, when water will rise in the tube; or by attaching a manometer, when the pressure will be registered. The magnitude of root pressure rarely exceeds one atmosphere and is generally much less, while in many plants it is not observable at all.

The existence of root pressure has often been regarded as difficult to explain, but such a pressure would naturally develop if the liquid in the xylem vessels and tracheids possessed an osmotic pressure greater than that of the soil solution surrounding the root hair. The suction pressures of the intermediate cells would adjust themselves automatically to produce a gradient between these two extremes. Since the soil solution has in general, as we have seen, a very low osmotic pressure, the concentration of solutes in the xylem elements need not be very high for the incidence of root pressure to be explained. Now Atkins found that the liquid in the xylem had a higher osmotic pressure than the soil solution, and although the osmotic pressure of the cells of the cortex was actually higher, this does not necessarily introduce any difficulty, since the actual capacity of the cells to absorb water depends, not on the osmotic pressure itself, but on the suction pressure, which is less than the osmotic pressure by the amount of the turgor pressure. The turgor pressure may undoubtedly be considerable, and will increase rapidly with water absorption, especially in a compact tissue, so that we should reasonably expect the actual suction pressures of the living cells of the root to be considerably less than their respective osmotic pressures. If then P_s is the osmotic pressure of the soil solution, P_x that of the liquid in the xylem tracheae P_a, P_b, . . . those of the successive living cells of the root through which the water passes and T_a, T_b, . . . the respective turgor pressures of these cells, then the condition for the development of root pressure is that

$$P_s < (P_a - T_a) < (P_b - T_b) \ldots < P_v$$

In order to demonstrate this point Priestley and Armstead made determinations of the solutes present in the liquid contained in the xylem of the vine, *Vitis vinifera*, and showed that the liquid contained 0·272 per cent. of dry matter, of which 0·136 per cent. was hexose, 0·066 per cent. inorganic matter and the remainder other organic matter. These values agree quite well with earlier determinations made by Dixon and Atkins of the liquid in the vessels of the sycamore. Comparing these numbers with those found for the soil solution, in which the concentration of solutes may be as low as 0·05 per cent., we see that the conditions for root pressure are likely to be fulfilled.

It may be noted that an osmotic pressure equal to that of 0·1 per cent. potassium nitrate solution will give rise to a root pressure of about 0·36 atmosphere. The observed differences in concentration of soil solution and liquid in the xylem tracheae are thus such as to give rise to root pressures of the order of those actually observed. In some cases, indeed, the exuding sap from the xylem may be more concentrated than in the cases quoted, and in some earlier investigations on *Acer platanoides* and *A. saccharinum* the observed concentrations of sugar were exceptionally high, that of sucrose in the latter species being found to be 3·57 per cent. It is, however, not clear that in these cases the liquid examined was derived exclusively from the xylem tracheae.

That the water which exudes from cut stems and roots comes from the water-conducting cells of the xylem has been disputed by James and Baker. These workers point out that an examination of the literature reveals that there is practically no experimental evidence to indicate that exudation of water is from the vessels, while their own experiments with *Acer pseudoplatanus* indicate that the sap which exudes from cut roots of this plant comes from living cells, particularly from those of the cambial region. They would themselves explain the passage of liquid out of cut roots and stems as resulting from the fact that the living cells are in a high state of turgor; the contents of the cells, that is, exert a considerable hydrostatic pressure. On cutting the stem or root, water is therefore forced out across the cut surface by this pressure. While this would explain a momentary exudation of sap it does not explain the maintenance of a root pressure for a considerable time such as has been observed in many cases. Whether, in actual fact, the water exudes from living or dead cells, the main difficulty in finding an explanation of root pressure is the same. In either case it would appear that there must be a mechanism by which osmotically active material is continually supplied to the exuding cells, so that a gradient of suction pressure is maintained and continued absorption of water by the root hairs made possible.

With the conditions under which root pressure is usually demonstrated, the passage of water through the root will, of course, stop when the hydrostatic pressure developed by the column of water above the cut end of the stump is equal to the exudation pressure. But if the liquid is allowed to flow away through the cut end of the stump, exudation will only continue if a supply of solutes to the xylem is maintained. The problem of root pressure therefore resolves itself into explaining the supply of solutes to the xylem. That this supply may continue for a long time is clear from the observations made by Priestley and Armstead on the vine. The roots of one large plant which had been decapitated a little above the soil level exuded on the average more than 500 c.c. of liquid per day for 36 days.

Atkins suggested that the sugars pass into the xylem tracheae simply by diffusion from living wood parenchyma cells bordering the tracheae, and Priestley and Armstead recorded that certain cells within the vascular cylinder have especially permeable protoplasts, a condition which they think may account for the presence of solutes in the xylem vessels. Having regard to the fact that the living cells of the xylem often contain a great deal of carbohydrate it is readily understandable that they may provide the solutes present in the vessels and tracheids.

It should be mentioned that Pfeffer regarded the difference in osmotic pressure between soil solution and liquid in the tracheae as insufficient to account for the active exudation of water such as is exhibited in the phenomenon of root pressure, and he concluded 'with comparative certainty' that even when the escaping liquid was rich in solutes the exudation pressure was yet due to the active excretion of water from living cells by means not yet known. Although he suggested several possible mechanisms, no evidence for these exists, and it seems doubtful whether Pfeffer was justified in his conclusions of the insufficiency of differences in osmotic pressure to explain root pressure.

THE ASCENT OF WATER THROUGH THE STEM

We can thus understand the absorption of water by the root hair and its passage through the cortex of the root into the xylem. The mechanism of its further transport from the root to the leaves is a problem which has given rise to considerable discussion and the formulation of many theories. It is clear that the rate of passage of water osmotically through the living cells of the cortex is too slow to account for the rate at which water is actually conveyed through the plant, and, indeed, it has been recognized for 200 years or more that the path taken by the water in its passage from root to leaves is the xylem of the vascular bundles. This was shown by experiments of two kinds carried out by Stephen Hales and others in the early part of the eighteenth century. By ringing experiments in which a ring of the outer tissues of the stem was removed, thus leaving a zone in which only tissues internal to the phloem were left, it was shown that the removal of these tissues had little effect on the movement of water, which must therefore take place in the xylem. The same conclusion was drawn from the second type of experiment in which the path taken by the water was demonstrated by supplying the roots or cut ends of stems with aqueous solutions of a dye. The xylem vessels and tracheids are stained by the dye, thus indicating that they are the water-conducting elements throughout the plant.

While the path of rapid water transport can thus be localized in the tracheae of the xylem, there is still the question whether the water travels in the cavities of these elements or in their walls. After all, the experiments with dye solutions result in the staining of the walls, and although this is no evidence that the water travels in the walls there is at least the possibility that it might do so. That the water is conveyed in the walls of the tracheae was the view warmly supported by Sachs, the movement of water up the stem being attributed to imbibition of water in the cell-walls and its movement through the walls to places where there is a saturation deficit, a consequence of evaporation from the surface of the leaves. The movement would thus be somewhat comparable to that of water through a vertical strip of blotting-paper supplied with water at its lower end and losing water by evaporation at the upper.

Attempts to disprove this imbibition hypothesis of Sachs were made by Elfving and Vesque, who showed that when the cut ends of twigs were placed in melted cocoa-butter which was then allowed to solidify and block the lumina of the tracheae, the movement of water through the plant was almost stopped. Similar experiments, with a like result, were made by Errera and Strasburger, gelatin being used instead of cocoa-butter. To these experiments the objection may be made that the cocoa-butter or gelatin may enter the walls and so alter their water-holding and water-conducting properties. To meet this objection Dixon carried out experiments with gelatin and paraffin wax of low melting-point. Different branches of lime (*Tilia europaea*) were placed with their cut ends in respectively (*a*) water at 50° C., (*b*) melted paraffin at 50° C., (*c*) gelatin coloured with Indian ink at 50° C. and (*d*) gelatin coloured with haematoxylin at 50° C. After transference to water at 13° C. the end of each branch was thinly pared and all were transferred to fresh water and left for 15½ hours. At the end of this time the untreated control was still quite fresh; on the branches treated with paraffin the leaves exhibited considerable wilting, while the leaves of the gelatin-treated branches had wilted to some extent though not so much as those treated with paraffin. The branches were next transferred to a strong solution of safranin where they remained for 1¼ hours, after which sections of the branches were cut at different levels. The examination of these sections showed that all the lumina of the xylem tracheae were blocked by paraffin in the branch treated with this material, while only portions of the lumina were blocked in the twigs treated with gelatin. It is important to note that in both cases the walls were stained with safranin. The conclusion to be drawn from this experiment is that when the cavities of the tracheae are blocked with paraffin or gelatin the passage of water up the stem is largely prevented, so that wilting takes place. Nevertheless the capacity of the walls for absorbing and transmitting water is not affected, at any rate not noticeably, as safranin, which travels in solution, is still able to pass through and stain the walls. It therefore appears that the channels through which water passes up the stem are the lumina of the xylem tracheae, and that the walls play at most only a negligible part in the transport of water.

While the movement of water has thus for long been generally recognized as taking place in the xylem, very diverse views have been put forward to account for the mechanism by which water can rise through the plant against the action of gravity. It is clear that root pressure is totally inadequate to account for the ascent of water to the heights to which it rises in many woody plants. Capillarity, the rise of liquids in narrow tubes, will account for the ascent of water in the vessels through a short distance, but is likewise insufficient to account for the rise of liquids through many feet. Reduced pressure in the upper part of the xylem tract, resulting from evaporation of water from the leaves, could not account for the rise of water to a height greater than that of the water barometer, namely, about 10 metres, and in many trees water is conveyed to heights greatly exceeding this. The theories devised to account for the ascent of water in plants are generally grouped as vital and physical theories according as they utilize, or dispense with, the activity of the living cells of the wood, in attempting to explain the rise of water. The chief of these theories will now be briefly reviewed.

Vital Theories. The principal theories of the ascent of water which invoke the activity of the living cells of the wood are those associated with the names of Westermaier, Godlewski and Janse. Westermaier, whose views were published in the years 1883 and 1884, thought that the passage of water upwards took place through the xylem parenchyma cells, the tracheae having the function of reservoirs in which the water is held by capillarity. This theory is, however, untenable because the passage of water through living cells could not take place at the observed rates of water movement up the stem, and the theory is, in general, unsupported by experimental evidence.

Godlewski, on the other hand, regarded the living cells in the wood as serving as pumps forcing the water up the tracheae with which they are in contact. To explain this action he supposed the living (medullary ray) cells of the wood to undergo periodic changes in osmotic pressure. Thus, if the osmotic pressure of the cells should increase, water will be drawn into them from the adjacent tracheae. If, after the cell has absorbed much water in this way, the osmotic pressure should fall owing to changes in chemical composition of the solutes in the cell-sap, water will pass out of the cell into the neighbouring tracheae. Godlewski supposed that the water which passed into the tracheae in this way rose in them to a certain extent, when it was drawn into a parenchymatous (medullary ray) cell at a higher level, the water rising because the pressure of air was supposed to be less at a higher than at a lower level. From the medullary ray cell the water would pass as already described into another trachea in which it would again rise. The water was thus supposed to take a staircase-like path up the stem. According to the theory, the actual pressure forcing water up the stem is a difference in air pressure. Now, as already pointed out, this cannot exceed one atmosphere, so that Godlewski's theory could not account for the water rising to a height of more than 10 metres.

Janse's theory of the ascent of water in stems has much in common with Godlewski's, but the former assumed a uni-directional pumping action of the medullary ray cells, that is, the water was always absorbed from a trachea on one side of the cell and given off to a trachea on the other side. To explain this action Jánse hypothesized a mechanism involving the transport of water in the cell by protoplasmic streaming. Apart from the fact that there is no evidence for such uni-directional pumping, such action could only be effective in raising water through the plant if, as Dixon has pointed out, the conducting tracts were at intervals completely interrupted by parenchymatous cells, which is not the case. Further, Dixon has shown that for water to pass upward at the rate of 7 cm. per hour, a rate observed by Ewart in stems of conifers, the rate of protoplasmic streaming in the parenchymatous cells required by Janse's theory would be more than 4 cm. per minute. No rate of protoplasmic streaming approaching this value has ever been observed.

The chief evidence in favour of the view that the living cells of the wood play a part in the ascent of water is that if the living cells of the lower part of a branch are killed, as, for example, by surrounding the branch with steam, the leaves on the upper part of the stem sooner or later wilt, thus indicating a cessation of, or considerable diminution in, the upward passage of water.

It has, however, been well argued by Dixon that this effect is not the result of the removal of the vital activity of the living cells of the wood, but is due to the blocking of the lumina of the tracheae by substances formed from the killed cells. When the living cells are killed they are rendered permeable, and solutes contained in them will diffuse into the water in the xylem tracheae. The high temperature will also bring about changes in cell proteins and other thermolabile substances. In any case the actual blocking of the cavities of the tracheae has been observed as a result of killing a part of the stem.

On the other hand the well-known experiment of Boucherie in 1840 provided a very important piece of evidence against all vital theories of the ascent of water. In this experiment a tree, cut across at its base, was there supplied with a poisonous solution. This was transported through the plant to the uppermost leaves, and would presumably kill any living cells on the way. Yet, when supplied with a second quantity of solution, this liquid was also carried to the top of the tree. This could scarcely happen if living cells were involved in the rise of water. In more recent years similar experiments have been performed with the same result by a number of different workers and there can be no doubt that stems are able to absorb water after treatment with a poison in the way described.

Physical Theories. It has already been noticed that capillarity, and reduced air pressure in the upper part of the plant, are both inadequate to account for the rise of water in plants, at any rate through tall trees. At one time it was thought that the rise of water might be accounted for by the liquid being in the form of a 'Jamin chain', that is, a chain of alternating air bubbles and water columns.

There appears, however, to be no justification for supposing that such chains are in any way effective in the way suggested, or that bubbles of air are present in the tracheae.

A number of possible ways in which water might ascend the xylem without the participation of living cells in the process were considered by Dixon and Joly, and they came to the conclusion that this rise of water could only be accounted for by its tensile strength or cohesion. As a result they propounded the 'cohesion theory' of the ascent of water. By cohesion is meant the mutual attraction of the molecules of a substance to one another, while adhesion refers to the attraction of the molecules of one substance to those of another. Thus it may be said that the molecules of water in the xylem tracheae cohere together, but adhere to the tracheal walls. According to the theory of Dixon and Joly, the water molecules in contact with the walls of the tracheae adhere to these walls, while the rest of the water coheres to the film of water adhering to the walls. Owing to the adhesion of the water to the walls so that it is not free to change its shape, and to its cohesion, the columns of liquid in the xylem behave to some extent like solid rods. We may, with advantage, at this point consider the forces operating on the water in the conducting tracts. As we shall see later, evaporation of water from the surface of the leaves results in the water in the xylem tracts in the leaves, that is, the water at the upper ends of the columns, being subjected to an osmotic pressure tending to pull water out of the xylem at these places. At the lower end of the plant the comparatively feeble root pressure, if present, is tending to send water up the stem. The pull on the water from above is likely to be greater than the push from below, and consequently the water will be in a state of tension, so long as the molecules of water cohere together and so long as the water cannot change its shape.

Measurements of the cohesive force, or tensile strength, of water show that it is very great. Dixon's own observations gave a minimum value of 150 atmospheres for the cohesion and adhesion to glass of water containing air, which means that a pressure of more than this magnitude would have to be applied against the cohesive force before the molecules of water would separate. Later determinations of Renner and Ursprung have yielded even higher values, up to 350 atmospheres. Observations on the tensile strength of xylem sap adhering to plant tissues indicate that the cohesion here is still greater.

We may therefore suppose, on the cohesion theory, that the ascent of water is brought about in the following way. Transpiration of water from the surface of the mesophyll cells will result in a continuous pull on the water in the xylem tracheae owing to the increase in osmotic pressure of the mesophyll cells resulting from loss of water. This pull on the water, amounting to a few atmospheres, is quite insufficient to break the water columns in the tracheae owing to the great cohesion of the water. The pull is thus transmitted to the whole of the water and this will be drawn through the plant much as a solid bar is drawn through, because the molecules cohere together.

On the theory of Dixon and Joly involving, as it does, the maintenance of a complete column of liquid throughout the xylem, a break in the water column by a layer of air would entail the cessation of movement upwards, at any rate above a height which would be determined by the pressure of air in the xylem, and which in any case could not exceed 10 metres. At one time it was thought that the water in the xylem contained small air bubbles and Dixon showed that where these were below a certain size they would not interfere with the upward movement of water, while above it they would expand, becoming rarefied, and might, by rendering the elements containing them functionless from the point of view of water transport, reduce the cross-section of the effective conducting tract. More recent work, however, by Holle and Bode indicates that air is absent from the conducting tracts, and they state that even in completely wilted plants the conducting tracts are free from air and contain unbroken columns of water. Only in the older parts of the wood, the elements of which have ceased to conduct water, or in younger wood which has suffered injury, does air occur. The observations afford striking support for the cohesion theory. It must, however, be mentioned that Haines has recently claimed to have demonstrated the presence of air in the xylem of a number of woody plants.

The rate at which water can move through the xylem can be measured in cut pieces of stem by forcing the water through these under pressure. Some measurements made by Dixon on the wood of *Taxus baccata* show that within limits the velocity of movement of water through the wood is proportional to the pressure, and that to obtain a velocity of water movement equal to that actually determined for living plants from observations of transpiration rate, a pressure equal to that of a water column twice the length of the stem is required. From what has already been stated it is obvious that this pressure is negligible in comparison with that of cohesion. Hence there is no question of the forces necessary to raise the water breaking the water columns.

Since the structure of the wood differs considerably in different species it is to be expected that the resistance offered by friction to the passage of water through the wood will vary from species to species, and that the efficiency of the wood for the conduction of water will thus vary with the species. An investigation of this question in about 60 species of woody plants was made by Farmer in 1918. The method adopted was to measure the quantity of water which passed through a standard length (15 cm.) of stem under a pressure of 30 cm. of mercury in 15 minutes. This quantity divided by the area of cross-section of the conducting tract gives a value which can be taken as a measure of the efficiency of the wood for water conduction and was called by Farmer the 'specific conductivity' of the wood in question. While there is a certain range in the value of the conductivity in the wood of a single species, it is clear that there are wide differences between the average conductivities of the wood of different species. The specific conductivity of evergreens is relatively low while that of deciduous plants is relatively high. Some of Farmer's results are summarized in Table XLIX. These numbers indicate the normal range of specific conductivity of the wood of the various species tabulated, numbers regarded as outside the normal range being excluded.

Table XLIX

Specific Conductivity of Wood of Different Species

Species	Range of Specific Conductivity
Podocarpus milanjiana	9 ± 2
Taxus baccata	12 ± 2
Ligustrum vulgare	37 ± 8
Prunus laurocerasus	10 ± 2
Hedera helix (juvenile)	25 ± 5
,, ,, (adult)	60 ± 10
Ruscus aculeatus	1 ± 0·3
Quercus ilex	32 ± 12
Quercus robur	75 ± 15
Fagus sylvatica	65 ± 10
Acer campestris	45 ± 15
Euonymus japonicus	12 ± 6
Euonymus europaeus	47 ± 8
Pyrus malus (var. Cox's Orange Pippin)	45 ± 15

It has been stated that the pull on the water in the xylem is exerted by the suction pressure of the mesophyll cells of the leaf, this pressure being maintained by evaporation from the leaf which prevents the cells from becoming fully turgid. Since the water has to be raised a greater distance the higher the insertion of the leaf, it might be expected that the higher the leaf above the ground the greater the osmotic pressure in the leaf cells. A number of workers have sought to show this relation between osmotic pressure and height above the ground, but published results of different observers are inconsistent enough to leave the matter in some doubt. The results of Harris, Gortner and Lawrence, however, seem quite definite. These workers determined the osmotic pressure of the sap expressed from leaves of 12 species of trees, and found an increase in osmotic pressure with height of the leaf insertion above the ground. A selection of their results is shown in Table L.

Harris, Lawrence and Gortner also found that the osmotic pressure of the sap of different species of plants of Arizona deserts varied in general according to the habit of the plant, the groups in order of increasing osmotic pressure of sap being (1) winter annuals, (2) perennial shrubs, (3) dwarf shrubs and half-shrubs, (4) shrubs and trees.

And again, Harris and Lawrence found that the osmotic pressure of the leaves of woody plants of the Blue Mountains of Jamaica was higher than that of the leaves of herbaceous plants of the same region, while Harris, Gortner and Lawrence found the same relation with plants of the Arizona deserts and of the north shore of Long Island. These observations support in general the idea that the farther the leaf from the ground the higher the osmotic pressure.

Table L

Osmotic Pressure of the Sap of Leaves inserted at Different Heights above the Ground

Species	Height in Feet	Osmotic Pressure in Atmospheres
Betula lutea	66	15·55
	52	16·01
	39	15·12
	25	14·11
	11	12·63
Juglans cinerea	52	18·31
	44	18·33
	38	17·18
	32	18·19
	21	17·85
	8	16·81
Quercus prinus	47	20·23
	36	20·08
	30	19·72
	19	19·57

TRANSPIRATION

It has already been noted that the pull of water through the plant results from the osmotic pressure of the leaf mesophyll cells which thus exert a suction on the water in the conducting elements of the vascular bundles of the leaf. The continued ascent of water accordingly depends on the maintenance of this suction by the leaf cells, and this will itself depend on the continuance of water loss by the leaf cells. By the continued evaporation of water, either through the cuticle or into the intercellular spaces of the mesophyll and so into the outer air through the stomata, a gradient of suction pressure will be set up in the leaf between the surface and the xylem tracheae similar, but in the reverse direction, to that between the absorbing cells and xylem tracheae in the root. The more rapid the evaporation, the greater will be the suction exerted on the water in the xylem, and so, if the rate of root absorption permits, the greater the rate at which water will be drawn through the plant. If root absorption cannot keep pace with the rate of evaporation the content of water of the leaf cells must fall and the leaves will eventually become flaccid and wilt, while the rate of evaporation itself must fall.

For the evaporation of water from the aerial parts of plants the term transpiration is used, and while some writers on the subject have emphasized the essential identity of evaporation and transpiration, others have stressed the necessity for distinguishing transpiration from simple evaporation owing to the complex structure in the former case of the body losing water, and the consequent influence of internal factors of the plant on the rate of water loss.

The importance of a supply of water to plants has already been noted, and it is generally held that transpiration is necessary in order to provide a sufficiency of nutrient salts to the plant and to distribute them more rapidly through the plant than would be the case if they diffused through living cells only, and it has been generally assumed that mineral salts absorbed by the root hairs are carried thus passively in what is called the transpiration stream. There is, however, evidence that the absorption of water and of mineral salts are largely independent processes. It has also been urged that transpiration prevents a plant from overheating when subjected to strong insolation, for much of the solar energy incident on the leaf under such conditions is utilized in the evaporation of water, but for which the temperature of the leaf would rise. It is doubtful, however, whether much importance can be attached to this consideration, for some desert plants exposed to very strong insolation are provided with devices which reduce the rate of transpiration to very small proportions, and while, in such plants, the temperature may rise some degrees above that of the surroundings, they suffer no ill effects from this. All the same, the temperature of wilted leaves in sunlight may be considerably higher than that of similar turgid leaves under the same conditions of insolation. But in any case it would appear that for the land plant transpiration is inevitable. Both for purposes of photosynthesis and respiration a structure making possible the interchange of gases between the plant and the atmosphere is necessary, and such a structure must involve the evaporation of water from the plant into the atmosphere. The actual amounts of water transpired from most plants are very large, having regard to the size of the plants, and appear to be very much larger than would be required for supplying the plants with mineral salts. Further, we find a multiplicity of devices throughout aerial plants for reducing transpiration, especially in the plants, known as xerophytes, exposed to conditions in which the rate of absorption of water may be low or that of evaporation high.

In many climates transpiration may be a problem, not only of scientific interest, but also of economic importance, for the development of crop, as of other, plants may be largely determined by the rate of water loss at different stages of development.

The measurement of transpiration can be made in various ways, involving (*a*) the determination of the vapour given off from the plant, (*b*) the loss of weight of the plant under conditions in which absorption of water is prevented, or (*c*) the rate at which water is absorbed by the plant. For determination of the rate of water vapour evolution the usual method is to use cobalt chloride (or nitrate) paper which is blue when dry and pink when damp. The time taken for the standard paper, when placed on the leaf, to change from its initial colour to a standard tint gives a value inversely proportional to the rate of water vapour evolution. This method, first introduced by Stahl in 1894, has been refined by American workers, and has been successfully used for quantitative determinations of rates of transpiration. A comparable principle is used in the horn hygroscope of F. Darwin. In this instrument the vapour given off from the leaf causes a strip of hygroscopic horn to curl, and the time taken for a definite degree of movement of the strip of horn gives a value inversely proportional to the rate of transpiration. The chief objection to these methods lies in the fact that placing either paper or horn on the leaf may itself considerably affect the rate of transpiration from the part of the leaf actually examined, as the external conditions are thereby greatly altered.

The measurement of transpiration by determining the loss of water from the transpiring object can be used with potted plants, plants in water culture, cut branches or even single leaves. With potted plants it is necessary to prevent evaporation of water from the soil by the use of a covering impermeable to water, while in the other cases, where the material is provided with a reservoir of water, it is necessary that no water should be lost by direct evaporation from the water itself. This method has been much used with potted plants and, where practicable, is probably the most satisfactory for determining the rate of transpiration of whole plants. It is, of course, not applicable to plants growing in the field. Also, it is necessary to bear in mind that in experiments of long duration, unless water is added to the soil from time to time, the water content of the soil will fall and may greatly influence the rate of water absorption and so limit the rate of transpiration.

The measurement of the rate of *water absorption* is carried out by means of the well-known instrument called the potometer, in which the absorption of water by a plant in water culture, a cut branch or a single leaf, is observed by the rate at which a bubble of air or a length of coloured liquid is drawn through a capillary tube filled with water and connecting the vessel holding the transpiring material with the water reservoir. A very simple form of potometer was used by Lloyd, consisting simply of a graduated tube forming a water reservoir connected to the transpiring branch of a plant by rubber tubing. The great disadvantage of potometric methods of measuring transpiration is that they assume the rate of absorption to equal the rate of evaporation, which quite definitely is not necessarily the case.

It will thus be clear that all methods of measuring transpiration rate have certain disadvantages, and while those depending on the determination, by weighing, of the water loss from the material are the most satisfactory, the choice of a method must depend on the type of plant material to be investigated and the particular problem to be examined.

THE PHYSICAL LAWS OF EVAPORATION

The rate of evaporation of water from a water surface of unlimited extent is proportional to the area of the surface considered, to the gradient of vapour pressure of water over the surface, and inversely to the atmospheric pressure. This relation can be expressed by the equation

$$\frac{dE}{dt} = \frac{kA}{P} \cdot \frac{dv}{ds}$$

where $\frac{dE}{dt}$ is the rate of evaporation, A the area of the surface, P the atmospheric pressure and $\frac{dv}{ds}$ the gradient of water vapour pressure. It is more usually expressed in the form, due to Dalton,

$$E = \frac{kA(V - v)760}{P}$$

where E is the rate of evaporation, V the saturation pressure of water vapour at the temperature of the evaporating surface, v the pressure of water vapour in the surrounding space, P the atmospheric pressure in mm. of mercury, while the constant k is called the evaporation coefficient.

Inspection of these formulae will make it clear that the rate of evaporation will increase both with rise in temperature and with movement of air over the surface. With rise in temperature the value of the saturation water vapour pressure, V, increases, while movement of air over the surface, by removing the evaporated water vapour, lowers the value of v. Hence in either case the value of $V - v$ increases, or, according to the mode of expression in the first formula, the diffusion gradient $\frac{dv}{ds}$ is increased.

The relation given above refers to a water surface of unlimited extent over which evaporation proceeds uniformly at right angles to the surface. With a relatively small area we may expect the rate of evaporation per unit area to be greater at the margins than in the middle part of the surface, for at the margin the vapour can diffuse not only vertically outwards from the surface, but also laterally as well. The position is, in fact, similar to that which we have already considered in dealing with photosynthesis where it was pointed out how marginal diffusion becomes relatively more important the smaller the aperture, so that with very small apertures the rate of diffusion is proportional to the linear dimensions of the aperture and not to its area. The same considerations obviously hold in regard to evaporation, so that with a very small surface the rate of evaporation into still air is proportional to the linear dimensions of the surface. Stefan's law for evaporation from a small circular area expresses this relation thus:

$$\frac{dE}{dt} = 4kr \log \frac{P - v}{P - V}$$

where r is the radius of the surface, and the other symbols have the meaning already assigned to them. If V is small in comparison with P the equation approximates to

$$\frac{dE}{dt} = 4kr.\frac{V - v}{P}$$

When, however, the air outside the surface is in motion, Stefan's law no longer holds. With moving air it has been shown from actual experimental observations by Thomas and Ferguson, and mathematically by Jeffreys, that for surfaces of so-called 'medium dimensions', the actual size depending on the air velocity, the rate of evaporation from a plane surface is proportional to $r^{1\cdot5}$ rather than to either r or r^2, where r is proportional to the linear dimensions of the surface. Thus for the simplest case of a circular surface, Jeffreys found the following relation

$$\frac{dE}{dt} = 3{\cdot}95\rho V_0(kur^3)^{\frac{1}{2}}$$

where ρ is the density of the air, V_0 the fraction of this due to water vapour, u the velocity of air blowing over the surface in a direction parallel with it, k the *effective* coefficient of diffusion of the water vapour, r the radius of the surface and $\frac{dE}{dt}$, as before, the rate of evaporation. It is assumed that r is large compared with $\frac{k}{u}$. The two important facts relating to evaporation indicated by this formula are that the rate of evaporation is proportional to (1) the square root of the velocity of the air, and (2) the cube of the square root of the radius of the surface. As regards the interpretation of the term 'medium dimensions', if the velocity of air movement is 400 cm. per second and k is 1,000 sq. cm. per second, the given relation holds for surfaces possessing a radius between 10 cm. and 250 metres. In a place in which the air movement is slight, the radius of the surface must be between 1 cm. and 25 metres. We may thus expect the relation here given to be approximately correct for areas of the size of leaves or collections of leaves (leaf mosaics).

The ratio of the rate of evaporation in a wind to that in still air will therefore depend not only on the wind velocity but also on the dimensions of the evaporating surface. Thus, the observation of Houdaille, quoted by Maximov, that the rate of evaporation in a wind with a velocity of 0·25 metres a second is three times that in still air, can only refer to a surface of one particular size and shape.

Cuticular and Stomatal Transpiration

In the cases of the leaves of mosses and a few other land plants the evaporating surface is approximately uniform over its whole area. In the vast majority of land plants, however, the problem of evaporation is complicated by the presence of stomata. While loss of water may take place through the cuticle, there seems little doubt, especially where the cuticle is well developed, that most of the water transpired by leaves is given off into the outer air through the stomata. Here the actual evaporating surfaces are the walls of the mesophyll cells bordering the intercellular spaces, and hence a gradient of vapour pressure will exist between the surface of these cells and the stoma.

The relative amounts of water transpired through the cuticle and stomata vary not only from species to species, but may be very different in the case of the same leaf at different times, owing to influences of external and internal conditions on the size of the stomatal opening. As is well known, the distribution of stomata may be very different on the two sides of the same leaf, and while it is most common for the upper surface of the leaf to have few, or even no stomata, whereas the lower surface possesses them in abundance, such a state of affairs cannot be regarded as sufficiently general to be called typical. Some observations of Garreau made more than 85 years ago, may be cited as indicating the relative amounts of cuticular and stomatal transpiration in some different types of leaves (Table LI). It will be seen that in ivy absence of stomata on the upper surface prevented all transpiration, whereas in the lime quite a considerable amount of water was lost through the upper surface although this was devoid of stomata. This difference in behaviour must be related to the development of the cuticle which is thick in the ivy but much thinner in the lime, and there can be little doubt that the physical and chemical characteristics of the cuticle are intimately related to the rate of transpiration through it.

Table LI

Transpiration from Upper and Lower Surfaces of Leaves

Species	Relative Numbers of Stomata, upper surface : lower surface	Relative Transpiration Rates, upper surface : lower surface
Atropa belladonna	10 : 55	48 : 60
Hedera helix	0 : 90	0 : 4
Syringa vulgaris	100 : 150	30 : 60
Tilia europaea	0 : 60	20 : 49

In the birch Stålfelt estimated that when the stomatal width is only 1 μ transpiration through the cuticle on both sides of the leaf only accounts for 8·6 per cent. of the total transpiration.

The question of diffusion of gases through the stomata has already been discussed in reference to the entrance of carbon dioxide into the leaf in photosynthesis. The evaporation of water vapour through the stomata must obey the laws of diffusion through small pores and multiperforate septa already enunciated. Thus, in still air diffusion through a single pore is proportional to the radius of the pore, while through a multiperforate septum, such as we may regard the epidermis of a leaf with scattered stomata to be, the evaporation of water may be practically as rapid as from a wet leaf, provided the distance between the stomata bears the right relation to their size. Thus, although the total area of stomatal apertures may be little more than 1·2 per cent. of the whole leaf area, evaporation may take place approximately as rapidly through the stomata as if the whole surface of the leaf were a wet evaporating surface.

In this connexion it is interesting to note some data regarding the distribution and size of stomata. In Table LII are shown the average numbers of stomata per unit area recorded by various observers.

Table LII

Species	Number of Stomata per sq. mm.		Observer
	Upper Surface	Under Surface	
Acer campestris	0–1	270	Salisbury
Acer pseudoplatanus	0	400	Weiss
Populus alba	0	315	Czech
Populus nigra	20	115	Salisbury
Quercus robur	0	450	,,
Quercus sessiliflora	0	400	,,
Taxus baccata	0	115	,,
Tilia vulgaris	0	130	,,
Corylus avellana	0	225	,,
Cytisus scoparius	90	90	,,
Hedera helix	0	125	,,
Ligustrum vulgare	0	300	,,
Prunus cerasus	0	216	,,
Apium graveolens	0	175	Lloyd
Arum maculatum	45	76	Salisbury
Carum petroselinum	0	150	Lloyd
Ficaria verna	21	75	Salisbury
Fouquieria splendens	160	160	Lloyd
Oxalis acetosella	0	45	Salisbury
Scilla nutans	55	51	,,
Verbena ciliata	100–175	100–175	Lloyd
Helianthus annuus	71	172	Muenscher
Zea mais	60	101	,,
Triticum sativum	21	46	,,
Hakea rugosa	70	70	Wood
Loranthus pendulus	140	140	,,
Billardiera scandens	124	0	,,
Pultenaea daphnoides	0	212	,,
Pultenaea floribunda	200	0	,,
Kochia triptera	14	14	,,

Salisbury has found, however, that the number of stomata per unit area varies not only with different leaves of the same plant, but on different parts of the same leaf, in general the frequency of the stomata increasing from the base of the leaf to its apex, and from the midrib to the margin. In dry conditions, and where the leaves are not too crowded, the stomatal frequency tends to be lower the larger the leaf, while the frequency tends to increase with height of the leaf from the ground. In general, stomatal frequency appears to run parallel with osmotic pressure, and, presumably, the suction pressure of the mesophyll cells in the neighbourhood of the stomata. Salisbury's observations were made on 25 trees, 29 shrubs and 150 'ground flora' species, and from these he concludes that stomatal frequencies tend to be higher for trees and shrubs than for herbaceous plants. This, again, may be compared with the observations already noted, that the osmotic pressure of the leaf cells of shrubs and trees is higher than that of herbaceous plants. Further, the drier the conditions under which the plant grows, the more frequent the stomata; indeed, from comparison of plants grown in dry and moist air Salisbury concludes that stomatal frequency is largely determined by the humidity of the environment. This conclusion may also be compared with the observation that osmotic pressure of the leaf cells of land plants tends to be higher the drier the conditions of life. Also Pisek and Cartellieri noted a greater stomatal frequency in sun plants than in shade plants of the same species. If from these observations we are to conclude that the higher osmotic pressure in leaves with more frequent stomata is due to the more rapid evaporation on account of the greater area through which transpiration can take place, we are faced with a difficulty, for we have already noted that according to Jeffreys' calculations the stomatal area can be reduced very much before any reduction of transpiration is to be expected.

Variation in Size of Stomata

The size of the stomata varies from species to species, while on the same plant, and even on the same leaf, variations in size of the stomata may occur. In Table LIII are shown some data obtained by Muenscher relating to the average size of stomatal aperture, presumably when the stomata are fully open. It will be observed that the variation is chiefly in the length of the pore, the width being approximately constant.

Table LIII

Size of Stomatal Aperture

Species	Upper Surface		Lower Surface	
	Length of Pore in micra	Width of Pore in micra	Length of Pore in micra	Width of Pore in micra
Helianthus annuus	19	4	14	3
Impatiens sultani	9	3	6	3
Phaseolus vulgaris	8	3	11	3
Ricinus communis	10	3	10	3
Triticum sativum	34	3	39	3
Zea mais	19	3	19	3

Stomatal Movement

By the opening and closing of stomata it has been supposed that the rate of transpiration from a leaf may be altered and regulated. These stomatal movements, as is well known, result from changes in shape of the guard cells owing to different regions of their walls possessing different elasticities. From the point of view of physiological anatomy stomata fall into several groups with regard to the mechanism of opening and closing, and descriptions of the various types are given in Haberlandt's *Physiological Plant Anatomy*. The important point for the plant physiologist is that the movements are brought about by turgor changes, increase in turgor resulting in the opening of the stomata, and a decrease in turgor resulting in their closure.

The early observation that the guard cells of the stomata contain chloroplasts while the ordinary epidermal cells are devoid of chlorophyll led to the view that changes in turgor in the guard cells are brought about by photosynthetic activity which increases the concentration of sugar and hence the osmotic pressure of these cells. Such a view was supported by the observation that in many plants the stomata open in the light and close in the dark.

There is no doubt that the changes in the degree of opening of the guard cells are related to changes in osmotic pressure. Iljin, indeed, in 1914, found that whereas the osmotic pressure of the guard cells when the stomata are closed is about the same as that of the neighbouring cells, when the stomata are open the osmotic pressure of the guard cells may be extraordinarily high, namely, from 45 to 90 atmospheres or 10 times as high as when the stomata are closed. Other authors have confirmed the considerable increase in osmotic pressure of the guard cells when the stomata open, although generally the observed increase is not nearly so great as that recorded by Iljin, being more generally in the neighbourhood of from 10 to 20 atmospheres.

That this increase in osmotic pressure is related directly to the photosynthesis of sugar is not now thought to be the case. Rather it is regarded as resulting from the formation of sugar by the hydrolysis of starch which is always present in the guard cells. As the stomata open the starch content decreases, and vice versa. Loftfield considers that light induces stomatal opening by initiating the conversion of starch in the guard cells to sugar. The rise in osmotic pressure conditions a flow of water into the guard cells which thus become more turgid and the stomata open. It has been suggested, and some evidence has been adduced for the view, that the concentration of hydrogen and other ions, by influencing diastatic action, can influence stomatal opening and closure.

Some workers as, for example, Scarth, consider that the rapidity with which stomatal movement takes place is too great for the process to be due entirely to the change of starch to sugar or the reverse.

Scarth suggested that closure of stomata in the dark might in part be attributed to the separation of water from the cell colloids and its loss to surrounding cells, as a result of the cell becoming more acid on account of the accumulation of respiratory carbon dioxide, while opening in light could result from the removal of this carbon dioxide in photosynthesis with increase in the imbibitional capacity of the cell colloids as the hydrogen-ion concentration decreases. These changes are, however, always accompanied or followed by hydrolysis or synthesis of starch, and the main point in Scarth's theory is that the carbohydrate balance depends strictly on the hydrogen-ion concentration which is chiefly determined by photosynthesis and respiration.

Measurement of Stomatal Aperture

Before the relationship of degree of stomatal opening to transpiration can be properly examined, or even the factors influencing stomatal opening and closing, it is necessary to have a method of measuring accurately the stomatal aperture. Four methods have been used for determining this. The method simplest in principle is to measure the aperture by microscopical observation of the living leaf. The other three methods require a few words of explanation.

A method due to Lloyd consists in stripping pieces of epidermis from the leaf and rapidly immersing them in absolute alcohol. This reagent dehydrates the cell-wall and hardens it so that it immediately becomes rigid; hence the guard cells retain their shape, and the stoma its degree of opening. Although the method seems drastic it is regarded as highly reliable by Loftfield, who checked the method by comparative observations on stomata so fixed and similar stomata remaining on the living leaf, and who also described various precautions which should be observed in using the method.

In the method developed by F. Darwin, the size of the stomatal aperture is estimated by drawing air through the leaf under a known pressure and observing the rate at which the air is thus drawn through. This rate depends on the size of the stomatal aperture and will be more rapid the wider the stomata are open. The instrument originally described by Darwin and Pertz was called by them the *porometer*; various forms of it have since been devised. The chief objection levelled against the porometer is that it involves unnatural conditions inasmuch as a current of air is pulled through the intercellular spaces of the leaf. Loftfield adds to this another objection to the effect that there is a doubt as to what porometer readings actually represent, while Maximov points out that with contraction of the intercellular spaces, which may be considerable during wilting, the rate at which air is drawn through the leaf under constant pressure will change independently of change in stomatal aperture. Indeed, Maximov regards the porometer as incapable of yielding exact data regarding the degree of stomatal opening. Later work by Nius indicates that, as an instrument for measuring stomatal aperture, the porometer is even more unsatisfactory than was formerly supposed, for he shows that during the course of 24 hours considerable variations can occur in the volume of the intercellular spaces of the leaf, and that the size of these has a much greater influence on the readings of the porometer than has stomatal aperture. As an instrument for measuring with precision the degree of stomatal opening, the porometer must be regarded with great suspicion.

An injection method for determining whether the stomata are open or closed (or nearly so) described by Molisch in 1912, consists in placing a drop of absolute alcohol or some other liquid on the surface of the leaf. If the stomata are open the liquid passes through the stomatal pores into the intercellular spaces which then appear as dark areas when the leaf is viewed by reflected light. If, however, the stomatal opening is below a certain size, the alcohol does not enter. The method can thus be used to determine whether the stomatal openings are greater or less than a certain size. Moreover, as this limit of size depends on the liquid used it is possible, by using a number of different liquids, to determine approximately the degree of opening of the stomata.

The Effect of Various Conditions on Stomatal Movement

Light. It has already been noticed that light in general induces stomatal opening while darkness leads to closing, so that frequently the stomata are open during the daylight hours and closed at night. This rule is, however, by no means general, and von Faber, indeed, in 1915 observed that in a number of species of plants of tropical rain forests the stomata remained open at night, while Stahl in 1919 observed similar behaviour in 25 out of 60 Central European species he examined. Molisch observed the stomata of *Cucumis sativus* and *Ranunculus acris* to be wide open at 1 a.m. in August, while those of a number of other species such as *Crambe maritima, Polygonum hydropiper* and *Scrophularia nodosa* were partly open. In two-thirds of the species he examined the stomata were then closed or nearly so

It has already been noticed that the effect of light in inducing stomatal opening was regarded by Loftfield as due to the initiation of enzyme action which brought about the hydrolysis of starch. Scarth, as we have seen, attributed the effect to changes in hydrogen-ion concentration resulting from photosynthesis and respiration, and the opening of some stomata in the dark he explained as ' acid ' opening, due to accumulation of carbon dioxide and corresponding high acidity. It would therefore appear that closure of stomata corresponds with a range of hydrogen-ion concentrations outside which, on either side, opening results.

Water Content. The water content of the leaf also affects the degree of stomatal opening, but here again many exceptions to the rule have been observed. Thus Molisch found that during wilting the stomata in about 40 per cent. of the species he examined remained open. These included *Cucumis sativus*, *Tropaeolum majus* and *Syringa vulgaris*. Many more cases were recorded by Burgenstein. The question is undoubtedly complex, for Laidlaw and Knight found that at the beginning of wilting the stomatal aperture may increase in size. Molisch and Burgenstein have further observed that after stomata have closed during wilting a subsequent opening may take place. It has been suggested that this is a post-mortal effect resulting from contraction of the dying tissues.

Iljin has recently shown that the opening of stomata during wilting is related to the starch-sugar equilibrium in the guard cells. Absorption of water by these cells leads to transformation of starch into sugar, while moderate withdrawal of water leads to the reverse action. If, however, the withdrawal of water is considerable, as in wilting, the starch disappears owing to the re-formation of sugar from it; consequently the osmotic pressure of the guard cells increases, water is absorbed by them and the turgor increases and the guard cells open.

According to Scarth's theory, loss of water, like light, will affect the hydrogen-ion concentration of the guard cells, and it may do this by reducing or inhibiting photosynthesis.

Temperature. Loftfield found that within limits the rate of stomatal opening in lucerne is increased by rise of temperature, the rate being approximately doubled for a rise of 10° C. between 0° C. and 30° C., so that the Van't Hoff rule appears to be obeyed. But in other species low temperatures tend to induce closure and high temperatures opening. Thus Scarth found in *Zebrina pendula* that temperatures between 0° C. and 8° C. prevent stomatal opening in light, whereas temperatures between 38° C. and 40° C. tend, in darkness, to bring about stomatal opening, or to prevent closure. In other cases, however, the behaviour is variable, while Kostychev and his co-workers observed that the stomata of plants on the shores of the Arctic Ocean remain wide open throughout the day, although the temperature during the 'night' may be as low as 4° C. There are yet other cases in which stomata close during the hottest period of the day.

Daily Periodicity of Stomatal Movement

From what has been stated above it would appear to be usual, though not universal, for stomatal movement to exhibit a daily rhythm characterized by an open condition of the stomata during the day and closure at night, while variations in the degree of opening might be expected with changes in light intensity during the daylight hours. Something approaching this behaviour was found by Lloyd in *Verbena ciliata*, as the curve in Fig. 26 clearly indicates. It will be observed, however, that on a summer day the stomatal aperture slowly increases during the early morning hours, the maximum degree of opening being reached by about 8 a.m. This condition is maintained only until about 1 p.m. after which the stomata gradually close, the minimum opening being reached at about 6 p.m. The course of stomatal movement in *Fouquieria splendens* was found to be similar, although the amplitude of movement was somewhat less.

The daily course of stomatal movement was examined by Loftfield in a large number of species. As a result of his observations he classified the stomata of these species into three groups typified by (1) barley, (2) lucerne and (3) potato. The barley or cereal type is characterized by stomata of peculiar structure and great sensitiveness. They never open at night and may only be open for a few hours during the day, though considerable variation is shown in respect to daylight opening. Thus, whereas the course of stomatal opening in wheat has been found to resemble that of the species examined by Lloyd, except that the stomata close completely at night, the stomata of barley close in the afternoon and on a hot dry day may only open partially for one or two hours in the early morning, and Loftfield

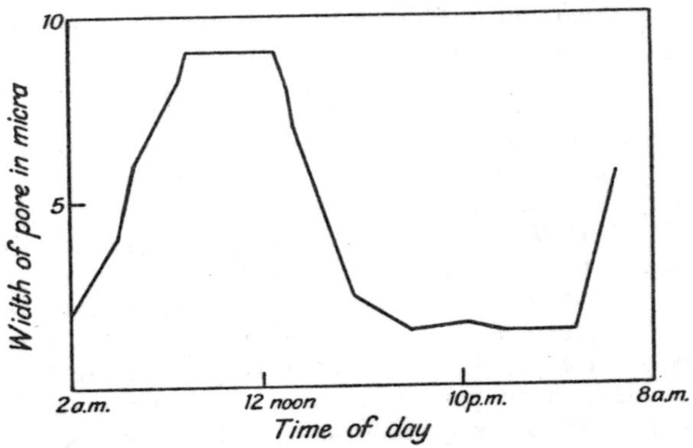

Fig. 26.—Curve illustrating the changes in stomatal aperture of *Verbena ciliata* during a summer day

(Constructed from the data of F. E. Lloyd)

states that under ' very favourable conditions ' (see below) the stomata of this type are wide open only for an hour or two.

In the lucerne group are included a large number of species including *Trifolium pratense* and *T. repens*, *Pisum sativum*, *Lathyrus odoratus*, *Beta vulgaris*, *Brassica napus*, *B. campestris*, *Cucumis sativus* and *Helianthus annuus*. With this type the course of stomatal movement depends largely on conditions. Under what are described as ' favourable conditions ', that is, with adequate water supply and medium temperature, the course of stomatal opening and closing is similar to that described for cereals, but as the conditions become more ' unfavourable ', that is, with lowering of the water content, high temperature, continuous bright sunlight and very dry air, the period during which the stomata remain open becomes shorter and shorter. Thus closure of the stomata at midday may take place, and with intensification of these conditions the stomata may remain closed all day. But with the incidence of midday closure the phenomenon of stomatal opening at night occurs, and this increases with increase in closure during the day. In the extreme case this results in complete closure of stomata during the day and partial opening at night. This is the typical behaviour of stomata in the xerophyte *Opuntia versicolor* as described by Edith B. Shreve, so that the plants of the lucerne group possess stomata which under mesophytic conditions behave like those of a typical mesophyte, but which under drier conditions behave like those of a xerophyte.

In the third group, which includes the potato, onion and species of *Cucurbita*, among others, the stomata remain open all day and all night under conditions of high water content. But if the climatic conditions are such as to lead to increased evaporation, the stomata tend to close. Some variation exists regarding the time at which this closure takes place. Loftfield would include *Verbena ciliata* in this group since in well-watered plants the stomata remain partially open all night (Fig. 26). Whereas the plants of the first group are cereals and those of the second group thin-leaved mesophytes, the third group includes plants with leaves of different morphological types such as *Portulaca oleracea, Cucurbita pepo* and *Allium porrum*.

Some doubt has been cast by Kostychev on the universality of the stomatal behaviour described by Loftfield. It appears that in the dry climate of Central Asia the stomata of cereals may remain open all day. In any case it is clear that no complete explanation of stomatal movement in relation to external conditions is yet available. Whether the theories of stomatal movement already mentioned afford a satisfactory explanation can only be decided after the various types have been studied in detail in relation to the theories. But, broadly speaking, there is no doubt of the profound influence of light and water conditions on the course of stomatal movement.

The Relation between Stomatal Aperture and Rate of Transpiration

It is clear that where transpiration takes place only through the stomata, complete closure of the latter must reduce transpiration almost to nothing. While Loftfield's researches show that in some plants complete closure does take place, it appears that in other cases complete closure is either never achieved or only rarely. For instance, complete stomatal closure appears to be comparatively rare in the species examined by Lloyd, and the variations in size actually found by him during a day in July in *Verbena ciliata* are shown in Table LIV.

The average width of the wide-open pore is thus about 0·009 mm., while during the period of greatest closure it is about 0·0015 mm. It is interesting to consider how far partial closure of the stomata of this plant should affect evaporation through them in view of Jeffreys' mathematical considerations of evaporation through multiperforate septa.

Table LIV

Variations in Stomatal Aperture in *Verbena ciliata*

Time of Day	Range in Pore Width in micra
1.30 a.m.	0 – 3
4.30 ,,	0 – 4
5.20 ,,	3 – 6
6.30 ,,	3 – 8
8. 0 ,,	6–12
9. 0 ,,	8–11
10. 0 ,,	6–12
12.30 p.m.	6–12
1.30 ,,	6
4.45 ,,	0 – 3
7.30 ,,	0 – 3
10.25 ,,	0 – 3
12.20 a.m.	0 – 4

When open the average diameter of the pore is about the same as that in *Helianthus annuus*. In the latter the average area of the pore of the open stoma is about 0·0001 sq. mm. (cf. p. 171). From Lloyd's measurements it would appear that on the average the length of an open pore in *Verbena ciliata* is about 2·2 times its width, whence, from the numbers given above, it can be calculated that, assuming the open pore is elliptical in cross-section, its area is roughly 0·00014 sq. mm. An actual set of measurements of stomatal aperture in *Verbena ciliata* made by Lloyd are shown in Table LV. The calculated areas of these

Table LV

Dimensions of Stomatal Aperture in *Verbena ciliata*

Stoma Number	Width of Open Pore in micra	Length of Open Pore in micra
1	6	15
2	5·8	9·8
3	8	18

pores are respectively 0·00007, 0·000045 and 0·00011 sq. mm. There thus appears to be considerable variation in the size of individual pores, but we may reasonably conclude that the size of the open stoma in this plant is about the same as in *Helianthus annuus*.

If it is assumed that the frequency of the stomata is only about 0·8 times that found in the sunflower it would mean that the distance apart of the stomata in *Verbena ciliata* is about 1·1 times that of the distance separating neighbouring stomata in *Helianthus annuus*. Now if Jeffreys' conclusion is correct that the stomata of the latter species can close until their diameter is only one-fiftieth of that of the open stomata before any reduction in diffusion takes place through them, it would appear that the stomata of *Verbena ciliata* can close until their diameter is reduced to one-fifty-fifth of that of the open stomata before the rate of diffusion of water vapour through them is appreciably affected. And even if Jeffreys' conclusions and the above calculations are only approximate, the opening and closing of the stomata of *Verbena ciliata* through the range indicated in Table LIV should be without appreciable effect on the rate of transpiration. This, indeed, is the conclusion which Lloyd drew from actual observations on the rate of transpiration and size of stomatal aperture in this plant, for he found the regulatory function of the stomata in transpiration to be negligible. This is opposed to the view that was at one time current, namely that the rate of transpiration from a leaf is affected by stomatal movement, and that indeed this serves to regulate transpiration. But already in 1919 Knight wrote: 'The conditions affecting the transpiration stream have been studied at every stage of its progress, and factor after factor has been successively added to the list until the stomata themselves cut but a poor figure in the array of regulating influences. In fact, in the opinions of some workers stomatal action is normally negligible compared with the controlling activities of other factors.'

Many workers have, indeed, attempted to determine whether the degree of stomatal opening influences transpiration rate, but as the quotation from Knight indicates, no general parallelism between the two has been made out, although a number of workers have appeared reluctant to accept as general the view, advanced by Lloyd in respect to *Verbena ciliata* and *Fouquieria splendens*, that 'stomatal regulation of transpiration does not occur'. It is, of course, clearly to be expected that throughout the course of the day the rate of transpiration, even if affected by stomatal aperture, will also depend on factors influencing evaporation such as temperature and relative humidity, so that the fact that transpiration rate and degree of stomatal opening do not run parallel is not in itself an argument against the view of stomatal regulation. From his work with a large number of species Loftfield concluded that when the stomata are wide open the rate of transpiration is determined by the factors influencing evaporation, but that after the stomata have closed to 0·5 or less of their full aperture, the size of the stomatal aperture may be a factor in determining the rate of transpiration. 'When closure is almost complete, the regulation of water loss by the stomata is very close.' This conclusion agrees qualitatively with that of Jeffreys, although Loftfield appears to suggest that stomatal opening would become a significant factor long before closure had proceeded as far as Jeffreys suggested it must before evaporation through the pores would be appreciably reduced.

There appears, then, to be considerable justification for accepting the view that the stomata can close to a certain extent before the closure has any effect on the rate of diffusion through them, and hence on the rate of transpiration, although we may hesitate to accept Jeffreys' conclusion as to the degree to which closure can proceed before affecting the transpiration rate. How far the stomata can close before this happens will depend mainly on the size of the stomata and their distance apart, and as both these, as we have seen, vary considerably, it is obvious that the extent to which the stomata can close before thereby bringing about a reduction in transpiration rate will vary from plant to plant. This, and also the failure to separate the effects of stomatal aperture and of environmental and perhaps internal factors influencing rate of evaporation, may account for the inconclusive results obtained with regard to the relation between stomatal aperture and rate of transpiration.

Relative Transpiration

If the stomatal aperture and other internal factors are without effect on the rate of transpiration, then the latter will be affected by environmental factors in the same way as evaporation. This being so, if the rate of transpiration from the plant is compared with the rate of evaporation from a free water surface under the same conditions, the ratio of the two rates should be a constant. Such comparisons have frequently been made, although usually the rate of evaporation is measured not from a free water surface, but from a porous surface saturated with water, such as a disk of filter paper, or a porcelain surface of a certain shape, such as a cup, a flat circular disk, or a sphere. While any instrument designed for the determination of rate of evaporation may be termed an *evaporimeter*, those in which the free water surface is replaced by a solid porous surface holding water are usually called *atmometers*.

The ratio of the rate of transpiration per unit area from a plant to that from unit area of the surface of an evaporimeter was called by Livingston the *relative transpiration*. Determinations of the relative transpiration show that its value may vary considerably in the same plant. Thus Livingston found that it varied between 0·371 and 0·029 in the desert plant *Allionia incarnata*, and between 0·193 and 0·009 in *Tribulus brachystylis*, another species of the Arizona desert. Hence it appears that the evaporating surface of the plant does not behave simply as a wet surface, and that some internal factor must be operative as well as the environmental factors. No doubt stomatal aperture can be effective in this way, but, as we have seen, probably only after closure has proceeded some considerable way. Even before the work of Jeffreys and Loftfield, it was suggested by Livingston and Brown that the stomata play only a minor part in regulating transpiration, the principal internal factor governing which they consider to be *incipient drying*.

According to this view, when transpiration is active the mesophyll cells of the leaf may lose water more rapidly than they receive it. The cell-walls bordering the intercellular spaces will then cease to be saturated with water and the surface of the water columns in the capillary spaces of the walls will retreat from the cell surface. The atmosphere in the intercellular spaces will then no longer be saturated with water vapour at the surface of the cells, and since the vapour pressure of water at the evaporating surface falls, the diffusion gradient of water vapour will be less steep, and consequently the rate of evaporation will be less. Thus, this incipient drying will lead to a falling off in transpiration rate.

However, the concept of relative transpiration has been much criticized. In the first place changing wind velocity may have different effects on evaporation from the surface of an atmometer and through the stomatal system, and secondly, solar radiation may influence evaporation from a leaf and from different types of atmometer differently, presumably because different proportions of the total incident radiation are transformed to heat by the different surfaces. So Livingston and Maximov found the maximum transpiration rate was reached earlier than the maximum rate of evaporation from an atmometer, while Briggs and Shantz found the maximum transpiration rate was reached after the maximum evaporation rate. The difference is ascribed by Maximov to the different types of evaporimeter used by these respective workers, that of Livingston being less sensitive to solar radiation than the leaf, that of Briggs and Shantz more so. It thus seems likely that values of relative transpiration are determined in part by the type of evaporimeter used. Consequently, the theory of incipient drying as a regulator of transpiration, being based on determinations of relative transpiration, still requires experimental evidence for its support.

From observations on the march of transpiration by maize plants, watered and unwatered, Maximov has outlined a theory of regulation of transpiration which involves both the theories of regulation by stomatal movement and by incipient drying, but which depends in the first place on the water content of the soil. He found that the transpiration from plants continually watered increased through the day until about 3 p.m., whereas in unwatered plants transpiration commenced to decline 2 to 4 hours earlier. By 5 p.m. the rate of transpiration from the unwatered plants was on the average less than one-third that from the watered plants. A shortage of water in the soil thus appears to bring about a considerable lowering of transpiration rate. Maximov argues that the drier the soil the less readily will water be absorbed by the root and the greater will be the retardation of movement of the water of the transpiration stream in the lower part of its course. Owing to the cohesion of water, the result of this retardation in water movement in the lower part of the plant is that water will not be able to pass into the mesophyll cells as rapidly as it is lost by transpiration. As a result, the water content of the leaf cells falls, and also the surface of each water column in the capillaries of the walls of the evaporating cells retreats from the cell-wall surface so that the vapour pressure of water vapour at the surface of the intercellular spaces becomes lowered, and the rate of transpiration falls as hypothesized by Livingston. The decrease in water content of the cells, that is, their loss of turgor, results as we have seen in closure of stomata, which, when a certain degree of closing has been reached, further retards transpiration, and with complete closure may stop it altogether. With renewed vigorous water absorption by the root the reverse chain of processes will be induced, so that regulation of transpiration, on this theory, may be traced back to the water content of the soil.

Influence of External Conditions on Transpiration

From what has been stated above it will be clear that internal conditions, which include stomatal aperture and water content of the cell-walls, may have a profound influence on transpiration rate. External conditions which normally affect rate of evaporation from a wet surface will likewise influence rate of transpiration. From the discussion on p. 223 it will be clear that both the humidity of the atmosphere and the temperature will influence rate of transpiration. Light

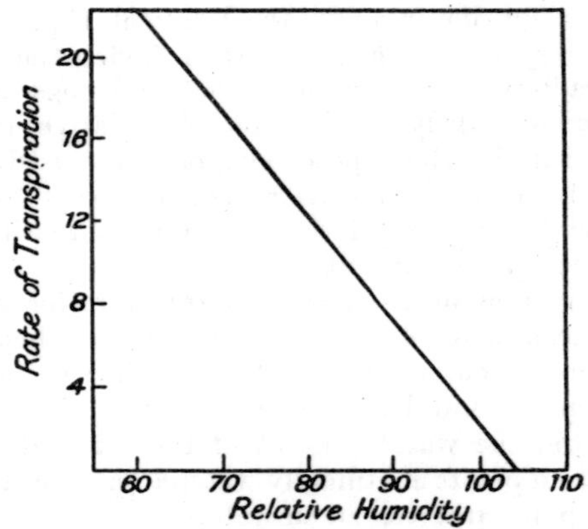

FIG. 27.—Graph showing the relation of transpiration to humidity (After F. Darwin)

(solar radiation) may affect transpiration not only by influencing the temperature of the evaporating surface, but also indirectly by its effect on stomatal movement, while the same holds, though less markedly, with regard to humidity.

Humidity. To eliminate the indirect effect of humidity in inducing stomatal closure, F. Darwin adopted the device of blocking the stomata on the under side of the leaf of cherry laurel (*Prunus laurocerasus*) with cocoa butter or vaseline, and then making incisions in the upper surface. Since in this leaf the stomata are confined to the under surface, the intercellular spaces are then connected with the outer air only through the artificially produced incisions. Potometric measurements of transpiration were then made under different conditions of humidity. The results of an experiment carried out in the dark are shown graphically in Fig. 27 and are typical of the results in general. They indicate an essentially linear relationship between relative humidity and transpiration rate, as would be expected from Dalton's formula, but it will be observed that by extrapolation transpiration would only be reduced to zero in a supersaturated atmosphere of about 105 per cent. relative humidity. This anomaly might be due to defects in the method, as, for example, the use of a potometer for measuring transpiration, or it could be explained as due to the surface of the evaporating cells being slightly higher in temperature than the surroundings. Darwin calculated that the actual difference in temperature necessary to give this result would be 0·8° C., which he considered might easily be produced by the respiratory activity of the leaf.

Later, Henderson confirmed Darwin's general conclusions, but the value he obtained for the relative humidity found by extrapolation for zero transpiration was about 102 to 103 per cent., corresponding to a leaf temperature of about 0·4° C. above the surroundings, as compared with the 0·8° C. calculated by Darwin.

Light. Like humidity, light may affect transpiration indirectly by its effect on stomatal opening and closure. By the method just described, F. Darwin investigated the effect of light on transpiration from leaves of *Hedera helix* and *Prunus laurocerasus* in which the factor of stomatal movement was eliminated. He found that transpiration from ivy was on the average 36 per cent. higher in diffuse light than in the dark, and in cherry laurel 32 per cent. higher.

The reason for this effect of light on transpiration is not yet clear. Two explanations have been suggested. Light may be absorbed by the leaf and transformed to heat and so increase the rate of evaporation, or it may have some effect on cell permeability which may lead to the readier passage of water through the cell membranes.

More recent work by Henderson, in which Darwin's work was repeated and extended, suggests that the normal effect of light on transpiration may be much less than that attributed to it by Darwin, for the considerable increase in transpiration rate in light appears immediately after making the incisions in the upper surface of the leaf, and may largely disappear after an hour or two, when the excess of transpiration in the light may only amount to 4 or 5 per cent., a quantity which would be readily explained as due to slight heating of the leaf.

Temperature. It is clear that, as mentioned above, temperature of the air will influence the rate of transpiration, by increasing both the water vapour pressure gradient and the coefficient of diffusion of water vapour. It is well known that on hot days some plants lose so much water that they wilt, and measurements of transpiration throughout the day leave no doubt that temperature may often be the chief factor determining changes in transpiration rate.

Wind. We have noted earlier that evaporation from a wet surface in a wind may be several times as rapid as in still air, but that the magnitude of the effect depends on the shape and size of the surface. The actual effect of wind on transpiration was examined by Knight in 1917. This worker compared the rate of transpiration from leaves with the rate of evaporation from a filter paper evaporimeter when both were subjected to a wind having a velocity of 7 metres per minute. In this case evaporation from the atmometer was about 73 per cent. higher than in still air, whereas transpiration from the leaf was only about 50 per cent. higher.

In these experiments the stomata remained open, but sometimes wind may induce closure of the stomata which, if carried far enough, will lead to a reduction of transpiration rate. On the other hand strong wind, by causing movements of the leaf lamina, may lead to movement of air in the intercellular spaces and its expulsion and absorption as the spaces contract and expand. When this occurs, owing to the expulsion of saturated air and the absorption of comparatively dry air, the rate of transpiration may be greatly increased; indeed, according to Wiesner, the rate of water loss under such conditions may be as much as 20 times that in still air.

The Daily March of Transpiration

Since transpiration is dependent on so many factors, internal and external, it is not improbable that throughout the day different factors may play the predominant part at different times in determining its rate. Notable observations of the daily march of transpiration in different plants have been carried out by Briggs and Shantz in Colorado, and by Maximov at Tiflis. In both cases continuous records were taken of transpiration and of climatic factors (air temperature, solar radiation, relative humidity and wind velocity). Although different species exhibit minor differences in the march of transpiration it may be stated that in general the rate of transpiration is low during the night and rises rapidly during the morning hours, attaining a maximum between noon and 2 p.m. During the afternoon the transpiration rate falls steadily to the minimum of the hours of darkness. In *Medicago sativa* on a clear day in June, Briggs and Shantz found this minimum transpiration rate only 2 per cent. of the maximum, while Maximov found a similar relation in *Helianthus annuus* at the same time of year. In *Stipa*, however, the minimum found by Maximov was 12 per cent. of the maximum rate. Under the conditions of these experiments, solar radiation, and also air temperature and saturation deficit, which are largely determined by the solar radiation, all reach a maximum in the middle hours of the day. Under other external conditions it is only to be expected that the march of transpiration during the day will deviate considerably from such a regular course.

The Relation of Water Absorption to Water Loss

It is obvious that if water loss by the plant is not equal to absorption the water content of the plant will alter. That changes in the water content of plants do take place has been shown by many investigators, and in some cases the fluctuations over a period of 24 hours may be very great. Thus in leaves of the desert plant *Amaranthus palmeri* Livingston and Brown recorded variations between 601·9 and 366·6 in the water content expressed as percentage of the dry weight, while considerable, though less wide, variations were observed in other plants of the Arizona deserts. The maximum decrease expressed as a percentage of the maximum water content was found to vary in these plants from 39·1 in the species mentioned above to 14·3 in *Prosopis velutina.* Values of this so-called water deficit of the same order were found by Mme Krasnoselsky-Maximov in plants of a definitely xerophytic type at Tiflis. Expressed as percentages of fresh weight of the leaves the maximum diurnal range in water content found by Livingston and Brown was about 8 per cent. and that by Mme Krasnoselsky-Maximov about 7·7 per cent. Similar values were found by Maximov and Krasnoselsky-Maximov for mesophytes such as *Helianthus annuus* and *Atriplex hortensis* in the temperate climate of Leningrad, but in England Knight found a maximum of only 1·3 per cent., and Yapp and Mason of 2·77 per cent. with an average of 1 per cent. These diurnal variations in water content are due to transpiration exceeding absorption during the morning and afternoon hours of the day, this excessive loss of water being made good during the night when absorption exceeds transpiration.

If the water deficit resulting from excess of transpiration over absorption becomes so great that the cells lose their turgor, the leaves become flaccid and the condition known as wilting results. From what has been stated above with regard to observed ranges in water content of leaves, it is to be expected that the loss of water necessary to produce wilting will exhibit considerable variation in different species. In general, wilting in thin-leaved shade plants sets in when the plants have lost comparatively little water ; indeed, in *Eupatorium adenophorum* Knight found that a loss of water of only 1 per cent. of the weight of the turgid leaf produced very definite wilting. In some plants, on the other hand, much more water has usually to be lost before wilting results. According to Mme Krasnoselsky-Maximov, these differences are related to the different changes in size between turgid and plasmolysed cells of the two types of plant. In shade plants she found the reduction in volume of turgid cells when plasmolysed was rarely as much as 3 per cent. whereas in sun plants a reduction of as much as 30 per cent. was observed. It would appear that the modulus of elasticity of the cell-walls is much less in the leaves of sun plants than in those of shade plants so that in the former the same pressure produces more stretching of the walls than in the latter.

Plants generally recover from moderate wilting, but if subjected to water shortage in the soil so that water can no longer be absorbed by the roots, the condition known as permanent wilting sets in from which the plants may derive permanent injury. In this condition the water content of the leaves is much lower than that observed in the transient wilting which may occur during the hotter hours of the day as a result of excessive transpiration alone.

Xeromorphy

It is well known that plants growing in dry habitats exhibit various characteristics which tend to retard transpiration. Such characteristics are a thick cuticle, the development of a layer of wax over the surface of the leaf, sinking of the stomata below the level of the epidermis, the development of hairs over the surface of the leaf, and the reduction of the surface area of the shoot in relation to the total volume of the plant.

Of these characters, the reduction in surface area does definitely result in a low transpiration rate. This is seen in such plants as the succulents, of which the Cactaceae exhibit the most highly developed examples. The low transpiration rate in these plants is aided by closure of the stomata during the day. But the succulents only constitute one group of many among xerophytes, and Maximov has emphasized that a low transpiration rate is not to be regarded as characteristic of xerophytes in general, and in groups of xerophytes other than succulents transpiration may take place as rapidly as from mesophytes. However, where the stomata are protected by hairs or papillae, or where they are sunk below the level of the epidermis, the rate of transpiration in wind will be less than it would be in an exactly similar leaf without these so-called protective devices and there will be little difference between the rate of transpiration in still and moving air. There is thus in such cases a diminution in the effect of wind in increasing transpiration rate, but so long as the stomata remain open transpiration will continue, and in those xerophytes in which the stomata are normally exposed we should expect transpiration to be as rapid as in mesophytes under similar conditions.

With closure of the stomata, however, a thick cuticle or a covering of wax may be very effective in retarding transpiration, and it is probably this that enables xerophytes to remain alive for a long time under very dry conditions. So Maximov comes to the conclusion that xerophytes are characterized, not by a generally low transpiration rate, for this may be as high as, or higher than, that of mesophytes when water is abundant, but by the capacity to restrict, by closure of stomata and the effective retardation of cuticular transpiration, the loss of water in times of drought.

GUTTATION

Water may not only be given off from the plant as vapour in transpiration, but also in liquid form, when the phenomenon of its excretion is called guttation. Much variation exists among plants as to the extent to which such excretion of liquid water may take place, but it has been observed in plants of more than 330 genera belonging to 115 families. In plants where it occurs it is induced by conditions which further root absorption and retard transpiration; it is thus not surprising that it should be more frequent at night than during the day, and commonest in humid tropical areas where the relatively high temperature of the soil is a factor favouring rapid water absorption and the moist atmosphere one tending to retard transpiration. Among plants of temperate regions species of *Impatiens* and grasses, including the cereals, readily exhibit the phenomenon. In the tropical plant *Colocasia antiquorum* it is said that during a single day 200 c.c. of water may be exuded by a single leaf.

The exudation of liquid water generally takes place through structures known as hydathodes which may be specialized epidermal cells unconnected with the vascular system, or may comprise a group of loosely arranged cells known as 'epithem' in contact with a vascular bundle and opening into a 'water stoma' which connects with the outside of the leaf and which differs from the ordinary stoma in the absence of power of movement in the cells bordering the pore.

The mechanism of excretion of water by guttation is not very clear. It is generally supposed that the excretion is produced in one of two ways. Where hydathodes provided with water stomata are present the exudation is attributed to root pressure or a similar pressure set up in the xylem of stems. As already explained, if solutes diffuse into the water in the xylem this may result in the passage of water into the vessels and tracheids, and if the process continues the water so drawn in will be forced out of the conducting tracts at the point which offers least resistance to the passage of water, and this will be where the tracheids abut on the epithem. Here the water is forced into the intercellular spaces and so out through the water stoma. The continuance of guttation in this way will thus depend on the continued supply of osmotically active solutes to the xylem, for the water of guttation contains a certain amount of these solutes which will therefore be constantly passing out of the plant. The loss of water from cut branches, the phenomenon known as 'bleeding', is generally ascribed to the same mode of action.

The simpler epidermal hydathodes are usually supposed themselves to play an active part in water excretion. We may suppose that osmotically active substances present in these cells pass through the membranes to the outside and set up an osmotic pressure outside the cell. Since the suction pressure inside the cell is less than the osmotic pressure so long as the cell is turgid, water will pass osmotically through the cell membranes to the outside. The maintenance of water excretion by this means will thus depend on the continued production of osmotically active material in the hydathode and the permeability of the cell membranes on the outer side to this material.

If this is a correct view of the mode of action of hydathodes of this type they may be compared with nectaries, where it has been shown that excretion of liquid ceases if the surface is washed with water, this having the effect of removing the osmotically active material, while addition of a small quantity of sugar will immediately bring about a renewal of the excretion.

It is obvious, then, that whatever the type of hydathode, the liquid excreted will contain solutes. Analyses have shown that these solutes may comprise a wide range of substances both inorganic and organic, varying from roughly 0·05 to 0·5 per cent. of the exudate, and including potassium carbonate, sodium chloride, calcium chloride, magnesium chloride, and other salts, while in species of *Saxifraga* and in many members of the Plumbaginaceae considerable quantities of calcium carbonate are present in the exudate.

REFERENCES

Burgenstein, A., *Die Transpiration der Pflanzen.* Jena, 1904.

—— *Die Transpiration der Pflanzen,* Zweite Teil (Erganzungsband). Jena, 1920.

Dixon, H. H. *Transpiration and the Ascent of Sap in Plants,* London, 1914.

—— *The Transpiration Stream.* London, 1924.

Maximov, N. A. *The Plant in Relation to Water.* English Translation edited with notes by R. H. Yapp. London, 1929.

Seybold, A. *Die physikalische Komponente der pflanzlichen Transpiration.* Berlin, 1929.

CHAPTER X

NITROGEN ASSIMILATION

ALTHOUGH compounds of nitrogen are present in plants in much less quantity than carbohydrates, they are of first importance, for some of them form an essential part of the protoplasm itself. The sequence of events between the absorption of comparatively simple nitrogenous substances from the surrounding medium and the formation of proteins and other complex nitrogenous substances is, however, very imperfectly known, and our information on this matter rests on a very much less secure foundation than our knowledge of the production of carbohydrates.

It is well known that the nitrogen absorbed by vascular plants, with the exception of the Leguminosae, is taken up by the root in the form of compounds in aqueous solution. Although 80 per cent. of the atmosphere consists of gaseous elemental nitrogen, the only plants capable of utilizing this are the Leguminosae, which are able to do so on account of the presence in their rootlets of nitrogen-assimilating bacteria. Although the relationship between these organisms and the host plant is still far from being completely understood, it is clear that they absorb nitrogen from which compounds are formed which pass into the tissues of the rootlets so that the source of nitrogen for the leguminous plant is actually combined nitrogen supplied to the roots as it is in autotrophic plants generally, although the actual compounds are possibly different.

The inability of the ordinary green plant to absorb elemental nitrogen was indicated by experiments of Boussingault nearly a hundred years ago. Plants were grown in (*a*) sand, (*b*) sand, ash and nitrate and (*c*) sand, ash and carbonate. Only in the cultures provided with nitrate did normal growth take place; in the others growth was negligible as was the increase not only in nitrogen, but also in organic matter. Confirmatory results were obtained shortly afterwards by Lawes, Gilbert and Pugh, and in more recent years by Otto and Kooper. The position of affairs in leguminous plants was made clear in 1888 by the work of Hellriegel and Wilfarth, who showed that whereas non-leguminous plants were dependent for their supply of nitrogen on compounds containing this substance in the soil, leguminous plants could absorb and utilize elemental nitrogen, but that the power to do so depended on the presence of micro-organisms, for the power is lost when the soil is heated to 70° C. and the micro-organisms thereby killed. The isolation of the bacteria growing in the tubercle-like rootlets of Leguminosae was effected about the same time by Beyerinck, and to these bacteria, to which a variety of names including *Pseudomonas radicicola* have been given, it was found the fixation of nitrogen by leguminous plants was due. *Pseudomonas radicicola* appears to be an aggregate species, comprising a number of different strains living in symbiosis with the respective members of the Leguminosae.

Apart, then, from the Leguminosae, the green autotrophic plant absorbs its nitrogen in combined form. In natural soils this nitrogen is nearly always derived from organic material; from the dead bodies of plants and animals or from the excreta of animals. Rocks containing nitrogen are rarely met with. What is known as the ' nitrogen cycle ' in nature is thus of supreme importance.

The nitrogenous substances of organic origin undergo degradation in the soil so that from the proteins are formed various decomposition products such as albumoses, proteoses, peptones and ultimately amino-acids. From the latter ammonia is produced, and may appear in the soil in the form of ammonium salts. Usually the ammonia or ammonium salts are absorbed by bacteria of the genus *Nitrosomonas* with the production of nitrous acid. The latter is then absorbed by the nitrating bacteria, *Nitrobacter*, which oxidize the nitrous acid to nitric acid and the latter, reacting with bases in the soil, produces nitrates. These are regarded as forming the chief source of nitrogen in the soil for the growth of plants, but it is clear that in addition there may be present nitrites, ammonium salts and organic compounds of nitrogen.

The nitrogenous substances to be met with in natural soils are thus various organic and inorganic compounds, the latter being ammonium salts, nitrites and nitrates. Analyses of such soils indicate that the organic nitrogen is generally present in greater quantity than the inorganic, while there is generally more ammonia than nitrate, which may, indeed, be present in only negligible quantity. A very considerable amount of work has been done on the absorption of these various nitrogen compounds by plants, by growing them in sand and water cultures supplied with the various compounds in question. The general conclusion may be drawn that ammonium salts, nitrites and nitrates are all absorbable and utilizable by plants. It appears, however, that different species exhibit differences as regards the relative utilizability of different inorganic nitrogen compounds. Thus, for example, Mazé concluded that maize can absorb and utilize ammonium salts equally as well as it can nitrates, while a similar conclusion has been drawn for numerous other crop plants by different workers. The results of other experiments are on record which suggest that in some plants, as in *Beta*, growth is more rapid when nitrogen is supplied in the form of nitrate than as an ammonium salt, while with other plants, as the potato, the reverse appears possibly to be the case. There also appears to be the possibility that the relative effectiveness of nitrate or ammonia as nutrient may vary with the stage of development, or may be different for different organs. Plants are also capable of absorbing and utilizing nitrites, although in higher concentrations these have a toxic effect.

As regards the absorption of organic compounds containing nitrogen, experiments made with sterile cultures, in which there is no possibility of the conversion of the organic compounds into ammonium salts or nitrates, indicate that such substances are absorbed and utilized by green plants, but much more slowly than inorganic nitrogen compounds are absorbed.

In natural soils the total nitrogen rarely reaches 1 per cent., and is generally very much less and may only amount to 0·1 per cent. Nearly all of this is organic nitrogen, the inorganic nitrogen being usually less than 1 per cent. of the whole. Since the nitrogen is absorbed principally as ammonium salts and nitrates, it thus appears that the actual solution of inorganic nitrogen is very weak. From this solution the nitrogen compound passes into the root hair, either by diffusion or by some more complex mechanism, the laws governing its intake being presumably those which govern the passage of dissolved substances into and out from cells in general, and which have already been discussed in Chapter IV.

The ultimate fate of the nitrogen after it enters the root is very varied. In some cases the nitrogen remains in inorganic combination, for de Vries found that potassium nitrate forms an important constituent of the vacuole in cells of *Helianthus tuberosus*, and that, indeed, the osmotic pressure of these cells is largely due to this substance. But the typical function of nitrogen in the plant is found in the production of proteins, and particularly the proteins which form part of the protoplasm. In addition to these, reserve proteins are common in plants, while numerous basic compounds containing nitrogen, including amines, betaines, alkaloids and purine bases are found throughout the plant kingdom, some widely distributed, others restricted to a single species. There are also those compounds of nitrogen bases with fatty acids, and containing phosphorus, which have already been mentioned in an earlier chapter, and which we have already seen appear to play so important a part in determining the characteristics of protoplasm.

While the nitrogenous substances entering the plant are known, and the final products of nitrogen assimilation are for the most part well recognized, the chain of actions connecting the raw materials and the final products lies largely in obscurity.

Where nitrates occur dissolved in the vacuolar sap, it may be concluded that they diffuse as such from the root hair through the cortical cells of the root, and so pass ultimately to the cells in which the accumulation of nitrates takes place. Part of their journey thither may, presumably, involve passive movement in the transpiration stream as well as active diffusion. This function of nitrogen, namely, the provision of osmotic pressure giving rise to turgor, is, however, a minor one only rarely met with.

Undoubtedly the principal problem of nitrogen assimilation is presented by the formation of proteins. The obscurity which surrounds this problem is evident when it is realized that some uncertainty still exists as to where protein synthesis actually takes place. Two main views have been held on this matter; according to one, protein formation takes place in the leaves, light energy being required for the synthesis, while according to the other, protein formation is independent of light and can take place in many other parts of the plant and perhaps in every cell, provided the materials necessary for the synthesis are present.

Schimper in 1888 found that nitrates accumulate in the leaves of plants maintained in darkness, these nitrates diminishing on exposure to light. This accumulation of nitrate in leaves was shown more recently, in 1922, by Chibnall in leaves of *Phaseolus multiflorus*. Further, Sapoznikow, about the year 1895, reported that during exposure of leaves to light their protein content increased as well as the content of carbohydrate. The more recent work of Chibnall on *Phaseolus multiflorus* has shown definitely that the protein content of leaves is higher during the day than at night. These observations suggest that protein synthesis from nitrates takes place in leaves during exposure to light. However, Sapoznikow himself, and later Zaleski, Sazuki and others, showed that protein synthesis still occurred in leaves in the dark if they were provided with carbohydrate. Sazuki, for example, transferred barley seedlings which had grown in a nutrient solution in the dark to a sugar solution, and found that protein synthesis resulted in darkness. It would thus appear that the formation of proteins in leaves in the light is not actually a photosynthetic process itself, but depends on the presence of carbohydrates and nitrates. In the dark photosynthesis of carbohydrate does not occur, and so the material necessary for protein synthesis is wanting. The matter is, however, not absolutely clear. As Maskell and Mason have pointed out, it is difficult to understand why nitrates should accumulate in leaves at night if protein synthesis is purely chemosynthetic, for although there is a reduction in sugar concentration at night, there may be a much greater reduction in the rate of entrance of nitrates in the leaf if these are conveyed in the transpiration stream, for, as we have seen in the last chapter, the upward movement of water is generally greatly reduced at night.

If, however, it is correct to assume that protein synthesis itself is not a photosynthetic process, it might be expected that this synthesis can take place in every living cell provided the requisite nitrate and carbohydrate are present, although it must be borne in mind that other conditions, such as the presence of certain enzymes, may also be necessary. There is a certain amount of evidence that root cells can manufacture protein provided they contain carbohydrate and nitrate, but more experimental evidence on the point is required definitely to establish it.

If we assume that normally the synthesis of proteins from nitrate and carbohydrate takes place largely in the leaves, it follows that the absorbed nitrate passes as such from the root hair to the mesophyll cells. It was earlier assumed that the nitrate was carried into the leaves in the transpiration stream through the xylem. In 1923, however, Curtis brought forward evidence suggesting that the nitrate might actually pass into the leaves by way of the phloem. Stems of privet, peach and lilac attached to the tree or bush were 'ringed', that is, a ring of tissue external to the xylem was removed, so that the phloem and all tissues outside it were wanting over a short length of the stem. The ring was covered with warm paraffin wax at the time of ringing. Immediately after ringing, sodium nitrate, or, occasionally, calcium nitrate was added to the soil. The nitrogen content of leaves above the ring was determined after various times in different cases, and the value obtained compared with that of leaves on unringed branches. The results obtained show that leaves on the ringed stems always contain less nitrogen than those of unringed stems, and this is so whether the ring is made in spring before the leaves open, and so before new xylem is formed, or in the summer when new xylem has been formed. It is also true whether the nitrogen quantities are expressed in terms of dry weight, leaf surface, or as absolute quantities.

The later observations of Maskell and Mason on the movement of nitrogen compounds in the cotton plant are at variance with those of Curtis. From analyses of the different parts of ringed and unringed plants Maskell and Mason concluded that the movement of inorganic nitrogen into the leaves is not apparently affected by removal of a ring of phloem and external tissues. They found little difference in nitrogen content of the leaves of ringed and unringed plants, the nitrogen content in both cases rising during the day and falling at night, so that in the ringed plants, as in the unringed, nitrates were able to reach the leaves, although in the former plants the phloem was missing over the ringed portion of the stem. Maskell and Mason suggest that the difference between the results obtained by them and by Curtis may be related to the different length of time which elapsed between ringing and analysis of the tissues. The analyses of Mason and Maskell were carried out over the next 24 hours or so after ringing, whereas in Curtis's experiments several weeks usually elapsed between ringing and analysis of the tissues. Hence changes in the wood resulting from ringing may have developed considerably by the time Curtis's analyses were made, and it is extremely possible that reduced transpiration resulted, as, indeed, Curtis found. The reduced conduction of nitrogen found by Curtis could therefore be explained as due to this reduction in the amount of nitrate carried passively in the stream. We may, therefore, suppose that inorganic nitrogen finds its way to the leaves largely by means of the transpiration current. Diffusion of the salts from cell to cell will probably also take place, but this is a very much slower process and will constitute a very subsidiary channel of supply of nitrogen to the seat of protein synthesis.

PLANT PROTEINS

In considering the problem of protein synthesis in the plant it is necessary to recall the molecular structure of the proteins. They are essentially condensation products of amino-acids, although other groups may enter into their composition. The amino-acids are substances containing both an acid group $-COOH$ and a basic group $-NH_2$, the simplest being aminoacetic acid $CH_2(NH_2).COOH$. In the proteins the condensation of the different amino-acids is effected by elimination of water, the combination taking place between the $-COOH$ and $-NH_2$ groups thus:

$$\underset{H_2N.CH.COOH}{\overset{\text{(X)}}{|}} + \underset{H_2N.CH.COOH}{\overset{\text{(Y)}}{|}} = \underset{H_2N.CH.CONH.CH.COOH}{\overset{\text{(X)}\qquad\text{(Y)}}{|\qquad\quad|}} + H_2O$$

Theoretically, at any rate, the number of amino-acids which can combine together is unlimited, and in naturally occurring proteins there are undoubtedly a very large number of amino-acid units involved in a single molecule. Thus from determinations of the rate of diffusion Sutherland concluded that egg albumin possessed a molecular weight of nearly 33,000, while Sörensen, from determinations of osmotic pressure, obtained a value of about 34,000. Such determinations indicate that the protein molecules contain several hundred amino-acid groupings. When it is realized that there are a large number of different amino-acids which may enter into the composition of the proteins, it will at once be evident that the number of possible proteins is practically unlimited. Indeed, a very large number of different proteins have been isolated from plants. These, while not varying greatly in respect to the proportions of different elements in the molecule, may differ considerably with regard to the relative amounts of the various amino-acids contained in them. The chief of these amino-acids which enter into the composition of plant proteins, and which are yielded by them on hydrolysis, are summarized in Table LVI.

The complete hydrolysis of proteins results in the production of the constituent amino-acids, but if the hydrolysis is only partial, compounds of similar constitution to the proteins, but possessing smaller molecules, are produced. These are described as metaproteins, albuminoses, peptones and polypeptides according to the size of the molecules, the metaproteins being the most complex and the polypeptides the least. Actually Emil Fischer succeeded in synthesizing polypeptides by combining amino-acid groups in chains. Commencing with combinations of two amino-acid molecules, called by him dipeptides, such as glycyl-glycine and alanylalanine, he proceeded to produce tripeptides, tetrapeptides and so on. The more complex of these artificially prepared substances, such as those containing 14 and 18 amino-acid groups, closely resemble proteins in their physical and chemical properties, and

Table LVI

Principal Amino-acid Constituents of Plant Proteins

Amino-acid	Formula
amino-acetic acid (glycocoll, glycine)	$CH_2(NH_2).COOH$
α-amino-propionic acid (alanine)	$CH_3.CH(NH_2).COOH$
α-amino-isovaleric acid (valine)	$(CH_3)_2:CH.CH(NH_2).COOH$
α-amino-isocaproic acid (leucine)	$(CH_3)_2:CH.CH_2.CH(NH_2).COOH$
α-amino-succinic acid (aspartic acid)	$CH_2.COOH$ | $CH(NH_2).COOH$
α-amino-glutaric acid (glutamic acid)	$CH_2.CH_2.COOH$ | $CH(NH_2).COOH$
arginine	$NH:C(NH_2).NH.(CH_2)_3.CH(NH_2)COOH$
lysine	$CH_2(NH_2).(CH_2)_3.CH(NH_2).COOH$
phenylalanine	$C_6H_5.CH_2.CH(NH_2).COOH$
tyrosine	$HO.C_6H_4.CH_2.CH(NH_2).COOH$
tryptophane	$C_8H_6N.CH_2.CH(NH_2).COOH$
cystine	$S.CH_2.CH(NH_2).COOH$ | $S.CH_2.CH(NH_2).COOH$

it is impossible to draw a hard-and-fast line between the true proteins and the simpler substances of protein character derived from their hydrolysis. They, that is the albuminoses, peptones and polypeptides, are thus sometimes spoken of as derived proteins.

The true proteins of plants are classified as albumins, globulins, prolamins or gliadins, and glutelins. Of these only the albumins are soluble in water. The globulins are soluble in dilute salt solutions, in which respect they differ from the prolamins and glutelins. The two latter groups are distinguished by their solubility in alcohol, the prolamins being soluble in 70–90 per cent. alcohol, while the glutelins are insoluble. All plant proteins investigated contain a small amount of sulphur, attributable to the presence in the molecule of a small proportion of cystine or some other amino-acid containing sulphur.

In addition to the simple proteins, more complex substances termed conjugated proteins are present in plants. Thus compounds of proteins and nucleic acids are probably present in the protoplasm of all cells, and the chromatin of the nucleus is supposed to be composed of such substances. On hydrolysis these compounds yield their constituent proteins and nucleic acids, the latter being themselves complex compounds containing phosphorus.

Mention has already been made of the fact that a large number of different proteins occur in plants, and it is possible, and even probable, that each species is characterized by its own specific proteins, although in nearly related species the proteins may be of so nearly similar composition as to go by the same name. Thus an albumin called legumelin has been isolated from a number of species of the Leguminosae, including soya bean (*Soja max*), lentil (*Ervum lens*), edible pea (*Pisum sativum*) and broad bean (*Vicia faba*) among others. The properties of these various preparations are closely similar, but minor differences are observable, as, for example, the nitrogen content of the protein, which varies in the different preparations from 16·1 to 16·3 per cent. and which suggests slight differences in the molecules of this protein in the different species.

Most of our information with regard to plant proteins has reference to storage proteins in seeds and tubers. Of late years, however, a certain amount of data regarding the cytoplasmic and vacuolar proteins of leaves have been obtained by Chibnall and his collaborators. From their work it appears that the cytoplasmic protein of leaves consists either of a glutelin, or of a mixture of glutelins with similar properties.

PROTEIN SYNTHESIS

The problem of protein synthesis is, then, to determine the series of reactions by which the absorbed nitrate or ammonium salt is converted into protein. We have already noted that the evidence appears to indicate that for this synthesis to occur the inorganic nitrogen compound must be brought into contact with carbohydrate, and that while this appears to take place chiefly in the assimilating organs, there seems no reason to suppose that protein synthesis cannot take place elsewhere provided the essential materials are present.

Since the nitrogen in proteins is contained in the $-NH_2$ and $=NH$ groupings, it might be supposed that ammonium salts would constitute a more readily utilizable form of nitrogen compound than nitrates. We have already seen that this is not generally so, and that nitrates in general constitute as good a source of nitrogen as ammonium salts or even a better one. It is, nevertheless, generally supposed that the absorbed nitrate is first reduced to nitrite and that this is further reduced to ammonia, which then reacts with non-nitrogenous acids, formed ultimately from carbohydrate, to give amino-acids. The latter then give rise to the proteins by condensation. On this view protein synthesis from nitrates follows essentially the reverse course of nitrification in the soil.

This theory of protein formation is supported by some facts. Reference has already been made to the existence of a nitrate-reducing enzyme in many plants giving rise to nitrites. There is, however, no evidence that the distribution of either this enzyme or of nitrite is universal, although it may be that further investigation may show the enzyme to be more widely distributed than it is supposed to be at present. No satisfactory explanation of the production of ammonia from nitrite in plants has so far been put forward.

The method of production of the amino-acids is also in doubt, both as regards the production of the non-nitrogenous acids from carbohydrate and the method by which the amination of these acids takes place. But the synthesis of the proteins from the amino-acids can be more readily explained, as the proteases are of very general occurrence, and if their capacity for synthesis is accepted, they would, under the appropriate conditions, effect the synthesis of proteins from amino-acids.

It is, however, possible that proteins are not produced in the manner suggested above. Schultze, in a series of investigations commenced in 1878 and extending over more than 30 years, showed that although in the earlier stages of development of seedlings a considerable number of amino-acids could be identified, in older seedlings these mostly disappear, while asparagine, $CO(NH_2).CH_2.CH(NH_2).COOH$, the amide of aspartic acid, accumulates in considerable quantity. This accumulation is particularly striking in various members of the Leguminosae, but also occurs, though to a less extent, in other plants. A second amide, glutamine, $CO(NH_2).CH_2.CH_2.CH(NH_2).COOH$, the amide of glutamic acid, was observed to accumulate in some plants, as, for example, in *Ricinus* and *Cucurbita*. In *Lupinus albus* the quantity of asparagine in etiolated seedlings 18 days old was found to amount to 25·7 per cent. of the dry weight. This quantity is, however, exceptionally high, but it appears that in general the amount of asparagine present cannot be explained as arising from the aspartic acid constituent of reserve proteins present in the seed, and it would therefore appear that the asparagine must arise in some other way, and that the nitrogen contained in it is derived from other amino-acids arising from reserve protein hydrolysis. These are supposed to undergo further decomposition with production of ammonia which then reacts with succinic acid to produce asparagine, although the mode of origin of the succinic acid from carbohydrate is not explained. The asparagine so produced is supposed to be an intermediate product in the formation of the proteins utilized in growth. If this view of Schultze's is correct, it means that in the transformation of reserve protein to protoplasmic protein the reserve proteins are broken down to amino-acids, these are then deaminated and the ammonia so produced then enters into combination with succinic acid formed from carbohydrate. The asparagine so produced provides the nitrogen, and so serves as an intermediary in the production of protoplasmic proteins. On this view it would follow that in the formation of protein from ammonium salts or nitrates and carbohydrates, asparagine might well constitute an intermediate stage. It must, however, be noted that Prianischnikov, while confirming Schultze's observations on the accumulation of asparagine in germinating seedlings, regarded this substance as a by-product of nitrogen metabolism, although one which might serve as a reserve of nitrogen, rather than as a general intermediate product in protein synthesis.

The same conclusion with regard to the part played by asparagine was reached by Chibnall from a consideration of the information obtained by him on the nitrogen changes in leaves of *Phaseolus multiflorus*. He found that in leaves placed with their petioles in water there is a large decrease in protein nitrogen with a corresponding increase in non-protein nitrogen, the increase being chiefly due to asparagine and free amino-nitrogen. Chibnall pointed out that one of the main products of protein oxidation is ammonia, which, in the free state, is toxic. As Prianischnikov suggested, the formation of the relatively harmless asparagine removes this toxic substance and has the result of storing the nitrogen in a form in which it can be subsequently utilized.

That the production of asparagine in leaves kept with their petioles in water is not simply a pathological effect resulting from abnormal conditions, is indicated by the fact, already mentioned, that in normal leaves attached to the plant the protein content increases during the day and decreases at night, suggesting that normally there is a continuous degradation of protein which is masked during the day by the excess of protein synthesis.

From the evidence at present available it may therefore be concluded that the inorganic nitrogen absorbed by the roots is largely conveyed as such to the leaves where in presence of carbohydrates it is worked up through a series of stages into proteins. That protein synthesis should take place chiefly in the leaves appears to be related to the fact that it is here that carbohydrate is synthesized, but there appears reason to suppose that proteins may be synthesized elsewhere. Thus Nightingale and others, from observations on the nitrogen nutrition of apple trees, have come to the conclusion that simple proteins are synthesized in the fine roots.

The protein manufactured in the leaf, like the carbohydrate, is, for the most part, utilized elsewhere, and so is removed away from the seat of synthesis. The large protein molecule is relatively immobile, with a low coefficient of diffusion, and consequently is converted into more readily diffusible substances for translocation. Here we may suppose that protease enzymes are active and that through the action of these the proteins are converted into amino-acids and asparagine, the latter substance being particularly significant in this respect. The observations of Chibnall on the change in the content of various nitrogenous substances in leaves of *Phaseolus multiflorus* during the day and night bear out this conclusion, there being a definite fall in the protein content of the leaves at night. Maskell and Mason have also found that in the cotton plant there is an increase in nitrogen content of leaves during the day and a decrease during the night, which can only be accounted for on the view of conveyance of nitrogen away at night.

In the mobilization of protein nitrogen for translocatory purposes, or in the breaking down of reserve protein, as a preliminary to the building up of protoplasmic protein, there is thus reason to suppose that, at any rate in some plants, asparagine is produced after deamination of amino-acids produced by protease action from proteins. Mrs. Onslow has pointed out that such a deamination is brought about by the oxidase of the potato tuber, and it may be that the catechol oxidase is the deaminating agent in plants. Other suggestions have, however, been made to explain the formation of asparagine from amino-acids, but as these lack much in the way of experimental support they need not be discussed here. A discussion of the whole matter will be found in Mrs. Onslow's book referred to at the end of this chapter. In some plants, it may be noted, the nearly related glutamine appears to take the place of asparagine.

According to Ruhland and Wetzel, only in some plants, which they designate as ' amide ' plants, does asparagine or glutamine result from the deamination of amino-acids. In another group this latter process results in the production of ammonia and organic acids, such as malic acid and oxalic acid ; among such plants are *Begonia semperflorens* and rhubarb (*Rheum hybridum*). Ruhland and Wetzel differentiate these ' acid ' or ' ammonia ' plants as a distinct physiological type from the ' amide ' plants investigated by Boussingault, Schultze, Prianischnikov and Chibnall. That the organic acid present in these plants arises from deamination, and not in respiration as in succulents, is indicated from a number of observations. Thus in *Begonia*-leaves the proportion of the total nitrogen in the form of ammonia is 5 to 10 times that in a non-acid plant, and the ratio of the ' residual ' nitrogen (that is, amino-acid and organic base nitrogen) to the ammonia nitrogen is low in *Begonia* (about 2 to 3) as compared with the ratio in non-acid plants (about 50). Amides, including asparagine and glutamine, are practically absent. Both the degree of deamination of amino-acids and the extent of protein formation run parallel with the increase in free oxalic acid. For these and other reasons Ruhland and Wetzel conclude that the production of organic acid in *Begonia* and other ' ammonia ' plants is related to the deamination of amino-acids and the synthesis of protein from the ammonia so produced. In the petioles of *Rheum*, where acid accumulates much as it does in *Begonia*-leaves, the formation of oxalic acid is preceded by that of succinic and malic acids. Since asparagine is the amide of α-amino succinic acid, it would seem possible that in ' acid ' plants asparagine is formed as an intermediate between storage proteins and the final products of deamination. It does not follow, however, that protein breakdown and synthesis follow the same course in all species.

PLANT BASES

Although proteins are no doubt the most important nitrogen compounds synthesized by plants, there are many other kinds of nitrogen compounds of different degrees of complexity formed in plant metabolism. Apart from proteins and amino-acids, which, as we have seen, no doubt arise by the hydrolysis of proteins and are possibly formed as intermediate products of their synthesis, the principal nitrogen compounds formed in plants are simple bases such as amines and betaines, purine and pyrimidine bases and the alkaloids. Another nitrogen compound of the first importance is chlorophyll, but as this has been considered in an earlier chapter, it will not be further dealt with in this place.

The part played by these various substances is not always clear, nor is their mode of formation in the plant. The simple amines which are sometimes found in plants include methylamine ($CH_3.NH_2$), isobutylamine ($C_4H_9.NH_2$) and aminoethyl alcohol (or colamine)

$$\begin{array}{l} CH_2NH_2 \\ | \\ CH_2OH \end{array}$$

It is possible, though by no means certain, these arise by the splitting off of carbon dioxide from the related amino-acids. Thus methylamine might arise from glycocoll, and aminoethyl alcohol from serine:

$$CH_2(NH_2).COOH = CH_3NH_2 + CO_2$$
$$CH_2(OH).CH(NH_2).COOH = CH_2(OH).CH_2(NH_2) + CO_2$$

Betaine occurs in many plant organs, including the root of the red beet. To it the formula

$$\begin{array}{l} COO \\ CH_2 \end{array}\!\!>\!N\vdots(CH_3)_3 \quad \text{or} \quad \begin{array}{l} O - N(CH_3)_3 \\ | \qquad\quad | \\ CO - CH_2 \end{array}$$

is ascribed and, like methylamine, it can be regarded as a derivative of glycocoll and so may possibly result from protein decomposition. While, like the simpler amines, its function in the plant is obscure, it may possibly be an intermediate in the production of the phospholipin lecithin, which is generally regarded as an important constituent of protoplasm. At any rate, the base choline which enters into the composition of lecithin is nearly related to betaine, its formula being given as

$$\begin{array}{l} HO-N\vdots(CH_3)_3 \\ \qquad | \\ \quad CH_2.CH_2OH \end{array}$$

Since it is easily produced in the laboratory from colamine, it seems likely that in the plant also these two substances are connected in origin, and it would thus seem possible that colamine, betaine and choline are all involved in one metabolic process. Choline is widely distributed in plants, having been found in many seeds and fruits, including those of wheat, barley, hop, beech and coco-nut among others.

The purine bases are derivatives of purine, a substance with a heterocyclic ring structure, its formula being

$$\begin{array}{l} N = CH \\ | \qquad | \\ CH \quad C-NH \\ \| \qquad \| \qquad\; >CH \\ N - C - N \end{array}$$

The principal purine bases found in plants are xanthine, caffeine or theine, theobromine, guanine and hypoxanthine and adenine. Of these xanthine occurs in the leaves of tea (*Thea sinensis*), in the root of *Beta*, and in various young seedlings; caffeine in the leaves of tea, in the leaves and seeds of coffee (*Coffea arabica*) and in some other plants; theobromine principally in the fruit of cacao (*Theobroma cacao*), guanine and hypoxanthine in a number of young seedlings, including sycamore and barley, and adenine in a few plant organs such as leaves of *Thea sinensis* and of *Trifolium repens*.

The mode of origin of these substances in the plant, and their function, are both in doubt. They may be protein decomposition products and either end-products of metabolism, in which case they are to be regarded as waste products which are not eliminated, or temporary storage substances, forming a reserve of nitrogen. It has also been suggested that some of them may serve as a protection against the attacks of animals and fungi. On the other hand, it has also been suggested that they may be synthesized in the plant from carbon dioxide and ammonia. According to this view by the synthetic action of urease (cf. Chapter V), urea is produced from these two simple substances, and the purine then synthesized from two molecules of urea and one molecule of methyl glyoxal which, as we have seen, is hypothesized as an intermediate product of plant respiration. This synthesis may be represented thus:

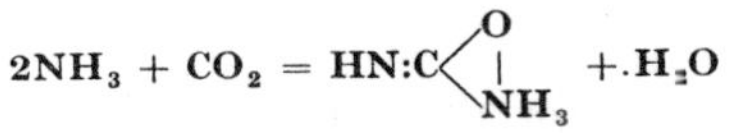

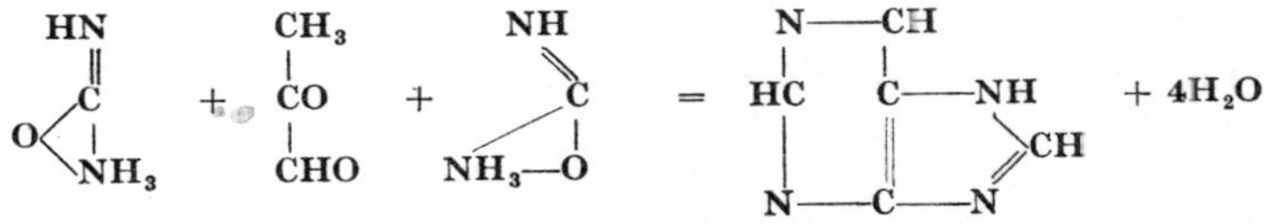

$$2NH_3 + CO_2 = HN{:}C\left\langle\begin{matrix} O \\ | \\ NH_3 \end{matrix}\right. + H_2O$$

```
  HN          CH3         NH              N───CH
  ||          |             \\            |    |
   C     +    CO     +       C      =    HC    C───NH        + 4H2O
O< |          |            / |            |    ||    \
   NH3        CHO      NH3──O             |    ||     >CH
                                          |    ||    //
                                          N────C────N
```

The alkaloids are a group of nitrogen-containing basic substances rather difficult to define. They may be regarded as nitrogen bases of plant origin derived from pyridine, pyrrolidine, tropane, quinoline or isoquinoline. The molecules thus all involve a ring structure with the nitrogen forming part of the ring. They are not widely distributed in plants, but occur chiefly in certain families, notably in Solanaceae, Papaveraceae and Ranunculaceae. Among the first named, nicotine occurs in tobacco (*Nicotiana tabacum*), atropine in deadly nightshade (*Atropa belladonna*), hyoscyamine in henbane (*Hyoscyamus niger*) and strychnine in *Strychnos nux-vomica*. Morphine in *Papaver somniferum* is the best-known alkaloid occurring in the Papaveraceae, while in the Ranunculaceae occur nigelline in the seeds of *Nigella sativa*, damascinine in those of *N. damascena*, delphinine, delphinoidine, delphisine, ajacine and ajaconine in various species of *Delphinium*, and aconitine in *Aconitum napellus*. Noteworthy alkaloids from other plants are quinine from the bark of species of *Cinchona*, a genus of the Rubiaceae, cocaine from the leaves of *Erythroxylon coca*, a member of the Erythroxylaceae, and coniine from hemlock (*Conium maculatum*), one of the Umbelliferae.

As examples of the chemical composition of alkaloids it will be sufficient here to refer to coniine, a comparatively simple pyridine alkaloid, and quinine, a somewhat more complex compound of the quinoline group. Their structural formulae as recorded by Haas and Hill are :

```
        CH2
       /   \
    CH2     CH2
     |       |
    CH2     CH.(CH2)2CH3                CONIINE
       \   /
        NH
```

```
                     CH2----------CH
                      |          /  \
                      |       CH2    CH.CH:CH2
                      |        |      |
                      |       CH2    CH2
                      |          \  /
   CH     CH--CHOH--CH-----------N
  //  \  /  \
 CH    C     HC--OCH3
 |     ||     |
 CH    C     CH                              QUININE
  \\  / \   //
   N     CH
```

Where they do occur the alkaloids may be present in solution in the vacuole, but frequently, especially in older tissues, they are present in the solid condition.

Like the purines, the mode of origin of the alkaloids is obscure. On the one hand, it has been held that they are formed along with proteins under conditions favouring assimilation, and on the other that they arise in the disintegration of proteins, by the combination of decomposition products of the latter with other substances in the plant. Their function is also obscure, but they are most usually regarded as waste products of metabolism. As already noted, they are of limited distribution in the plant kingdom, so that no universal rôle in plant metabolism can be attributed to them.

REFERENCE

ONSLOW, MURIEL W., *The Principles of Plant Biochemistry*, Part I, Chapters IV and V. Cambridge, 1931.

CHAPTER XI

TRANSLOCATION AND STORAGE

In the unicellular plant, and in multicellular plants composed of uniform cells, such as many filamentous algae, every cell is, physiologically, a self-contained system, and movement of dissolved material from one cell to another is negligible. But in the vascular plant the state of affairs is very different. Here there are specialized organs of assimilation in which carbohydrates, proteins and possibly other substances are produced of which only a small proportion is required for the maintenance of the activity of the assimilatory organs themselves. The great proportion of the assimilates are required for the building of fresh tissues either at once or in the future. Hence most of the material formed as a direct or indirect result of photosynthesis is conveyed away from the seat of its manufacture either to actively growing parts or to places where it is stored for future use. The actively growing parts are the apices of main and side shoots and roots, and, in plants with secondary growth in thickness, the cambium and phellogen. The places of storage are very varied. In annual plants the chief storage organ in general is also the organ of reproduction, the seed, which, when conditions for its germination are fulfilled, is thus provided with food material for growth until, with the development of assimilatory tissue and roots, it can provide for itself. In perennials, in a similar way, the perennating parts are usually also storage organs. Thus in woody plants much material is stored in the xylem parenchyma, medullary rays and cortex of stem and root. In perennials the aerial parts of which die down annually, storage takes place in the subterranean rhizomes, tubers, corms and bulbs, which generally serve also as organs of vegetative propagation. With renewal of growth the reserve materials are required for the production of new tissues, and consequently a transference of these materials must take place from their places of storage to the seat of growth.

It is reasonable to suppose that for a substance to be readily translocated it should be soluble in water or cell-sap, and that its molecules should not be too large. As we have seen in an earlier chapter, with increasing molecular size the coefficient of diffusion tends to decrease and, although the rule is by no means absolute, the penetrability of the molecules through the cell membranes also tends to decrease. For storage, on the other hand, we should expect a substance to be relatively immobile, and hence storage products are likely to be either solid, in which case they obviously cannot pass from the cell containing them, or, if soluble, to possess large molecules or molecular aggregates of low diffusivity and penetrability. Actual observations show that these suppositions are correct. The utilization of the products of assimilation thus involves four processes : the translocation of nutritive substances from one part of the plant to another, the storage of these substances as reserves for future use, the mobilization of these reserves as they are required for growth and the building up of the material into the plant body. As respiration is an accompaniment of all life activity, some of the material is also utilized in this process.

THE TRANSLOCATION OF ASSIMILATES

In relatively small multicellular, but non-vascular plants, such as liverworts and fern prothallia, movement of manufactured food material must be slight since there is little differentiation of function in such plants. What movement of material there is in such plants will take place from cell to cell by diffusion. In non-vascular plants there frequently occur, however, cells more elongated than the rest, which, by the consequent reduction in the number of cross-walls per unit length, may be slightly more adapted for the movement of dissolved substances than are the rest of the cells. Such cells among non-vascular land plants reach their highest development in the stem of the gametophyte of the moss *Polytrichum*, where a rudimentary presumed conducting system of elongated cells is present as a cylinder occupying the middle of the stem, comprising a central portion of so-called hydrom which undoubtedly conducts water and a surrounding region of leptom, which is supposed to serve for the conduction of food material. It has also been supposed that in the larger brown algae a tissue specialized for the conduction of food material is developed.

It is, however, in vascular plants that differentiation of function reaches its highest expression, and where, as part of this differentiation, tissue for conduction of material reaches its highest development in the elongated cells of the vascular bundles. While, as we have seen, there is now general agreement that the main channels of conduction of water are the tracheids and vessels of the xylem, there have been divergent views expressed as regards the conducting channels by which manufactured food material is translocated, and a favourite view of the older writers, including Nägeli, Haberlandt and Schimper, was that nitrogenous and non-nitrogenous food even travelled through the plant by separate paths. While the nitrogenous substances were supposed to move through the sieve-tubes of the phloem, Haberlandt in 1882 from anatomical considerations, and Schimper three years later by microchemical tests, concluded that sugar was transported from the leaf by way of the parenchymatous cells which form a sheath round the vascular bundles of the leaf. This view was combated in 1897 by Czapek, who considered that the sieve-tubes constitute the path of translocation of manufactured food in general, although sugar might travel by diffusion through the cells of the bundle sheath as through any other cells of the plant. Sugar, indeed, appears to be generally present in sieve-tubes, often in considerable quantity, and Mangham found this to be the case even when it appeared to be absent from neighbouring tissues. The work of more recent investigators has brought considerable support to Czapek's view.

The experimental determination of the path of transport of assimilates involves two methods of investigation: that of observing the results of ringing experiments, and chemical examination of cells and tissues. As explained in an earlier chapter, by ringing is meant the removal of a ring of tissues from the stem so that a break in certain of the more external tissues results at the place of ringing. In the comparatively early experiments of Hanstein in 1860 it was found that removal of a ring of the outer tissues, as far as the cambium, near the lower end of a cut woody twig, resulted, if the twigs were placed in water, in the production of adventitious roots *above* the ring, even when there were nodes below it, whereas in unringed twigs, or in only partially ringed twigs in which a vertical strip of the outer tissues was left intact, the roots formed at the lowest node. The experiment is easily repeated with twigs of willow and many other plants. Ringing experiments subsequently made by Lecomte in 1889 and Czapek in 1897 showed that in actively growing plants there may occur much greater vegetative development above the ring than below it, and that a much greater production of callus may result at the upper margin of the ring than at the lower. These experiments all suggest that the material required for growth is conveyed in the tissues external to the xylem, since removal of them from a short length of stem appears to prevent or retard the movement of substances necessary for growth across the ringed region. Ringing, as shown by Lecomte, can also lead to the accumulation of carbohydrates in the cells above the ring, thus indicating that these substances are included in those transported through the outer tissues. Later work supports these findings. Thus Curtis in 1920 found that the removal from a defoliated stem of *Philadelphus pubescens* of a ring of tissue up to the xylem led to cessation of growth, thus indicating that the *upward* movement from below of substances necessary for growth is also stopped by removal of outer tissues. Curtis points out that these substances may not only serve to provide material for new cells, but may also be necessary for producing the osmotic pressure in the cell-sap as a result of which the cells absorb water. He also found that the passage of carbohydrates is practically stopped by removal of a ring up to the cambium, and hence, although much carbohydrate is stored in the xylem, this is not translocated in a longitudinal direction through the xylem but has to be first conveyed in a radial direction to the phloem before it can be transported upwards or downwards. Whether the flow of carbohydrate is upwards or downwards must depend on a variety of factors, as, for example, whether the twig is leafy or defoliated, the position of food reserves and the place of ringing. Indications were also obtained of a similar state of affairs with regard to nitrogenous substances.

In the cotton plant also, Mason and Maskell, by a long series of chemical analyses, have found that ringing a stem prevents the flow of carbohydrates and of organic nitrogen across the region of the ring.

Ringing experiments in themselves generally only indicate that the movement of manufactured food materials in stems takes place in tissues external to the cambium. More exact information with regard to the location of the conducting tracts is forthcoming from chemical investigations of the tissues, and such investigations may be on normal unringed stems or, as in the case of much of Mason and Maskell's work, combined with ringing experiments.

Mention has already been made of the fact that Lecomte found that ringing led to the accumulation of starch above the ring. He found this accumulation in the parenchymatous cells of cortex, phloem, medullary rays and even in the outer part of the pith, but not in the sieve-tubes, from which observation he concluded that the latter did not form the path of transport of carbohydrates, which he supposed moved in the parenchymatous tissue. Czapek, however, held the view that sugars were transported through the sieve-tubes, his conclusions being based partly on the demonstration of the presence of sugars in the sieve-tubes by means of microchemical tests. Mangham also, with an improved technique, thus demonstrated the close association of sugars with the sieve-tubes.

The most definite evidence in favour of the view that the sieve-tubes form the conducting channels of carbohydrates is provided by the work of Mason and Maskell on the cotton plant. By means of chemical analyses of the different regions of the outer tissues they found that the quantity of sugars in the sieve-tubes varies markedly with variations in the supply of sugars to the outer tissues as a whole, whereas variations in the amount of sugar in the parenchyma are small.

Mason and Maskell, from their chemical analyses, were able to draw certain conclusions with regard to the movement of sugars through the cotton plant. They determined the changes of sugar concentration in the leaf and outer stem tissues (bark) over 24-hour periods and found that such changes in the leaf-sap were followed by similar changes in sugar concentration in the bark, while determinations of sugar concentration at different levels of the stem showed that a gradient of sugar concentration exists in the stem, the concentration being higher at the higher level. The movement of sugar through the stem thus resembles physical diffusion, inasmuch as the movement is from a place of higher to one of lower concentration. When the normal direction of movement of sugar is reversed, as, for example, by removing leaves from the upper part of the stem and so eliminating the formation of carbohydrate in that region of the plant, the movement of sugar up the stem is accompanied by a reversal of the concentration gradient, the concentration of sugar at a higher level being less than that at a lower one. Also, by lessening the area of cross-section of the path of conduction by partial ringing, the rate of movement is also reduced.

Determinations of the various carbohydrates present in different parts of the leaf of the cotton plant at different times were made by Phillis and Mason, values being obtained for glucose, fructose, total hexoses, sucrose and polyglucosides. From their determinations these authors conclude that carbohydrate is conveyed from the assimilating cells to the phloem chiefly as sucrose and that it is in this form that it also travels through the phloem of leaf and petiole to the stem. There is a positive concentration gradient [1] of sucrose in the phloem of leaf and petiole, but the concentration of this sugar is about 10 times as great in the phloem of the petiole as in the assimilating cells. Since it follows from the existence of the positive gradient that the concentration in the phloem of the fine veins is higher than in that of the petiole, there must be a sudden change in concentration in passing from the mesophyll to the phloem of the bundle ends. At these bundle ends the veins contain a large proportion of phloem, consisting of small sieve-tubes and large companion cells, and undivided mother cells resembling the companion cells. The two latter kinds of cells are grouped together as transition cells. They form a large proportion of the total cross-section of the fine veins and it is suggested that they are the cells in which sucrose accumulates and from which it passes into the sieve-tubes. This supposition leads to another suggestion, namely, that the function of companion cells is also to accumulate sucrose from neighbouring parenchyma cells and pass it on to the accompanying sieve-tubes. The mechanism of the accumulation is, however, not understood.

So far it has been mainly sugar, the transport of which has been considered, but it has already been pointed out that organic nitrogenous materials are also in all probability conveyed in the phloem. Thus Maskell and Mason found that in the cotton plant, if a stem is ringed so as to remove a band of phloem, nitrogen accumulates in the stem above the ring and decreases below it, thus indicating that the movement of nitrogen is stopped by removal of a ring of phloem. It is significant that ringing does not affect the movement of inorganic nitrogen into the leaf, a finding which is in agreement with the generally accepted view that the nitrogen compounds absorbed by the root travel in the xylem dissolved in the water of the transpiration stream.

The observations of Maskell and Mason on the transport of nitrogen in the cotton plant show, however, that the nitrogen relations in the phloem are by no means simple. These authors found that, contrary to what might be expected, the vertical concentration gradient of nitrogen in the outer tissues of the stem is such that there is a higher concentration of organic crystalloidal nitrogenous compounds at a lower than at a higher level. With leaves and wood, however, the

[1] That is, the concentration is higher towards the tip of the leaf than towards the base.

gradient is in the opposite direction, and the authors hold this to support the view that the gradient of *mobile* nitrogen in the phloem is in the direction of downward movement, for the mobile nitrogen in the leaves and wood of the stem should be in dynamic equilibrium with the mobile nitrogen in the phloem and other outer tissues (bark). That the gradient of total organic nitrogen should be in the opposite direction might be due either to chemical masking, that is, the masking of the positive gradient of mobile nitrogenous substances by a negative gradient of immobile crystalloidal nitrogenous substances, or by regional masking, in which case the actual channel of nitrogen transport is limited to one part of the bark, and the positive gradient in this channel is masked by negative gradients of nitrogen in the outer tissues of the bark. But Maskell and Mason found firstly that there are definite negative gradients of amino-acids and asparagine in the bark as a whole and that these would more than account for the observed negative gradient of organic crystalloidal nitrogen, and secondly that both the total organic crystalloidal nitrogen and the asparagine and amino-acid fractions exhibit a steeper negative gradient in the inner part of the bark which contains the sieve-tubes than in the outer part which consists chiefly of medullary ray cells and cortex. There is a possibility of masking by negative gradients in the companion cells, while it is also possible that part of the amino-acids and asparagine in the sieve-tubes is not really mobile but is rendered immobile in some way, as, for instance, by adsorption.

The analyses of different layers of the bark indicate that amino-acids and residual nitrogen (that is, crystalloidal nitrogen which is not accounted for by asparagine, amino-acids, nitrate and ammonia) are chiefly to be found in the sieve-tubes, while asparagine is present in highest concentration in the medullary rays, although it is also present in high concentration in the sieve-tubes. This suggests that organic nitrogen is transported chiefly as amino-acids or as the substances of unknown composition comprising the residual nitrogen, while asparagine is a storage substance. It is important to note that the positive gradient observed in the leaf and petiole is due chiefly to residual nitrogen and to a less extent to amino-acids, while the gradients of nitrate, as we should expect, and of asparagine, are in the reverse direction.

That the observed negative gradient of crystalloidal nitrogen is the resultant of two components, a static negative gradient of storage nitrogen and a dynamic positive gradient of mobile nitrogen, is supported by observations on defoliated stems ringed near the base. In stems subjected to this treatment, which has the effect of stopping translocation of dissolved material, there is still a marked negative gradient of crystalloidal nitrogen, as well as of protein, in the outer tissues of the stem; indeed, the gradients are steeper than when translocation is proceeding. This behaviour is in marked contrast to that of the sugars, for with cessation of movement of these substances the positive gradient of them disappears.

The problem is further complicated by the fact that the conversion, as the result of conditional changes in the bark, of one nitrogen compound into another may be very rapid. By a series of analyses Maskell and Mason showed that such changes are brought about, for example, as a result of ringing, or by alterations in humidity, desiccation bringing about a conversion of crystalloidal nitrogen to protein. A change in the reverse direction results from a decrease in sugar concentration and perhaps of hydrogen-ion concentration. Since nitrogen compounds in the bark thus appear to be very labile it would seem that all the nitrogen fractions in the sieve-tubes may be involved in the translocation of nitrogen through the stem, since immobile nitrogen can be transformed to mobile nitrogen, and vice versa, with such rapidity.

While carbohydrates and nitrogenous substances no doubt constitute the bulk of manufactured food material conveyed through the stem, the various mineral elements enter into the composition of tissues and so must find their way to the growing points as well as carbohydrates and nitrogen. Mason and Maskell have made preliminary observations on the transport of phosphorus, potassium and calcium on the same lines as their researches on the translocation of carbohydrates and nitrogen. They have come to the general conclusion that the ash constituents ascend the stem in the wood, and that phosphorus and potassium at any rate, like carbohydrates, are translocated downwards in the phloem, although there is no evidence of a movement of calcium in this tissue. There appears to be more of the ash constituents moving downward than is necessary for growth, having regard to the quantities of carbohydrate moving downward, and this applies to nitrogen as well as potassium and phosphorus. It is suggested that this excess may pass back into the tracheae, analyses of the contents of which indicate a comparatively high content of ash constituents as compared with that of sugar, while organic nitrogen may amount to one-third of the total nitrogen present. Potassium, like carbohydrates, exhibits a positive vertical concentration gradient in the bark, and it is therefore assumed that potassium is not stored in this region of the stem. With phosphorus, on the other hand, there is in the earlier stages of development a negative gradient as with nitrogenous substances, and it is supposed that this is also due to storage as in the case of nitrogen. The negative gradient of phosphorus ultimately becomes reversed to a positive one, which suggests that during development the stored phosphorus is utilized and translocated away.

It will be noticed that these observations suggest that the complex organic substances containing phosphorus and potassium are also manufactured in the leaf, at any rate, in great part.

Mason and Maskell were able to calculate the effective coefficient of diffusion of sugars through the stem. Their analyses of different levels of the stem provided data for calculating the concentration gradient, while a comparison of analyses of ringed and unringed stems at different times provided data of the rate of movement of the sugars. For since ringing stops the downward transport of sugars, the excess of sugar in a ringed stem above the ring over that in a similar portion of an unringed, but otherwise similar, stem exposed to the same conditions, can be taken as a measure of the rate of movement of the sugars. The calculations led to the surprising conclusion that the effective coefficient of diffusion of sugar through the phloem is about 20,000 to 40,000 times that of sucrose in water, which means that the sugar travels through the phloem 20,000 to 40,000 times as fast as it diffuses through water.

To account for this rapid movement of sugar Curtis has suggested that streaming of the protoplasm of the sieve-tubes is responsible, the sugar being conveyed by mass movement in the protoplasmic stream.

A different mechanism to account for the rapidity of translocation has been hypothesized by E. Münch, according to whom also the movement of dissolved substances in the phloem is not brought about by diffusion, with each substance travelling more or less independently down its own concentration gradient, but by a mass movement of liquid. In Münch's theory the assimilating tissue as a whole may be compared with an osmotic cell attached to a capillary tube, the phloem elements corresponding to the latter. If such a cell, separated from water by a semi-permeable membrane, develops an osmotic pressure, water will be drawn into the cell and its contents will be forced up the capillary. The rate of movement through the capillary depends, among other factors, on the ratio of the area of the surface through which water is absorbed to the area of cross-section of the capillary, and if this ratio is large the rate of movement of the liquid through the capillary may be very rapid. On the Münch hypothesis the absorption of water from the xylem by the assimilating cells leads in a similar way to a mass movement of liquid into and through the sieve-tubes of the phloem.

Although the phloem elements at their far end terminate in a similar cellular osmotic system, the mass movement of liquid will yet proceed towards it, provided the pressure tending to send water in at this end is less than that at the other. It will be observed that the theory will involve the movement of all dissolved substances through the phloem at the same rate, whereas if the movement is governed by the laws of diffusion each substance will move at a rate determined partly by its own coefficient of diffusion and its own diffusion gradient.

Münch's theory has been criticized by O. F. Curtis and H. T. Scofield. These authors point out that the theory necessitates a gradient of turgor pressure, chiefly dependent on osmotic pressure, from the supplying and receiving tissues, and examined the theory from the point of view of this requirement. But in material of various kinds, which they examined, including sprouting potatoes, sprouting onion bulbs and seedlings they found the osmotic gradient was always in the reverse direction from that required by the theory and was in fact from the receiving tissues to the supplying tissue. On the other hand, Dixon and Gibbon measured the osmotic pressure of the sap exuding from punctures of the phloem of the ash (*Fraxinus excelsior*) at different levels and found osmotic pressure gradients of 2·2 to 8·9 atmospheres per meter were present in the direction required by Münch's theory. A gradient of osmotic pressure in the same direction has also been recorded by Mason and Maskell for the cotton plant.

Another theory of the movement of organic substances in plants has been put forward by A. S. Crafts according to which translocation is accomplished by a mass flow of solution through the cell-walls of the whole phloem, but Crafts' work has been destructively criticized by F. C. Steward and J. H. Priestley mainly on the ground that the use made by Crafts of Poiseuille's formula for the rate of flow of a fluid through a tube involved assumptions which would lead to quite incorrect conclusions.

STORAGE

It may thus be concluded that the nutritive substances out of which new tissue is built, and which are utilized in respiration, are manufactured mainly in the leaves and are translocated away from the leaves in the phloem. Some of this material will be thus conveyed direct to the growing tissues, while excess of it beyond that required for immediate activities of the plant is conveyed to places where it is stored. Such stored reserve material is later drawn upon for subsequent activity and it may be concluded that the translocatory channels for the movement of material from the place of storage to the place of utilization are again the sieve-tubes of the phloem.

It has already been pointed out that whereas materials during their translocation must be relatively mobile, and so soluble in water and crystalloidal with molecules of comparatively small size, for storage these substances will be converted into others which are immobile and so either insoluble in water or possessed of large molecules or molecular complexes with perhaps colloidal characteristics. Thus carbohydrate which is translocated as sugar is largely stored as insoluble starch, while other storage carbohydrates are the complex polysaccharides known as hemicellulose and inulin, which with their large molecules are relatively indiffusible. On the other hand sucrose, which as we have seen is regarded by Mason and Maskell as the form in which carbohydrate is translocated, at any rate in the cotton plant, is a not infrequent storage substance, as in the roots of biennials such as mangold and beet (*Beta vulgaris*), carrot (*Daucus carota*) and parsnip (***Pastinaca sativa***). It may be that here the diffusion of the sucrose out from the storage cells is prevented by the relative impermeability of the cell membrane, but there is still the problem of how the sucrose can accumulate, for if sucrose is both translocated and accumulated in the storage cells, such accumulation involves movement of the molecules against the concentration gradient. That such accumulation, either real or apparent, does actually take place in plant tissues there can be no doubt (cf. Chapter V), but it may also be that in such cases as beet and carrot the sugars of translocation are not sucrose but hexoses.

The commonest storage substances in seeds are fats, and evidence has already been presented which leaves no doubt that on mobilization of such reserves sugars are produced, so that a transference of immobile fat to mobile carbohydrate is effected.

Nitrogen is for the very great part stored as protein, the number of individual storage proteins being very numerous. Indeed, not only do different species contain different storage proteins, but one species may contain several different proteins. Actually in a seed the protein would appear to consist of at least two kinds, reserve protein and active protein forming an essential part of the protoplasm. Thus the total protein as determined by chemical analysis generally consists of a mixture of the reserve protein or proteins and the protoplasmic proteins, and the separation of these different components is generally not easy. In the wheat grain, for example, there are, according to Osborne, at least five different proteins which are present in various quantities. In Table LVII these quantities are shown as percentages of the total weight of the grain. Of these proteins glutenin

Table LVII

Proteins of the Wheat Grain

	Spring Wheat	Winter Wheat
Glutenin	4·683	4·173
Gliadin	3·963	3·910
Globulin	0·624	0·625
Leucosin	0·391	0·359
Proteose	0·231	0·432

and gliadin are to be regarded as reserve materials, while the globulin, leucosin and proteose appear to be constituents of the protoplasm, for whereas these latter three proteins are present in the cells of the embryo, glutenin and gliadin are obtained from the endosperm and are absent from the embryo.

Although the same protein, or very similar proteins, may occur in nearly related species, these substances from all but closely related species as a rule differ. Thus a protein excelsin is obtained from the Brazil nut, while a different protein, edestin, is present in hemp seeds, and another, legumin, is found in a number of Leguminosae (for example, *Pisum sativum*, *Vicia faba*, *Ervum lens*). These various proteins differ constitutionally in the number, arrangement and kind of amino-acids which compose them.

The mobilization of each reserve substance is brought about by the action of its appropriate enzyme as described in Chapter V. Thus starch is hydrolysed into glucose by the action of the diastatic enzymes, fat is hydrolysed by lipase into fatty-acids and glycerol, inulin into fructose by the action of inulase and proteins into amino-acids by proteases. It is generally supposed that the formation of the storage substances from the mobile substances of translocation is brought about by the synthetic action of the same enzymes, there being evidence, as we have seen, that enzyme action is reversible.

The places of storage in plants are very varied. But since the stored materials are chiefly required when a plant comes to renewed activity after a period of rest and its assimilatory function is imperfect or out of action, the places of storage have most frequently a definite relation to this necessity. For a similar reason reproductive bodies are nearly always also storage organs or possessed of storage organs or tissue. Thus in annual plants the chief storage organs are the seeds, where reserve material occurs either in the cells of the embryo itself or in the endosperm surrounding the embryo. Similarly, in vegetative reproductive bodies, such as bulbs, corms and tubers, the cells are rich in stored material. In these, as in seeds, material for the development of new tissue is thus available until the new plant has provided itself with assimilatory organs. In perennial plants generally, there is usually a rest period which in temperate regions is in the winter, and in most cases the assimilatory organs of such plants disappear, either by the shedding of the leaves, as in the case of deciduous shrubs and trees, or by the dying down of the whole subaerial portion of the plant to leave only a subterranean rhizome. In the case of woody plants we thus find much storage of material in the parenchyma cells and medullary rays of the wood and phloem, where it is easily available for the development of the cambium or for conduction to the growing points. In rhizomes, as in bulbs, corms and tubers, with which they may be included as perennating organs, the parenchymatous cells are usually very rich in stored material which is mobilized when activity is renewed in spring.

REFERENCE

CURTIS, O. F., *The Translocation of Solutes in Plants.* New York and London. 1935.

CHAPTER XII

THE FUNCTION OF THE MINERAL CONSTITUENTS OF PLANTS

THE MINERAL ELEMENTS PRESENT IN PLANTS

LARGELY owing to the water culture experiments carried out by Sachs and by Knop about the year 1860 it was for long regarded as an established fact that ten elements only were essential for the nutrition and growth of most plants. These ten elements were carbon, hydrogen, oxygen, nitrogen, sulphur, phosphorus, potassium, calcium, magnesium and iron. It was found that plants could be grown with their roots in a nutrient solution containing salts involving only the last seven of these elements. Many nutrient solutions of slightly different composition have been used among which two of the most popular are those of Knop and von der Crone, the former of which is constituted thus:

Calcium nitrate	0·8 g.
Potassium nitrate	0·2 g.
Potassium dihydrogen phosphate	0·2 g.
Magnesium phosphate	0·2 g.
Ferric phosphate	trace
Water	1000 c.c.

and the latter thus:

Potassium nitrate	1·0 g.
Calcium sulphate	0·5 g.
Magnesium sulphate	0·5 g.
Calcium phosphate	0·25 g.
Ferrous phosphate	0·25 g.
Water	1000 c.c.

The hydrogen and oxygen required by the plant are, of course, provided by water, and the carbon by the air. As a result of these water culture experiments it was concluded that only the ten elements noted above were essential constituents of all plants, but analyses of plants have shown that many other elements may be present in them.

It has also been stated that calcium is not necessary for many fungi and for some of the simpler algae.

The quantitative determination of the elements present in plants is generally carried out by burning away the organic substance when the ash, which contains the mineral elements, is left and analysed.

This ash contains all the elements which, in the form of salts or their ions, were absorbed from the soil, with the exception of nitrogen Including carbon, hydrogen, oxygen and nitrogen, actually some 38 elements have been identified as occurring in plants, although some of the rarer elements such as vanadium and caesium, appear to occur very rarely. Other elements usually regarded as non-essential for most plants may actually occur in considerable quantity in some species. Thus in plants of *Equisetum maximum* silica has been found to account for 71 per cent. of the ash, while Richardson found it provided 49·4 per cent. of the ash of *Equisetum hyemale.* It may also constitute 50 per cent. or more of the ash of the cereal straws. Not only silicon, but a number of other elements, including aluminium, manganese, boron and chlorine, are present in all or most plants.

It may be that elements essential for the life of one plant are not essential for that of another. Certainly the amount of the different chemical elements present in plants of different species varies markedly, even when growing on the same soil. Analyses of plants of the same species growing on different soils show that the composition of the soil may influence considerably the relative amounts of the various mineral elements taken into the plant.

Many ash analyses of plants are on record, both of lower and higher plants. As regards the former, it will be sufficient to give the data for a fungus (*Tuber cibarium*), an alga (*Cladophora glomerata*) and a liverwort (*Fegatella conica*) selected from a number collected by Czapek (Table LVIII). The low concentration of potassium in the ash of *Cladophora* may be noted. This is a characteristic of many water plants besides algae.

Table LVIII

Percentage Composition of the Ash of Various Plants

Ash Constituent	*Tuber cibarium*	*Cladophora glomerata*	*Fegatella conica*
K_2O	25·15	0·35	36·6
Na_2O	1·0	6·38	6·4
CaO	9·40	59·18	15·0
MgO	0·20	1·84	8·6
Fe_2O_3	3·20	0·53	—
Al_2O_3	—	—	3·3
P_2O_5	30·25	4·14	7·8
SO_3	4·65	13·33	13·3
SiO_2	10·0	10·60	5·0
Mn_2O_3	—	—	trace
Cl	0·2	1·05	6·4

As an example of the amounts of the different elements in an angiosperm it will be sufficient to quote the results of analyses of maize plants published by Latshaw and Miller in 1924. It will be noticed that their numbers (see Table LIX) are not comparable with those given above for lower plants since the numbers given for the latter represent analyses of ash, whereas the analyses of Latshaw and Miller give the percentage of each element in the total dry matter. It will be observed that carbon and oxygen each account for from 41 to 46 per cent. of the total dry matter of the plant, leaving only about 10 to 15 per cent. for all the other elements together. Of these it will be observed that silicon, aluminium, manganese and chlorine

Table LIX

Percentage of Various Elements in Total Dry Matter of Different Parts of Plants of *Zea mais*

Element	Leaves	Stems	Grain	Cobs	Roots
Carbon	41·27	44·51	44·72	45·75	42·31
Oxygen	43·86	43·90	45·30	45·89	43·58
Hydrogen	5·86	5·90	6·96	6·36	5·72
Nitrogen	1·30	0·84	2·15	1·38	1·27
Phosphorus . . .	0·207	0·089	0·34	0·094	0·120
Potassium . . .	1·48	1·23	0·42	0·46	0·48
Calcium	0·47	0·17	0·025	0·022	0·61
Magnesium . . .	0·21	0·16	0·20	0·11	0·17
Sulphur	0·24	0·16	0·14	0·21	0·25
Iron	0·070	0·052	0·043	0·025	0·52
Silicon	2·59	0·42	0·016	1·33	4·44
Aluminium . . .	0·074	0·013	0·023	0·052	0·98
Manganese . . .	0·043	0·017	0·037	0·031	0·066
Chlorine	0·222	0·224	0·033	0·12	0·11

as well as the ten elements universally accepted as essential, are present throughout the plant. It will also be clear that the relative amounts of all the ash constituents vary considerably in the different organs, the variations in the percentage of calcium, sulphur, iron and silicon being particularly noteworthy. The low percentage of calcium in the fruits of maize, a condition confirmed by analyses by other investigators, is not general. Thus, in a number of analyses of plant ash by various workers, collected by Wolff, the percentage of calcium expressed as CaO in the ash of maize grains varied from between 0·6 and 3·8, whereas in rape (*Brassica rapa*) it varied between 10·4 and 17·3, while in the fruits of *Lithospermum officinale* Hornberger found that calcium oxide accounted for 59 per cent. of the total ash. Similar variations, though not always so wide, are met with in the cases of the other mineral elements. This is illustrated by the analyses of the ash of the leaves of two species, *Fraxinus excelsior* and *Delphinium geyeri*, the analyses being due respectively to Ramann and Gossner and to Heyl, Hepner and Loy (Table LX).

While there are these wide variations in the relative amounts of each of the mineral constituents, it may be said that of the metals potassium tends to be present in greater amount than the others, although sometimes it is exceeded in amount by calcium. In leaves the amount of potassium expressed as K_2O generally constitutes 30 to 50 per cent. of the ash although lower values tend to occur in water plants, while the calcium expressed as CaO varies very greatly. Percentages of lime as high as 66·5, 65·2 and 61·1 have been recorded in the ash of the leaves of *Abies pectinata*, *Sedum album* and *Carpinus betulus*, while in *Ajuga reptans*, *Briza media* and *Tripsacum dactyloides*

Table LX

Percentage of the Various Constituents of the Ash of the Leaves of Two Species

Ash Constituent	*Fraxinus excelsior*	*Delphinium geyeri*
K_2O	19·0	22·42
Na_2O	1·2	1·88
CaO	49·6	24·49
MgO	10·6	0·05
Mn_3O_4	0·15	
Fe_2O_3	0·15	5·36
P_2O_5	7·07	1·26
SO_3	2·98	3·26
SiO_2	5·50	11·88
Cl	1·41	0·42

percentages of lime in the ash as low as 2·1, 2·0 and 1·64 have been found. The calcium content of the leaf increases with age, and this holds both for deciduous and evergreen plants. Thus in *Robinia* leaves an increase in the lime content of the leaves from 21 per cent. of the ash in May to 73 per cent. in September has been recorded. On the whole magnesium and iron are present in leaves in less quantity than potassium and calcium, but here again the amounts vary greatly, a percentage of magnesium oxide as high as 28·47 per cent. having been observed in leaves of *Solanum tuberosum* and as low as 0·05 per cent. in *Triticum repens*, while as regards iron, although the amount of iron, expressed as ferric oxide, is generally only about 1 to 4 per cent., in exceptional cases percentages as high as 18 or 19 per cent. have been found.

Of other metals sodium is a general ash constituent of leaves, and often comprises from 1 to 3 per cent. of the ash. In some non-halophytes the content may be greater than this, while in halophytes the sodium content may rise to very high values, often 30 or 40 per cent. of the whole ash calculated as NaO. Aluminium and manganese also are generally present in appreciable quantity.

Of the other elements of the ash, phosphorus is generally present in comparatively large amount, particularly in fruits. Thus in leaves 8 to 15 per cent. of the ash can often be accounted for by P_2O_5, while percentages as high as 30 per cent. have been recorded. While similar variations occur in fruits, values for P_2O_5 approaching 50 per cent. of the whole ash have been found in some cases.

The amount of SO_3 in the ash of leaves is generally about 3 to 6 per cent., but values as low as 0·5 and as high as 18 per cent.. are on record. Many Cruciferae possess an exceptionally high content of sulphur.

The amount of silicon in plants is very variable, some plants possessing the merest trace, while in others silica may comprise 80 per cent. of the ash. The high content of silica in some grasses and in *Equisetum* is well known, but other species also exhibit a high content of this element. Frequently a high content of silicon accompanies a low calcium content and vice versa. The actual ratio of lime to silica often depends on soil conditions, plants growing on a soil rich in lime exhibiting a higher Ca : Si ratio than plants of the same species growing on a soil poor in lime. But, as is well known, many plants only grow on soils containing a sufficiency of lime, while others are intolerant of even a moderate amount of calcium in the soil, a condition of affairs which presents plant ecology with one of its most outstanding, and for the most part, unsolved problems.

Chlorine, like silicon, varies much in amount in plants, from a trace to about 25 per cent. of the ash, or even more, in some halophytes.

THE ESSENTIAL ELEMENTS

It has already been noted that the earlier experiments with water cultures led to the very general acceptance of the view that only ten elements are as a rule necessary for plant growth. Work during the last twenty years has been tending more and more to indicate the incorrectness of this view. As far back as 1905 Bertrand had pointed out the favourable effect of small quantities of manganese on the development of plants, while Mazé, in 1914 and succeeding years, as a result of carefully controlled water culture experiments with maize and some other plants, came to the conclusion that for normal growth this plant requires not only the generally accepted ten essential elements, but also small quantities of zinc, aluminium, manganese, boron, silicon and chlorine.

As regards manganese, McHargue has produced further evidence that this element is essential for the development of plants. In 1919 he showed that when grown in water cultures free from manganese, wheat plants became etiolated and otherwise developed abnormally as compared with their normal development when the culture solutions contained a small quantity of manganese. Later he showed that members of the Leguminosae are more sensitive to lack of manganese than are other plants, but he evidently regards manganese as an essential element for all plant life. Among plants for which he has shown manganese essential for normal development are radish, soya bean, cowpea (*Vigna*), field pea (*Pisum*) and maize.

Several workers have suggested that manganese is essential for the normal development of species of *Aspergillus*, and McHargue and

Calfree found that cultures of a species of this fungus of the *flavis* group grew more rapidly on a medium containing low concentrations not only of manganese, but also of copper and zinc. The maximum growth was actually reached only in presence of all three. While there seems to be now no doubt that manganese is essential for the life of many plants there is still controversy regarding the necessity of this element for all plant species. Thus the need for manganese for the development of *Lemna major* has been disputed recently by Clark and Fly.

The importance of zinc and copper in the development of *Aspergillus* has just been mentioned. In 1926 Sommer and Lipman found that zinc was absolutely essential for the development of barley and dwarf sunflower plants, but there is at present little evidence of the universal necessity for this metal in plants. More recently Sommer, and also Lipman and Mackinney, have decided that copper is essential for the growth of some plants. Sommer worked with plants of flax, sunflower and tomato grown with and without a small supply of copper, and found a striking retardation of growth of these plants in absence of the element. Lipman and Mackinney examined the growth of barley in water cultures with and without a small supply of copper and found that vegetative growth of the plants was less in complete absence of copper, while normal seed production was completely inhibited, although the presence of copper in a concentration of $\frac{1}{5} \times 10^{-6}$ to $\frac{1}{16} \times 10^{-6}$ was sufficient to enable the seed to develop normally.

As regards boron, in 1923 Miss Warington found that small quantities of boron were essential for the growth of the broad bean (*Vicia faba*) and possibly also for *Phaseolus multiflorus* and *Trifolium incarnatum*. Boron did not, however, appear essential for the growth of the cereals barley and rye. Sommer and Lipman, however, in 1926 concluded that boron was essential, not only to *Vicia faba* but also for the development of barley, buckwheat, flax, mustard, cotton and *Ricinus communis*. Boron has also been shown to be necessary for the development of a number of other plant species including tomato, tobacco and species of *Citrus*.

It thus seems evident that the view that has been accepted for more than half a century to the effect that only ten elements are necessary for the growth of plants requires modification. The work of the last twenty years on the mineral nutrition of plants indicates very definitely that in many plants, and possibly in all, small quantities of manganese and boron are necessary, while evidence is steadily accumulating that small quantities of other elements such as zinc, copper and possibly aluminium, sodium, chlorine and others are also necessary, at any rate in certain species. Undoubtedly small quantities of these elements frequently lead to increased growth even where it has not been proved that growth ceases without them. For healthy development of the plant the concentration of several of these substances in the external medium must be very low, for they exert a toxic action in very moderate concentrations. That the necessity of boron, manganese and other elements for plant growth was overlooked is explained by the fact that the small quantities necessary might be supplied (1) by the seed itself, (2) by impurities in the salts normally used for culture experiments and (3) by the bottles used to contain the culture solutions, small quantities of compounds of the elements concerned being possibly dissolved out of the material of the bottles by the culture solutions.

THE ABSORPTION OF THE MINERAL NUTRIENTS

The mineral constituents of the plant are all absorbed from the medium surrounding the absorbing organ, whether this is the photosynthesizing cell of a submerged alga or the root hair attached to the root of a vascular plant. The laws governing the absorption of solutes by plant cells have been discussed in Chapter IV. Two points relating to the absorption of salts may be recalled here in particular. In the first place the ions of salts are not absorbed towards a position of equilibrium which is one of equality of concentration inside and outside the cell. Actually with dilute solutions there is always, as far as is known, an accumulation of ions inside the cell so that a concentration of the ion is reached inside the cell which may be very many times the concentration outside. The significance of this in absorption of solutes by root hairs from a solution as dilute as that of the nutrient salts in the soil is obvious. The rate of absorption towards a position of equilibrium (which may never actually be reached) in which the internal concentration of the solute is many times that of the solute in the external medium will be much greater than would be the case if the position of equilibrium were one of equality of concentration inside and outside the cell.

The second point which calls for comment here is the observed connexion between respiration and salt intake to which attention has been drawn in recent years. The importance of soil aeration for healthy development of plants, and the increased rate of growth which has been observed in water cultures of a number of plants kept constantly aerated, can thus be related not only to the general necessity for respiration in living cells, but also to the absorbing function of the root hairs. Aeration, by providing a constant supply of oxygen, enables aerobic respiration to proceed more vigorously than it would with deficient oxygen supply, and so promotes the most vigorous absorption of salts.

For long there was controversy regarding the interdependence or otherwise of the absorption of water and salts by roots. On general principles it is to be expected that the absorption of water and dissolved substances will be independent of one another, the entrance of water depending on the difference between the osmotic pressure of the outer solution and the suction pressure of the absorbing cells, while the absorption of any dissolved substance will depend on the concentrations of the substance in the outer solution and in the cell interior and on the position of equilibrium in the distribution of the substance between the outside and inside of the cell. The permeability of the cell membranes to water and to dissolved substances must also influence the relative rates at which water and various solutes are absorbed. In the extreme case of a solute to which the cell membranes are completely impermeable water will be absorbed, but no solute. It is thus clear that the soil solution or other external solution is not absorbed as such by the absorbing cells.

That the two processes may not be completely independent in vascular plants will, however, appear from the following considerations. As shown in a previous chapter, the rate of absorption of water is in part due to the rate of transpiration. Now if we suppose an absorbed solute to diffuse from the root hair through the cortex to the conducting elements of the xylem, a gradient of concentration of the solute through the root tissues will be set up. The steepness of this gradient will depend on the concentration of the solute in the liquid in the xylem conducting cells. If transpiration is rapid the mass movement of the liquid upwards in the xylem will also be more rapid, consequently the concentration of the solute will be lower in the tracheae of the roots than when transpiration, and the mass movement of water in the xylem, are slow. But probably over a wide range of transpiration rates there is little difference in the concentration of the solute in the xylem of the root, this concentration probably approximating to zero for quite moderate rates of transpiration.

Actual attempts to determine a relationship, or absence of one, between water and solute uptake by vascular plants, confirm these considerations. Thus when tobacco plants were grown under different conditions for transpiration, Hasselbring in 1914 found that actually more ash was present in the plants which had absorbed less water. This lack of any proportionality between water absorption and salt intake has been found also for a number of cereals by other workers.

THE UTILIZATION OF THE MINERAL ELEMENTS IN THE PLANT

The inorganic ions absorbed by the root find their way to various and remote parts of the plant, as analyses of different plant organs show. The mechanism of this transference must in part be similar to that by which the ions enter the cell, but no doubt a certain amount, perhaps a very considerable amount of some elements, is carried in the transpiration stream in the xylem. This is certainly indicated by the accumulation of calcium salts in the leaves to which attention has already been called. Further, the observations of Mason and Maskell, to which reference has been made in the last chapter, indicate that potassium and phosphorus are brought to the leaves in the transpiration stream and are then translocated away in the phloem. The failure of Mason and Maskell to find evidence of movement of calcium away from the leaves falls into line with the observed accumulation of that element in the leaves.

As already indicated in this chapter, not all the elements absorbed are necessary for the plant, and even in the case of essential elements it does not follow that the whole amount absorbed is actually utilized. The absorbed elements may be directly employed for manufacture of cell-wall material, or protoplasm, or they may accumulate in the vacuole (cf. Chapter IV) and so provide the necessary concentration of solutes for producing the osmotic pressure exerted by the cell. The actual function of the various elements is, however, far from clear in many cases, as the following remarks regarding function of the individual mineral elements indicate.

Potassium. Examination by microchemical tests and chemical analyses of the distribution of potassium in plants indicates that meristematic tissues are particularly rich in this element, but it is stated to be absent from nucleus and plastids, although it appears to be plentiful at or near the surface of these bodies. Thus it would appear that potassium is essentially a constituent of the cytoplasm.

Quite a considerable number of observations have been made on the effect of potassium starvation on plants, and from these observations conclusions have been drawn regarding the function of potassium. Shortage of potassium often results in some form of damage to the leaves. Thus those of fruit trees exhibit the condition known as ' leaf scorch ', in which there is an appearance resembling partial withering owing to deficient water content. A low water content of leaves of plants deficient in potassium has been recorded in other plants, but some workers have recorded the reverse. It has generally been concluded that shortage of potassium leads to a reduction in photosynthetic activity, a result particularly emphasized by Stoklasa. Thus, in 1916, in collaboration with Matoušek, he recorded that the leaves of sugar-beet plants growing in a complete nutrient solution photosynthesized more than 3 times as rapidly as the leaves of similar plants growing in a solution without potassium, but otherwise similar. This effect of potassium has been well established by a number of other workers during the last 65 years, but it may be noted that Briggs found a similar reduction in the rate of photosynthesis as a result of deficiency in magnesium, iron or phosphorus as well as of potassium. Other effects of potassium shortage have been described, but these appear to be different in different species. Thus Miss Hartt in 1929 mentioned deficient development of chlorophyll as one of the effects of potassium starvation in the sugar cane, but observations by Reed on moss protonema and by Janssen and Bartholemew and by Nightingale, Schermerhorn and Robbins on tomatoes indicate that in these plants, when they are deprived of potassium, the chloroplasts develop at least as much chlorophyll as those of plants with an adequate supply of this element. It appears impossible, therefore, to ascribe the influence of potassium on photosynthesis to any general effect on the development of chlorophyll.

A favourite view of the function of potassium in the plant is that it acts as a catalyst. Recently James pointed out that Briggs had already shown that potassium shortage affected both the photochemical and dark stages of photosynthesis while that starch formation is furthered by potassium had been shown by Reed for moss protonema, by Briggs for sunflower seedlings and for the potato plant by Maskell. James himself concluded from these results and his own investigations on the potato plant that one important effect of potassium on leaves is to further catalytic activity in the case of the three processes leading to starch formation mentioned at the beginning of this paragraph, while James and Miss Cattle actually showed that the rate of breakdown of starch by potato diastase was accelerated by 5 per cent. potassium chloride.

In spite of the lowered rate of photosynthesis in plants suffering from potassium starvation, there has several times been observed an accumulation of carbohydrate in such plants. This has been attributed by Nightingale, Schermerhorn and Robbins to retardation of the earlier stages of nitrogen metabolism, with the result that carbohydrate which would normally be involved in the synthesis of organic nitrogen compounds remains unutilized. Janssen and Bartholomew, and Nightingale and his collaborators agree that potassium is essential for cell division and the latter writers suggest that the reason for this may be that potassium is necessary for the synthesis of the proteins of meristematic material, a conclusion which falls into line with the observed comparatively high concentration of potassium in meristematic tissues.

Another function attributed to potassium in grasses is that it affects in some way the mechanical strength of the straw by leading to an increase in the amount of the strengthening tissues and in the strengthening quality of the cell-walls.

The evidence derived from chemical analysis of plant tissues indicates that the potassium is in a soluble and principally inorganic form. In some plants it appears, indeed, to exist in the cell-sap in so high a concentration in the form of simple salts as to play an important part in the maintenance of cell turgor (cf. p. 35).

Calcium. Although various suggestions have been made regarding the function of calcium in plants, the part played by this element in the life of the plant is no clearer than the part played by potassium. The theory of Schimper was at one time popular, though it appears to meet with little support at the present time. On this view it is pointed out that oxalic acid is produced in many plants, perhaps as a by-product in the formation of proteins. It was supposed that calcium combined with this oxalic acid to form calcium oxalate which, being insoluble, was thus removed from any active participation in cell activities. In this way the poisonous effect of the oxalic acid was removed. The objections to this theory are that oxalic acid does not appear to be formed in some plants to which, all the same, calcium is necessary, while oxalic acid does not appear to be a very strong poison to some plants, which will, indeed, grow in nutrient solutions in which the calcium is provided in the form of oxalate. It may, however, be that in some plants calcium does serve to neutralize acids formed during certain phases of plant metabolism. Thus in 1920 Kelley and Cummins found that in *Citrus* trees (orange, lemon and grape-fruit) suffering from calcium starvation a mottling of the leaves was associated with a higher acid content and a lower calcium content than were found in normal leaves. Although the mottled leaves contain a higher total content of acid than normal leaves, the hydrogen-ion concentration is about the same, which indicates that excess of acid in the mottled leaves is non-ionized and hence attributable to some little-dissociated organic acids. Parker and Truog, also in 1920, pointed out that in a series of 34 plant species the nitrogen content and calcium content ran parallel, and they saw in this fact evidence of the acid-neutralizing function of calcium. It was argued that if acids are formed as a by-product of nitrogen metabolism, the higher the nitrogen content of the plant the more organic acid is formed and hence the more calcium necessary to neutralize the acid. That the calcium content should be higher under such conditions would presumably be explained on the ground that the greater utilization of the element would result in a steeper concentration gradient and so in a more rapid absorption. That there is indeed some connexion between nitrogen metabolism and calcium is indicated by recent work by Nightingale and his collaborators, who found that tomato plants suffering from calcium starvation were unable to absorb or utilize nitrates to any appreciable extent. On the acid-neutralization theory this could be explained by supposing that under conditions of calcium starvation acids accumulate in the plant and inhibit the normal processes involved in nitrogen metabolism. But other explanations are possible.

Parker and Truog regarded calcium as playing two parts in the life of the plant, that of an acid-neutralizer as we have seen, and that of a constituent of certain plant proteins, thus accepting the view advanced by Loew in 1903. In this connexion it is worth noting that Chibnall and Channon in 1929 found calcium phosphatides present in leaf cytoplasm. Loew, it may be noted, thought that the calcium protein compounds were essential constituents of the nucleus and plastids, and Sorokin and Sommer, from observations on meristematic cells in roots of calcium-starved plants of *Pisum sativum*, found that shortage of calcium results in a decrease in the amount of cytoplasm and a failure of the nuclei to divide mitotically. There is thus some evidence to suggest that calcium forms an essential part of the protoplasm, although it may be noted that in the plants examined by Reed in 1907 the nuclei divided mitotically in calcium-deficient plants.

It has been mentioned in Chapter II that one of the views of the composition of the middle lamella of the cell-wall is that it is made of calcium pectate, and holders of this view would see in this one function of calcium in the plant. In support of this can be cited Reed's observation that in plants deprived of a supply of calcium new cell-walls are either not formed or are abnormal, but Sorokin and Sommer in their work cited above could find no indication in calcium-starved plants of separation of cell-walls owing to the breaking down of the middle lamella or its failure to form, and the same conclusion was drawn by Day who also examined the effect of calcium starvation in roots of *Pisum sativum*.

Magnesium. One function of magnesium in green plants is obvious, since this element enters into the composition of the chlorophyll molecule. It seems likely, however, that magnesium plays some other essential part in the life of the plant, for it appears to be necessary for non-green, as well as green, plants. It has been suggested that magnesium is connected in some way with phosphorus metabolism, since there is some evidence that higher magnesium contents are characteristic of those parts of the plant where higher phosphorus contents also occur, as, for example, the seeds and root and shoot apices. According to Loew, the content of magnesium in seeds storing fat is much higher than in those storing starch. Thus it has been thought that magnesium may be concerned in some way in the formation of lipoid substances containing phosphorus, the phospholipines, which undoubtedly form an important part of protoplasm. The evidence, and views based on the evidence are, however, rather vague.

Iron. Although not entering into the composition of chlorophyll, a small quantity of iron is necessary for the production of the green pigment, and when grown in absence of this element the shoots of normally green plants are pale yellow in colour and exhibit the condition known as ***chlorosis***. It is most generally thought that iron acts as a catalyst in a number of reactions in the plant, including some of the reactions leading up to the production of chlorophyll, while Warburg regarded iron compounds as functioning in this way in aerobic respiration. It is certainly true that only a small quantity of iron is necessary for the healthy development of the plant. Little, however, is really known of the actions in which iron actually participates in plant metabolism.

Manganese. While it is not yet clear whether manganese is an essential element for all plants it obviously cannot be stated that there is a definite function which it universally fulfils in plants. But in some cases it appears to behave in a similar way to iron, inasmuch as in complete absence of manganese the shoots of a number of species exhibit chlorosis. It has also been supposed that, like iron, it plays a part as a catalyst in other metabolic processes, such as those involved in respiration and nitrogen metabolism. As in the case of iron, our knowledge of the manner in which manganese is involved in reactions in the plant is still vague.

Sodium. Although sometimes present in considerable quantity this element is not regarded as essential for the growth of most plants, but Osterhout in 1912 found it was necessary for a number of marine algae, since replacement of the sodium chloride of sea-water by some other compound not containing sodium always led to injury of the plants. What its particular function is in these plants is not clear. From experiments with various species of flowering plants, experiments mostly concerned with crop yield under various conditions of mineral nutrient supply, it appears that sodium can partly, but not wholly, replace potassium as a nutrient, since the lowering of the amount of dry matter produced in a crop on account of shortage of potassium can be partially made good by the addition of a sodium salt. While our knowledge of the part played by potassium in plant metabolism is as slight as it is at present, it is obviously not possible to say in what way sodium can replace potassium in the plant.

Phosphorus. It has already been mentioned that the concentration of phosphorus in plants is particularly high in fruits and seeds. This is partly accounted for by the high concentration of this element in all meristematic, including embryonic, cells, due to their high content of protoplasm. As already noted, complex phosphorus-containing lipoid substances such as lecithin and other lecithides or phospholipines enter into the composition of protoplasm, while the nucleoproteins, which, as mentioned in an earlier chapter, contain phosphorus, are also supposed to be constant constituents of the cytoplasm and nucleus.

As pointed out in Chapter VI phosphates are regarded as taking part in an early stage of yeast fermentation and hence probably of both anaerobic and aerobic respiration. If this should be so, a second universal function of phosphorus in plant cells is indicated.

Other functions of phosphorus in the plant have also been suggested, particularly in relation to nitrogen metabolism. Thus both MacGillivray and Kraybill found that tomato plants suffering from a shortage of phosphorus contained an abnormally high proportion of nitrogen, but that this was due to soluble nitrogen compounds such as amide nitrogen, amino nitrogen and ammonia nitrogen, whereas protein nitrogen was less than normal, whence it seems possible that phosphorus may play some part in one or other of the processes leading to protein formation.

Sulphur. This element is a constituent of plant proteins which contain the amino-acid cystine (cf. p. 251). As the amount of this amino-acid in different proteins varies considerably so does the proportion of sulphur in the protein. In proteins containing much cystine, sulphur may comprise as much as 7 per cent. of the molecule, whereas in others it may amount to considerably less than 0·01 per cent. of the protein. Sulphur is thus an essential element since it forms part of the molecules of protoplasmic proteins.

In 1932 Nightingale, Schermerhorn and Robbins recorded that tomato plants grown in a medium without sulphate possessed yellowish-green leaves, hard and woody stems and extensive roots, although the diameters of both roots and stems were abnormally small. They point out, however, that all these characteristics are exhibited by tomato plants starved of nitrogen, phosphorus or potassium, while the fact that the plants deficient in sulphur had abnormally high percentages of carbohydrates and nitrate, indicates that, as with phosphorus, deficiency of sulphur retards the processes leading to protein formation. Since sulphur enters into the composition of plant proteins this is to be expected.

The failure of sulphur-starved plants to produce an adequate supply of chlorophyll had also been noted by a number of earlier workers and the conclusion drawn that sulphur is connected in some way with chlorophyll formation. The relation may be an indirect one, and it may be that the formation of chlorophyll is linked in some unknown way with protein formation.

In Leguminosae, sulphur, both in elemental form and as sulphate, has been found to bring about an increase in the number of root tubercles. The mode of action of the sulphur is not known, but the indication is that it brings about an increase in the number of the bacteria inhabiting the tubercles.

As already noted, some members of the Cruciferae, particularly species of *Brassica*, contain relatively large quantities of sulphur. The same is the case with species of *Allium* among the Liliaceae. In these plants the sulphur is contained in glycosides such as sinigrin (p. 106), a condensation product of glucose, oil of mustard (allyl thiocyanate) and potassium hydrogen sulphate. The significance of the formation of this type of glycoside is not evident.

Silicon. Although as we have seen earlier in this chapter the ash of some plants contains a very large proportion of silica, numerous water culture experiments have made it clear that silicon is either a non-essential element or is required in only very small quantities.

At one time it was thought that the silicon present in the stems of *Equisetum* and various grasses aided in the maintenance of the rigidity of these organs, but it now appears that these stems are as rigid when grown without silicon as when containing a high percentage of it.

According to Palladin, although silicon is not essential for plants grown in the laboratory, it may be necessary for plants grown under natural conditions, where it serves the purpose of protecting them from the attacks of fungi, the hyphae of which cannot readily penetrate cell walls containing much silica, and of animal parasites such as aphides as well.

Experiments carried out at Rothamsted over more than 40 years have shown that the addition of silicates to the soil has the effect of increasing the yield of plants supplied with only a little phosphorus, and this result has been confirmed by other workers. Brenchley, Maskell and Warington suggest that this effect is not due to a replacement of phosphorus by silicon but that it results from an action within the plant which has the effect of rendering mobile phosphorus which is otherwise immobile and so rendering it available for translocation to regions of active development.

Boron. It has already been noted that evidence is accumulating which indicates that boron is an essential element for plant growth, although it must be supplied in low concentration to avoid a toxic effect. The function of boron is, however, obscure. Where it has been proved that boron is an essential element for plant growth its absence leads to a cessation of cell division in the meristematic regions while the older tissues exhibit various pathological conditions. The xylem may be feebly developed; the phloem and other tissues may disintegrate. In *Vicia faba* Brenchley and Thornton found that a supply of boron brings about an increase in the number of root tubercles, while in plants without boron the tubercles are abnormal, being small and with vascular tissue either absent or very poorly developed.

Chlorine. Although chlorine is generally present in plants, sometimes in considerable quantity, it is doubtful if it is ever essential to the life of the plant, at any rate in more than very small quantity. Yet for many years it was thought that buckwheat required an external supply of chloride for its growth. More recent water culture experiments have, however, shown that this plant will develop normally with no supply of chloride beyond that present in the seed.

The chlorine in plants is apparently always present as chloride, and although a number of observations are on record indicating that the addition of chloride to the nutrient medium may influence the development of various plants in certain ways, it does not seem possible to draw any general conclusions with regard to the function of chloride in the plant.

CHAPTER XIII

HETEROTROPHIC PLANTS

THE essential characteristic of the green plant, in which it differs from non-green plants and animals, is that it is able to absorb water and inorganic salts from the soil, and carbon dioxide from the air, and from these simple compounds synthesize all the various and complex substances which make up the plant body. The energy required for these syntheses is derived in the first place from the light energy of the sun and transferred to the plant by the process of photosynthesis of carbohydrates. This process, as we have seen, requires the green chlorophyll pigments, and plants devoid of chlorophyll are thus unable to absorb light energy and to manufacture carbohydrates. Such plants must therefore absorb organic material from the outside world, for only in such a way can they provide themselves with organic compounds of adequate energy content to form a basis for further syntheses in the plant, and generally for carrying on the maintenance and development of the organism. In contrast with green plants which, with their normal mode of nutrition, are termed autotrophic, plants in which organic material is absorbed directly from the environment are termed heterotrophic. The distinction between autotrophic and heterotrophic plants is by no means absolute. In completely heterotrophic plants the whole of the organic food material of the plant is absorbed from outside, but in quite a number of plants in which a certain amount of organic material is obtained from without in this way, chlorophyll is developed and some photosynthesis of carbohydrate takes place. Sometimes the term mixotrophism has been used to describe this condition.

Heterotrophic plants are differentiated into parasites and saprophytes according as the organic material is derived from the bodies of living organisms or from dead organic matter, that is, the dead bodies of animals or plants or products of their decomposition. Where some photosynthesis takes place as well, the plants are either partial parasites or partial saprophytes. Further, it has been shown that some autotrophic plants can absorb and utilize organic materials to a certain extent, and so when kept in the dark and provided with such organic nutrients they exhibit a degree of heterotrophism. Similarly, the difference between saprophytes and parasites also is not absolute. Some fungi, for example, can develop either saprophytically or parasitically, while others which are normally parasitic have been cultivated on artificial media as saprophytes. A distinction is often made between obligate and facultative parasites, the latter being able to grow without obtaining material from a host plant, whereas without a living host as a source of organic nutriment the obligate parasite is unable to develop.

Related to the heterotrophic condition is that of symbiosis in which two organisms of different species live together in a close association apparently to their mutual advantage, or at least without any indication of disadvantage to either of the associates or symbionts. A number of cases exist of symbiosis in which two species of plants are concerned, while cases are known in which the association is between a plant and an animal.[1]

SAPROPHYTES

The majority of saprophytes are found in the groups bacteria and fungi, among which the moulds (including Zygomycetes and a number of Ascomycetes) and many Basidiomycetes grow on non-living organic material. There are a few saprophytic algae, such as *Polytoma*, a member of the Chlamydomonadaceae, and some supposed saprophytes are also met with among higher plants. These are plants which grow in soils rich in decaying organic matter or humus and include the yellow bird's nest (*Monotropa hypopitys*) and some orchids, namely, the bird's-nest orchid (*Neottia nidus-avis*), the coral root (*Corallorhiza innata*) and *Epipogon aphyllum*. Associated in symbiotic relation with the roots of these plants are fungal mycelia to which further reference will be made later.

By far the vast majority of observations dealing with the nutrition of saprophytes concern the fungi. While, as noted in the last

[1] The term symbiosis was originally used by de Bary to include all kinds of intimate association of two organisms, and is still used in this wide sense by some writers. Thus C. S. Gager (*Fundamentals of Botany*, Philadelphia, 1916) distinguishes four types of symbiosis: (1) disjunctive or social symbiosis, (2) epiphytism, (3) mutualism, (4) parasitism. Disjunctive or social symbiosis may be nutritive or non-nutritive. An example of nutritive social symbiosis is afforded by the association of certain species of ant and fungi. The ants sow fungus spores on leaves removed from trees; the mycelium of the fungus grows on the leaves and is harvested by the ants as a crop. Pollination of flowers by insects constitutes non-nutritive social symbiosis between the flower and insect. In epiphytism one plant merely serves as a mechanical support for the other and there is no nutritional relationship between the two. The growth of mosses and ferns on the trunks of trees affords well-known examples of epiphytism. In mutualism the association is a close nutritional one in which the two organisms derive mutual advantage from the association, or at least apparently suffer no disadvantage. The association is so close that the two organisms often appear externally as one, as in the case of the lichens, and certain algal-containing worms. It is, perhaps, more usual to exclude epiphytism and parasitism from the usual conception of symbiosis, which thus includes disjunctive symbiosis and mutualism, or conjunctive symbiosis.

chapter, it is held that calcium, at least in any but very small quantity, is unessential for the development of some fungi, the other elements generally regarded as essential for the growth of higher plants are also necessary for that of saprophytic fungi. As source of organic food a very large range of carbon compounds has been found to be utilizable by saprophytic fungi, although some individual differences between species have been observed in this respect. Of organic compounds not containing nitrogen, sugars appear to be most readily absorbed by the majority of moulds, while other compounds forming a source of carbon supply for saprophytic fungi are mannitol, quinic acid, glycerol, citric acid, tartaric acid, lactic acid and fats among many others. There are even saprophytic organisms which can utilize paraffins as a source of carbon. Among such are a species of *Penicillium* and the Actinomycetes, a group of organisms of doubtful systematic position exhibiting relations with both bacteria and fungi. Among nitrogen-containing organic compounds peptones and amino-acids can frequently provide a satisfactory source of carbon for these plants.

With regard to a supply of nitrogen many saprophytic fungi can absorb this element in an inorganic form, particularly as ammonium nitrate. Others require a supply of organically combined nitrogen, as, for example, that provided by amides or amino-acids, peptones or even more complex proteins, while a number can absorb their nitrogen in the form of inorganic salts but grow more vigorously when supplied with organic nitrogen compounds.

As regards the capacity of various saprophytes to utilize different organic nutrients, there is considerable variation. Some species are capable of readily utilizing a wide range of substances, others exhibit more limitation in regard to substances they can make use of readily. For this reason an attempt is sometimes made to distinguish between omnivores and specialists. Some of the common moulds, such as *Aspergillus niger* and *Penicillium glaucum*, undoubtedly come within the first category, but even here there may be very marked differences in the nutritive value of different substances. Thus the moulds in question absorb *d*-tartaric acid very much more readily than its stereo-isomer *l*-tartaric acid, while *Aspergillus fumigatus* absorbs the two forms equally well. On the other hand, it is doubtful whether there actually exist true specialists, in the sense of saprophytic organisms which can utilize only one or a few nutrients, although no doubt there are many which utilize some few substances better than any others. Certainly as a rule saprophytic fungi can utilize very many different substances, but the metabolic processes in the organism may vary greatly with the nature of the substrate on which the organism is growing, and the end-products of the metabolism of the same organism may be varied very considerably according to the nature of the organic nutrient with which the plant is provided, as the work of Raistrick and his collaborators has shown for a number of saprophytic fungi, including many species of *Aspergillus* and *Penicillium*. But both the utilization of material and the metabolic changes within the organism depend on other external conditions, including the hydrogen-ion concentration of the medium and other substances present. This is certainly the case with regard to the utilization of nitrogen. Attempts have, indeed, been made to arrange saprophytic fungi and bacteria into groups according to the most favourable source of nitrogen supply, namely, elemental nitrogen, nitrates, nitrites, ammonium salts, amides and peptones, yet it must definitely be regarded as doubtful whether such a classification can have any rigidity.

Little is known regarding the immediate fate of the organic compounds absorbed by saprophytes or of the subsequent changes which lead to the building up from these substances of the various constituents of the plant body. Clearly enzymes play their part in the life of such plants as they do in autotrophic plants, and a large number of enzymes with similar characteristics to those from higher plants have been extracted from fungi. Thus diastase enzymes, maltase, sucrase, lipase, protease and hemicellulase are all found in fungi. Many fungi appear, indeed, to possess the capacity to develop enzymes in relation to the composition of the organic nutrient. Thus, according to Knudson, *Aspergillus* only develops the enzyme tannase when provided with tannic acid or gallic acid. In other cases, where the enzyme is formed in the plant whether the enzyme substrate is provided or not, the enzyme is formed in greater quantity when its particular substrate is present. These phenomena are described respectively as qualitative and quantitative regulation of enzyme formation.

In the higher autotrophic plants the cell membranes are very generally impermeable to enzymes, but a number of fungi can utilize starch as a nutrient, from which it is evident that diastase must pass out of the plant in order to bring about the transformation of the starch into soluble sugar.

Where saprophytes absorb their carbon as carbohydrate and their nitrogen in the form of a nitrate or ammonium salt, the synthesis of the cell materials may quite possibly follow the same lines as in autotrophic plants, but where the carbon is absorbed in some form other than carbohydrate, and where the nitrogen is absorbed as an organic compound, clearly the course of the vital syntheses must be more or less modified. But when it is considered how much the nature of these syntheses is a matter of conjecture in the normal autotrophic plant, it is not surprising that little is known of them in heterotrophic plants.

HUMUS PLANTS

The flowering plants generally regarded as saprophytes are all included in a few families, Orchidaceae, Burmanniaceae and Triuridaceae among monocotyledons, and Monotropoideae and Gentianaceae among the dicotyledons. They are essentially plants growing in humus, and apart from the absence of chlorophyll they possess other characteristics. Important among these from the physiological point of view are the feeble development, or complete absence, of root hairs, and, except perhaps in one species of orchid, *Wullschlaegelia aphylla,* the association with the subterranean absorbing organs of fungal mycelia. To such an organ, consisting of a root with associated fungus, the name mycorhiza was given by Frank in 1885.[1] The spelling mycorrhiza is now more usually adopted. In *Monotropa* the fungus is external to the root and the mycorrhiza is therefore described as ectotrophic, while in the orchids and other members of this group of heterotrophic plants the fungus is present inside the cells and the mycorrhiza is therefore said to be endotrophic. The subterranean organs themselves are sometimes roots, but in other cases true roots are absent and the fungus is developed in the cells of the rhizome. Frequently the development of the roots is arrested and they present a clustered appearance, the so-called ' coralloid ' arrangement.

It would seem certain that the associated fungi must be connected in some way with the nutrition of these plants. The latter, possessing no chlorophyll, must obtain their supply of carbon from organic compounds present in the humus surrounding the roots or rhizomes. There are obviously two possibilities. Either the plants themselves can absorb these organic nutrients, or the latter are first absorbed by the fungal mycelium and passed on, either changed or unchanged, to the cells of the root or rhizome. In some plants the poor development of root hairs, and the presence of a fungal investment covering the root, favour the second view, though much variation exists in this respect. There is, at any rate, some likelihood that the absorption of carbon compounds takes place by means of the fungus. Moreover, it is likely that water and mineral salts are absorbed through the same channels while it is possible that organic nitrogen compounds are as well. If this is a correct interpretation of the relationship between flowering plant and mycorrhizal fungus the question arises whether it is correct to regard the former as saprophytes, and it has been suggested they would be more correctly regarded as parasites upon their associated fungi, since the angiospermous humus plant receives its food supply from the fungus and gives it nothing in return. Moreover, the condition of the fungus in *Neottia* suggests that the fungus suffers damage as is usually the case with the host when attacked by a parasite. As described by Magnus in 1900, the fungus in this

[1] Frank defined mycorhiza in the following terms : ' a union of two different individuals into a single morphological organ, which can probably be conveniently denoted as fungus-root, mycorhiza '. Since then many authors, as, for example, M. Skene (*The Biology of Flowering Plants*, London, 1924), have used the term mycorhiza to denote, not the organ itself, but the *condition* of association of root and fungus. The spelling ' mycorrhiza ' is now in almost general use. Although frequently done, there appears to be no justification for the use of ' mycorhiza ' or ' mycorrhiza ' as a plural as well as a singular noun.

orchid is found in the cells of the outer cortex, it is not present in the inner part of the cortex and only to a slight extent outside the root. The infected region of the cortex appears to form three distinct layers; there is an outer 'digestive' layer, a middle layer of so-called 'host' cells, and an inner 'digestive' layer. In the host cells the mycelium is differentiated into thick-walled hyphae lining the cell-walls, and finer hyphae which penetrate the cytoplasm and vacuole, and which are supposed to be absorptive in function. The hyphae in these cells persist, but those in the digestive cells, which at first are thin-walled and rich in protein, disorganize into a structureless mass which becomes partially digested.

Observations on the mycorrhizas of *Neottia nidus-avis* described by Wolff in 1926 throw some light on the problem. Wolff isolated the fungus from the mycorrhizas of this plant and found it belonged to the genus *Orcheomyces* which had previously been isolated from the aerial roots of epiphytic orchids. The fungus can be cultivated in nutrient media containing glucose and inorganic salts, and the fact that nitrogen can be dispensed with in the culture solution indicates that molecular nitrogen from the air can be assimilated by this fungus. Analyses showed, indeed, the presence of nitrogen compounds in the culture when none were provided in the culture medium. Wolff further found that although the fungus will grow on a medium in which all the carbon is provided as glucose, it thrives much better on tannin. Apparently the fungus secretes the enzyme tannase which splits up tannin into glucose and gallic acid, and uses both constituents of the glucoside. The *Neottia* plant itself contains tannin, and Wolff thought the plant absorbed tannin from the humus, and the mycorrhizal fungus then absorbed the tannin from the cells of *Neottia*. If this view of the relations between angiosperm and mycorrhizal fungus is correct then the association between the two plants is rather one of conjunctive symbiosis than of fungal host and angiosperm parasite. How the nutritive conditions in *Neottia* and other plants with endotrophic mycorrhizas compare with those in *Monotropa*, the mycorrhizas of which are ectotrophic, is not clear, but it must be noted that the distinction between the ectotrophic and endotrophic condition is rather one of degree than an absolute one, for in the former there is generally penetration of the outermost layer of cells of the root with occasional penetration of deeper tissues.

While the *Neottia* plant thus appears to act as a parasite as far as the digestion of the outermost and innermost layers of the mycelium are concerned, there is a suggestion that the middle regions of the fungus may obtain something from the cells of the *Neottia* plant. There is thus at any rate the possibility that the relation between the two associates has some of the features of conjunctive symbiosis. McDougall indeed suggested the term symbiotic saprophytism to express such a relation, and if this mode of life is differentiated from true saprophytism, in which organic food is absorbed directly by the plant without the intermediary of a mycorrhizal fungus, truly saprophytic angiosperms would be limited to the single case of *Wullschaegelia aphylla*, which is the only non-chlorophyllous angiosperm which is definitely not a parasite and which has been reported as lacking mycorrhizas. The very fact that it should occupy such an isolated position from a physiological point of view has suggested the need for its re-investigation to make sure that it is indeed devoid of mycorrhizas. It will thus be clear that considerable doubt exists regarding the mode of nutrition of the so-called saprophytes met with among angiosperms.

It may be noted that a number of plants growing in humus and provided with mycorrhizas but developing chlorophyll have been regarded in the past as hemi-saprophytes. Direct evidence of absorption of organic food material by such plants appears, however, to be wanting.

PARASITES

As with saprophytes, the vast majority of parasitic plants are fungi and bacteria, while a comparatively small number of angiosperms are also parasites. Of the fungi, the rusts (Uredinales) and smuts (Ustilaginales) are all parasites, as are many of the Oomycetes and Ascomycetes and some of the Basidiomycetes. As these parasites produce disease in their hosts they are known as pathogenic fungi or pathogenes. The majority of them are parasitic on plants, but a number are parasitic on animals. Thus *Cordyceps militaris* is parasitic on caterpillars while the miscellaneous group known as dermatophytes are parasitic on the skin and hair of human beings and other mammals, and produce diseases such as ringworm. Bacteria include numbers of animal and plant pathogenes.

Complete parasites derive their nutriment entirely from the body of the host plant, but theoretically it should be possible to grow the parasite on a non-living medium containing the substances which it normally obtains in the body of the host plant, and so cultivate it saprophytically. In a number of instances it has been found possible to do this, and investigation of plant pathology has been greatly aided by the artificial cultivation of pathogenic forms in this way. Other pathogenic fungi it has not so far been found possible to cultivate in this way. Such may be termed obligate parasites. On the other hand, some fungi which are usually saprophytic can penetrate living tissues, sometimes by way of previously formed wounds, and adopt a parasitic mode of life in the tissues. Such forms are spoken of as facultative parasites.

With parasites a division into omnivores and specialists is much more justifiable than with saprophytes. Thus *Sclerotinia sclerotiorum* is parasitic upon, and causes diseases of, tomatoes, potatoes, mangolds, carrots, artichokes, beans, lettuce and *Antirrhinum*, while *Botrytis cinerea* also attacks a great variety of plants, including tomatoes, potatoes, lettuce, various bulbs, and also the young shoots of some woody plants such as gooseberries, vines and conifers. But many parasitic fungi attack plants of a single species or of a few nearly related species. This is especially the case with many of the rusts and smuts. Thus *Endophyllum sempervivi* attacks only species of *Sempervivum*, while *Phragmidium rubi-idœi* appears to confine its attack to the raspberry. In some cases what appears to be a single species morphologically is a collection of varieties each of which limits its attack to one particular species or variety of host plant. Such closely related parasitic forms are termed physiological or biological species. They appear to be most highly developed in rusts of cereals. Thus within the morphological species *Puccinia graminis*, parasitic on cereals, a number of varieties have been recognized, each of which attacks a different species of cereal, while within each variety of the fungus a number of biological forms have been identified each of which attacks a different variety of the host species. The phenomenon of heteroecism is also met with to a considerable extent in the rusts, but also to a slight extent in other groups of fungi. Here more than one host plant is involved in the life-cycle of one species of parasite, one stage in the life-cycle being carried out on one host, another stage on a different species.

The special physiology of the relations of host and fungal parasite is generally regarded as coming within the province of plant pathology. There are, however, a few general observations which may be made here. In the first place it will be noted that the parasitism is only possible if the fungus is able to invade the tissues of the host plant. Frequently invasion of the tissues of the host commences with the germination of the fungal spore on the surface of the host, the young hypha produced then either entering through a stoma or lenticel or penetrating directly into a surface cell. Where the invasion of the host is by way of stoma or lenticel, the cells of the host are entered sooner or later, but the mechanism of this entrance is not understood. It would be readily explained by the fungus secreting an enzyme or some toxic substance which digested, dissolved or weakened the cell-walls of the host plant, but although at one time it was assumed that this must be the case, more recent work has failed to produce evidence that entry of the fungus is effected in this way. It may be noted that corky cell-walls in general offer effective resistance to fungal penetration.

The attack of a parasitic fungus generally results in certain morphological changes in the host which constitute the symptoms of the disease. These include changes in colour resulting from a reduction in the quantity of chlorophyll or from its degeneration. Sometimes the presence of the parasite may result in feebler development of tissues and may bring about dwarfing, while conversely the fungal attack may bring about increased growth with hypertrophy of certain parts, the production of excrescences and abnormal morphological developments which can be described as malformations. Such developments as galls and witches' brooms on aerial parts of plants, and club root or finger-and-toe on roots of *Brassica* plants, are examples of such abnormal growths. These developmental changes must be connected in the first place with the physiological relations between host and parasite. Thus the hypertrophy of tissues resulting from fungal attack can be related to the stimulus to development resulting from wounding. The physiological problems of the relation between fungal parasite and host have, however, for the most part, still to be worked out.

ANGIOSPERMOUS PARASITES

Complete, or almost complete, parasites are not common among angiosperms. Such as exist fall into two groups, those parasitic on the roots and those parasitic on the stems of other vascular plants. Root parasites are represented in this country by two genera of Orobanchaceae, *Orobanche* and *Lathraea*. The species of *Orobanche* are parasitic on the roots of clover and other plants, while the native species of *Lathraea, L. squamaria* is often parasitic on the roots of hazel, but sometimes on those of other plants. In these plants the aerial parts consist of a shoot with scale leaves, terminating in the inflorescence, while in the tropical root parasites, members of the Rafflesiaceae, the whole plant is reduced to little more than the flowers, the flowering shoots arising as adventitious buds on what is little more than a mycelium. In root parasites the roots produce haustoria which penetrate the tissues of the roots of the host plant, and xylem and phloem of the parasite come into contact with those of the host, so that both water and solutes contained in the xylem, and manufactured food materials in the phloem, of the host are diverted in part into the parasite.

The only native stem parasites in this country in which parasitism is practically complete are the three species of *Cuscuta* (dodder) parasitic on heather, flax and some other plants. Here union of tissues of the host and parasite takes place in much the same way as with root parasites, haustoria from the twining stems of the parasite penetrating the tissues of the host so that here also the xylem and phloem of the parasite come into contact with those of the host.

Apart from the method of obtaining the food supply the nutritional physiology of these parasites is presumably the same as that of autotrophic plants.

Cuscuta and some species of *Orobanche* are stated to contain a little chlorophyll, so presumably they are capable of weak photosynthesis. No green coloration is, however, obvious to the naked eye, and these plants are no doubt virtually total parasites. They are unable to live without their respective hosts, and indeed, in *Orobanche*, for example, the seeds can only germinate in the neighbourhood of the roots of the host plant.

HEMI-PARASITES

In addition to these practically total parasites there exist among flowering plants a number of partial parasites which are green in colour, but which are also parasitic to some extent on other flowering plants. They include both root and stem parasites. The best known of the former are the Rhinantheae, a sub-family of the Scrophulariaceae including *Rhinanthus*, *Bartsia*, *Melampyrum* and *Euphrasia*. Other examples are met with in the Santalaceae, represented in this country by *Thesium humifusum*. The best-known stem hemi-parasite is the mistletoe, *Viscum album*, but the family to which this plant belongs, the *Loranthaceae*, contains some hundreds of semi-parasitic species, while other stem semi-parasites occur in the Santalaceae and Myzodendraceae.

In the Rhinantheae the degree of parasitism varies greatly. Thus the tropical *Striga orobanchoides*, according to Heinricher, is a total parasite, while *Tozzia* spends most of its life underground as a complete parasite, but later develops a slight photosynthetic activity in aerial shoots which persist for only a few weeks. Another species of *Striga*, namely *S. lutea*, is a total parasite in the early stages of its development, but later develops chlorophyll. The seeds of these species, like those of *Orobanche*, only germinate in the neighbourhood of the roots of the host plant. In our native Rhinantheae, however, the seeds can germinate in absence of the host plant and can develop to a certain extent independently. *Euphrasia rostkowiana*, according to Heinricher, is only capable of feeble independent development, while *E. odontites* and *E. minima* can complete their development from germination to seed formation independently of a host plant. Most of the Rhinantheae occupy a position intermediate between these two extremes.

The exact nutritional relations of the root hemi-parasites are in some doubt. Their photosynthetic activity, except in the two or three extreme cases of parasitism noted, appears to be normal, and it is usually supposed that the chief advantage they derive from their parasitic habit is the absorption of water and salts from the host, although the possibility of absorption of some organic material from the host is not excluded, particularly in the early stages of development of the parasite.

The same appears to be the case with the stem hemi-parasite *Viscum album*. Here the leaves are green and photosynthesize actively, while according to Pierce only the xylem of host and the haustoria of the parasite come into contact. Ringing the stem of the host above and below the haustoria of the parasite, so as to remove all tissues outside the xylem, has no adverse effect on the parasite, which also suggests that the mistletoe plant does not depend on its host for a supply of organic nutrients.

If this is the correct view of the relation between host and parasite in the case of these chlorophyll-containing hemi-parasites, the latter clearly differ from typical heterotrophic plants which are incapable of manufacturing their own organic food material. They differ in their nutrition from typical autotrophic plants in absorbing their water and mineral salts wholly or in part by way of the host plant instead of directly from the soil, but apart from this their nutrition appears to be normal.

INSECTIVOROUS (CARNIVOROUS) PLANTS

Insectivorous or carnivorous plants constitute a group of heterotrophic plants which may be regarded as semi-parasitic. Their common characteristic is that by means of some special device insects or other small animals are trapped and organic material from their bodies, after decay or digestion, is absorbed by the insectivore.

About 400 species of insectivorous plants are known, but they are all included in a very few families: Droseraceae, Lentibulariaceae, Sarraceniaceae, Nepenthaceae and Cephalotaceae. The trapping devices are of three main types which are met with respectively in (1) the Droseraceae and in *Pinguicula*, a member of the Lentibulariaceae, (2) in *Utricularia*, another genus of the Lentibulariaceae and (3) in Sarraceniaceae, Nepenthaceae and Cephalotaceae.

Droseraceae. This family contains six genera with about 100 species, all of which are insectivorous. The most widely distributed genus is *Drosera* (sundew) while the other five genera contain only seven species altogether. The best known of the latter are *Dionaea* and *Aldrovanda*.

In *Drosera* the upper surface of the leaf bears numerous glandular hairs or tentacles, those towards the margin of the leaf having long stalks, those in the middle shorter ones. When an insect alights on a leaf, as will be described in a later chapter, the outer tentacles bend over the insect and the glands, both of the incurved outer tentacles and of the short erect middle ones, all secrete a fluid in which the insect becomes partly or entirely submerged, this being helped by an incurving of the leaf margin which renders the leaf somewhat cup-shaped. The secretion contains proteases by which the nitrogenous substances of the small animal are converted into amino-acids, which are most probably absorbed by the cells of the glands. This process may take several days, the length of time depending on the size of the animal digested, but after the digestion of the animal is complete the tentacles return to their original position and are then ready to respond again to the contact of another small animal. There is always a certain amount of sticky secretion on the surface of the leaf and it is possible, though not definitely established, that insects may be attracted by this secretion.

In *Drosophyllum*, *Roridula* and *Byblis* the conditions have generally been regarded as similar except that the tentacles have no power of movement. Recently, however, Lloyd has examined the case of *Roridula* and concludes that this plant is not carnivorous, the secretion of the glands being resinous and not digestive.

In the two other species of Droseraceae, *Dionaea muscipula* and *Aldrovanda vesiculosa* the trapping mechanism is different. When an insect alights on the upper surface of the leaf the two halves of the bilobed leaf move upwards and close together with the insect between them, the midrib acting as a hinge. The movements are discussed in a later chapter. In *Dionaea*, glands exude a digestive fluid as in *Drosera*, but there is some doubt regarding the secretion of a digestive fluid in *Aldrovanda*.

In *Pinguicula* also, insects are caught by the sticky secretion on the leaves and are then partially digested by the proteases secreted by the glands of the leaf as in the Droseraceae.

Utricularia. The species of *Utricularia* are either, like the British species *U. vulgaris*, *U. major*, *U. intermedia* and *U. minor*, aquatic plants, or, like many tropical species, are terrestrial species growing in a wet substratum. The plants are devoid of roots and the leaves are much divided, some of the leaf segments being modified to form bladders. Small aquatic animals are trapped in these bladders, and the mechanism of this trapping has been the subject of observations by a number of observers, notably by Darwin, Goebel and Lloyd, the last-named of whom has examined between 75 and 80 species.

The bladders are borne on short stalks and in *U. vulgaris* are about 2·5 mm. long, but in *U. cornuta* and *U. capensis* only about 1 mm They are oval in shape but rather more pointed at one end than the other. At the more pointed end is an entrance into the interior closed by a door of tissue which acts as a valve, the door opening inwards and being attached to the wall of the opening above and resting on the lower margin of the opening when closed. A number of long hairs, often branched, arise from the outer surface of the door. In some species slight opening of the door appears to result from a small animal pushing against it, while in other cases contact with the hairs attached to the door may bring this about. When the door has been slightly opened in this way water enters the bladder carrying with it the small animal; the door then closes and the animal is trapped inside the bladder. The explanation usually given of this action is that the walls of the bladder are practically impermeable to water, while attached to the inner wall of the bladder, and so projecting into its cavity, are a number of hairs which withdraw water from the bladder cavity, the wall of which therefore tends to collapse inwards. According to Lloyd, the histology of the door is such that it tends to move outwards; hence it will rest closed against the margin of the entrance unless pressure is exerted on it from the outside. According to Lloyd, the door when closed is practically watertight. Now if the door is opened slightly there will result a rush of water into the bladder, since as a result of the absorption of water by the internal hairs combined with the practical impermeability to water of the bladder-wall and the watertightness of the door, the hydrostatic pressure inside the bladder will be somewhat less than that of the water outside. Even before the pressure inside and outside has equalized, the door closes.

The whole process is extremely rapid, and for this reason Lloyd describes it as a snap-action mechanism. By taking cinemicrophotographs this investigator showed that the whole action takes place in less than $\frac{1}{16}$ second, the opening of the door occupying less than $\frac{1}{160}$ second and the closing of it about $\frac{1}{40}$ second.

No evidence has so far been produced that any enzyme is excreted by *Utricularia* into the bladders, and it is usually supposed that the trapped animals die in the bladder and then form the substrate of bacteria, as a result of which they decay and their decomposition products are then absorbed, either by the cells of the bladder-wall or, more probably, by the internal glandular hairs.

As well as in the 250 species of *Utricularia*, the same mechanism occurs in the related genera *Biovularia* with one species and *Polypompholyx* with three species.

Pitcher Plants. Members of the third group of insectivorous plants are characterized by the possession of certain leaves the apical portions of which are modified into the form of a pitcher capable of holding liquid. The size of the pitcher varies greatly from species to species, so that its content is no more than a cubic centimetre or so in some, while in other species it may be more than a litre. In *Nepenthes* the leaf-blade proper is continued into a tendril-like extension which terminates in the pitcher, and the tendril, by coiling round a support, maintains the pitcher in a vertical position. The pitcher is provided with a lid, which is at first closed but which later opens and remains so. In the Sarraceniaceae the whole leaf develops as a long slender pitcher with a terminal lid opening as in *Nepenthes*.

In *Nepenthes* the inner wall of the pitcher from the bottom to halfway up or higher bears numerous multicellular glands which secrete a fluid which passes into the pitcher. The upper part of the wall, which is devoid of glands, is covered with wax. It is supposed that insects are attracted to the pitcher by nectar secreted by glands situated on the outer surface of the pitcher, on the inner surface of the lid and on the inner edge of the margin of the pitcher opening. If the animal enters the pitcher it fails to find a foothold on the slippery wax-covered internal wall of the pitcher and slips into the liquid in the lower part of the pitcher. In this liquid the insect is digested and products of its digestion are absorbed by the cells of the pitcher walls, possibly by the glands.

The usual view held with regard to the means by which digestion of the insect is effected, and the one supported by Goebel and Vines, is that the pitcher liquid itself contains protease enzymes which are instrumental in effecting the digestion. Hepburn found that the pitcher liquor possesses no proteolytic activity until it has received an animal or some nitrogenous material, after which it is capable of digesting a number of proteinaceous materials. A distinction is thus made between 'stimulated' and 'non-stimulated' pitchers. The difference between them appears to be due to the fact that the presence of an insect stimulates the glands to increased secretion, the latter containing the protease. In the presence of a weak acid, however, digestion of some substances is effected by liquor from non-stimulated pitchers; such is the case with carmine fibrin. With liquor from stimulated pitchers digestion is accelerated by addition of weak acid.

Hepburn further found that in open pitchers bacteria are fairly numerous, but that when isolated from the pitcher liquor they usually digest protein very slowly. His general conclusion is, therefore, that the chief factor in the digestion of the insects is the protease present in the pitcher liquor after stimulation by the entrance of an insect. Any part played by bacteria in the digestion is a very secondary one. The enzymes present in the body of the insect may play some part by bringing about autolysis of material of the insect tissue.

In a later investigation on pitcher plants belonging to the family Sarraceniaceae, Hepburn, St. John and Jones, found the same state of affairs in species of the genus *Sarracenia.* In these plants the pitcher liquid appears to exert either a narcotic or wetting action on insects so that their movements are retarded. The pitcher liquor, like that of *Nepenthes,* contains proteases and also proteolytic bacteria, which no doubt play similar parts to those performed by them in *Nepenthes.* But in a second genus of Sarraceniaceae represented by the single species *Darlingtonia californica,* the pitchers do not appear to secrete any proteolytic enzymes, although proteolytic bacteria may be present.

Hepburn and his collaborators showed experimentally that the pitcher walls of both *Sarracenia* and *Darlingtonia* absorb water from the pitcher cavity. They also introduced aqueous solutions of various nitrogen compounds into the pitchers of plants of *Sarracenia,* and showed by chemical analysis that these substances (acetamide, urea, asparagine, peptone) were absorbed by the cells of the pitcher; so also were phosphate and lithium ions. There can be little doubt that in *Sarracenia,* as in *Nepenthes,* insects are digested mainly by proteases secreted into the pitcher liquid from glands on the internal walls of the pitcher, while proteolytic bacteria and autolysis of the insect tissues may play subsidiary parts in the digestion. What the position is with *Darlingtonia* is not so clear. Liquid is secreted as a result of stimulation by pieces of raw beef, though not appreciably by pieces of cooked beef, and the absence of protease in the liquid secreted suggests that the supply of nitrogenous material to the plant by the bodies of trapped insects must depend in this plant on the slower processes of bacterial action and autolysis.

The West Australian pitcher plant, *Cephalotus follicularis,* is the single species of the Cephalotaceae, a family related to the Saxifragaceae. The small pitchers show the same general peculiarities as those of

Nepenthes, and according to Dakin the pitcher liquor contains protease, which digests blood fibrin in presence of weak acid. The conditions in this plant have not been fully worked out, but it appears possible that they are generally similar to those in *Nepenthes*.

The usual view taken of the significance of the insectivorous or carnivorous habit is that it provides the plants concerned with a supply of nitrogen additional to that which is absorbed from the medium in which the plants grow, and it would certainly appear to be the case that many of these plants grow in soils or water deficient in nitrates. Thus a number occur in swamps and bogs in which deficient aeration limits the development of nitrifying bacteria. It has been suggested that the absorption of salts apart from nitrogen compounds is important, while Dakin further suggested that in *Cephalotus*, where the conditions under which the plants grow are unfavourable for transpiration, the development of pitchers provides for a water current in the plants by secretion of water by the glands, but whether a current of water to the pitcher-walls is really of value in the life of this plant depends on the significance of transpiration in itself, a matter which is in some doubt, and which has been discussed in an earlier chapter.

SYMBIOSIS

A number of cases of symbiosis, used in its restricted sense of the close association of individuals of different species for their mutual advantage, exist in the plant kingdom. The best known are the lichens, in which the symbionts consist of an alga and fungus; leguminous plants with bacteria living in the root tubercles; and many forest trees and other plants in which there is mycorrhizal development. There are also a number of examples of symbiosis between plants and animals. Among these are the chlorophyll-containing worms (*Convoluta roscoffensis* and *C. paradoxa*), in which an alga lives in the body of the worm. Less intimate examples of animal and plant symbiosis are those in which ants play a part. These include the case of myrmicophilous plants in which ants live in cavities of the pulvini of the leaves and protect the leaves from attack by leaf-cutting ants, and the case of fungi cultivated by ants for food material. Still less intimate is the connexion between plants and pollinating insects and between plants and the animals which, by various means, disseminate fruits and seeds.

Lichens. The thallus of the lichen consists of two distinct constituents, an alga and a fungus. In some lichens it has been possible to isolate the two constituents and cultivate them independently. Thus the algal constituent of three lichens, *Physcia parietina*, *Evernia furfuracea* and a species of *Cladonia*, was found to develop exactly as the unicellular green alga *Cystococcus humicola*, while the algal constituent of the lichen *Peltigera canina* appears to be identical with the alga *Gloeocapsa monococca*. Similarly the fungal constituents of *Lecanora subfusca*, *Xanthoria parietina*, and other lichens have been cultivated independently of the alga in artificial media.

In some few lichens a state of affairs has been described which has suggested to the observers that in these particular cases the fungus may be to some extent parasitic on the alga, but whether such is a correct interpretation of the observed facts has been disputed. Thus the presence of dead cells in the lichen thallus may be due to natural decay, and if material of these dead cells is absorbed by the fungus the condition is not one of parasitism. At any rate, there seems no doubt that in at least the vast majority of lichens both algal and fungal constituents maintain a healthy life and the condition is one of symbiosis in the narrow sense sometimes described as mutualism.

The algal symbiont, being green, manufactures carbohydrate by the ordinary photosynthetic process. The carbon dioxide for this it obtains from the air. But the water, as well as mineral salts, which it would normally absorb from the medium in which it lives it must obtain from the fungus, for the algal cells are surrounded by fungal mycelium and do not generally come into direct contact with the substratum. There is some evidence that the algal symbiont of the lichen *Physcia parietina* can absorb peptone, and Chodat found that in artificial cultures of the algal symbionts of lichens, these forms developed much more satisfactorily when provided with nitrogen in the form of glycine or peptone than when the nitrogen was presented as potassium nitrate. There is thus at least a possibility that the alga obtains a supply of organic nitrogen from the fungus as well as a supply of water and mineral salts. There is also some evidence that the alga can absorb carbohydrate in the same way. This has been shown by cultivating lichens on artificial media containing organic substances and mineral salts and depriving them of carbon dioxide. Under these conditions the alga developed satisfactorily, even in the dark, especially when the medium contained glucose. In *Physcia parietina* it has similarly been shown that the alga can utilize for its nutrition oxalates excreted by the fungus. But in general the conclusion is drawn that the alga photosynthesizes carbohydrate in excess of its own requirements, the excess being absorbed by the fungus.

Leguminosae. It is well known that members of the family Leguminosae develop on their roots small tubercles or nodules as a result of rootlet infection with a species of bacterium known under a number of different names including *Bacillus radicicola* and *Pseudomonas radicicola*, which absorbs nitrogen from the soil atmosphere. From this nitrogen certain compounds are manufactured by the bacteria and excreted into the cells of the rootlet, but what these substances are has not been established. Culture experiments have clearly shown that the power of the leguminous plant to fix atmospheric nitrogen is entirely due to the bacteria of the root tubercles. In this symbiotic relationship the bacteria thus provide the leguminous plant with nitrogen compounds while the bacteria receives from the root water and salts and perhaps carbon compounds.

Root tubercles of similar appearance are met with in species of *Alnus, Myrica, Eleagnus* amd *Ceanothus*, and these tubercles also contain bacteria of *Pseudomonas radicicola* or some nearly related species.

Mycorrhizas. The existence of the mycorrhizal habit in humus plants has already been mentioned, but it is not by any means confined to them and is, indeed, of very wide distribution throughout the plant kingdom. Endotrophic mycorrhizas occur typically in the Orchidaceae and Ericaceae, while they are also present in many species of Gramineae and other plants. Ectotrophic mycorrhizas are present in forest trees such as beech, birch, oak, chestnut, aspen, pine and spruce, and shrubs such as hazel. Fungal mycelium is also met with in association with many Bryophyta and Pteridophyta. Much has been written with regard to the nutritive relations of the mycorrhizal fungus and its associated plant, and very diverse views have been held on this matter, but the more generally accepted view is that the condition is one of symbiosis in which both members derive benefit from the association. As regards the ectotrophic mycorrhizas of coniferous trees, a number of the fungal species concerned have been grown independently of the host, but under such conditions their growth is much slower than when grown in association with their hosts, and it is supposed that the fungi obtain carbohydrates and possibly other organic substances from their hosts. On the other hand, the roots are thought to obtain a supply of nitrogen and mineral salts by way of the fungal mycelium, for it has been found that the mycorrhizas in culture can utilize organic nitrogen compounds much more readily than uninfected seedlings. Thus seedlings of *Picea abies* when supplied with nitrogen in the form of peptone or nucleic acid show characteristics of nitrogen starvation, whereas similar seedlings infected with the mycorrhizal fungus show no sign of such starvation. It is also significant perhaps that woodland soils in which these trees with mycorrhizas flourish are generally poor in nitrates. The fungal constituents of the endotrophic mycorrhizas of the green Orchidaceae have been isolated and found to grow readily on artificial media in independence of their normal orchidaceous associate. On the other hand, the view is generally held that the fungus is necessary for germination of the orchid seed and subsequent development of the seedling, so that for the orchid the association is obligatory. While in nature this may be so, it may be noted that Knudson has not merely germinated orchid seeds in absence of a fungus, but has cultivated a *Laelia-Cattleya* hybrid on agar in absence of a fungus for 8 years, when the plant flowered. Orchid seeds, however, do not germinate when supplied only with water and nutrient salts, and Knudson found that for germination to take place a supply of organic food material (sugar) must be provided.

The nutritional relations between orchid plant and associated fungus are in much doubt. The fungal constituents of orchid mycorrhizas are mostly related to parasitic forms, and the view is generally held that the present association has been evolved from a condition of parasitism of the orchid root by the fungus. The condition is now generally regarded as one of symbiosis largely because neither fungus nor orchid appears to suffer damage from the association, and also because the association, as far as the germination and development of the orchid is concerned, is of an obligate nature.

Knudson's view regarding the association of the fungus in the germination of orchid seeds appears a reasonable one. Having shown the necessity for a supply of organic food material for the germination of the orchid seed, he concluded that the part played by the fungus in the germination is to bring about a transformation of insoluble organic food material in the medium to soluble material which can then be absorbed by the seed and seedling. He thus regards the orchid plant in the early stages of growth as saprophytic. The fungus he regards as playing the part of a parasite of a mild type, and there appears, indeed, to be no evidence that the mycorrhizal fungus in the adult orchid plant provides the latter with anything.

Among the Ericaceae the case of *Calluna* has been most completely investigated in regard to the relationships between mycorrhizal fungus and host. Here Rayner has shown that, although seeds will germinate in absence of the fungus, the seedlings fail to develop beyond an early stage. If the fungus is available, the germinating seed may be infected as soon as the radicle begins to elongate. As the seedling develops, the fungus spreads through the cells of the cortex of the root and into the shoot. Rayner pointed out that *Calluna* normally grows in humus soils very deficient in nitrates as well as other mineral salts, and she considers the relations between the associates to be similar to those in the ectotrophic mycorrhizas of forest trees. The fungus is thus supposed to enable the *Calluna* plant to draw on the nitrogenous material in the humus. Moreover, there is evidence that the fungus of the endotrophic mycorrhizas of *Calluna*, like *Pseudomonas*, can fix atmospheric nitrogen. However this may be, there is definite digestion of the fungal mycelium by the root cells of *Calluna*. It may then be supposed that the mycorrhizal fungus of *Calluna* provides nitrogen to the host plant and absorbs other materials from it. However, Rayner is of opinion that the obligate relation in *Calluna* is associated with infection and seedling development and that the plants would probably grow quite well if developed seedlings could be obtained free from the mycorrhizal fungus.

In the grass *Lolium*, McLennan found that the root fungus forms fat which ultimately finds its way into the cells of the host.

It will thus be clear that the relationships between mycorrhizal fungi and their hosts may be different in different cases, but in all of them the essential character of the relationship is nutritional.

REFERENCES

C. DARWIN, *Insectivorous Plants*. London, First edition 1875. Second edition 1888, and reprints.

M. C. RAYNER, *Mycorrhiza*. London, 1927.

BOOK III

THE PHYSIOLOGY OF DEVELOPMENT

CHAPTER XIV

THE COURSE OF DEVELOPMENT

It is one of the characteristics of the living organism that it undergoes development. Beginning as a single cell or a group of comparatively few cells, it increases in size and develops a form, in many cases very complex, which is determined by heredity and environment, the materials for the increase in size being furnished by this environment. It must be one of the aims of plant physiology to determine the laws governing this development.

It is convenient to divide reproductive methods in plants into three kinds: sexual, asexual, and vegetative.[1] In sexual reproduction the new individual arises from a single cell that has resulted from the fusion of two cells (sexual cells or gametes); in asexual reproduction a single cell, specially produced for the purpose, without a previous fusion, forms the commencement of the new generation; while in vegetative propagation a mass of ordinary body cells (rarely a single body cell) from the parent forms the starting-point of the new individual. Such a mass of tissue may either function solely as a reproductive body, as in the case of a gemma of *Marchantia*, a corm of *Crocus*, or a tuber of potato or *Dahlia*, or it may consist of ordinary organs of the individual, such as stems, roots, leaves, or portions of them, which under certain conditions, usual or unusual, grow into complete new individuals independent of the parent.

The course of development may be widely affected by the mode of origin of the individual. This can be illustrated very conveniently by certain ferns in which all three kinds of reproduction occur. In

[1] An examination of the text-books reveals that there is some diversity in the meaning attached to the terms 'asexual' and 'vegetative'. By some writers (e.g. Molisch) the terms are regarded as synonymous and refer to any kind of reproduction not involving a sexual fusion. By others (e.g. Scott) vegetative reproduction is regarded as a particular case of asexual reproduction, or, conversely, asexual reproduction is spoken of as a special case of vegetative reproduction (e.g. by Fritsch and Salisbury). Others, again (e.g. Tansley), regard the term asexual as referring only to those cases in which the new individual arises from a spore formed without a previous fusion, while vegetative reproduction signifies the arising of new individuals from cells of the vegetative body, occasionally from single cells but more usually from multicellular masses of tissue such as gemmae, whole shoots or parts of them, pieces of leaf or root tissue. The matter is, of course, chiefly one of definition, although perhaps not entirely, but to the writer the last method of classification appears the most convenient and so is adopted here.

Asplenium bulbiferum, for example, the asexually produced spore develops into the prothallus (gametophyte), the sexually produced fertilized egg cell develops into the sporophytic fern plant. On this are formed vegetatively the bulbils which develop into similar sporophytic fern plants, and asexually the spores which develop into prothalli. And although the organisms resulting from the fertilized egg and from the bulbil are similar, the early development must obviously be different in the two cases.

Development, to the eye at any rate, is not as a rule a continuous process. In most cases there is a period, or series of periods, during which the individual appears to remain unchanged. During these so-called 'rest periods' it is clear that in many cases some life processes continue, if only those connected with respiration, although the rate of the processes may be so slow as to be scarcely observable. On the other hand, there are many cases known in which developmental changes are actually proceeding during the apparent rest period.[1]

Frequently the asexually produced spore, and not the fertilized egg, undergoes, or may undergo, a period of rest before development of the individual commences. This is so where the spore is shed or otherwise becomes separated in space from the parent plant. In the Gymnosperms and Angiosperms this is obviously the case with the microspores, and here the rest period actually occurs after development of the spore has proceeded a little way towards the formation of a rudimentary prothallus. The sexually produced zygote develops into an embryo of generally considerable organization before a rest period occurs and the seed containing the embryo is separated from the parent.

Although the whole course of development must be regarded as continuous, it is nevertheless convenient to consider certain stages in the development separately. It is clear that germination, the emergence of the actively growing individual from the resting spore, zygote, seed, tuber or other organ, is a well-marked event or phase in the life history of most individual plants, but from what has been written in the preceding paragraph it is clear that the germination stage occurs at a different place in the life-history of different species.

Other periods of rest and subsequent sprouting may also occur, as in most perennial plants of temperate and arctic regions and many tropical plants as well, where, during a season with conditions unfavourable to

[1] Jost distinguishes between complete rest in which, consequent upon withdrawal of water, all activity ceases, and partial rest in which, with cessation of growth, various metabolic processes, particularly respiration, continue. Seeds and spores afford examples of the former, winter buds of trees examples of the latter. It is doubtful how far such a distinction is absolute, for it may be that in all cases where a structure remains alive respiration proceeds, albeit at so very slow a rate as to be unrecognizable. On the other hand, it may be that complete rest, with absolute cessation of all activity, yet with retention of vitality, is possible. The settlement of the question appears at present impossible.

growth, the whole plant may remain in an apparently dormant condition.

The production of reproductive bodies marks another stage in the development of the individual, but the position of this stage varies even more than that of germination.

The course of development, then, is not uniform throughout the plant kingdom. In annual flowering plants we may, however, conveniently consider the following stages as constituting the whole life-history :

1. Development of the zygote into the embryo, including seed formation.
2. Germination of the seed.
3. Vegetative development.
4. Reproduction, involving formation of flowers with asexual spores, the development of the gametophytes and structures in relation with them, and fertilization.

It is clear that the last two stages may overlap considerably, as in many species the production of a crop of flowers does not involve the cessation of vegetative growth.

In the perennial plant the course of development is generally complicated by the intercalation of rest periods with subsequent periods of sprouting as mentioned above. Here it may be some years after the commencement of the individual's life when the reproductive phase commences, and then quite usually vegetative development and reproduction proceed each year. There is in such plants an annual course of development largely comparable with the whole course of development of the annual plant, the stages being :

1. Renewal of activity of resting buds.
2. Vegetative development.
3. Reproductive processes.

Here again the vegetative and reproductive phases of development overlap as a rule, while not infrequently the formation of flowers may precede vegetative development in the annual cycle. Rarely flowers may be formed not in every season, but only once in a number of seasons, a year in which flowers form being succeeded by a number of years in which the development is purely vegetative, after which flowers are again produced. Very rarely in perennials the formation of flowers may terminate the growth and life of the plant altogether.

From the physiological point of view, then, various well-marked phases in development can be recognized, each presenting its own characteristics, the laws governing which it is the business of the plant physiologist to discover. These phases are : rest period, renewal of activity, vegetative development, reproduction. It has already been pointed out that the position of these phases in the life-cycle is not uniform throughout the plant kingdom but may vary from one species to another. These various phases of development will be considered in the next chapters.

CHAPTER XV

GERMINATION OF SEEDS

For a seed to germinate certain external conditions are necessary. A supply of water, a temperature within a limited range, a supply of oxygen, and, in some cases, certain light conditions, are recognized as necessary external factors for germination. But even when all these are provided, it does not follow that the seed will germinate. It may remain ' dormant ', and it has to be recognized that internal conditions of the seed may also be factors in determining its germination.

EXTERNAL FACTORS

Water Supply. Seeds, although to all appearances dry, contain a small quantity of water, but the proportion of water in the cells of the seed is very much less than that contained in the cells of the actively growing plant of the same species. Thus seeds of *Ricinus communis* were found to contain 6·45 per cent. of water, whereas the seedlings of the same plant contained 92·7 per cent. So long as the seed is kept dry it cannot germinate. The seed, either in the embryo or endosperm (rarely perisperm), contains food reserves which can as a rule only be transformed into mobile nutritive substances in the presence of water. The enzymes responsible for the transformation of the reserves may, in some cases, only be developed in active form in presence of water. The diffusion of the mobile substances so produced takes place through water and the actions which proceed in the cell appear to require for the most part an aqueous medium ; in short, an absorption of water by the seed is an essential for renewal of activity.

Provided the temperature and internal conditions are suitable, a seed brought in contact with water absorbs the latter, partly by imbibition, partly on account of the osmotic pressure of the cell contents. Possibly the absorption is at first largely or entirely due to imbibition, but as more water is absorbed the consequent production of osmotically active substances from the reserves, as, for instance, sugar from starch, may increase the osmotic absorption of water.

The quantity of water absorbed by germinating seeds depends on the species. In Table LXI are given the quantities of water observed to be absorbed by air-dry seeds and fruits of different species, the quantities being given as percentages of the original weight of the air-dry seed. These values give a good idea of the variation in the water requirements for germination of different species of seeds, but they must not be regarded as absolute values characteristic of the species, as not only seeds of different varieties of the same species, but different individuals of the same variety or species, differ greatly in regard to the quantity of water they can absorb. The quantity will also depend on the amount of water present in the air-dry seed, and this will itself depend on the time that has elapsed since the seeds were shed and on the conditions to which they have been subjected during that time.

Table LXI

Water Absorption by Seeds

Species	Water absorbed as percentage of original weight	Observer
Wheat	60·0	Nobbe
Barley	46	A. J. Brown and Worley
Oat	59·8	R. Hoffmann
Maize	39·8	Nobbe
Rye	57·7	R. Hoffmann
Buckwheat	46·9	R. Hoffmann
Castor oil (*Ricinus communis*)	42	Stiles
Poppy	91·0	R. Hoffmann
Rape	48·3	Nobbe
Sunflower (*Helianthus annuus*)	56·5	R. Hoffmann
Broad bean (*Vicia faba*)	157·0	Nobbe
Edible pea (*Pisum sativum* var. Thomas Laxton)	186	Stiles
Lathyrus maritimus	230	Stiles and Dellow
Xanthium pennsylvaticum	50	Shull
Pinus austriaca	35·8	Nobbe

Speaking generally, seeds of the Leguminosae are conspicuous for the large amount of water absorbed in germination; the cereal grains and seeds containing much fat absorb relatively much less water, although poppy is somewhat of an exception.

The time taken for seeds to attain their maximum swelling depends no doubt on the species or variety, but is dependent also on the temperature. Some observations by Dimitrievicz more than fifty years ago illustrate this point. Samples of seeds of red clover, rape and chick pea (*Cicer arietinum*) were soaked in water at four different temperatures: 0°, 10°, 15° and 35° C. and the absorption of water measured after the lapse of 6, 12, 24 and 48 hours. At 0° the absorption of water by rape was practically complete after 24 hours, but in red clover and the chick pea this stage was only reached after about 48 hours. In all cases absorption was more rapid at the higher temperatures. Accurate experiments carried out by A. J. Brown and Worley in recent years show that the total amount of water absorbed by barley grains is practically unaffected by temperature, but that increase in temperature speeds up the entry of the water. It was found that between 3·8° and 34·6° C. the rate of water absorption is almost exactly doubled for a rise in temperature of 10°, the actual temperature coefficient (Q_{10}) being

1·99. Shull, from a very careful series of experiments carried out on seeds of a variety of the edible pea and of the cocklebur (*Xanthium pennsylvaticum*), concluded that the effect of temperature on water absorption in these cases was less than that found by Brown and Worley, the temperature coefficient over the range 5° to 35° C. being 1·6 in the case of the pea and 1·55 in one set of experiments with *Xanthium pennsylvaticum* and 1·83 in another.

Brown and Worley concluded that the rate of water absorption is an exponential function of the temperature, the relation between temperature and rate of absorption being expressed by the equation $v = ae^{k\theta}$, where v is the rate of water intake, θ the temperature and a and k are constants. Because this relation is the same as that between vapour pressure of water and temperature, and because of the high value of the temperature coefficient, which is that characteristic of chemical and not physical actions, Brown and Worley suggested that the semi-permeable coat of the barley grain permitted only simple water molecules (' hydrone ') to penetrate it and that raising the temperature increased the proportion of hydrone molecules at the expense of the more complex water molecules of which water, on this view, is supposed to be partly composed. Shull pointed out that if this idea were correct it should apply to all semi-permeable membranes, including those of *Xanthium pennsylvaticum*. As the temperature coefficients actually found in this case were less, being between those characteristic of physical and those characteristic of chemical reactions, Shull's conclusion was that absorption is not conditioned by a single chemical change such as the simplification of water molecules with rise in temperature, but that both physical and chemical changes affecting water absorption take place as the temperature rises, the chemical changes occurring in the colloidal material in the seed. This would also explain the different temperature coefficients obtained with different samples of seed of the same species, as, on account of a different history prior to the experiment, the colloidal condition of the seeds of the two samples might be different.

Because a seed can absorb water it does not follow that it is capable of germination. The materials of dead seeds may absorb water by imbibition, while in many cases the seed coat is a semi-permeable membrane so that a solution contained within it will absorb water by osmosis whether the cells containing the solution are themselves possessed of semi-permeable membranes or not. Such semi-permeable coats have been described in barley, horse-chestnut, *Xanthium pennsylvaticum* and a number of other species. On the other hand, failure of a seed to absorb water does not necessarily mean that it has lost its viability, that is, its capacity to germinate. Many seeds possess hard coats which for a considerable length of time remain completely impermeable to water. The Leguminosae include many such species, among which may be cited those of *Ulex, Robinia, Acacia, Gleditschia, Trifolium, Medicago* and *Lathyrus*. As examples among plants of other families may be particularly mentioned *Polygonum, Spergula, Canna* and many palms. Varieties of the same species may exhibit differences in respect of impermeability of the seed coat to water, a fact which will be familiar to growers of sweet peas (*Lathyrus odoratus*).

An experiment made by Nobbe illustrates the length of time that such seeds may remain viable while the seed coat remains impermeable to water; it also illustrates the great variability among different individuals of the same species. In this experiment 400 seeds of *Robinia* were allowed to remain in contact with water. After about a year ten of these seeds had swollen and germinated, after another year one more seed, and during a third year two more seeds germinated. The experiment was continued for 32 years and at the end of that period seeds were still occasionally swelling and germinating.

Such hard-coated seeds absorb water at once when a portion of the seed coat is removed by filing or chipping it. In some cases immersion of the seeds in concentrated sulphuric acid for a certain time renders the seed coat permeable to water, and, if the immersion has not lasted too long, the seeds are still capable of germination. In other cases treatment with hot water will bring about the same result. It is extraordinary what exposure to high temperatures such seeds are sometimes capable of withstanding. Thus Schneider-Orelli found that seeds of *Medicago denticulata* and *M. arabica* are not injured by immersion in boiling water for 7½ hours. On the other hand, Honing found that the hard-coated seeds of *Crotolaria striata* are injured by exposure to water at temperatures above 50° C, while seeds of *Albizzia moluccana* may be damaged by treatment with water at 80° to 100°, although the percentage commencing to germinate may be higher than at lower temperatures. For this species 60° appears to be the best temperature for obtaining germination without damage, while 70° appears to be the best temperature for seeds of *Pithecolobium saman* and *Mimosa invisa*.

Temperature. It is a matter of common knowledge that seeds only germinate within a certain range of temperature. If the temperature is too low or too high germination does not take place even if other conditions are suitable. What determines the limits of the temperature range for the germination of any particular species is not always clear. Frequently the lower limit is in the neighbourhood of 0° C., and it might be supposed that withdrawal of water by freezing reduces or removes the supply of this necessary substance and so fixes the lower limit of temperature. This may be the explanation in some cases, but it certainly does not explain all, as quite frequently the lower limit of temperature at which germination takes place is considerably higher than the freezing-point of water; for example, in some species of the Cucurbitaceae it is as high as 18° C. The upper limit of the temperature range is probably determined by some effect of comparatively high temperature on some cell constituent, as, for instance, denaturing of proteins. This upper limit of temperature, like the lower limit, shows considerable variation among different species, a fact which is not surprising as the proteins and other constituents of different species show considerable differences in their resistance to higher temperatures.

Within the range of temperature over which germination takes place the rapidity of germination is much influenced by temperature. As with other plant processes, the conception of three cardinal points, the minimum, optimum and maximum temperatures, has been applied to germination, the minimum temperature being the lowest at which germination takes place, the maximum the upper limit of the temperature range for germination, and the optimum that temperature at which germination is most rapid. The cardinal points for temperature in regard to the germination of a number of species, as determined by Haberlandt, were recorded by Detmer. A selection of these results is given in the accompanying table. It has already been pointed out that the conception of an optimum is not very helpful in the consideration

Table LXII

Cardinal Points of Temperature in regard to Germination of Seeds in Degrees Centigrade

Species	Minimum	Optimum	Maximum
Wheat	0 – 4·8	25–31	31–37
Rye	0 – 4·8	25–31	31–37
Barley	0 – 4·8	25–31	31–37
Oat	0 – 4·7	25–31	31–37
Maize	4·8–10·5	37–44	44–50
Millet	4·8–10·5	37–44	44–50
Hemp	0 – 4·8	37–44	44–50
Buckwheat . .	0 – 4·8	25–31	37–44
Sunflower . .	4·8–10·5	31–37	37–44
Flax	0 – 4·8	25–31	31–37
Red Clover . .	0 – 4·8	31–37	37–44
Pea	0 – 4·8	25–31	31–37
Lucerne . . .	0 – 4·8	31–37	37–44
Melon	15·6–18·5	31–37	44–50
Pumpkin . . .	10·5–15·6	37–44	44–50
Cucumber . . .	15·6–18·5	31–37	44–50

of comparatively simpler processes such as photosynthesis where it is clear that an apparent optimum temperature results from the combination of a pure temperature effect on the process and a secondary effect of higher temperature on some constituent or constituents of the cell. Germination is a very much more complex matter, but undoubtedly the apparent optimum here also results from the combination of a secondary temperature effect on the cell constituents with the pure temperature effect on germination. It is therefore to be expected that at high temperatures there will be a falling off in the rate of germination with time. If, however, we define the optimum temperature for germination as the highest temperature at which the process continues unaffected by a time factor, the term 'optimum' may have a definite value, provided no other condition is acting as a limiting factor. The values obtained by Sachs, Haberlandt and others may be such, as in the majority of species limiting factors are much less likely to be operative in laboratory experiments on germination than in experiments on some other plant processes.

At the same time it must be emphasized that the cardinal points must of necessity be approximate, as was realized fully by Pfeffer fifty years ago. It is not easy to determine rates of germination as distinct from rates of swelling. In practice rates of germination at different temperatures are compared by counting in a sample the number of seeds in which the radicles have appeared through the testas after different periods of time. The method is clearly very approximate, but it serves for determining cardinal points. On the other hand, there is very considerable individual variation among seeds belonging to the same species or even variety, while according to various workers the minimum temperature for germination depends on the conditions under which the plant from which the seeds were obtained grew. The possibility of the influence of other conditions, as already mentioned, must not be overlooked.

It will be observed from Table LXII that the temperature range for germination varies for different species, both in regard to its position and its extent. On the whole the cardinal points are lowest for plants of arctic and alpine habitats and highest for tropical species.

Within the temperature range for germination the rate of germination increases with increase in temperature provided no other condition is limiting the rate. At temperatures higher than the optimum, as defined above, the rate of germination is almost certainly initially higher the higher the temperature, but falls off with time, more rapidly the higher the temperature. Owing to the practical difficulty of measuring germination rate this view is difficult of exact experimental confirmation, but there is no reason to doubt its truth.

It is easy to understand that germination should be accelerated by increase in temperature. Among the processes involved in germination are diffusion of water into the seed, enzyme actions and diffusion of mobile substances from one part of the seed to another. All these are processes accelerated by rise in temperature.

Light. The influence of light on germination is very varied. With possibly the majority of species it would appear that light is without influence on germination, and it was at one time thought that this applied to all seeds, a view championed by Nobbe even after exceptions to the rule had been recorded. Seeds the germination of which is uninfluenced by light may be described as light-indifferent ; so far as we know at present most seeds of agricultural importance belong to this category. However, as long ago as the year 1860 Caspary recorded that the germination of seeds of *Bulliarda aquatica* (*Tillaea aquatica*), a member of the Crassulaceae, was strongly favoured by light, and 18 years later Wiesner found that the germination of the seeds of mistletoe was favoured by light. Since then a very considerable number of species with such 'light-sensitive' seeds have been recorded by various workers, particularly by Kinzel. Later it was discovered that there were seeds the germination of which was retarded or prevented by light, the first recorded case of which appears to be that of the seeds of *Phacelia tanacetifolia* described by Remer in 1904; such seeds may be termed 'light-hard'. Among light-sensitive seeds are those of various species of *Epilobium* and *Rhododendron, Pinguicula vulgaris, Ficus aurea, Lythrum salicaria* and many members of the Gesneriaceae. Whereas nearly related species often show similarity of behaviour in regard to light sensitivity, this is not necessarily so. Thus, although as noted, above, the seeds of *Viscum album* are sensitive to light, those of tropical species of *Viscum* do not require light for germination. Again, most species of *Veronica* possess light-sensitive seeds, but at least one species of this genus, *V. tournfortii,* has seeds which are light-hard. Other species described as having light-hard seeds are *Nigella sativa, Phlox Drummondii, Nemophila insignis, Whitlavia grandiflora, Allium spp.* and many members of the Amarantaceae. Thus Lehmann found that at 21° C. about 75 per cent. of a sample of seeds of *Nemophila insignis* germinated in the dark within 8 days, whereas only 1 or 2 per cent. germinated in the light (of intensity about 150 metre candles) in the same-time.

In some cases the quantity of light necessary to bring about the germination of a light-sensitive seed is very small. The seeds of tobacco are regarded as light-sensitive, but an exposure of the fully swollen seeds to diffuse light for one hour is sufficient to promote germination. Even more striking is the case of *Lythrum salicaria* investigated by Lehmann. In darkness at 30° C. only 6 or 7 per cent. of a sample of the fully swollen seeds germinated within 24 hours and this percentage was found not to have increased appreciably after a period of 7 days. But exposure to a light-intensity of 730 metre candles for only 0·1 second was sufficient to give a germination of 50 per cent. of the seeds of a sample within 24 hours.

It is very clear that the influence of light on germination must be considered in relation to other factors. Among those which can profoundly influence the effect of light in such cases are the age of the seed (the degree of 'after-ripening'), temperature, temperature variation during germination, and the nature of the substratum. Other factors which may have an effect in some cases are the condition of the seed coat (for example, whether ruptured or not), pressure of oxygen and pressure of carbon dioxide.

With regard to the influence of after-ripening, as a rule the percentage of germination increases for a time with age of the seed, both in light and in darkness, and the percentage of seeds of a light-sensitive species germinating in the dark may approach more nearly the figure for those germinating in the light in the case of older seeds than in that of freshly harvested seeds.

As regards temperature, Lehmann concluded from a number of experiments carried out with both light-hard and light-sensitive seeds, that within the temperature range of germination a higher temperature furthers the germination of light-sensitive seeds in the dark, while with light-hard seeds a lower temperature favours germination in the light. Two examples, one of a seed light-hard at ordinary laboratory temperatures, the other of a seed light-sensitive at such medium temperatures, may be quoted from among Lehmann's observations. These are *Nemophila insignis* (light-hard) and *Epilobium hirsutum* (light-sensitive). The results are summarized in the accompanying table:

Table LXIII

Influence of Light and Temperature on the Germination of Light-hard and Light-sensitive Seeds

Species	Temperature in degrees Centigrade	Light intensity in light experiments in metre candles	Percentage germination in	
			light	dark
Nemophila insignis . .	10–11	—	81	91·5
,, ,, . .	10–11	—	87	87·5
,, ,, . .	21	145–155	1·5	74·5
,, ,, . .	22–24	160	0	34·5
,, ,, . .	31	160	0	0
Epilobium hirsutum . .	20	145–155	78	7·5
,, ,, . .	22–24	145	60·5	3
,, ,, . .	22–24	175	58	—
,, ,, . .	31	145	67·5	53·5
,, ,, . .	31	175	61·5	—

However, the relation between light and temperature in regard to the germination of light-sensitive and light-hard seeds cannot be stated generally in such simple terms. From the point of view of temperature influence, light-sensitive seeds alone appear to comprise at least five different types. These are as follows:

1. Light brings about the germination of the seeds in low temperatures, while in the dark germination can be brought about by subjecting the seeds to comparatively high temperature. It is important to note that with this type of seed germination will also take place in the dark if the high temperature is not constant but if instead the seeds are subjected to alternations of high and low temperature. *Epilobium hirsutum* and *Veronica longifolia* are species with seeds of this type.

2. The second type of seed is represented by those of *Poa pratensis*. These seeds are light-sensitive, germinating well at comparatively low temperatures, but unlike the seeds of the first type germination in the dark is not improved by exposure to constant high temperature. On the other hand, the seeds, like those of the first type, germinate in the dark if subject to alternation of relatively high and relatively low temperatures.

3. Seeds of the Gesneriaceae constitute a third type. These seeds germinate in the light at low temperature, whether kept constant or not, and also if exposed to alternation of high and low temperatures. They appear to germinate not at all in the dark, neither with constan temperature nor when subjected to temperature variations.

4. The seeds of *Ranunculus sceleratus*, investigated by Gassner, behave differently from any of those already described. So long as the temperature remains constant the seeds of *Ranunculus sceleratus* do not germinate either in the light or in the dark. With variations of temperature, however, the seeds appear to be light-sensitive, germination then being very considerably furthered by light. Moreover, it was found that the exposure to light and to varying temperature need not be contemporaneous in order to produce maximum germination. If the seeds were first exposed to light at constant temperature and then to alternations of high and low temperatures in the dark, germination resulted as satisfactorily as if the light and varying temperatures had synchronized. Moreover, it was found that the seeds could be dried between the period of illumination and the exposure to alternations of temperature, and the germination capacity remain unimpaired.

5. Lastly, among light-sensitive seeds is to be mentioned the curious case of the grass *Chloris ciliata*, also investigated by Gassner. When seeds of this species are supplied with distilled water only, light furthers germination at higher temperatures (33°–34° C.), but retards germination at temperatures below 20° C. At intermediate temperatures, namely, those about 22° C., the seeds are indifferent to light as regards germination.

Light-hard seeds have been less fully investigated than have light-sensitive seeds, but it seems probable that the temperature-light relations may be as varied with them as with the latter.

The behaviour of light-sensitive and light-hard seeds may be profoundly influenced by the nature of the substratum. A number of cases have been reported in which light-sensitive seeds can be made to germinate in the dark if treated with certain substances, notably proteolytic enzymes, weak acids, nutrient solutions and particularly the nitrogen-containing constituents of these. At one time it was stated that the action of light could be replaced by that of these substances, but it is extremely doubtful if such a simple method of expression represents the actual state of affairs. Gassner found treatment with Knop's solution could bring about germination of seeds of *R. sceleratus* exposed to alternation of high and low temperatures, indicating that the nutrient solution acted in the same direction as light. The effects of light and the nutrient solution were, however, additive. This result is clearly shown in Table LXIV. A similar additive effect of light and nitrate was observed with *Chloris ciliata* at higher temperatures. At lower temperatures, however, where light reduces germination, nitrates further it, so that here the two factors work against one another. Lehmann found similarly that in the case of

Table LXIV

Germination of Seeds of ***Ranunculus sceleratus*** on Different Substrata (The seeds exposed daily to temperatures of 19° for 4 hours and 28° for 20 hours)

Substratum	Lighting Conditions	Percentage of Germination
Distilled water	Dark	0·7
,, ,, ...	Daylight	28
Knop's solution	Dark	55
,, ,, ...	Daylight	86

Veronica tournfortii, whereas light retards germination, potassium nitrate favours it.

Experiments, chiefly by Lehmann and his collaborators, show that dilute solutions of acids (N/200 N/1000) will further germination of light-sensitive seeds in the dark.[1] They find, however, in the case of ***Lythrum salicaria*** that in complete darkness and *constant* temperature, treatment with acid has little or no effect on germination. If, however, the seeds are subjected to temperature variation, or to light, then the germination is much increased by treatment with acid. In this case, therefore, acids do not replace the action of light or of varying temperature, but act as sensitizers, increasing the effect of these factors. Nitrates are supposed to act in the same way as acids. This view cannot, obviously, be regarded as of general application, for it cannot possibly hold for those cases in which nitrates and light act in opposite directions.

What the action of light is in the germination of light-sensitive and light-hard seeds is not at all clear, but it seems extremely probable that no one single explanation can be found for all cases. Three theories of the action of light in germination have been put forward. On the first view light acts as a stimulus, on the second as a catalyst, while on the third view it is supposed that during germination an ' inhibition principle ' forms, and this is suppressed by light, so that, in the case of light-sensitive seeds, germination can only proceed when light or some other factor suppressing the inhibiting principle is present. On this view, due to Gassner, the favourable influence of light on germination in certain seeds is the result of an inhibition of an inhibition.

Lehmann originally regarded the action of light as a stimulus, that is, as releasing an action or chain of actions within the seed. Against this view Gassner has marshalled a number of arguments. In the first place the relation between the magnitude of the quantity of light and the germination resulting does not follow the Weber-Fechner law (see Chapter XIX) characteristic of many stimulus actions. But it is well known that in the case of many stimuli the Weber-Fechner law

[1] Care has to be exercised in carrying out experiments on the germination of seeds in presence of acid, as a condition known as ' false germination ' may be produced in which, owing to swelling of the seed and rupture of the seed coat, the radicle comes to project through the testa, although germination has never commenced. Often in such cases the seed has actually been killed by the action of the acid, although the appearance is that of a germinating seed.

is not obeyed and at best it is only an approximation which holds over a very limited range. It does indeed seem very possible that in the case of *Lythrum salicaria,* where a very small light quantity is sufficient to release the germination process, the light acts as a stimulus. On the other hand, it seems fairly clear that in the cases of *Ranunculus sceleratus* and *Chloris ciliata* the reaction of the seeds to light cannot be the result of a stimulus alone. The fact that the action of light can remain latent, even while the seeds are dried, and then have effect in the dark when varying temperature is applied, indicates, as Gassner pointed out, that some other explanation must be sought in these cases.

That light acted as a catalyst in certain cases of light-sensitive seeds was held by Lehmann. Enzymes play a considerable part in germination, and since in some cases it had appeared that the action of light could be replaced by enzymes or by acids which also act catalytically in the hydrolytic actions taking place in germinating seeds, it seemed likely that light might also act catalytically. This explanation obviously will not suffice as it stands for those cases in which the action of the acid is not to replace light, but merely to increase its action. Also in the cases of *Ranunculus sceleratus* and *Chloris ciliata* acids do not act as releasers of the germination processes, although nitrates do. But Gassner showed that in these cases the nitrate does not enter the seed, the testa forming a layer impermeable to the salt. If the nitrate acted as a catalyst in these cases it could only do so by penetrating into the seed. Gassner also pointed out that Lehmann found removal of the seed coat rendered the light-sensitive seeds he investigated capable of germination in the dark. It is thus possible that the action of acid is to alter the condition of the testa so that it is more permeable. This was found to be the case with the seeds of *Alisma,* where the favourable action of acids in regard to germination was traced by Crocker and Davis to an action on the testa.

Gassner's view of the action of light on the germination of the seeds of *Chloris ciliata* is that in germination some inhibiting principle is produced outside the seed proper (embryo and endosperm). Evidence for this is found in the fact that nitrates and light act in the same direction (at higher temperatures) and the nitrates do not enter the seed. How light and nitrates destroy or damage this inhibiting principle, or, indeed, what is the nature of the inhibiting principle, is not clear. It might be the presence of a layer possessing some characteristic chemical or physical property, or some change in the substratum produced by the diffusion out from the seed of certain substances. That substances do diffuse out from germinating seeds and seed coats is certain.

It is quite possible that the mechanism of the action of light is different in different cases, just as the result of the action of light is different. It may be noted that in general those seeds the germination of which is affected by light are small ones, but the significance of this, if any, is not clear. It is true that for the processes of growth involved in germination food materials stored in the seed are required and in small seeds these are small in amount and consequently the seedling has to assimilate its own carbon dioxide at a very early stage. Kinzel observed that during the germination of the light-sensitive seed of *Poa pratensis* chlorophyll develops even before the rupture of the seed coat, and thought the necessity for light might be traced back to this. But the interaction of light, temperature and composition of the substratum make it clear that the need for light cannot be connected in general with the production of chlorophyll and need for early photosynthesis, since in a number of cases so-called light-sensitive seeds can be brought to germinate in the dark, and the seeds of *Poa pratensis* are among these. Generally it is only after the germination stage can be considered over that the resulting seedling has to fend for itself and light becomes a necessity for nutrition.

It has been suggested by Kummer that the content of fatty acid in the seed may have something to do with its light-sensitivity. He found that a number of grass seeds possess a relatively high content of fat (4 to 26 per cent.) and that of these seeds those with a high content of fatty acid (e.g. *Trisetum flavescens, Alopecurus pratensis, Holcus lanatus* and *Aira flexuosa*) germinate readily in the dark, whereas those in which the fatty acid content is low, namely, *Koeleria cristata, Anthoxanthum odoratum, Dactylis glomerata, Poa trivialis* and *P. pratensis*, generally germinate in the dark with difficulty. Moreover, other light-sensitive seeds containing fat, such as those of *Nicotiana tabacum, Lythrum salicaria* and *Epilobium roseum* also possess a low content of fatty acid. Kummer therefore thinks that light-sensitivity is related to the quantity of fatty acid in such seeds, although he admits that other factors are doubtless operative. If there is anything in Kummer's suggestion it would appear that the effect of light in the case of certain fat-containing light-hard seeds might be to bring about the formation, or increase in concentration, of lipase, and so further the mobilization of reserves.

Oxygen. A supply of oxygen is a necessity for germination. Thus no germination takes place as a rule in the case of seeds submerged in water, particularly if the water is free from air; if the latter condition is maintained, even the seeds of water plants for the most part fail to germinate.

The capacity of seeds for germinating under water has been examined by Morinaga, who tested seeds belonging to 78 different genera. He found that in two genera seeds germinated better under water than on damp filter paper, in 18 genera germination under water was as good as on filter paper, and that altogether seeds of 43 of the 78 genera were capable of germinating under water. If the water was kept in contact with pure oxygen instead of air, so that the concentration of the oxygen in the water was increased, the number of genera in which seeds could germinate in the water was somewhat increased. On the whole small seeds were found to germinate more readily under water than were large ones.

The germination of seeds at too deep a level of the soil may be prevented because of the extremely slow diffusion of oxygen from the upper soil level and atmosphere to the seeds.

The necessity for oxygen for germination is readily understood. During this process growth takes place rapidly, inasmuch as cells swell and divide actively and consequently respiration, involving intake of oxygen and output of carbon dioxide, proceeds actively.

It is curious that in some cases a reduction in the partial pressure of the oxygen in the air leads to a more rapid germination. Thus Morinaga found that when 40 to 80 per cent. of the air was replaced by hydrogen or nitrogen about 96 per cent. of a sample of seeds of *Typha latifolia* germinated within 4 days, whereas in normal air no germination took place at all within this time and only 3·8 per cent. in 10 days. In complete absence of oxygen, however, no germination took place. This result is all the more curious as, when the seed coats are removed, germination takes place readily over a wide range of partial pressures of oxygen (from 1 to 90 per cent. of the normal oxygen partial pressure). Possibly oxygen reacts with some constituent of the seed coat when this is wet to produce a substance or layer which retards germination.

Various Substances. From what has already been said with regard to the influence of the substratum on the germination of light-sensitive seeds, it is obvious that various substances can have a profound effect on germination. This is so, quite apart from the special case of light-sensitive seeds. A few cases of special interest may be mentioned here.

(*a*) *Water Plants.* A. Fischer observed that mature seeds of a number of water plants such as *Sagittaria sagittifolia, Alisma plantago-aquatica, Hippuris vulgaris,* and various species of *Potamogeton, Scirpus* and *Sparganium,* do not germinate in pure water but do so readily in impure water containing bacteria. He thought that excretions from the bacteria might lead to the germination of the seeds and so examined the influence of various acids and alkalies on germination of the seeds concerned. As a result he found that a number of acids (lactic, malic, oxalic, hydrochloric, sulphuric, nitric) and alkalies (potassium and sodium hydroxides) in dilute solution bring about the germination of these seeds.[1] Thus out of a sample of 1,400 seeds of *Sagittaria sagittifolia* there was no single case of germination during several months in water, whereas in 0·045 per cent. lactic acid 84 per cent. of a sample germinated within 29 days. It is interesting to note that in the case of rice Nagai found germination best in distilled water, acids and alkalies having no beneficial effect.

(*b*) *Parasites.* It is stated that seeds of various root parasites, *Orobanche, Lathraea* and *Tozzia* (a hemi-parasite) germinate only in the neighbourhood of the roots of their respective hosts. The conclusion appears inevitable that certain substances diffuse from the

[1] W. Crocker thinks it extremely doubtful that the slightly favourable action of fungi and bacteria is due to acid action.

roots of the host, possibly from dead or dying outer cells, and further the germination of the seeds of the parasite.

(*c*) *Orchids.* The germination of orchid seeds has been a subject of interest to botanists and horticulturists for more than a century. The view is generally held that these seeds only germinate, or at any rate only produce seedlings, when a fungus, of a species depending on the species of orchid, is present. This question has been dealt with earlier (Chapter XIII) and will therefore not be further discussed here.

DORMANCY

In discussing the influence of light on germination reference has already been made to a number of cases of seeds which, given the normal conditions requisite for germination, such as a suitable temperature, a supply of water and access of oxygen, nevertheless fail to germinate unless subjected to a certain amount of light or to some other special condition or conditions such as variations of temperature or the presence of certain substances in the substratum. Failing these special conditions the seeds remain alive but dormant. Such a condition of 'dormancy' is met with fairly frequently and may often not be related to conditions of illumination. In such cases the freshly harvested seeds do not germinate under ordinary conditions for germination, but do so after the lapse of a certain time. Indeed, in species growing in temperate climates it appears to be commoner for the seeds to undergo a rest period before becoming capable of germination, than for them to have the capacity for germinating immediately they are shed. In other cases the seeds will germinate if sown immediately after maturing, but if not sown then they only germinate after a more or less prolonged rest period. Thus seeds of *Taxus* will germinate if sown at once; otherwise germination is delayed for from two to four years. We can thus distinguish between primary and secondary dormancy. Dormancy is absent as a rule from seeds of cultivated species of agricultural importance.

The problem of dormancy in seeds has been the subject of very considerable investigation by Crocker and his school. It is clear from their researches that the dormant condition may be brought about by a variety of causes. Crocker considers that a condition of primary dormancy in seeds may arise in at least five different ways, and that in some cases two or more types of dormancy may be combined. These five types of primary dormancy are as follows:

1. Immaturity of the embryo. In some seeds the embryo when the seed is shed is immature and it has to undergo further development before germination can take place. This further development takes place when the seeds are brought into conditions favourable for germination and may involve a period of a few days or many months. This type of dormancy is found in most of the large groups of seed

plants including Gymnosperms (*Ginkgo biloba, Gnetum gnemon*), Dicotyledons (*Ranunculus ficaria, Corydalis cava*) and Monocotyledons (*Paris quadrifolia, Hymenocallis speciosa*). The accompanying table indicates the variation in time required by the embryos of different species exhibiting this kind of dormancy to complete their development under conditions favourable for germination. The times given are naturally approximate as they will vary with temperature and possibly other conditions, and also to some extent with the individual.

Table LXV

Time required for Immature Embryos to complete their Development

Species	Time
Caltha palustris . . .	10 days
Clematis vitalba . . .	17 ,,
Anemone hepatica . .	2 months
Fraxinus excelsior . .	4 ,,
Corydalis cava . . .	10 ..

2. The seed coat is impermeable to water. The phenomenon presented by hard-coated seeds in which the testa is impermeable to water has already been noted when the question of water absorption by seeds was discussed. It was there pointed out that many hard-coated seeds are met with among the Leguminosae, although they occur also in other families, notably in Cistaceae and Malvaceae. The coats of such seeds gradually become permeable to water with storage, very slowly as a rule when stored dry, less slowly when kept under conditions favourable to germination, and apparently most rapidly when subjected to the very varying conditions of the normal environment. The length of time taken for the impermeability to disappear also varies with the individual as well as with the species ; the individual variation in this respect is indicated by the results of the experiment recorded by Nobbe described earlier. Means by which this kind of dormancy may be removed have already been mentioned. According to Verschaffelt, in the Leguminosae the micropyle remains open in the seed and water fails to enter by this channel because it cannot wet the walls of the micropyle. Treatment with alcohol renders certain of such seeds capable of absorbing water, but as the action is negligible or non-existent in many members of the Leguminosae Crocker thinks Verschaffelt's explanation can only apply to some, and not all, species. The impervious layer (or layers) apparently varies in different species ; information on this point is not definite and opinions expressed are controversial.

3. The force exerted by the swelling seed is insufficient to rupture the seed coat. This type of dormancy must be sharply distinguished from the last type. The seed coat is readily permeable to water and, indeed, water absorption takes place readily up to a point. If the coat is removed the embryo grows readily, never at any time exhibit-

ing a dormant period. Such seeds, investigated by Crocker and Davis, are those of *Alisma plantago-aquatica* and *Amaranthus retroflexus*. Seeds of the former, when kept dry, will remain dormant for years, whereas those of the latter under similar conditions lose their dormant condition in a few months, no doubt on account of changes in the testa, although when remaining moist in the soil and so partially or fully imbibed with water, they may remain dormant for many years. The methods suitable for removing dormancy from hard-coated seeds are for the most part applicable to seeds exhibiting this third kind of dormancy: that is, mechanical injury to the seed coat or treatment with sulphuric acid. A variety of acids and alkalies in various concentrations may also be effective. It is possible that the seeds of *Sagittaria sagittifolia* already mentioned in connexion with the action of substratum on germination may be seeds exhibiting this type of dormancy.

4. The seed coats are relatively or completely impermeable to oxygen. The seeds of this type which have been most fully investigated are those of species of *Xanthium*. If the oxygen supply to the seed is limited, respiratory processes which play so prominent a part in germination are also limited. Along with the limitation of oxygen may also go a failure of respiratory carbon dioxide to leave the seed, and this will also tend to repress respiration. Removal of the testa allows germination to proceed readily, while the same result may be obtained by increasing the oxygen pressure outside the intact seed. When stored dry, or under natural conditions, the permeability of the testa to oxygen gradually increases.

It is possible that the dormancy of the seeds of *Chloris ciliata*, already noticed as a light-sensitive seed, may, in part at least, be due to a limited supply of oxygen conditioned by the seed coverings. Gassner found that the glumes retarded the entrance of oxygen. After-ripened seeds (see below) of *Chloris* before removal of the glumes, at high temperatures (33° C.) germinate much better in the light than in the dark, whereas after removal of the glumes the seeds germinate equally well in the light and dark. It would appear that here light has an effect on the glume, rendering it more permeable to oxygen. It is possible that the effect of light may be to lower the minimum oxygen pressure necessary for germination as was found to be the case with the action of high temperature on the seed of *Xanthium*. But the relations in the case of *Chloris* are obviously very complex, for at medium temperature (22° C.) seeds from which the glumes have been removed germinate better in the dark than in the light.

The curious case of *Typha* in which removal of the seed coat leads to germination, but in which *reduced* oxygen pressure has the same effect on the intact seed, has already been noted.

5. After-ripening of the embryo is necessary. In seeds exhibiting this kind of dormancy the embryo fails to germinate after removal of the seed coat even when supplied with all the normal conditions

for germination. Certain changes have to take place in the embryo or part of it before germination can take place. This type has to be distinguished from that in which dormancy is due to morphological immaturity of the embryo, for in the latter case the embryo has to undergo morphological development, whereas in the type of dormancy under discussion the embryo appears to be fully formed, but has to undergo certain chemical and physical changes. The seed of this type that has been most fully investigated is that of *Crataegus mollis*, investigated by Davis and Rose and by Eckerson. The temperature most favourable for after-ripening is 5° in this case, but the time required for after-ripening depends very greatly on the presence or absence of seed coverings. Thus instead of the 3 or 4 weeks required for after-ripening in the case of the detached embryo, as many months are required when the testa only surrounds the seed, and more than a year when the stony endocarp is present as well. Apple seeds also belong to this class although the dormancy is much less pronounced.

It was found that after-ripening could be hastened and so the dormant condition removed, by treatment with acids, this treatment apparently bringing about a change in reaction of the hypocotyl in the acid direction such as is characteristic of normally after-ripened seeds.

It is possible that in the case of some light-sensitive seeds we have to do with dormancy of this type in which light brings about or hastens after-ripening processes. On the other hand in some cases, at any rate in that of *Chloris ciliata*, the action of light is on the seed coats and the type of dormancy is one of the others previously described or some different kind altogether.

Secondary Dormancy. It was mentioned above that some seeds will germinate when they mature, or if taken from the fruit before they are shed and brought into conditions favourable for germination, but that if this time is allowed to pass they will remain dormant for long periods. Seeds of some species or varieties of *Primula*, for example, behave in this way. The seeds pass over into a condition of so-called secondary dormancy. Such a condition may be produced in some seeds artificially. Thus Kidd and West found that when seeds of *Brassica alba* are exposed to high concentrations of carbon dioxide not only is the germination of the seeds inhibited, but the initial inhibition is followed by prolonged dormancy after removal of the carbon dioxide. Crocker and his co-workers found that secondary dormancy can be produced in seeds of *Sorghum halepense* and *Amaranthus retroflexus* by exposure to temperatures below the minimum necessary for germination along with all other conditions required for germination. According to Crocker, when secondary dormancy occurs in seeds capable of immediate germination, it results from exposing the seeds to the conditions necessary for germination with the exception of one essential condition or in presence of a substance inhibiting germination or hardening the seed-coat colloids. In many cases the dormancy results from the seed coats becoming hard or impermeable to gaseous exchange. In the case of *Brassica alba*, however, Kidd and West think that the secondary dormancy is due to the passing of the tissues of the embryo into a stable condition and that a stimulus is needed to bring about a change in the state of this tissue equilibrium before germination can take place. In other cases, where the dormancy results from modification of the seed coats, rupture of the coats is sufficient to end the period of dormancy. This is, indeed, also the case with most of the seeds of a sample of fully swollen *Brassica alba* seeds, but Kidd and West regard the renewal of activity in these seeds as due to the stimulus of injury or shock involved in the removal of the coats with a fine sharp needle; if the seeds are not fully swollen the secondary dormancy persists after the removal of the coats.

RETENTION OF VITALITY

The question of dormancy in seeds is related to another, namely, that of the length of time during which a seed can remain dormant and yet retain its viability (see p. 310). This problem concerns not only seeds exhibiting dormancy, but all seeds, for it is well known that seeds of agricultural importance such as those of wheat and other cereals can be kept dry for a term of years and germinate at any time they are exposed to the ordinary conditions necessary for germination. The question resolves itself into two parts: how long can the power of germination be retained, and how is it that seeds can retain their viability for long periods?

With regard to the first question it is evident that the duration of the power of germination differs very greatly in different species. The seeds of some species of *Oxalis* (e.g. *O. rubella* and *O. pentaphylla*) will germinate when fresh immediately after their release from the capsule, but drying kills them. Seeds of various species of *Salix* germinate within 12 hours if sown immediately after ripening, but if kept for a few days germination takes longer, while if kept for a longer time the power of germination is lost altogether. The seeds of *Populus* retain their viability for a longer period, but not for more than a few weeks. The seeds of many trees remain viable from the time of shedding in the autumn until the following spring; among such are those of the birch, elm, beech, and probably the oak. The seeds of some other trees retain their power of germination for longer periods; thus lime, maple, alder, *Sorbus torminalis*, hornbeam, larch and Weymouth pine (*Pinus strobus*) possess seeds which retain their viability for from one to three years, while seeds of ash and Scots pine may remain viable for periods up to four years and spruce seeds up to five years. Seeds of the commoner cereals can remain alive for longer periods still. Burgerstein found that samples of wheat and barley grains which had been kept in a drawer in paper packets germinated to the extent of from

70 to 90 per cent. after 10 years' storage in this way. Grains of rye, however, kept for the same time were found no longer capable of germination.

The most striking record of the longevity of seeds is that furnished by a report made by Becquerel in 1907.. He used seeds from the Paris Natural History Museum which had been collected and dried at various known dates, the youngest being 25 and the oldest 135 years old. Seeds of about 500 species belonging to 30 families were examined. It was found that on the whole the seeds of Leguminosae retained their vitality longest; the other families providing viable seeds were Nelumbiaceae, Malvaceae and Labiatae. The longest-lived seeds of all were those of *Cassia bicapsularis*; three seeds out of a sample of ten 87 years old germinated. Other long-lived seeds were those of *Cytisus biflorus* (84 years), *Stachys nepetaefolia* (77 years), *Trifolium arvense* (68 years), *Ervum lens* (65 years), *Lavatera pseudo-Olbia* (64 years), *Nelumbium codophyllum* (56 years), *Nelumbium asperifolium* (48 years). In 1935 Becquerel made a second examination of the germinative capacity of these seeds. He found that seeds of *Cassia bicapsularis* 115 years old germinated, as did seeds of *Leucaena leucocephala* 99 years old, seeds of *Dioclea pauciflora* 93 years old and those of *Mimosa glomerata* 81 years old. He also examined the viability of seeds in the Adanson herbarium made during the reigns of Louis XV and Louis XVI. Among these he found two seeds of *Cassia multijuga* collected in 1776 which germinated readily when 158 years old. Other records of the germination of seeds more than a century old were recorded by Ewart who succeeded in germinating seeds of *Goodia latifolia* and *Hovea heterophylla* 105 years old. A great number of similar tests were also made by Ewart in Australia on old seeds mostly belonging to a collection sent out from Kew to Melbourne in 1856. He also found that the Leguminosae provided the majority of the cases of longevity. As long ago as 1850 R. Brown observed the germination of the seed of *Nelumbium speciosum* which had been stored for 150 years.

More recently Ohga has reported the finding of seeds of Indian lotus (*Nelumbium nucifera*) in a drained lake-bed in Manchuria which gave 100 per cent. germination. The lake was drained at least 160 years, and probably more than 250 years, ago, and the seed must have been produced before the lake was drained. This appears to constitute the case of longest viability so far recorded for which the evidence appears sound. Cases of greater longevity of seeds are recorded from time to time, but the evidence is open to serious criticism. Tales of the germination of wheat found in Egyptian tombs and even in tombs a thousand or two thousand years old may be dismissed as erroneous; in such grains the embryo is always found to have suffered fatal injury.

The seeds so far mentioned as exhibiting unusual longevity have one characteristic in common: they all possess hard coats which are impermeable to water and, as Becquerel showed, are in some cases, at any rate, impermeable to gases. Although undoubtedly longevity is exhibited principally by hard-coated seeds there is evidence that some other seeds without impermeable coats can, under certain conditions, remain dormant for many years. Thus there are many records of seeds buried in the soil at various depths being able to germinate when brought near the surface after the lapse of years. Perhaps the best-known observations of this kind are those of Peter who took samples of soil at various depths down to 24 centimetres from soil that had been woodland for a long time (20 to 46 years), but which had previously been arable. When such soil samples were brought into conditions favourable for germination a number of weeds of arable land appeared and the obvious conclusion is that the seeds of such plants had remained dormant in the soil from the time when the land had been cultivated. Although various arguments have been advanced against accepting this conclusion, such as the possibility of the seeds being carried from neighbouring land by various agents and the possibility of the species remaining growing in the woodland as dwarf forms and so being overlooked, yet there seems no doubt that seeds of certain species, not necessarily hard-coated, can remain dormant and viable in the soil. Among such are various species of *Brassica*. Peter found that seed of *Brassica sinapistrum* (charlock) remained dormant in this way for 25 years; Kidd reported with regard to *Brassica nigra* (black mustard) that in Sussex ' every farmer and labourer along the northern slope of the South Downs will give examples from his experience of the seeds sprouting in newly-ploughed land after they have lain dormant for years, while the land has been under pasture or hay '. This species is also included in a list of eight, seeds of which were found by Beal and Darlington in America to remain viable in soil for 30 years. Other species on this list are *Capsella bursa-pastoris*, *Setaria glauca* and *Stellaria media*.

We may conclude therefore that many seeds can remain dormant for a very considerable time, but sooner or later the power to germinate is lost. In seeking an explanation of longevity we have really to consider two cases. The majority of long-lived seeds are ' hard ' seeds with coats impermeable to water and perhaps gases; these are seeds which remain viable in the dry condition. But also there are seeds of the type of *Brassica nigra* in which the coats are not impermeable and which retain their vitality in the soil and hence in a damp environment in which they must be more or less swollen with water. With hard seeds maintained in the dry condition the question at once arises whether any change whatever is taking place in the seeds during storage. Obviously if the state of the seed remains absolutely unchanged it should retain its power of germination indefinitely. But is such a state of static equilibrium possible with living matter? Can respiration and the changes bound up with it cease altogether, to reappear when the seed is brought into conditions suitable for the development of the embryo? Or does respiration proceed, although extremely slowly, in such seeds? A definite answer to this question is scarcely possible. Becquerel did indeed show that the testa of these hard seeds is impermeable to oxygen and carbon dioxide at a certain stage of desiccation, and that in consequence interchange of material between the embryo and the outer environment is prevented. Nevertheless, respiration might proceed slowly within the seed, the oxygen contained inside the testa being utilized for this. To eliminate this possibility Becquerel perforated the coats of the seeds, dried them as far as possible and kept them for two years *in vacuo*. In these circumstances aerobic respiration should be suppressed, yet the seeds remained viable. All the same, the possibility of slow *anaerobic* respiration is not ruled out.

To settle the matter Becquerel prepared a number of samples of seeds *in vacuo* the germination of which is to be tested after various intervals of time. In this way it will be possible to determine whether the seeds are actually in a state of static equilibrium, or whether chemical changes, which, in the absence of constructive metabolism must lead to the death of the seeds, actually continue throughout the period of dormancy.

Certain experiments on the freezing of seeds appear of the greatest significance in this connexion. It is found that freezing in liquid air (– 190° C.) or liquid hydrogen (– 250° C.) does not affect the viability of dry seeds. Under these conditions we must suppose the whole of the contents of the seed form a solid block in which the space relations of the molecules and molecular aggregates of the living tissues are preserved, as the rapidity of the freezing would prevent the small amount of water present separating out as ice crystals. The chemical changes taking place in living cells are those which characteristically take place in an aqueous medium. Hence, when frozen the seeds are surely in that condition of static equilibrium which we have postulated as one of the possible states of long-lived dormant seeds.

Of course, whether chemical changes stop completely, or proceed at so slow a rate that they are indeterminable after a long term of years, may be regarded as an academic question. It does seem clear, however, that in all cases the power of germination is lost sooner or later. Whether this will be the case with the seeds used in Becquerel's samples time alone can show. It is, however, interesting to note that loss of vitality in seeds in ordinary circumstances is not due to utilization of material by respiration. After the loss of viability the quantity of respirable material is not appreciably different from that in the viable seed. Nor is the loss of viability due to loss of enzymes which are necessary for transformation of stored food reserves into mobile material. Some years after seeds have lost their viability these enzymes can still be obtained in an active state from some seeds. Crocker and Groves are of the opinion that the loss of viability of seeds stored dry is due to slow denaturing or coagulation of proteins of the embryo protoplasm. A certain amount of evidence for this was obtained by comparing the coagulation time of proteins and the length of life of wheat grains at various temperatures between 50° and 100° C. The relationships are similar in the two cases. At low temperatures the rate of denaturing of the proteins would be very slow if the law found for temperatures above 50° C. holds for temperatures below that level. Ohga has actually applied this theory to the case of the seeds of *Nelumbium nucifera* and calculates from data obtained with regard to the length of life of the seeds at high temperatures that in soil with an average temperature of about 10° C. the length of life of such seeds should be about a thousand years. Such estimates are naturally only crude approximations at best.

We have yet to consider the case of longevity exhibited by seeds buried in the ground and containing more than minimal quantities of water. Here it would seem at first sight that the conditions are favourable for continued respiration. Kidd found in white mustard (*Brassica alba*) that such a dormant condition could be produced by subjecting the seeds to high partial pressure of carbon dioxide. Soil air may possess a high partial pressure of carbon dioxide resulting from the decay of organic matter, and Kidd holds that the dormancy of seeds like those of species of *Brassica* in the soil results from the action of carbon dioxide which depresses both aerobic and anaerobic respiration and retards or inhibits the germinative processes accordingly. The depression of respiration produces no permanent disorganization and as soon as the high partial pressure of carbon dioxide is removed the seeds, other conditions being satisfactory, can germinate unless a condition of secondary dormancy, to which reference has already been made, should ensue.

CHEMICAL CHANGES IN GERMINATION

Morphological study has revealed that a considerable variety of form exists among seeds of different species, while chemical study has shown that these differences in form are frequently associated with differences in chemical composition. It may be well to recall that the essential part of the seed contained within the seed coat consists of the embryo, which may or may not be surrounded by endosperm derived from the secondary nucleus of the embryo-sac. Exceptionally, a layer of perisperm, persistent nucellar tissue, may surround the endosperm. In dicotyledons the embryo, with very few exceptions, consists of the two cotyledons lying close together side by side with the plumule between them and passing over at their point of junction with one another and with the plumule into the hypocotyl which is continued into the radicle. In the monocotyledons the embryo presents a variety of forms. Thus the seeds of most orchids contain an embryo consisting of a small mass of cells undifferentiated into cotyledon, plumule, hypocotyl and radicle. In the grasses, on the other hand, the differentiation is considerable.

Coulter and Chamberlain distinguished four types of monocotyledonous embryos, the Alisma, Pistia, Lilium and Orchid types, while among the dicotyledons, although the type of the well-known *Capsella* appears to prevail, there seem to be considerable variations from it, and the range in structure and development among the dicotyledons appears to be not less than that in the monocotyledons.

As seeds thus exhibit a wide range of morphological structure it is only to be expected that the chemical reactions and changes involved in the germination of seeds should also show a wide range of variation. But in all cases the first chemical changes involved in germination following upon absorption of water from without, must consist in the mobilization of food reserves contained in the seed, either in the endosperm, perisperm, or in the embryo itself. Consequently the processes involved must depend mainly on the chemical nature of the reserve.

It is estimated that fat forms the chief food reserve in the seeds of 80 per cent. of the species of flowering plants, while starch forms the principal reserve in only about 10 per cent. Protein occurs in considerable quantity in still fewer species. In other cases complex carbohydrates other than starch, and which are loosely spoken of as cellulose or hemi-cellulose, constitute the main reserve of food, as in the date palm, plantain and *Tropaeolum*. Frequently not one substance but several different substances which may serve as food, are present in considerable quantity in seeds. Some analyses of seeds are given in Table LXVI.

Table LXVI

Analyses of Various Seeds

(The values are given as parts per 100 of dry matter.)

	Triticum vulgare	*Pisum sativum*	*Quercus robur*	*Brassica rapa*	*Cocos nucifera*
Carbohydrate	68·65	52·68	46·83	24·41	12·44
Fat	1·85	1·89	3·08	33·53	67·00
Protein	12·04	23·15	3·26	20·48	8·88

Whatever the food reserve, this must be utilized for the building up of new cells as the embryo develops into the seedling. This developmental activity, as we know, involves respiration and a loss, therefore, of some carbon in the seed as carbon dioxide. Until the seedling develops a certain amount of chlorophyll and is capable of photosynthesis, the seed and its contained embryo or the seedling which develops from it, actually lose in dry matter although they may show considerable increase in bulk and weight owing to absorption of water.

How great this loss in dry matter may be is shown by the numbers in Table LXVIII which gives, among other data, the dry weight of seeds and seedlings of *Zea mais*.

This loss of dry matter must be ascribed to respiration since in this process carbon is lost to the plant in the form of carbon dioxide. The actual loss of food reserve substance must be greater than the loss of dry matter, since some of the reserve material must be utilized in the production of new tissues involving both protoplasm and cell-wall. In Table LXVII are given the amounts in grams of various substances in 100 grams of dried pea seeds (*Pisum sativum*) and in the seedlings developed from a similar quantity of pea seeds, the data having been obtained by Sachsse.

Table LXVII

Change in Composition of Edible Pea Seeds on Germination

	Seeds	Seedlings. 114 hrs. old	Seedlings. 184 hrs. old
Total dry matter	100	96·58	92·54
Starch	42·44	38·10	33·43
Dextrin	6·50	5·03	5·41
Cellulose	7·13	7·87	8·10
Fat	2·27	2·24	2·03
Protein	23·84	23·84	23·71
Undetermined matter	13·76	15·36	15·74
Ash	4·08	4·08	4·08

From this it will be observed that the actual loss of dry matter over the first period of germination is about 3·5 grams, whereas the loss in starch and dextrin together amounts to nearly 6 grams. The formation of new cellulose walls accounts for some of the difference, while undetermined matter, which may include complex protoplasmic material, accounts for more. More striking are some analyses of maize seeds and seedlings made by Boussingault. Twenty-two seeds of a sample were analysed and the same number of similar seeds were allowed to germinate in the dark for a period of 20 days. The amounts of various substances found in the seeds and seedlings are given in Table LXVIII.

Table LXVIII

Change in Composition of Maize on Germination

	Seeds	Seedlings
Total dry matter	8·636	4·529
Starch and dextrin	6·386	0·777
Glucose and sucrose	0·0	0·953
Cellulose	0·516	1·316
Fats	0·463	0·153

From this it will be observed that the actual loss in dry matter is about 4·1 grams, whereas the loss in starch and dextrin amounts to 5·6 grams. The difference is largely accounted for by the formation of fresh cell-walls in the form of cellulose. It will also be observed that mobile carbohydrate has increased from nothing to 0·953 grams. No doubt a prelude to the utilization of the reserve is its mobilization. From the analyses it also appears that fat has been utilized, although the amount present in the first place is small.

The changes taking place during the germination of seeds containing much fat appear very different from those taking place in starch seeds. As an example of such a seed we may take hemp (*Cannabis sativa*). In Table LXIX are shown the results of an analysis made by Detmer of 100 grams of hemp seed and of two sets of seedlings from the same quantity of seeds after germinating in the dark for periods of 7 and 10 days respectively.

Table LXIX

Change in Composition of Hemp Seeds on Germination

	Seeds	Seedlings 7 days old	Seedlings 10 days old
Total dry matter	100	96·91	94·03
Fat	32·65	17·09	15·20
Sugar	0·00	0·00	0·00
Dextrin	0·00	0·00	0·00
Starch	0·00	8·64	4·59
Protein	25·06	23·99	24·50
Undetermined substances	21·28	26·13	26·95
Cellulose	16·51	10·54	18·29
Ash	4·50	4·50	4·50

Here it will be observed that the fat decreases greatly in amount whereas protein remains constant, indicating that in this seed, where sugar and starch are absent, fat is the principal respiratory substrate. It is to be observed that a loss of cellulose occurs at first, suggesting that complex carbohydrates may form part of the food reserve and may be utilized in the metabolic processes of the germinating seed. With further development of cell tissue, however, the amount of cellulose increases.

As an example of a seed containing reserve protein may be taken that of the yellow lupin, the chemical changes in which during germination were studied by Beyer. The quantities in grams of various substances in 1,000 seeds and the same number of seedlings are shown in Table LXX. Analyses of the seedlings in two stages of development are given, the first when the seedlings were about 1 and 1½ inches, the second when they were about 2 to 3 inches long.

Table LXX

Change in Composition of Seeds of Yellow Lupin on Germination

	Seeds	Seedlings. 1st stage	Seedlings. 2nd stage
Fat	4·832	4·603	3·439
Protein	49·075	46·281	43·097
Asparagine	0·00	0·745	2·612
Sugar and alkaloids	8·498	17·091	15·698
Gum	5·542		
Cellulose, Starch, Pectin	8·869	7·715	9·253
Ash	3·384	3·493	3·633

It will be observed that the chief loss on germination is in the protein, although there is at first a decrease in the polysaccharides and also a quite definite disappearance of fat, in spite of this being present in relatively small quantity. The amount of sugar actually increases, although a similar decrease in the amount of more complex carbohydrate, followed by an increase, as is the case in germinating hemp, is also to be noticed.

As mentioned in the chapter on respiration, earlier in this work, it is not known whether, in the case of respiratory substrates other than sugar, the latter is first formed, or whether fat, protein or more complex carbohydrate is utilized more or less directly. In any case we must suppose that the actual katabolic processes of respiration are intimately linked with the anabolic processes of development, since the two run parallel.

The fact, however, that in both fat and protein seeds sugar either makes its appearance or increases in amount during germination, even when the seeds are kept in the dark and photosynthesis excluded, suggests that sugar is always produced whatever the substrate. In starch seeds no doubt the first phase of germination from a chemical point of view is development of diastatic enzymes and consequent production of glucose which is then translocated to the actively growing regions. In fat seeds lipase is developed and the substrate converted into fatty acids and glycerol. What the immediate fate of these substances may be is not known. The fatty acid can be recognized in considerable quantity in many germinating seeds, but the glycerol disappears at once and its presence can rarely be demonstrated. It is possible that both the fatty acid and glycerol are converted into sugar before translocation takes place; on the other hand glycerol, being a substance to which cells are readily permeable, may be translocated and utilized at the actual growing points. It has even been suggested that the fatty acids may also be translocated in the form of a very fine emulsion. This is not as improbable as might at first sight appear, for both the protoplasm and cell-wall contain fatty constituents and the fine fatty acid particles might be able to pass from cell to cell through these fatty components of the cells. Whether, then, the constituents of fresh cells are built up directly out of fatty acids and glycerol or out of sugar formed from either of these substances or both, we do not know. The processes are no doubt intimately linked with those of respiration, and even here, although it is generally suspected that hexose sugar is first formed, we do not really know the course of events.

The utilization of material in the germination of protein seeds presents similar or even greater difficulties. Here, also, the increase in glucose content of germinating seeds suggests that sugar is formed from protein, or rather from the amino-acids which result from the action of proteases, although we have seen earlier there is no very definite evidence that protein is utilized in respiration. In the case of all seeds it would seem likely that some protein reserve must be present, since the building up of fresh protoplasm in the course of cell multiplication during development of root and shoot necessitates a supply of nitrogenous material.

The mobilization of the protein reserve of seeds is no doubt effected by the proteases which are present in germinating seeds, with production of amino-acids. A number of the latter have been identified in seedlings of one species or another. Thus phenylalanine, tyrosine, valine and leucine have all been found in *Lupinus luteus* and *L. albus* seedlings, while the same amino-acids, with the exception of valine, have been found in seedlings of *Cucurbita*. An amino-acid very commonly present in young seedlings is asparagine, a substance (see p. 253) with the formula

$$\begin{array}{l} NH_2.CH.COOH \\ \quad\;\;\; | \\ \quad\;\; CH_2.CONH_2. \end{array}$$

Schulze found in *Lupinus luteus* that the asparagine content increased with the age of the seedling, not reaching a maximum until the plant was several weeks old. Some of the results of Schultze and Umlauft are summarized in Table LXXI. The accumulation of asparagine

Table LXXI

Content of Asparagine in *Lupinus luteus* Seedlings

Age of Seedling in days	Asparagine Content in percent. of dry weight of seed	Asparagine Content in percent. of dry weight of seedling
4	3·12	3·3
7	9·78	11·2
10	15·24	17·3
12	18·22	22·3
15	19·43	25·0
16		25·7

takes place chiefly in the young shoot, and this can be regarded as evidence that the asparagine is largely used in the building up of new protein. On the other hand, asparagine has been regarded as an end-product of metabolism, which can, however, be used in metabolic processes after being subjected to the action of light. In fact, in spite of a number of observations and much speculation, we do not know what is the fate of the amino-acids, and what changes take place in the rebuilding from them of the proteins of the new tissues formed in germination. It is possible that amino-acids are further acted upon and that even simpler derivatives of them form the basis for the formation of fresh protein. In this connexion it is possible that the enzyme urease may play some part in the nitrogen metabolism of germinating seeds as it is known to be actively present in those

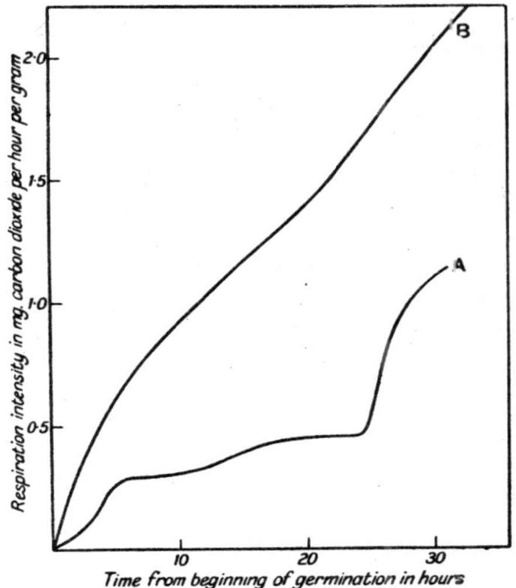

Fig. 28.—Curves to show the course of respiration of seeds of *Lathyrus odoratus* during germination

A, intact seed provided with testa ; B, seed deprived of testa

(From W. Stiles and W. Leach)

of many Leguminosae, as well as in *Ricinus*. The evidence at present available, however, does not suggest that it is universally present in seeds, so that it can scarcely have a general significance in germination.

As regards the actual course of respiration during germination, it has been usual to suppose that there is a ‘ grand period of respiration ’, during which the rate of respiration rises from approximately zero to a maximum and then falls again. Recent work by Stiles and Leach on the respiration of sweet pea (*Lathyrus odoratus*) seedlings supports the view that respiratory activity does rise to a maximum in this way, but that the very considerable falling off observed by earlier workers is probably to be explained as chiefly related to the conditions under which the experiments were carried out. In older seedlings enclosure in a chamber or exposure to air saturated with water vapour has the effect of reducing the respiration rate, probably because the movement of mobile respiratory substrate from the organs of storage to the growing points is retarded under such conditions.

It naturally follows that the course of respiration of a seedling kept in the dark is likely to differ from that of a seedling exposed to light, for as soon as chlorophyll develops in the latter the concentration of respiratory substrate in the young shoot may be increased by photosynthesis. The respiration rate of the whole seedling is therefore likely to increase rapidly on this account, although its increase will sooner or later be actually masked by photosynthesis with its gaseous exchange in the opposite direction.

From what has been said earlier regarding the retarding influence on germination exerted by the seed coats of many seeds, it is to be expected that the course of germination and of respiration may be more or less considerably affected by removal of the testa. In the case of the sweet pea the rate of respiration of the seed increases from the time the seed begins to absorb water until it reaches a rate which remains approximately constant until the seed coat splits, upon which the rate again rises to a maximum. Removal of the testa eliminates this constant period. The course of respiration during germination of sweet pea seeds with and without the seed coat is shown in Fig. 28.

CHAPTER XVI

VEGETATIVE DEVELOPMENT

THE development of the vegetative body of an organism is one of the most obvious characteristics of living things. In the simplest unicellular organisms, it is true, this development may consist to external view of no more than an increase in size of a single cell without any other apparent change, but in the majority of multicellular plants, and particularly in those with a vascular system, this development involves a very considerable increase in complexity, parts with different functions and of different form developing, so that there is what has been called a 'physiological division of labour' in which each function is carried out by a particular set of organs. At the same time these organs do not function independently of one another; there is a correlation between the activities of the various parts so that the whole plant develops as a single unit.

This development involves growth, and indeed the two terms are often used synonymously. The term development, however, involves the idea of increase in complexity, whereas growth rather suggests enlargement only. The simple definition of growth as increase in size, or volume, or mass, is, however, unsatisfactory, for this may result solely from increase in water content due to an absorption of water which may be lost again by evaporation, and such fluctuations in water content may be of quite considerable magnitude. Increase in dry weight has therefore sometimes been taken as a definition of growth, but objections can be raised to this also, as dry weight increase can be due to an increase in the amount of cell inclusions which are dead material and have no real concern with the growth of the plant. Moreover, during germination the seedling may be growing and yet actually losing in dry weight, although increasing in size and complexity. Other definitions of growth are increase in size involving development, and increase in volume which is irreversible. Both these definitions are fairly satisfactory.

But although it may be difficult to find a satisfactory definition of growth, there is rarely any doubt as to its meaning. In the multicellular plant its essential character is the utilization of the material formed in photosynthesis and other anabolic processes to produce new tissue. It is thus a necessary condition for growth over a length of time that the material introduced into a plant body by absorption from the environment should exceed that lost to the plant in respiration, although over shorter periods, as noted above, a plant might develop and yet lose in dry weight. Actually, during the growing season the photosynthesis is many times the respiration. For example, Boysen Jensen found that a plant of *Sinapis* during the 24 hours of a summer's day absorbed a net amount of 84·6 mg. carbon dioxide (that is, total assimilation less total respiration) through the leaves, and lost only 4 mg. of carbon dioxide through respiration from the non-green parts, making a total excess of assimilation over respiration of 80·6 mg. of carbon dioxide corresponding to 49·5 mg. increase in dry weight of the plant if this is calculated as starch, or 55 mg. if calculated as hexose sugar. The former of these values is about 17·8 per cent. of the actual dry weight of the plant.

The parts of the plant where growth is actually taking place, that is, where new tissue is being formed and the increase in size is located, may be remote from the assimilatory organs where carbohydrates, and possibly proteins, are formed. Thus, as pointed out in a previous chapter, these synthesized materials have to be conveyed in a mobile form to the regions of growth. Hence the rate at which new tissue is formed, that is, the rate of growth, depends not only on the rate at which photosynthesis takes place, but also on the rate of diffusion of the synthesized material to the growing parts. There may also be other internal factors in the growing cells themselves as well as external factors which influence the rate of growth.

In unicellular organisms vegetative growth may consist only of enlargement of a single cell which may after a time pass over into its reproductive stage. In filamentous algae such as *Spirogyra* vegetative growth can be regarded as involving two stages, cell division and cell enlargement. In the higher plants Sachs pointed out long ago that three stages of growth can be recognized, namely, cell division, cell enlargement and the internal differentiation of the cell. So long as a cell is meristematic its growth activities are confined to the first two of these stages, but each cell that forms part of a permanent tissue goes through the third phase and so takes on that special form which determines its function and which characterizes the tissue or part of a tissue of which it is a component.

In an organ of limited growth, such as a leaf, when the adult permanent condition is reached every cell has gone through these three phases of growth. Consequently, as the increase in size of the whole organ is the sum of the increases in size of the individual cells, it is to be expected that the organ itself will, to some extent, show the same sequence of phases as the cells of which it is composed. It is true that some cells will reach the second and third stages before others. However, on the whole in the earliest stage of leaf development, while the leaf is in the bud in a rudimentary condition, most of the cells are meristematic and still dividing. This early stage is followed by a period of enlargement and unfolding, and this again by the phase of internal differentiation which it overlaps. With organs of unlimited growth, such as roots and shoots, a zone at the apex remains meristematic, but cells formed by division of these meristematic cells on the side remote from the apex pass in time into the phases of enlargement and internal differentiation and ultimately cease growth. But while these changes are taking place the meristematic cells at the apex have continued to divide and cut off on the inside more cells, which gradually pass over into the permanent condition. Thus in passing from the apex towards the older parts of the organ we find the youngest parts chiefly in the first stage of growth, the next principally in the phase of elongation while the oldest parts have ceased to increase in size and may be completely differentiated. This condition of affairs was originally pointed out by Sachs who published a number of measurements illustrating it. In Table LXXII is shown the relation between rate of elongation and distance from the apex of a primary root of *Vicia faba* growing in moist air at about 20·5° C., the numbers in the second column giving the increase in length, during 24 hours, of successive lengths of root, each, at the beginning of the period, 1 mm. long. In Table LXXIII are given similar numbers for

Table LXXII

Relation between Rate of Growth and Distance from Growing Apex in Primary Root of *Vicia faba*

Original mean Distance of Zone from Root Apex in mm.	Increase in Length during 24 hours in mm.
0·5	1·5
1·5	5·8
2·5	8·2
3·5	3·5
4·5	1·6
5·5	1·3
6·5	0·5
7·5	0·3
8·5	0·2
9·5	0·1

Table LXXIII

Relation between Rate of Growth and Distance from Growing Apex in Shoot of *Phaseolus multiflorus*

Original mean Distance of Zone from Root Apex in mm.	Increase in Length during 40 hours in mm.
1·75	2·0
5·25	2·5
8·75	4·5
12·25	6·5
15·75	5·5
19·25	3·0
22·75	1·8
26·25	1·0
29·75	1·0
33·25	0·5
36·75	0·5
40·25	0·5

a shoot of *Phaseolus multiflorus*, the increases in length recorded being in this case that of successive lengths originally 3·5 mm. long and the period of time over which the increases in length occurred, being 40 hours. The data in both tables were obtained by Sachs.

The matter can be looked at in another way. If a small zone of tissue just behind the apex is considered, it will thus increase in length as the cells within it go through the stage of elongation. In this zone those cells more remote from the apex will cease growing before those nearer the apex, but sooner or later all the cells in the zone will have ceased enlarging, the whole zone by this time having not only greatly increased in size, but become much further removed from the apex on account of the growth of the cells between it and the apical meristem. Thus, if such a small zone behind the apex is considered it will go through a series of changes similar to those undergone by an organ of limited growth such as a leaf, that is, it will grow slowly at first, then more rapidly, until the rate of growth reaches a maximum, after which it will gradually fall to zero. In reference to this phase of enlargement of the cell, organ or portion of an organ, during which the rate of growth increases from practically zero to a maximum and then falls again to zero, Sachs introduced the term ' grand period of growth '. In a case given by Sachs himself, that of a zone of the primary root originally occupying 1 mm. behind the apex of a seedling of *Vicia faba*, growing in moist air at a temperature of about 20° C., this period lasted 7 days, the growth on successive days being that shown in Table LXXIV. There are also shown here similar numbers for the growth on successive days of a zone just behind the apex of the growing shoot of *Phaseolus multiflorus*.

Table LXXIV

Growth of a Zone of Primary Root of *Vicia faba* originally 1 mm. long and of a Zone of Shoot of *Phaseolus multiflorus* originally 3·5 mm. long

Day	Increment of growth: *Vicia faba* root	Increment of growth: *Phaseolus multiflorus* shoot
1	1·8	1·2
2	3·7	1·5
3	17·5	2·5
4	16·5	5·5
5	17·0	7·0
6	14·5	9·0
7	7·0	14·0
8	0·0	10·0
9	0·0	7·0
10	0·0	2·0

If these results are expressed graphically the curve of total growth is of the form usually termed sigmoid, while that illustrating the relation between rate of growth and time shows a maximum. Such curves for the growth of *Phaseolus multiflorus* are shown in Figs. 29 and 30.

The curve giving the relation between rate of growth and time has been termed, not very happily, the 'grand curve of growth'.

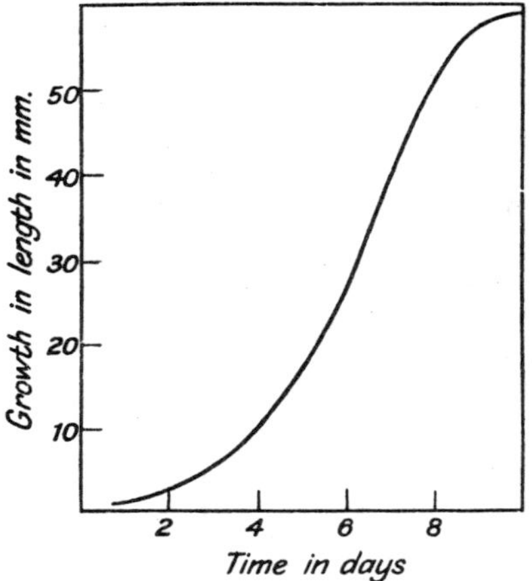

FIG. 29.—Curve of total growth of an internode of *Phaseolus multiflorus*
(Constructed from the data of J. v. Sachs)

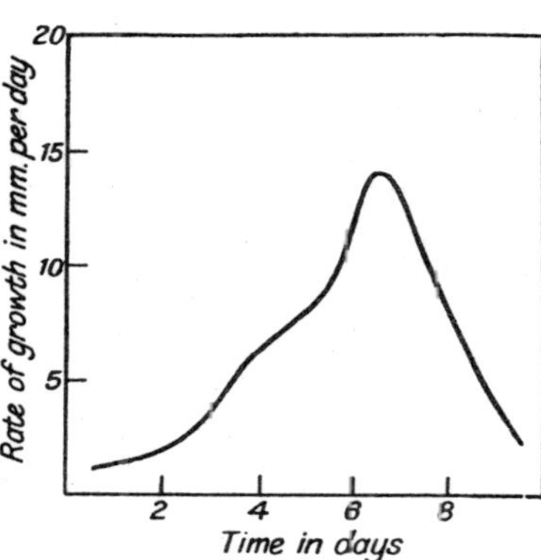

FIG. 30.—Grand curve of growth of an internode of *Phaseolus multiflorus*
(Constructed from the data of J. v. Sachs)

The growth of an organ or zone of an organ does not always follow the simple course indicated by Fig. 30. The best-known exception to this is the curve of the grand period of growth of the sporangiophore

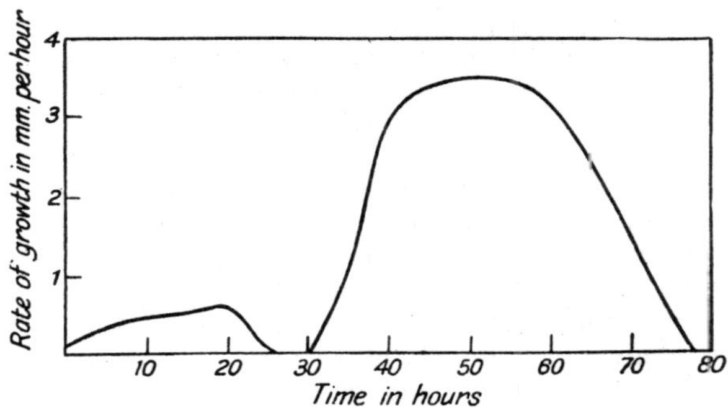

FIG. 31.—Grand curve of growth of the sporangiophore of *Phycomyces blakesleeanus*
(After L. Errera)

of *Phycomyces blakesleeanus* [1] determined by Errera in 1884 and shown in Fig. 31. Here the rate of growth reached a maximum, under the conditions of observation, after 19 or 20 hours, and then fell rapidly

[1] Generally known as *Phycomyces nitens*.

to zero, after which growth recommenced and reached a second, and much higher maximum, at the end of the second day. The explanation of the cessation of growth at the end of the first day is that the material available for development is then used in the development of the sporangium, after which growth of the sporangiophore itself is resumed. Other members of the Mucorineae behave similarly, but in the case of *Pilobolus* growth in length of the sporangiophore terminates with the production of the sporangium.

A somewhat similar grand curve of growth was observed by Miyake for the inflorescence axis of *Taraxacum* (cf. Fig. 32). The explanation is probably similar, the cessation in growth of the axis corresponding to the period of enlargement of the inflorescence.

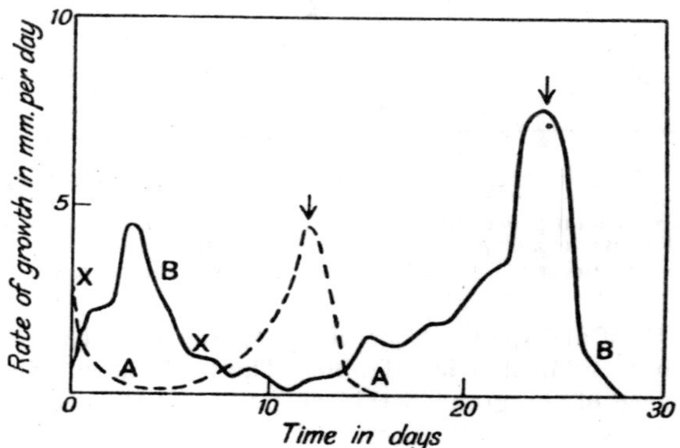

Fig. 32.—Growth-rate curves of the inflorescence axes of two varieties of *Taraxacum officinale* during flowering and fruiting

A, var. *glaucescens*; B, var. *albiflorum*. The time of opening of the flowers is indicated by ×, that of the expansion of the fruiting heads by ↓

(After K. Miyake)

The grand period of growth is generally comparatively brief, but sometimes it may last for months. This is the case with the sporangiophore of *Pellia* in which the rate of growth is extremely slow for several months during which time the total growth in length may not exceed 1 or 2 mm. This period of slow development is followed by an exceedingly rapid growth of the sporangiophore which may reach its final development of sometimes as much as 80 mm. in a few days.

The length of the growing zone of a shoot or root varies greatly among different species. Some measurements made of the length of the growing zones of the shoots of a number of species are summarized in Table LXXV. The number of internodes involved may vary from one to five or more.

With roots the growing zone is generally much shorter than in shoots. Thus the numbers shown in Table LXXIı indicate that the growing region of the primary root of *Vicia faba* may not exceed a centimetre. In aerial roots, however, the growing zone may be much longer, and in *Cissus*, a genus closely allied to, and sometimes included in, *Vitis*, Blaauw recorded that the growing zone might be as long as 100 cm.

Table LXXV

Species	Length of Growing Zone in cm.	Observer
Elodea candensis	2·5	Askenasy
Galium mollugo	3	,,
Fritillaria imperialis	8	Sachs
Polygonum sieboldi	15	,,
Asparagus asper	20	,,
Hippuris vulgaris	25	Askenasy
Allium cepa	30	Sachs
Cephalaria procera	35	,,
Allium porrum	40	,,
Aristolochia sipho	45	Askenasy
Aristolochia atropurpureum	50	Sachs

The actual rate of growth in length of different organs varies exceedingly. Thus among fungi A. Möller, according to Jost, recorded a maximum growth rate of the stipe of *Dictyophora phalloidea* of 5 mm. per minute, an extraordinarily high value, while hyphae of *Botrytis* were observed by Reinhardt to grow at the rate of 0·034 mm. per minute. Among higher plants a stem of giant bamboo (*Dendrocalamus giganteus*) was observed by A. M. Smith to elongate at a rate of nearly 0·4 mm. per minute. Some workers have thought a better method of expressing growth rate is in terms of the increase in length in unit time per unit length of the growing zone, as this will give a truer idea of growth activity. When so expressed it is found that very great variations in growth rates exist, and that pollen tubes can exhibit particularly high growth rates. The numbers in Table LXXVI recorded by Buchner give some idea of the possible variation in growth rates between different species.

Table LXXVI

Growth Rates of some Plant Organs

Plant Organ	Growth rates in mm. per minute per mm. growing Zone
Impatiens Hawkerii pollen tube	2·20
Impatiens balsamia pollen tube	1·00
Mucor stolonifer hyphae	1·18
Botrytis hyphae	0·83
Bambusa shoot	0·0127
Bryonia shoot	0·0058

In unicellular organs the rate of growth is often taken as the inverse of the time between one division and the next. In the case of bacteria this is often about 20 to 30 minutes. This may be regarded as the time taken for the growing material (or zone) to double in size, in which case a value for the growth rate of such bacteria comparable with those given in Table LXXV would be only of the order of 0·05.

INFLUENCE OF EXTERNAL CONDITIONS ON RATE OF GROWTH

It has already been pointed out that in an autotrophic plant the external conditions of light and temperature will affect the rate of growth by their effect on photosynthesis and hence on the supply of the material out of which new tissue is formed. It is of interest to inquire whether, apart from this indirect effect, there is also a direct effect of external conditions on the rate of growth. This growth, as we have seen, really involves two stages, cell division and cell enlargement, but it is most frequently impracticable to separate these two phases, at any rate in multicellular plants, when investigating the effect of external conditions. We are also faced with a similar complexity regarding the interaction of factors as in photosynthesis, for if increase in temperature, for example, should directly bring about an increase in the reactions involved in growth, we can understand that the rate of supply of available material to the growing zone might impose a limit on the rate of growth and that increase in temperature of the growing zone in such cases could not bring about a rise in the growth rate simply because of insufficiency of material.

The exact relation between growth and internal and external factors has, however, not yet been investigated sufficiently to enable the relations to be stated as definitely as in the case of photosynthesis, and so there still obtains in regard to growth the conception of cardinal points (minimum, optimum and maximum), a conception which, it is now realized, does not help to explain the relation between photosynthesis and its conditioning factors. As regards temperature, the minimum and maximum temperatures for growth of a number of species are given in Table LXXVII, but it must be realized that these tem-

Table LXXVII

Minimum and Maximum Temperatures for Growth in Centigrade Degrees

Species	Minimum Temperature	Maximum Temperature
Pisum sativum (roots)	−2	44·5
Ulothrix zonata	0	24
Triticum vulgare	0	42
Lepidium sativum	2	28
Mucor racemosus	4	33
Pinus sylvestris	7	34
Acer platanoides	7	26
Zea mais	9·5	46
Cucurbita pepo	14	46
Aspergillus fumigatus	15	60

peratures may vary from individual to individual and from organ to organ and from time to time in the same individual. Generally speaking, the range of temperatures over which tropical plants grow is higher, though not necessarily wider, than that for temperate plants, while the range of temperatures over which arctic plants grow is lower. The range of temperatures may also vary in extent from one species to another, and the same may be the case with different organs of the same species. Thus Wiesner found the limits of temperature for the germination of spores of a species of *Penicillium* to be 1·5° and 43° C., for growth of the mycelium 2·5° and 45° C., and for spore formation 3° and 40° C. Again, in *Vaucheria repens* the minimum and maximum temperatures for vegetative growth are respectively 0° and 30° C., but for zoospore formation 3° and 26° C. According to Setchell, vegetative growth in the marine angiosperm *Zostera marina* occurs between 10° and 20° C., whereas reproduction only takes place between 15° and 20° C. In spring-flowering plants, as, for example, *Crocus*, the minimum and maximum temperatures for flowering are often lower than those for vegetative growth.

It is to be noted that the temperatures to which a plant is exposed may often lie outside the range within which it grows. The minimum and maximum temperatures for growth do not by any means correspond with the range of temperatures over which a plant can remain alive.

These minimum and maximum temperatures for growth are dependent to some extent on external conditions and in some cases, at any rate, are susceptible to modification by what has been called 'accommodation'. Thus, by gradually lowering the temperature the minimum temperature for the growth of *Bacillus anthracis* has been lowered from about 14° to 10° C., and by gradually raising the temperature the maximum temperature for growth of *Bacillus fluorescens* has been raised from 35° to 41·5° C.

As regards the direct effect of temperature on growth Miss Leitch in 1916 examined this question in the roots of *Pisum sativum*, while Miss Talma in 1918 recorded results obtained with roots of *Lepidium sativum*. Miss Leitch measured the rate of growth of pea roots at temperatures from − 2° to 45° C. (Between − 2° C., which appears to be the minimum temperature for the growth of pea roots, and 29° C., there is a regular increase in the rate of growth with rise in temperature (see Fig. 33) although Miss Leitch's determinations show some irregularity due, no doubt, to individual variations, that is, to internal conditions which are, of course, not under control nor measurable. The relationship between rate of growth and temperature was found not to be a logarithmic one, the temperature coefficient (Q_{10}) decreasing with increasing temperature, much in the same way as it does in the case of the respiration of these roots (cf. p. 138) only much more definitely. The values of the temperature coefficients, calculated from Miss Leitch's data for two series of experiments, are shown in

Table LXXVIII. The average coefficients for the whole range 0° to 28° in the two series are respectively 3·56 and 3·89.

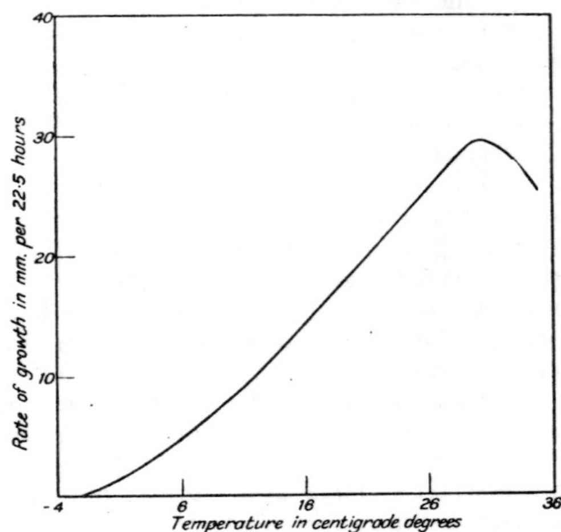

FIG. 33.—Curve illustrating the relation between temperature and the rate of growth of roots of *Pisum sativum*

(After I. Leitch)

Table LXXVIII

Temperature Coefficients for the Growth of Roots of *Pisum sativum*

Temperature Interval	Temperature Coefficient. Series I.	Series II.
0–10	10·0	8·2
5–15	3·4	4·1
10–20	2·3	2·9
15–25	1·9	2·4
18–28	1·7	2·2

Above 29° C. the rate of growth does not remain constant with time, but shows that falling off in value which is attributed to a so-called 'time factor', although up to 35° the regular fall is interrupted by a temporary rise. At 40° C. the fall in rate with time is continuous. Above 30·3° C. the initial rate of growth was actually less than at lower temperatures. Miss Leitch applies the expression 'optimum temperature for growth' to the highest temperature at which growth is maintained at a constant rate, this temperature, in her experiments, being about 29° C. It is true an initially higher rate was observed with increasing temperature up to 30·3° C., but the initial rate, as we have seen, is not maintained. For the temperature at which the initial rate of growth is highest she proposed the term 'maximum rate temperature'.

The observations of Miss Talma on the effect of temperature on the growth of roots of *Lepidium* show a general agreement with those on pea roots just described. Miss Talma's results are shown graphically in Fig. 34, the three curves giving the amount of growth at different temperatures after 3·5, 7 and 14 hours. The recession of the

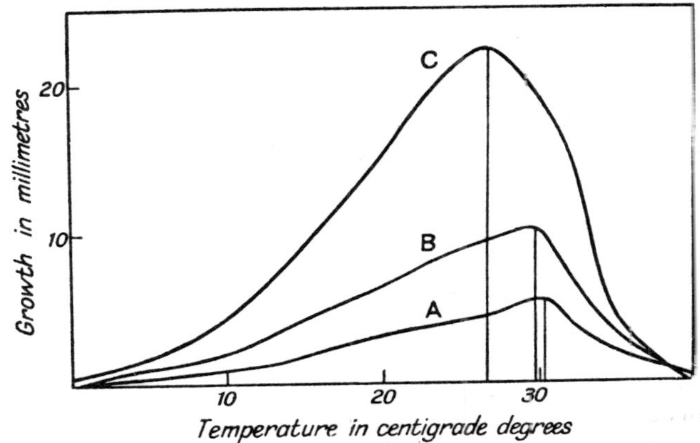

FIG. 34.—Curves showing the growth of roots of *Lepidium sativum* at different temperatures after (A) 3·5 hours, (B) 7 hours and (C) 14 hours
(After E. G. C. Talma)

apparent optimum temperature with time is accounted for by the action of the time factor at higher temperatures. The optimum temperature in the sense in which it was defined by Miss Leitch would here appear to be about 26° or 27° C. The temperature coefficients of growth rate, calculated from Miss Talma's data are summarized in Table LXXIX.

Table LXXIX

Temperature Coefficients of Growth of Roots of *Lepidium sativum*

Temperature Interval	Temperature Coefficients over Different Periods		
	3·5 hours	7 hours	14 hours
0–10	—	—	7·9
5–15	3·6	4·7	5·2
10–20	3·4	3·43	3·45
15–25	2·25	1·9	2·27
16–26	1·86	1·88	2·1
17–27	1·81	1·83	1·9
18–28	1·86	1·79	1·64
19–29	1·71	1·68	1·43
20–30	1·59	1·57	1·2

The effect of temperature on the rate of growth of the sporangiophore of *Phycomyces* was examined by Miss Graser, and the observed results were also in general agreement with those found for the roots of higher plants. The minimum and maximum temperatures were found to be 0° and 34° C. respectively but the falling off in growth rate at temperatures above 29° C. was very marked. Within the range of 0° to 29° C. the fall in the value of the temperature coefficient with rising temperature is as marked as in the cases observed earlier. Some of Miss Graser's values are summarized in Table LXXX.

Table LXXX

Temperature Coefficients of Growth of Sporangiophore of *Phycomyces blakesleeanus*

Temperature Interval	Temperature Coefficient
5·1–16·0	7·5
6·2–17·4	4·8
6·3–19·4	2·5
9·8–23·4	1·52
10·8–25·7	1·34

There is evidence that in some cases a *change in temperature* leads to changes in the rate of growth which are related to the change in temperature itself as distinct from the actual temperature level. For example, True in 1895, found that roots of *Vicia faba*, *Pisum sativum* and *Lupinus albus*, when subjected to a sudden change of temperature from about 1° to about 20° C., as extremes, exhibited rapid elongation followed by a period of depressed growth, while if the temperature change took place in the reverse direction a depression in growth rate occurred. Attention has more recently been called to this phenomenon by Silberschmidt and by Erman, the former of whom has shown a similar behaviour in coleoptiles of oat, barley and wheat, as well as in the hypocotyl of flax, the results with oat being confirmed by Erman. Some of Miss Graser's observations on *Phycomyces blakesleeanus* suggest that the same phenomenon occurs in the growth of the sporangiophore of this fungus.

An example of Silberschmidt's results is given in Fig. 35 which shows in graphical form his results with wheat. The rate of growth of a plant of *Triticum vulgare* at room temperature (14·5° C.) was observed for a period of 40 minutes after which the plant was transferred to a temperature of 0·5° C. for 45 minutes. It was then transferred back to room temperature and the rate of growth again observed. The initial rise in growth rate above the normal with the subsequent fall are well shown in the figure. Results with oat are shown in Fig. 36. Here also it will be observed that transference from a temperature of 5·5° C. to one of 15° C. brought about a temporary high growth rate.

As will be described later, a similar behaviour as a result of exposure to light for a limited time is termed the 'light-growth reaction'. In comparison with the latter, this change in growth rate as a result of exposure to a different temperature has been termed the thermo-growth reaction. Whether such a comparison of the two reactions is really justified is not clear. At any rate the so-called 'thermo-growth reaction' appears to be in the nature of a response to a stimulus, that of change in temperature (see Chapter XIX). Thus Silberschmidt records experiments in which the growth was observed at room temperature, then for a period during which the plant was maintained at a low temperature and then for a further period after it was re-transferred to room temperature. From the graphical representation of the results in Fig. 36 of such an experiment with *Avena*, it will be observed that the rise and subsequent fall in growth rate occur both after a change from a higher to a lower temperature, and after the reverse change from the lower to the higher temperature.

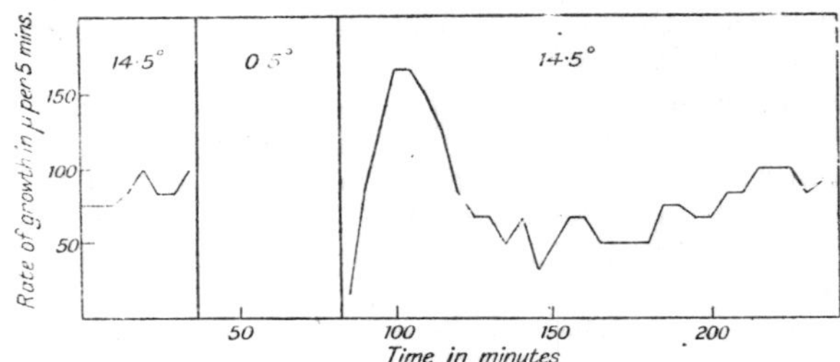

FIG. 35.—Curves illustrating the thermo-growth-reaction in *Triticum*
(Constructed from the data of K. Silberschmidt)

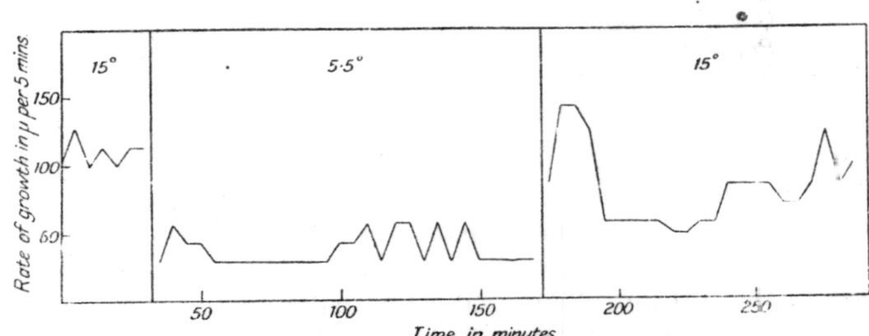

FIG. 36.—Curves illustrating the thermo-growth-reaction in *Avena*
(Constructed from the data of K. Silberschmidt)

Light. While many fungi can live through their whole life-history in darkness, it is obvious that light is necessary for the growth of autotrophic plants as it is an essential factor in the synthesis of material out of which new tissue is formed. As plants only grow within certain limits of temperature so similarly there are ranges of light intensity outside the limits of which different species do not grow. Wiesner determined these limits for a large number of species growing under natural conditions, expressing his results in two ways. By the term 'absolute light requirement' ('Lichtgenuss') he denoted the actual range of light intensities measured in Bunsen-Roscoe units; by 'relative light requirement' he denoted the range of light intensities expressed as factors of the intensity of full sunlight at the place where the plants were growing. Since the maximum intensity of full sunlight varies with the latitude and the height above sea-level it is to be expected that the relative light requirement will be different in different places. Wiesner found that the relative light requirement of plants increased with latitude, the values for *Betula nana*, for example, being from 0·294 to 1 in Oslo, from 0·455 to 1 in Tromsö, 0·77 to 1 in Troll Fjord and 1 only in Advent Bay, Spitzbergen. The comparison of the light requirements at different places is complicated not only by the different light intensity of full sunlight but also by the different length of day and again by the fact that possibly in colder places more of the heat rays of solar radiation are utilized.

Some of Wiesner's values for the minimum relative light requirements of plants at Vienna are summarized in Table LXXXI. The maximum relative light requirement is unity in all cases.

Table LXXXI

Light Requirements of Different Species at Vienna

Species	Minimum relative Light Requirements
Buxus sempervirens	0·0092
Fagus sylvatica (open form)	0·0118
Fagus sylvatica (close form)	0·0167
Carpinus betulus	0·0179
Acer platanoides	0·0182
Quercus robur	0·0385
Populus alba	0·0668
Betula verrucosa	0·111
Liriodendron tulipifera	0·133

Apart from the indirect effect of light on growth through photosynthesis it has been realized for long that light must have a direct effect on the growth of some organs. The phenomenon of etiolation, in which subjection of a plant to complete darkness involves certain changes in the form of plants, indicates that alterations in light intensity may bring about alterations in the growth rate of various organs, and that, in particular, feeble light intensity leads to increased rate of growth. Sachs and others showed that reduction in light intensity leads to increased rate of growth of stems and leaves, but more recent researches have shown that there is no general rule to express the relation between light intensity and growth rate. Thus in the coleoptile of *Avena*, Vogt and Sierp have both shown that the more intense the light the shorter are the coleoptiles, and Sierp found that with increasing light intensity, not only was the total growth reduced but that the grand period of growth was also shortened. But he further found that in experiments lasting one day only growth was more rapid in stronger light intensity. These relations are made clear by Fig. 37, showing the curves of the grand period of growth of *Avena* coleoptiles in different intensities of light.

The results of experiments on the influence of light intensity on the growth of roots present some contradictions. In 1902 Kny recorded that roots of plants of *Lepidium sativum*, *Vicia sativa* and *Lupinus albus* growing in water culture, grew more rapidly in darkness than in light, but the more recent short period experiments of Miss Leitch and Blaauw indicate that light is without effect on the rate of growth of most roots. The experiments of the latter with hypocotyls of *Helianthus globosus* showed that growth in these organs increased with light intensity.

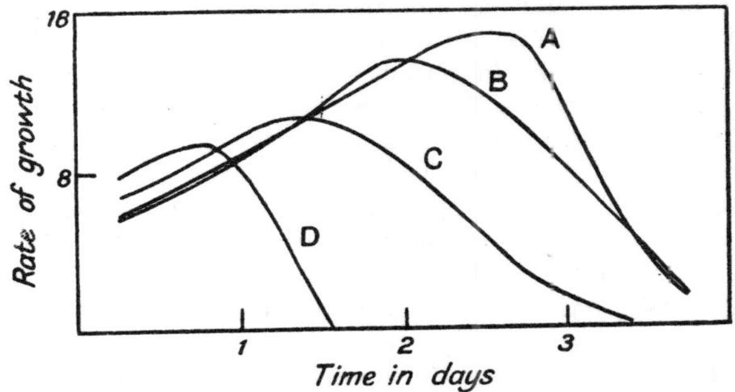

Fig. 37.—Grand curves of growth of *Avena* coleoptiles in different intensities of light A, weak red light; B, 16 MK; C, 500 MK; D, 4000 MK. The coleoptiles were kept in darkness for alternate half-days
(Constructed from the data of H. Sierp)

In 1897 Stameroff recorded that the rate of growth of pollen tubes and hyphae of *Mucor* is the same in the light and in the dark, but more recently, in 1918, Blaauw reported that the rate of growth of sporangiophores of *Phycomyces* is strongly affected by light, being greater in light intensities of 8 and 64 metre-candles[1] than in the dark, but being strongly retarded in a higher intensity of illumination, 4,000 metre-candles.

Some of the contradictory results obtained are probably to be traced to temperature variations and differences in the earlier experiments, for, as has already been explained, temperature has a considerable direct effect on growth rate. But from the more recent work, in which temperature has been carefully controlled, it is evident

[1] The metre-candle is the intensity of illumination from a standard candle at one metre distance from the source of light. Unit quantity of light is the quantity of light from this standard candle falling during one second on unit area in a plane at right angles to the direction of the light and at one metre distance from it. If the illuminated area in a series of experiments remains constant, the quantity of light is proportional to the product of the light intensity and the time, and can be denoted by the term metre-candle-second.

that light has, in many cases, a very definite effect on rate of growth, and it may be concluded that in such cases a photochemical reaction must be involved in addition to the purely chemical, or enzymic reaction, indicated by the effect of temperature on growth.

According to the observations of Klebs on the effect of light on the germination and development of fern prothallia it appears that light intensity, and also the wave-length of the light, can influence both the division phase and the enlargement phase of growth. In *Pteris longifolia*, for example, the prothallus in weak white light (from an electric lamp) with an intensity of 27 metre-candles grows into a filament which may reach a length of 2·2 mm. without any cell division occurring; the single cell thus formed may be more than 100 times as long as a normal prothallial cell. With increasing light intensity there are an increasing number of transverse cell divisions and with still higher light intensities divisions take place in two directions so that a heart-shaped prothallus is formed. With still higher light intensities, 500 metre-candles or more, divisions take place in three planes at right angles, so that a solid tissue results.

By using red light and blue light in place of white light Klebs further showed that in blue light cell division takes place, but enlargement is greatly retarded; red light, on the other hand, furthers growth in length but inhibits cell division. From his observations Klebs concluded that light has two actions on growth; a trophic action consisting in the production of nutrient material, and a blastic action, in which light acts as a catalyst, which shows itself as cell division and enlargement.

Considerable attention has been directed in recent years to what is termed the 'light-growth reaction', to which passing reference has already been made. This reaction was first described in 1914 by Blaauw as a result of observations on the sporangiophore of *Phycomyces blakesleeanus*, his observations being later extended to higher plants. Blaauw's experiments showed that when a sporangiophore previously grown in the dark is subjected to a certain amount of illumination provided equally on four sides, an increase in the rate of growth of the sporangiophore results, this increase being followed as a rule by a decrease below the original rate and then by slighter variations before the rate is again constant. In these experiments a very constant temperature was maintained, the variation throughout one experiment not usually exceeding 0·01° C. Both the light intensity and time of illumination were varied, so that different quantities of light ranging from four times 0·25 to four times 1,920,000 metre-candle-seconds (M.K.S.) were employed. The actual light intensity and period of illumination as such, at any rate within the limits imposed by Blaauw's experiments, do not appear to affect the change in growth rate resulting from exposure to light; the extra growth appears to be determined by the actual quantity of light. The light-growth reaction is characterized by certain quantities, namely, the time which elapses between illumination and the commencement of acceleration of growth, the maximum rate of growth reached, the time taken to reach the maximum and the amount of the extra growth. Since the normal rate of growth varies in different individuals the amount of the extra growth is expressed in terms of minutes of growth, the time in minutes which would be required for the plant to increase in length by the extra amount, when growing at its normal rate. The subsequent fall in growth and the time taken for the normal rate to be regained exhibit considerable variations. The mean of the values of the light-growth reaction characteristics for different quantities of light published by Blaauw for *Phycomyces blakesleeanus* are shown in Table LXXXII.

Table LXXXII

Mean Values of the Characteristics of the Light-growth reaction in *Phycomyces blakesleeanus*

Light quantity in M.K.S.	Time in minutes before growth rate changes. (Reaction time)	Time taken in minutes for maximum growth rate to be reached	Maximum value of growth rate in percentage of normal rate	Extra growth in minutes of growth
4 × 0·25	8·30	10·90	121	0·89
4 × 1	5·33	8·92	130	1·62
4 × 4	3·83	6·67	158	2·69
4 × 30	3·60	7·10	194	5·35
4 × 210	3·36	7·06	253	9·82

Within the limits of light quantity used in these experiments it will be observed that the light-growth reaction appears sooner, the greater the quantity of light, and the maximum growth rate is also reached more quickly. The extra growth produced by the reaction also increases with the light quantity, the extra growth being proportional to the cube root of the light quantity. With higher light quantities the reaction appears to be more complex, as the numbers given in Table LXXXIII show. It would seem that in addition to

Table LXXXIII

Mean Values of the Characteristics of the Light-growth reaction in *Phycomyces* with higher Light Quantities

Light quantity in M.K.S.	Reaction time in minutes	Time taken in minutes for maximum growth rate to be reached	Maximum value of growth rate in percentage of normal rate	Extra growth in minutes of growth
4 × 16,000	3·5	6·75	177	4·00
4 × 240,000	4·0	7·33	175	6·37
4 × 1,920,000	5·3	7·92	183	9·23

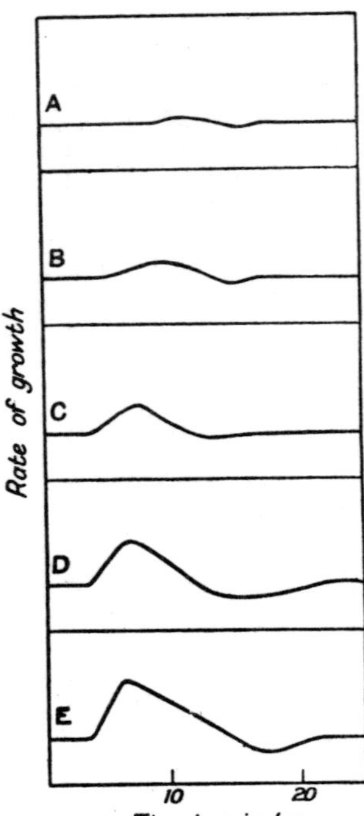

Fig. 38.—Curves illustrating the light-growth-reaction in the sporangiophore of *Phycomyces blakesleeanus*

A, 4 × ¼ M.K.S.; B, 4 × 1 M.K.S.; C, 4 × 4 M.K.S.; D, 4 × 30 M.K.S.; E, 4 × 210 M.K.S.
(After A. H. Blaauw)

the 'light-growth reaction' which brings about a temporary increase in growth, there is also what Blaauw calls the 'anti-reaction', which involves a *decrease* in growth rate. In lower light quantities this decrease appears after the light-growth reaction is completed (cf. Fig. 38), but in higher light quantities this anti-reaction is speeded up and increased in amount more than the light-growth reaction with the result that the two overlap and the resulting growth is dependent on two opposed reactions. An inspection of the curves shown in Fig. 39, representing the change in growth rate with time with higher light quantities, will make it clear that this is a possible way of interpreting the results.

The light-growth reaction in higher plants was examined in the hypocotyl of *Helianthus globosus* and in the roots of *Sinapis*, *Raphanus* and *Avena* by Blaauw himself, and in the coleoptile of *Avena* first by Vogt and more recently by a number of other experimenters. In the hypocotyls of *Helianthus*, the roots of *Sinapis* and the coleoptile of *Avena*, the light-growth reaction occurs, but results in a decreased rate of growth. Blaauw thus distinguished between a

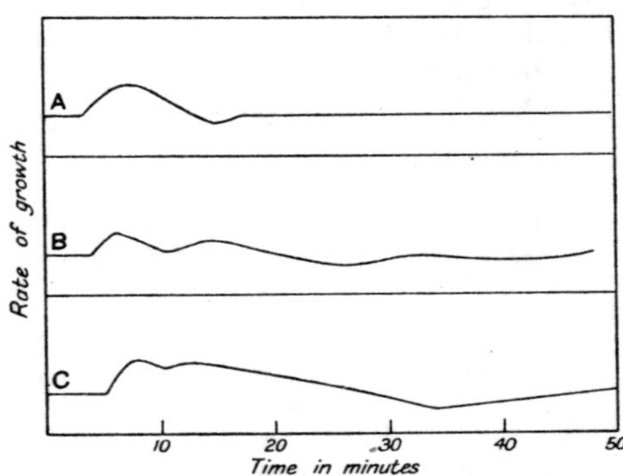

Fig. 39.—Curves illustrating the light-growth-reaction in the sporangiophore of *Phycomyces blakesleeanus* in high intensities of light

A, 4 × 6,000 M.K.S.; B, 4 × 240,000 M.K.S.; C, 4 × 1,920,000 M.K.S.
(After A. H. Blaauw)

positive reaction in *Phycomyces* and a negative reaction in these higher plants, while in some roots, as those of *Avena*, and *Raphanus*, there is no light-growth reaction. In these multicellular organs where the reaction occurs it resembles the reaction in *Phycomyces* in that the reaction is followed by an anti-reaction having the reverse effect on growth rate, the two effects being partly superimposed in higher light intensities. The course of the reaction in the hypocotyls of *Helianthus globosus* after illumination with different quantities of light is shown in Fig. 40.

In the coleoptile of *Avena*, one of the most popular objects of investigation in growth problems, there appear, from the work of F. W. Went, to be separate light-growth reactions in the basal and apical parts of the organ.

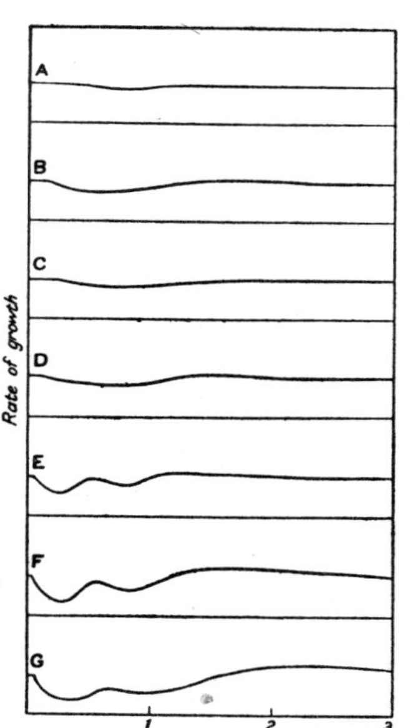

Fig. 40.—Curves illustrating the light-growth-reaction in hypocotyls of *Helianthus globosus*

A, 4 × 4 M.K.S.; B, 4 × 32 M.K.S.; C, 4 × 256 M.K.S.; D, 4 × 2,048 M.K.S.; E, 4 × 16,400 M.K.S.; F, 4 × 131,200 M.K.S.; G, 4 × 1,050,000 M.K.S.

(After A. H. Blaauw)

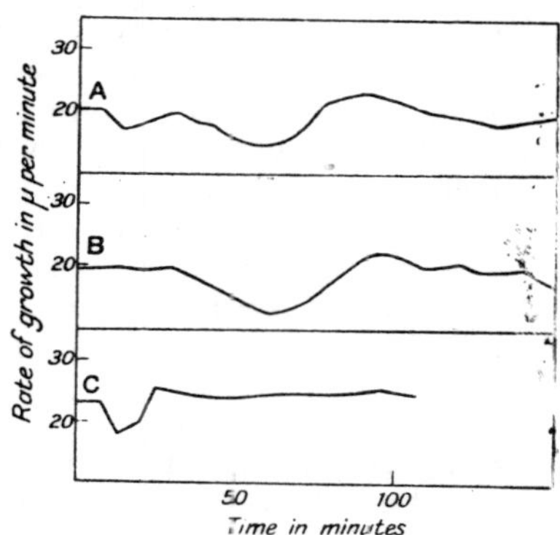

Fig. 41.—Curves illustrating the light-growth-reaction of the coleoptile of *Avena*

A, whole plant illuminated; B, tip only illuminated; C, basal part only illuminated

(After F. W. Went)

The curves in Fig. 41 show the change in growth rate when (A) the whole coleoptile is illuminated, (B) when the apex only is illuminated and (C) when the base only is illuminated with a light quantity of 500 M.K.S. It will be seen that the reaction in the first case is actually the sum of the reactions in the other two. Of these the reaction of the basal part of the coleoptile occurs sooner, and is less in amount, than that in the upper part of the coleoptile. It has been suggested that the two reactions can be related to different internal factors, that in the apex to the development of a substance influencing growth rate to which reference will be made shortly, that in the base to a change in the permeability of the protoplasm brought about by light, the evidence for which rests, however, on doubtful foundations.

In order to eliminate certain errors in measuring the light-growth reaction, Cholodny introduced a method of experimentation in which the coleoptile was maintained in a stream of running water. Isolated coleoptiles under this condition when suddenly illuminated from two opposite sides, exhibited a typical light-growth reaction. About 20 to 30 minutes after illumination a decrease in growth rate amounting to 30 to 40 per cent. of the original rate, took place. This was followed by a recovery of the original rate and subsequent variations as already noted. But if the coleoptiles, instead of being subjected to sudden illumination, were subjected to illumination of low intensity at first, but gradually increasing, a quite different effect resulted. The rate of growth now *increased* with increase in light intensity, and then, some time after the full intensity was reached, fell again to the original value. Cholodny therefore drew the conclusion from such experiments that the so-called light-growth reaction is due to an 'energetic shock', a shock, that is, due to the sudden exposure to illumination of an organ growing in the dark. In fact, the data that have been accumulating with regard to the light-growth reaction seem to point more and more to its being in the category of a response to a stimulus. The analogous thermo-growth reaction, and the fact that a dark-growth reaction has been observed when plants are transferred from light to a brief period of darkness, also suggest that the light-growth reaction is a response to the stimulus of a change in light intensity.

Wounding. When some of the cells of a tissue are damaged by pricking, cutting, high temperature or pressure, such treatment nearly always leads to the death of the cells affected and also to that of adjacent cells. But cells bordering these latter frequently exhibit renewed activity, so that not only meristematic cells, but adult cells in which division had ceased, once again divide and enter into the first phase of growth, and new tissues are formed. Sometimes this renewed activity leads only to the formation of a layer of cork which closes and heals the wound, but in other cases a considerable development of new tissue known as callus takes place which, in extreme cases, may lead to the regeneration of the whole plant from a piece of tissue as, for example, when plants of *Begonia* are obtained from pieces of leaf.

An examination of the conditions leading to this renewed activity following wounding was made by Haberlandt, his observations being made chiefly on kohl-rabi and potato. In tubers of these plants wounding leads to renewed cell division in the tissue near the wound. If, however, a tuber is cut and the surface of the wound washed well with water so as to remove the contents of the injured cells, the renewal of cell division is much reduced or altogether suppressed. But if the cut surface is then covered with a layer of macerated tuber of the same kind the renewed cell activity takes place. From these observations Haberlandt concluded that the cut cells contained or produced a substance which, on diffusing into the intact cells in the neighbourhood of the wound, induced cell division, and it seems a possibility that, if this is a correct interpretation of the experimental results, such substances may play a part in normal cork formation and in other cases where there is a renewal of activity. Moreover, if this is so, we may be allowed to suppose that in all meristematic activity cell division requires the presence of such a substance. As in Haberlandt's experiments the substance produced in one place is conveyed to others, Haberlandt described it as a growth hormone, and it appears likely that it is formed in, or carried by, phloem cells, as pieces of potato form no cork over a wound if they contain no phloem, while such pieces will form wound cork if the cut surface is covered with another piece of potato containing phloem.

The part played by the growth hormone in cell division is obscure. If its action is analogous to that of a catalyst, it appears, nevertheless, not to be enzymic in character, for it will withstand heating to 100° without any impairment of its activity in promoting cell division.

Oxygen Concentration. As their name implies, oxygen is necessary for the continued growth of aerobes. This necessity is no doubt due to the part played by oxygen in bringing about the release of energy in respiration, which, as we have seen, is an essential activity of living matter and which is no doubt intimately connected with the building up of materials of higher energy content required for growth. It is therefore not surprising that below a certain value, which varies with the species, decrease in oxygen concentration brings about decrease in the rate of growth. Increase in oxygen concentration above a certain value may also lead to reduced growth, but in most aerobic plants a considerable retardation is not obvious until a pressure of 3 or 4 atmospheres is reached. Some workers have attempted to define cardinal points for oxygen concentration in its relation to growth, and have attempted to determine minimum, maximum and even optimum values of oxygen concentration. When it is considered that the plant kingdom includes both anaerobes and aerobes it is no wonder that both the minima and maxima vary widely. Thus for the anaerobe *Bactridium butyricum* the maximum oxygen pressure for growth is 0·001 atmosphere, whereas in some aerobic plants it is stated to be as high as 9 atmospheres, though the possibility of high pressure itself having a deleterious effect must not be forgotten. The minimum also varies, being zero in the case of anaerobes, but generally from about 0·1 to 3 per cent. in aerobes, although some very much lower minimum values were found by Wieler in 1883. His results are summarized in Table LXXXIV.

Table LXXXIV

Minimum Oxygen Concentrations necessary for Growth

Species	Oxygen Concentration
Helianthus annuus	Between 19×10^{-11} and 29×10^{-4}
Vicia faba	19×10^{-11}
Lupinus luteus	Between $5{\cdot}2 \times 10^{-6}$ and $7{\cdot}8 \times 10^{-2}$
Brassica napus	Between 0·08 and 0·51
Cucurbita pepo	0·09
Ricinus communis	0·09
Bellis perennis	0·09
Coprinus lagopus	Between 0·09 and 0·58
Mucor mucedo	Between $2{\cdot}9 \times 10^{-4}$ and $4{\cdot}6 \times 10^{-3}$
Phycomyces blakesleeanus	Between 0·14 and 0·20

The range of oxygen pressures over which different species grow also varies, but an exact analysis of the relationship of oxygen concentration to growth is for the most part wanting.

Carbon Dioxide Concentration. Carbon dioxide concentration, as a factor in growth, acts chiefly through photosynthesis in providing material for the growing cells. But in high concentrations carbon dioxide exerts a toxic effect on the plant and so reduces growth rate. With some roots, carbon dioxide in as low a concentration as 5 per cent. may retard growth and in a concentration of 25 to 30 per cent. inhibit it altogether. Shoots are generally more resistant, retardation not resulting until the concentration is as high as 15 per cent., although inhibition of growth may occur in concentrations of 20 to 25 per cent.

INFLUENCE OF INTERNAL CONDITIONS ON RATE OF GROWTH

Water Content. While the external conditions of temperature and light both have an effect on the enlargement stage of growth, and while wounding induces renewed activity of the division stage, it is clear that internal factors may also be operative in affecting growth rate, although they may often elude experimental control to any extent. Since the enlargement stage depends very largely on absorption of water, it is clear that the rate of growth must depend on the water relations of the growing organ ; on the rate, that is, at which water is brought to the growing regions, and the rate at which it is lost, if at all, by transpiration. Clearly, then, this factor may depend not only on the effectiveness of the conducting channels, but also on certain external conditions, such as, in the case of an aerial vascular plant, the water content of the soil and the relative humidity of the air. Thus, according to Walter, the roots of young seedlings of *Lepidium* and *Pisum sativum* are only able to grow in an atmosphere with a relative humidity of 97·5 to 100 per cent., and even at the former value the growth is greatly reduced. The growth in length of the roots of the latter in atmospheres with different relative humidities is shown in Table LXXXV. Walter has also determined the minimum relative humidity for the growth of a number of fungi and bacteria. While some of these can grow in somewhat

Table LXXXV

Effect of Relative Humidity on the Growth of Roots of Young Seedlings of *Pisum sativum*

Age of Seedling in Days	Growth Rate in millimetres per day			
	100% R.H.	99% R.H.	97·5% R.H.	95% R.H.
3	16·0	6	0	0
4	12·5	5	1	0
11	22·5	14	1	0

drier media than can the roots mentioned above, on the whole this minimum of relative humidity is surprisingly high (cf. Table LXXXVI).

Table LXXXVI

Minimum Values of Relative Humidity for Growth

Species	Minimum Value of Relative Humidity in per cent.
Aspergillus glaucus	85
Penicillium glaucum	85
Mucor racemosus	90
Zygorhynchus exponens	92
Alternaria tenuis	93·5
Phycomyces blakesleeanus	95
Bacillus mycoides	95–97
Chaetocladium	98

It will thus be readily understood that growth can be retarded or inhibited by surrounding cells with solutions of a sufficiently high osmotic pressure, by which the absorption of water by the cell is hindered or prevented altogether. Some plants, however, have a power of accommodation to strong solutions, and this must be brought about either by absorption of the solute from outside, or by the development in the cell-sap of other osmotically active substances. Sometimes this accommodation can only be effected if the increase in external concentration is gradual.

According to Sachs, a turgid condition of cells is necessary for the enlargement phase of growth, as through the turgor pressure the cell-wall is stretched and then becomes thickened afterwards either by the deposition of further material on or in the walls, by apposition or intussusception. Although this simple explanation of growth in size is not now accepted, it is clear that growth cannot take place without an adequate supply of water, and that where this supply is feeble it may limit the rate of growth.

Growth Substance. It has already been noted under the effect of wounding that there is some evidence of the existence of a substance which induces or furthers cell division. During the last few years much information has been accumulating with regard to a substance which plays an important part in cell enlargement, although the first intimation of its existence came from work on phototropism by Boysen Jensen in 1910, to which reference will be made in a later chapter. The development of our knowledge on this subject has been largely the work of the Dutch school, including F. W. Went, Dolk, Van der Wey, Heyn and others, carried out mainly on the coleoptile of *Avena*. From the series of researches performed by these workers it would appear that in the apex of the coleoptile of *Avena* and of other members of the Gramineae there is produced a substance which diffuses to the base of the organ. Thus, if the apex (the uppermost 1 to 2 mm.) of the coleoptile is cut off, growth of the coleoptile ceases, but if the coleoptile apex is re-fixed to the decapitated stump by means of a thin layer of gelatin, growth is renewed. The explanation of this is that a substance is produced in the apex which is transported to, and furthers enlargement of, the cells in the coleoptile below the apex, and which is responsible for,

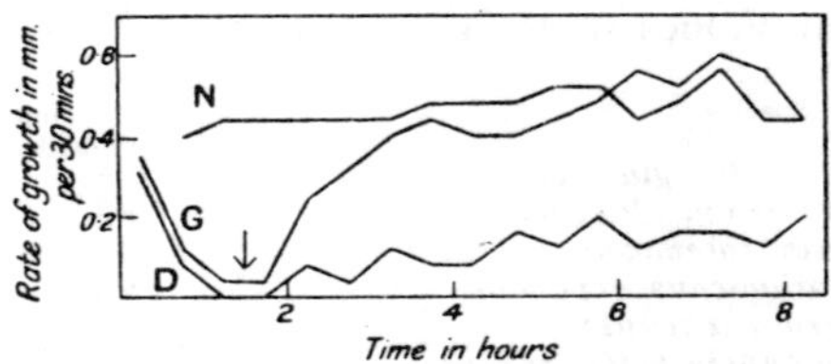

Fig. 42.—Curves illustrating the growth of coleoptiles of *Avena*
N, normal coleoptile; D, decapitated coleoptile; G, decapitated coleoptile treated with auxin at ↓
(Constructed from the data of A. N. J. Heyn)

or assists in, the growth in size of these cells. If the apex is not replaced on the base, a certain amount of growth does take place after a time owing to the formation of some fresh 'growth substance' at the top of the decapitated stump. That this is a correct explanation of the observed behaviour is indicated by the fact that the substance can be extracted from the isolated apex and applied to the decapitated base and produce growth in the latter. The method adopted to show this is to place the apex, or a number of apices, on a plate of gelatin or agar-agar, when the substance diffuses from the apex into the plate. A piece of the gelatin or agar, on being placed on a decapitated base, then induces growth in the latter, just as if the coleoptile apex itself had been placed there. Actually the growth substance can be transferred, not only from the coleoptile apex to agar, but from one agar block to another. The curves in Fig. 42, constructed from data published by Heyn, indicate the rate of growth in normal and decapitated coleoptiles, and in decapitated coleoptiles treated with growth substance in the manner described.

The growth substance does not appear to be different in every species, for that obtained from the coleoptile of one species will further growth in another. Thus growth substance from *Avena* will bring about growth in the decapitated seedlings of other Gramineae such as maize and barley, while Miss Uyldert has shown that the growth of the decapitated axis of *Bellis perennis* can be promoted, not only by the replacement of the young inflorescence, but also by treatment with the growth substance from *Avena*. Moreover, Nielsen showed in 1930 that growth substance capable of furthering growth in oat coleoptiles can be obtained from *Rhizopus suinus*, and subsequent workers have found that other fungi also yield a growth-promoting substance. Thus a preparation of growth substance obtained by Boysen Jensen from *Aspergillus niger* was shown not only to further the growth of the oat coleoptile, but also that of decapitated roots of *Vicia faba*. The presence of growth substance has now been demonstrated in many dicotyledonous seedlings including *Raphanus sativus*, *Lepidium sativum*, *Vicia faba* and *Lupinus* spp., in the pollen of many plants, in the shoots of both dicotyledonous and coniferous trees and in some algae. Malt extract and urine are other sources of the growth substance. Whether all the preparations so obtained contain the same growth-promoting substance or substances cannot be definitely stated, but it is at least a probability.

According to the researches of Kögl and his collaborators, there are at least three growth-promoting substances. The first of these to be isolated was a substance soluble in ether and able to withstand drying and also heating to 100° C. It has been obtained in a crystalline form and the formula $C_{18}H_{32}O_5$ has been assigned to it. It is described as an organic acid with the formula R.CHOH.CH_2.CHOH.CHOH.COOH, the R involving a five carbon atom ring with two branched side chains $CH_3.CH_2.CH(CH_3)$. This substance has been termed *a*-auxin. A second growth-promoting substance, termed *b*-auxin by Kögl, appears to be a closely allied substance with the formula $C_{18}H_{30}O_4$. This again is an organic acid, a keto-acid with the composition R.CHOH.CH_2.CO.CH_2.COOH. A third growth-promoting substance (hetero-auxin) has been found to be β-indolyl-acetic acid, $C_{10}H_9O_2N$.

All these substances behave similarly in furthering growth, and they can be regarded as examples of plant hormones. According to Went, the growth substance in the apices of maize coleoptiles is *a*-auxin, while that obtained from *Aspergillus niger*, *Rhizopus nigricans* and yeast is hetero-auxin.

Söding has obtained a growth hormone in agar from *Cephalaria tartarica*, a member of the Dipsaceae, in the same way as that obtained by Went from the *Avena* coleoptile. The *Cephalaria* hormone was recognized in the seedlings, leaves, and stalks of buds, flowers and unripe fruits. This hormone, however, induced no response in the oat coleoptile. Although it has been shown that many plants contain a hormone which will induce growth in the *Avena* coleoptile, Söding found that the growth substance from *Symphoricarpus* and *Rheum*, like that from *Cephalaria*, will induce growth in the latter but not in *Avena*. He suggests several possible explanations. Thus there may be other growth hormones besides the auxins of *Avena*, or the same growth hormone may be present in different conditions in different plants, or there may be impurities present in the hormone preparation from *Cephalaria* which inhibits its action on *Avena*.

It has been shown recently by Van Overbeck that in hypocotyls of *Raphanus* light reduces the sensitivity of the cells towards the growth substance, so that the same quantity of auxin has less effect in furthering elongation of the cells in the light than in the dark. This provides an explanation of the light-growth reaction in shoots in terms of the growth-promoting substance. The reaction is due to the action of light in reducing the sensitivity of the cells to the auxin.

Attempts have been made with some success to carry out quantitative investigations on the effect of the growth substance. For this purpose it is necessary to employ a unit of growth substance. The unit proposed by Kögl and Haagen-Smit is the quantity of growth substance (contained in a piece of 3 per cent. agar 2 mm. square and 0·5 mm. thick) which, when placed eccentrically on the apex of a decapitated coleoptile of *Avena* produces a curvature of 10° in 2 hours at a temperature of 22° to 23° C. and from 90 to 95 per cent. humidity. The unit proposed by Dolk and Thimann is the quantity of substance present in 1 c.c. of solution which, after mixing with 1 c.c. of 3 per cent. agar, produces a curvature of 1° when placed eccentrically on the decapitated coleoptile of *Avena*.

The mode of action of the growth substance in coleoptiles, according to the work of Söding, De Haas and Heyn, is as follows. After its production in the coleoptile apex the substance is transported to the growing region below. Here it probably acts in the first place on the protoplasm, but ultimately on the cell-wall, Its particular effect on this is to increase the plasticity, as a result of which the extension of cell-wall produced by the hydrostatic pressure of the turgid cell resulting from water absorption is rendered permanent. That the plastic extensibility of the cell-wall is increased by the action of the growth substance is easily shown by stretching, by means of an attached weight, intact and decapitated coleoptiles. The deformation produced in this way is greater in intact than in decapitated coleoptiles, but in the latter, after decreasing for 3 hours, the plastic extensibility increases, corresponding to the renewed formation of growth substance in the new apex. If the decapitated coleoptiles are treated with growth substance, the plastic extensibility is greater than in similar decapitated, but untreated, coleoptiles. Again, if coleoptiles are placed in water, the resulting absorption of water, and consequent increase in length of the organ, is considerably greater in intact than in decapitated coleoptiles, whereas the contraction in plasmolysis of the intact and decapitated coleoptiles is practically the same, which indicates that the greater increase in length of the intact coleoptiles in water cannot be referred to an increase in the elastic extensibility of the cell-wall.

This theory of cell enlargement thus differs from Sachs's view, mentioned earlier in this chapter.

Keeble, Nelson and Snow have found that if roots of maize and pea seedlings are decapitated and then washed in water for 12 or 20 minutes, their rate of growth for the next 22 hours or longer is more rapid than that of similar decapitated roots which have been submerged in water for a similar length of time before decapitation and not after it. From this they conclude that a substance is formed at the wound retarding growth, the substance being removed when the roots are washed after wounding.

THE GROWTH RATE OF WHOLE PLANTS

So far the factors influencing the growth of individual cells, or of those of a particular organ, have been subject to discussion, but it is clear that the rate of growth of a whole plant must be intimately related to that of the growth of its constituent organs. We should therefore suppose that, broadly speaking, the growth of whole plants, that is, the total increase in size or dry weight, would be influenced in much the same way by temperature, light and water content, as is the growth of single cells. This is no doubt the case, but it must be remembered that the growth of a whole plant depends on the rate of cell division as well as on the rate of enlargement of cells, whereas it is the latter phase of cell growth which has been the main subject of investigation in the growth of individual cells and organs. The rate at which the third phase, that of differentiation into permanent tissue, takes place may also affect the rate of growth of the plant as a whole, as the amount of permanent tissue will affect the production of new materials used in growth and the rate of transport of this material to the growing points.

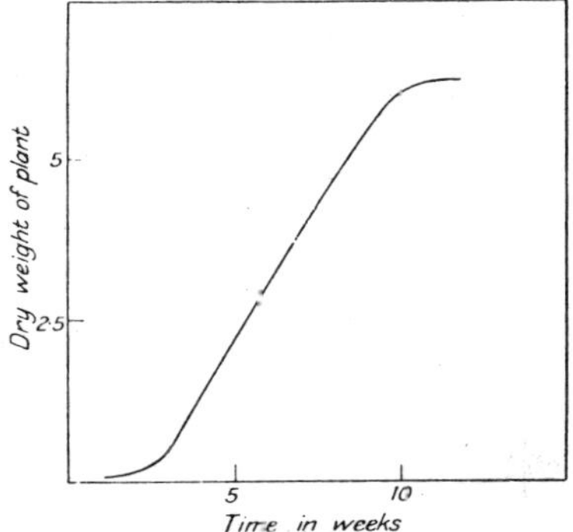

FIG. 43.—Curve of total growth of an annual plant (*Sinapis alba*)

(Based on the data for the growth of a whole crop obtained by R. Hornberger)

As a plant grows it is to be expected that, with the increase in the amount of leaf and root, the rate at which material is absorbed from outside will increase and so the rate of growth will increase. With the growth of most plants, assuming external conditions are satisfactory, this is, within limits, usually obvious. It is, for example, well recognized by most people that the increase in size of an annual plant is slow at first but proceeds at an increasing rate as the plant enlarges. This rate of increase, however, after a time slows down, and, with flowering, increase in size may stop altogether. The curve showing the relation between rate of growth of an annual plant with time is therefore of the so-called S-shaped or sigmoid type shown in Fig. 43.

It is not to be expected that the *rate of growth* of an annual plant, even if growing under approximately constant climatic conditions, will rise regularly to a maximum and then fall regularly to zero. Thus, from a consideration of available data, Briggs, Kidd and West decided that the curve of *growth rate* of a plant of *Zea mais* throughout its life has the form shown in Fig. 44, the two subsidiary maxima late in the life of the plant corresponding to the flowering and fruiting phrases respectively, during both of which growth activity rises to a maximum value and then falls.

It is clear that during the first part of the growth of an annual plant the rate of growth per plant increases with development. It has been pointed out by V. H. Blackman that at any stage in development the rate of growth per unit of dry weight of the plant gives an idea of the growth activity of the plant, which the rate of growth per plant does not. If it is assumed that the rate of growth is proportional to the dry weight of the plant we have the following relation:

$$\frac{dw}{dt} = rw$$

where w is the dry weight of the plant at any time and $\frac{dw}{dt}$ is therefore the rate of increase in dry weight. If r is a constant the integration of this equation between limits gives this relation in the form

$$\log \frac{W}{W_0} = rT$$

$$\text{or } W = W_0 e^{rT}$$

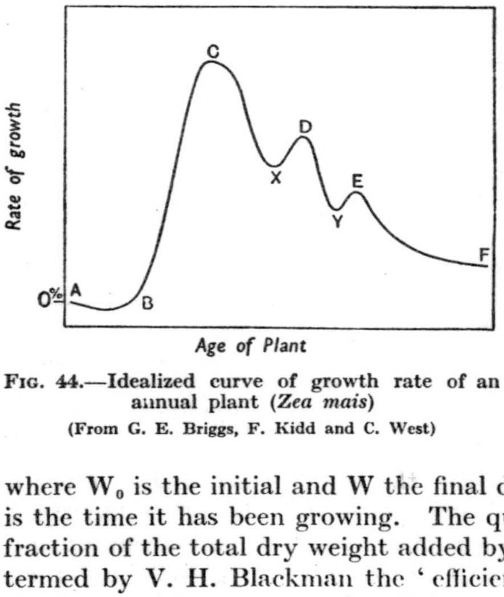

FIG. 44.—Idealized curve of growth rate of an annual plant (*Zea mais*)
(From G. E. Briggs, F. Kidd and C. West)

where W_0 is the initial and W the final dry weight of the plant when T is the time it has been growing. The quantity r, which represents the fraction of the total dry weight added by growth in unit time, has been termed by V. H. Blackman the 'efficiency index' of dry weight production, for 'it represents the efficiency of a plant as a producer of new material'. It will be observed that total growth will depend both on the initial dry weight of the plant (W_0) as well as on the efficiency index r. Thus, as V. H. Blackman has pointed out, for high production of vegetative material two factors are necessary, namely, large seeds and a high efficiency index.

It is clear that the efficiency index is not a constant value throughout the life of a plant, for it falls off after a time and ultimately reaches zero with cessation of growth. Also during the period of active vegetative growth external conditions may vary and so modify the efficiency index, and determinations of the latter over any particular period must necessarily give a measure of the *average* efficiency over the period of observation. It is to be noted that where r is itself a function of w, the integration given above is inadmissible so that the equation $W = W_0e^{rT}$ must be used with caution in comparing the efficiencies for growth of different plants.

West, Kidd and Briggs have defined 'relative growth rate' as the weekly percentage rate at which dry weight increases. If R is the relative growth rate and w the dry weight we have the relation

$$\frac{dw}{dt} = R\frac{w}{100}$$

It will thus be seen that the relative growth rate so defined is Blackman's efficiency index multiplied by 100 if the same units are used.

Other attempts have been made to find mathematical relations between the amount of growth and the age of an annual plant. It was pointed out by Brailsford Robertson that the sigmoid curve of growth resembles in form that of an autocatalytic chemical reaction, a reaction, that is, which is catalysed by one of the products of the reaction, but in which the amount of material available for the reaction is limited. In such a case, if m is the total quantity of material available for the reaction and x is the amount transformed after a time t, the rate of the reaction at any time is given by the equation

$$\frac{dx}{dt} = Rx(m - x)$$

which, on integration, gives the relation $\log \dfrac{x}{m - x} = Kt$. The curve of this equation is sigmoid in shape and a number of growth curves have been determined which show a general resemblance to it. In the case of a growing annual plant it would have to be supposed that the available material would be limited by the total amount of growth of which the plant is capable; a limit imposed by internal factors. In the equation given above, m would therefore represent total growth and x the amount of growth at any time. An agreement with this equation was recorded by Reed and Holland for the growth of *Helianthus annuus* in 1919, and Reed has attempted to apply the formula to the yearly growth of some perennials. Yet it must be doubted whether the superficial resemblance between the growth curve of a plant and the curve for an autocatalytic reaction has any fundamental physiological significance, or that attempts to stress the resemblance will lead towards the solution of the problems of growth.

The rate of growth of an organism, as pointed out by Briggs, is a function of the various factors of the environment and of the internal conditions which are at present unanalysed. Hence change in environmental conditions may not produce the same change in growth rate in two organisms, even when the two have the same rate of growth under one particular set of conditions, for the internal factor, a term we may use for the sum of the internal conditions, may differ in its components in the two cases, and may be affected differently by changes in the environmental conditions. Where organisms are grown under constant conditions we may regard the growth curve as providing information concerning what Briggs has called the 'drift' of the internal factor. When, however, an attempt is made to determine the effect of any particular external factor on the rate of growth it is obviously a difficult problem to determine a definite relationship between the value of the external factor and the rate of growth which shall be of general applicability, owing to this drift in the internal factor. As Briggs puts it, the problem is to determine both the drift of the internal factor, and the relationship between growth rate and an external factor, from actual measurements of growth rates under different external conditions. As a matter of fact, growth is dependent upon, and influenced by, such a complex of factors, external and internal, that an exact determination of the inter-relationship of the factors and of their respective relations to the growth rate is not to be expected with our present knowledge and with the methods of analysis which are at present available. Moreover, it seems necessary to regard the growth through the annual life-cycle as changing not only quantitatively, but also qualitatively, and the relationships of the various factors to vegetative growth may be quite different from their relationships to reproductive growth. Indeed, the drift of the internal factor for vegetative growth appears to involve a reduction in the vegetative growth rate, and to lead ultimately to the replacement of vegetative growth by reproductive growth. There is much evidence that this drift in the internal factor is largely determined by external conditions. These are discussed in the later chapter dealing with reproduction.

THE DEVELOPMENT OF ORGANS

In the simplest plants growth consists of the enlargement of a single cell which ultimately divides into two cells which separate and develop into two daughter cells similar to the parents. Examples of such are found in the Protococcaceae, the Volvocaceae and the Bacteria. In the filamentous algae, and many other Thallophyta, although the cells resulting from division do not separate, they remain morphologically and physiologically similar, and only exceptionally is there any differentiation between them. As, however, the plant kingdom is ascended, differentiation of the plant body into organs with different functions and composed of cells of different forms, becomes more and more evident. A particular feature of this differentiation is the early formation in the development of the individual of apical growing points of different kinds at the two ends of the organism, which thus has a polar structure with an axis. From these two growing points develop respectively the shoot and root, by the division, enlargement and internal differentiation of the cells of which they are composed. As is well known, by the unequal development of the cells of the shoot apex, outgrowths which develop into leaves and side branches are formed, the more rapidly developing cells being located in definite regions of the outer layer of the shoot apex. In the root, on the other hand, the lateral roots arise endogenously from cells in the pericycle and therefore have to make their way through the outer tissues to emerge from their axis of origin. By the repetition of these processes in side-shoots and side roots plants of great complexity may be produced. In this power of indefinite or open growth, resulting from the continuous activity of the apical meristems, the higher plant stands in the greatest contrast to the higher animal.

Since, in many cases, cuttings of either shoot or root can regenerate complete plants, it would appear that the meristematic cells of both shoot and root apices have the same potentialities, and if this is so, the difference in the modes of growth of shoots and roots must be related to the organization of the respective apices. While explanations of the relationship between the organization of the meristems and the mode of growth of the cells developing from them have been attempted, these rest at present on too hypothetical a basis to render a discussion advisable in this place.

THE INFLUENCE OF EXTERNAL CONDITIONS IN MODIFYING THE FORM OF PLANTS

While the major problems of the causal explanation of the normal course of development of root and shoot and their constituent parts still await attack by physiological methods, considerable information is available with regard to the influence of various factors in modifying the normal course of development. Changing conditions might result in a general reduction or increase in the rate of growth, all parts being affected in the same proportion, or there might result a differential change in growth rate in different organs, in which case a difference in the general form of the plant will result. In such cases the total amount of growth in organs of limited growth is generally altered, so that leaves may not merely grow more slowly or rapidly, but may, in their final condition, be smaller or larger, while they may also show differences in their internal structure. Obviously such results must be due to the interaction of the external factor under consideration with other factors of the environment, or with internal conditions, but only in relatively few cases is there much information regarding this interaction of factors.

Undoubtedly the most familiar example of the effect of external conditions on form is the phenomenon of etiolation, but other conditions besides light can exert an influence on the form of plants. Some of the effects of different conditions are described below.

Temperature. It was shown by Vogt that in the coleoptile of the oat not only is the rate of growth influenced by temperature, but that the total growth of the coleoptile in the dark depends on the temperature. The results of his experiments are summarized in Table LXXXVII. It will there be seen that up to a temperature of about 13° C. the final length of the coleoptile increases with rise in temperature, but that above that temperature the total growth is less the higher the temperature. The coleoptile, however, develops its full size more rapidly the higher the temperature.

Table LXXXVII

Final Length and Time of Growth of Coleoptiles of *Avena sativa* at Different Temperatures in the Dark

Temperature in Centigrade degrees	Final Length in mm.	Growth Time in Days
7·5	117·1	30
8·4	120·8	17
10·2	131·4	14
12·8	150·3	13
14·0	122·0	9
20·1	99·4	5
20·2	94·6	5
25·5	75·6	4
29·8	59·7	3·5
33·3	45·8	3
33·4	45·1	3
34·0	38·0	3
35·1	35·7	2·5
42·0	0	0

While the earlier observations of Popovici on the effect of temperature on roots of *Vicia faba, Cucurbita pepo* and *Phaseolus multiflorus* also suggest that the total amount of growth in length of cells behind the growing point may be affected by temperature, especially when this approaches the minimum or maximum for growth, such an effect of temperature on total growth does not appear to be a general phenomenon, for Miss Graser found the total length of the sporangiophore of *Phycomyces blakesleeanus* was unaffected by temperature, although the rate of growth was very definitely influenced, as described earlier.

Cases are on record in which temperature plays a very marked effect on the course of development. Among lower plants the alga *Stigeoclonium* passes at low temperatures into the palmella condition in which the cells are separate in a jelly, while at higher temperatures it forms filaments. A somewhat similar behaviour is recorded for *Bacterium pasteurianum*, the cells of which, when the bacterium is grown at 34°, form short rods, but elongate at 40° C. into slender filaments which may be as much as 150 times the length of the rods at 34° C. Among higher plants Vöchting found that a variety of potato called Marjolin, after exposure for 4 to 5 weeks to a temperature of 6–7° C., produced tubers in place of the foliage shoots normally produced at higher temperatures.

Light. As already mentioned, of all the factors of the environment which may affect the growth form of plants, light is the most obvious. Plants which are kept in complete darkness exhibit the phenomenon of etiolation which has certain very definite characteristics. Etiolated plants are devoid of chlorophyll and possess a pale yellow colour, and because of this absence of chlorophyll are unable to manufacture carbohydrate. For this reason an etiolated autotrophic plant with only a small reserve of food will be unable to grow, and will consequently exhibit a retardation of growth which is a secondary effect of the absence of light and not related primarily to the effect of light on growth. For the examination of pure etiolation it is therefore necessary to utilize material containing abundant food reserve, and for this reason sprouting potatoes provide a favourable object for the study of the phenomena of etiolation. In general, the absence of light results, in the case of vascular plants, in greater growth in length of stems, but in a reduction in the growth of leaves, with the result that etiolated plants have a straggly appearance. There are, however, exceptions to this rule. Thus in the Liliaceae the leaves of etiolated plants are longer and narrower than those of normally illuminated plants, while in *Beta vulgaris* the leaves of plants grown in darkness are not greatly reduced in size as compared with the normal leaves. Vöchting showed that in absence of light the stems of certain Cacti, *Phyllocactus* and *Opuntia*, which normally take the form of phylloclades, develop more or less radially.

In general it would seem, as Priestley has suggested, that in absence of light there is a different distribution of the growth activities of the apical meristem of the stem, from that in light, the increased growth in length and reduction in surface suggesting that in the dark the surface layers of the meristem make less growth, and the more deeply seated layers more, than in the light.

The characteristics of etiolation are shown, though to a less degree, in plants exposed to very brief periods of illumination, or to more prolonged illumination with light of very low intensity. In Fig. 45 are shown results obtained by Priestley with seedlings of *Pisum sativum* subjected to complete darkness and to brief periods of illumination (2 minutes, 10 minutes and 60 minutes) each day from an 80 candle-power electric lamp, the distance of the plant from the source of light varying from 70 cm. to 30 cm. as the plant grew. It will be observed that plants exposed for only 2 minutes each day do not show the complete etiolation of those grown in total darkness; on the other hand, chlorophyll apparently developed only in the plant exposed for 60 minutes each day. It would thus appear that since light exposures too short to induce chlorophyll formation reduce the characteristics of etiolation, the subsequent development of normal structures in etiolated plants cannot be due to products of photosynthesis, but rather to some photochemical reaction comparable, perhaps, to that which must be involved in the light-growth reaction.

The changes in external form characteristic of etiolated plants are accompanied by changes in internal structure. Among the latter, in *Vicia faba*, for example, Priestley has recorded that the diameter of the stele relative to that of the stem is smaller in etiolated than in normal plants, and both xylem and sclerenchyma are less developed. According to Priestley, one of the most striking differences between etiolated and normal stems may be the formation in the former of a typical endodermis with Casparian strips. In the etiolated shoot the stele is first limited by a starch sheath (never present in roots of *Vicia faba*) which the endodermis later replaces. Priestley holds the view that the cell-walls of an etiolated shoot differ from those of one grown in the light in that they contain more fatty substances, and that in the change from the starch sheath to the endodermis carbohydrates are replaced by fatty substances deposited in the Casparian strip. Priestley further thinks that the redistribution of growth at the apex of an etiolated shoot, to which reference was made above, is to be traced to the greater difficulty of the meristem in obtaining material from the vascular supply when the plants are grown in the dark, owing to the walls of the cells between the vascular tissue and the meristem becoming relatively impermeable on account of the presence in them of fatty substances and proteins.

It should be observed that not all plants are equally affected by absence of light as regards change of form. Thus, according to Priestley, etiolation effects are much less marked in *Phaseolus multiflorus* than in *Vicia faba*.

Not only does the complete absence of light, or intermittent illumination for very short periods, induce modification in the form of plants, but feeble light intensity for normal periods of illumination may also result in changes in form. The most familiar example of this is found in what are called ' sun ' and ' shade ' leaves, which are produced in a number of woody plants, including *Fagus sylvatica* and *Corylus avellana*. Leaves which have developed in weak light as compared with those which have expanded in the sun, are thinner and generally possess relatively less palisade tissue, while the epidermal cells are larger, with more irregular walls and a thinner cuticle. Here again, some species are much less affected by differences in light intensity than others, so that stable and labile species are distinguished according to the little or great effect of light intensity upon their form.

A large number of miscellaneous observations on the influence of light on the development of different organs of plants are on record. Thus, Stahl stated that in darkness *Adoxa* forms tubers instead of runners, while Vöchting recorded many changes in the production of flowers as a result of darkness, including the fact that while in some plants, as, for example, *Tropaeolum*, flower buds are able to form and develop in darkness, in others, such as *Mimulus luteus*, no flowering occurs even in weak light.

Water Content. After light, no factor of the environment has so great an influence on the form of plants as the water content of the environment. It has already been noted that the enlargement of the growing cell depends on a supply of water, and it is therefore not surprising that limitation of the water supply of plants may lead to

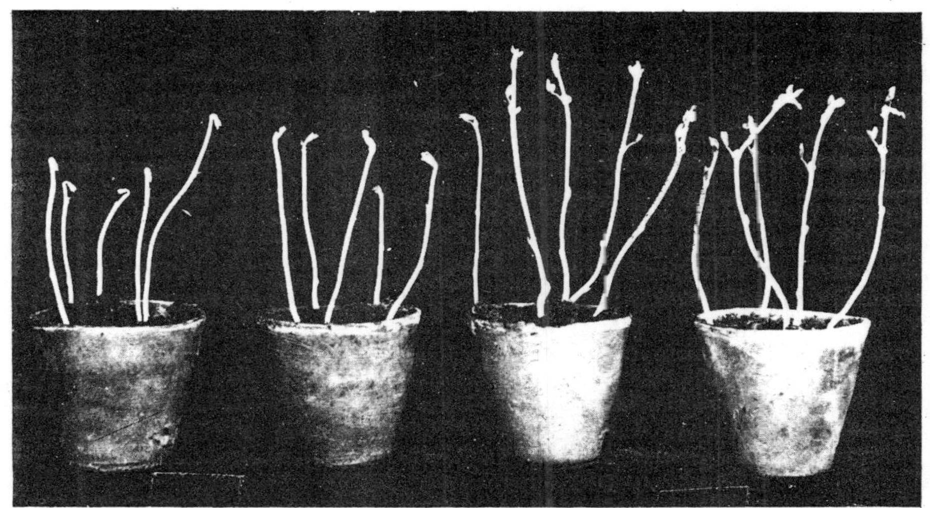

FIG. 45.—Seedlings of *Pisum sativum* subjected to different periods of illumination each day (complete darkness, 2 minutes, 10 minutes, 60 minutes)

(From J. H. Priestley)

FIG. 46.—Photographs of leaves of *Taraxacum dens-leonis* indicating the wide range of form

(From a collection made by late R. H. Yapp)

To face page 370

dwarfing. Generally speaking, plants of habitats with an abundant supply of water are characterized by a luxuriant growth, the foliage being well developed and often with a high water content, whereas plants of dry regions tend to be stunted in form with a leaf surface much reduced, while devices for the reduction of transpiration are developed. Further, not only are the normal forms of plants of different species related to the water content of their respective habitats in this way, but there is a direct effect of the water content of the environment on the form of plants. When grown in a dry habitat plants tend to develop more collenchyma and sclerenchyma, while the leaves have a thicker cuticle, whereas in an abnormally wet habitat collenchyma and sclerenchyma may be reduced or absent altogether, the xylem is feebly developed and the leaves develop a thin cuticle. Thus gorse (*Ulex* spp.) and broom (*Cytisus scoparius*), when grown in moist conditions lose their spiny character, and the former, at any rate, might be mistaken for a different genus. Among species which are particularly affected by differences in water content are the dandelion (*Taraxacum dens-leonis*), in which the leaves of specimens growing in wet habitats may be 5 times as long as those of plants growing in dry conditions (Fig. 46).

While under natural conditions it is usual for a high water content of the soil to accompany a high relative humidity of the air, this is not necessarily so, and it would appear that the water contents of the two parts of the environment of a land plant do not influence growth in the same way. It will be sufficient in this matter to refer to the observations of Kohl published in 1886. Plants of *Tropaeolum majus* were grown in different combinations of dry and damp soil with dry and moist air. Plants in a habitat both parts of which had a high water content produced leaves with 5 times the area of those grown in a dry soil and dry air. Leaves of an intermediate size were produced on plants when either the soil or the air was moist and the other part of the habitat dry. On the other hand, the thickness of the cuticle depended only on the humidity of the air; if this was high the cuticle was thin whether the plant was grown in wet or dry soil. While collenchyma was absent when both soil and air possessed a high water content and only feebly developed when relative humidity of the air was high and the soil dry, in dry air it was well developed. It thus appears that a restricted water supply to the roots may have a different influence on the development of the plant from a low water content due to excessive transpiration.

Nutrient Substances. The supply of the various necessary nutrients of the plant may influence, not only its rate of growth, but also its form. Particularly sensitive to this factor is the passing over from vegetative to reproductive growth, a matter to be dealt with in a subsequent chapter. Starvation in respect of any particular nutrient may result in deviation in form from the normal.

Poisons. Many poisons not only retard plant growth, but bring about abnormal development. Conspicuous among such effects are those produced by gases such as those present in smoke of various kinds. For example, in seedlings of *Vicia* exposed to the action of tobacco smoke the shoots remain short but are comparatively thick and grow in abnormal directions. Plants of *Impatiens sultani* exhibit a creasing of the leaves, while other results of the action of smoke are an abnormal development of lenticels and inhibition of anthocyanin formation. A very common result of the action of smoke is premature leaf-fall. For these effects produced by smoke the ethylene and acetylene present are probably chiefly responsible.

Many substances which, in even moderate concentrations, are poisonous to plants, bring about an increase in growth rate of whole plants when presented to the latter in sufficient dilution. This is notably the case with compounds of zinc, boron and manganese (cf. Chapter XII).

Parasitic Organisms. The attacks of parasitic organisms may result in profound changes in the growth forms of plants. The hypertrophy of certain organs is a frequent result of such attacks. The abnormal development of the roots of plants of the genus *Brassica* resulting from attack by *Plasmodiophora brassicae* is a familiar example, and the production of irregular swellings in the rhizomes and tubers of a number of varieties of potato by *Synchytrium endobioticum.* Other parasitic fungi produce deformation of leaves, such as the peach leaf curl due to attacks by *Exoascus deformans*, or abnormally shaped fruit as in the case of pocket plums caused by *Exoascus pruni.* Attacks by small animals produce even more marked modifications of form, hypertrophy of tissues in the neighbourhood of the attack giving rise to what are known as galls. The Cupuliferae, and particularly the genus *Quercus*, are most liable to attacks of this kind, but gall formation is met with in a number of other families, particularly Compositae, Salicaceae, Rosaceae and Leguminosae. Gall-like structures are also caused by some fungi, as, for example, by *Exobasidium vaccinii* on *Vaccinium vitis-idaea.* A description of galls and other abnormal developments of plant tissues will be found in Küster's books on the subject noted at the end of this chapter.

CORRELATION

In investigating the physiology of the higher plant it is not unusual to treat individual organs as if they were complete entities ; to investigate the photosynthesis of detached leaves, for example, or the water relations of detached organs or even isolated pieces of tissue. Such a method of treatment is often necessary in order to analyse the numerous vital processes which make up the plant's activity. At the same time it must be borne in mind that the various organs of a plant are interrelated in their activities, the roots absorbing material necessary for the assimilatory processes which take place in the leaves, while the development of the roots depends on the manufacture of assimilates in the leaf and their conduction to the growing points. But the various constituent organs of a plant are inter-dependent in their development not merely through these more obvious nutritional relations; a closer relationship exists between the parts than can be accounted for solely on these lines. Thus, to quote a familiar example of this inter-dependence, if the terminal bud of the shoot is removed, very frequently a lateral bud which previously remained dormant, now develops and carries on the growth of the shoot. In fact, every organ and, indeed, every cell, develops in such a way that all the organs or cells which make up the plant body form one harmonious whole in which the parts fit in with one another as regards form and function. If the system is disturbed, for example, in the way just mentioned, by the removal of the terminal bud, there is a tendency for development to take that direction which leads towards a restoration of the former conditions, in this case by the development of a lateral bud. For these inter-relationships of one organ with another and with the whole organism the term 'correlation' is employed, and we must suppose that these correlative relationships whether merely nutritional or of more subtle origin play a most important part in determining the course of development.

One of the best-known examples of correlation, the dormancy of certain lateral buds when the terminal bud is active, and the wakening into activity of one of these lateral buds on removal of the terminal bud, has already been mentioned. Among other comparatively simple examples are the three following observed by Goebel. In some cases, when a number of leaves are removed from a plant, the remaining ones develop larger. Removal of the larger leaves from a branch of certain trees will result in the expansion of the axillary buds without removal of the apical bud. Removal of the lamina of leaves of the broad bean results in enlargement of the stipules if the operation is performed sufficiently early in development. Vöchting showed that similarly shoots that would normally be reduced to thorns could be brought to develop into ordinary shoots. Other simple examples of correlation are observable in the potato. Here, if the aerial shoots are removed, buds which would normally form underground stems grow into foliage shoots above the ground, while in some varieties aerial shoots may be induced to form tubers if the formation of normal underground tubers is prevented.

No doubt the way in which correlation acts varies in different cases. Sometimes, perhaps, correlative effects depend solely on nutrition. Thus, in the case mentioned above, where removal of a proportion of leaf rudiments on a shoot leads to a greater development of the remaining leaves, it might be argued that more food is now available per leaf. It was at one time thought that the development of lateral buds on removal of the terminal bud was to be explained on somewhat similar lines, namely, by the accumulation of shoot-forming nutrients at the upper end of the stem. That the development of dormant lateral buds involved more than this is indicated by the work of a number of investigators of recent times, among whom Loeb and Snow deserve special mention.

Loeb's investigations were made chiefly with *Bryophyllum calycinum*. In this plant the arrangement of the leaves is decussate, there being, as usual, a bud in the axil of each leaf. If a piece of stem is separated from a plant and, after all the leaves are removed from the piece, it is placed horizontally with the line joining the morphologically uppermost pair of buds vertical, only this pair of buds will commence to develop into shoots, while the growth of the lower of the pair is soon retarded or even stops. No other buds develop. If the upper bud is removed, the lower bud grows rapidly, but one or both of the two buds of the node next below (morphologically speaking) also sprout.

Loeb explained these results by assuming that a bud, on beginning to grow out, produces substances which pass from it and travel towards the base of the stem. These substances inhibit the growth of other buds into which they diffuse. It was supposed that the substances travel in the conducting tissue in the side of the stem on which the bud is borne, but that they also tend to travel downward. Hence in the first case, where no buds are removed, the inhibitory substances produced by the morphologically uppermost bud on the physically upper side of the shoot find their way to all other buds on the stem and retard or inhibit their development; when, however, this upper bud is removed, the inhibitory substances from the remaining bud, which is on the lower side of the shoot, do not reach the neighbouring pair of buds, since, owing to the decussate arrangement of the leaves, these are situated above the developing bud producing the inhibitory substances. These latter do, however, succeed in reaching buds farther away from the morphological apex of the stem.

That the behaviour just described cannot be explained as due to the apical bud absorbing all the shoot-forming material because it is the first to grow out, is indicated by the fact that in a piece of stem treated in the same manner as that described above, but with the petioles (only) of the morphologically uppermost pair of leaves left attached to the stem, all the other leaves being entirely removed, the growth of the apical pair of buds is prevented or retarded, and the buds of the node below grow instead. The presence of the petioles has an inhibiting effect on the development of their axillary buds. But as soon as the petioles wither and fall off, the two apical buds develop and soon exert their inhibitory action on the growth of the two buds morphologically lower down the stem, as a result of which the development of these latter buds is soon retarded or stopped.

It would thus appear that not only does the growing bud produce growth-inhibiting substances which travel towards the base of the stem, but that the leaf also forms such substances which inhibit the growth of shoots. Thus, if a stem of *Bryophyllum calycinum* is placed horizontally in the way already described, and all the leaves removed except those of the morphologically uppermost node, no shoots are produced at all. Removal of one of the pair of leaves at once results in the sprouting of the corresponding axillary bud. That the leaf and developing bud should behave similarly in this respect is not surprising, since the development of the bud is so largely a development of leaves.

Although the leaf thus inhibits shoot formation lower down the stem, adventitious roots are formed in abundance at the nodes. It is possible that the leaf also produces substances which further root formation, or it may be that the same substances which inhibit shoot formation have a furthering action on the development of roots.

Numerous other experiments of a similar kind were performed by Loeb in support of his view of the production of inhibitors of shoot growth. They scarcely explain, however, why the morphologically uppermost bud develops first. Loeb's explanation of this is that when a piece of stem is cut out from the intact plant, the inhibiting substances which have travelled downwards from the apex of the shoot are distributed throughout the cut piece of stem. The inhibitors now continue to move towards the morphological base of the stem and so the buds at the morphological apex will be clear of inhibitors before those lower down, and so will grow out first and continue to inhibit the growth of the others.

It is to be noted that, by similar experiments with shoots of *Bryophyllum calycinum* in which leaves were removed from various parts of the stem, Loeb showed that a leaf also produces substances which travel towards the apex of the stem and further shoot development. It is, however, not clear whether these are simply sugars and other nutrient substances, or whether they are unknown substances of hormone character like the inhibitors.

A number of years before Loeb's conclusions were published Dostál had recorded that, in species belonging to 15 different families, after removal of the apical bud, the axillary buds developed more rapidly if their respective leaves were removed than if these leaves were left attached to the stem. Dostál concluded from these and later comparatively recent observations that the leaves produce a substance which inhibits the development of their own axillary buds. As the inhibitory action only occurs if the leaves are allowed to assimilate or are otherwise supplied with nutrient material, Dostál concluded that the substance inhibiting growth was formed in the leaves during their metabolism.

The inhibiting action of the leaf has been more recently investigated by Snow in young seedlings of the edible pea, *Pisum sativum*. Here, by removal of different leaves, it was shown that almost the whole of the inhibiting effect of the shoot upon the axillary buds comes from three (or four) of the leaves developing towards the shoot apex, namely, the two largest leaves which may be regarded as still forming part of the terminal bud and the leaf next below it. The leaf of *Pisum sativum*, that is, only inhibits during a certain stage in its development; it begins to inhibit when it reaches a length of 2 to 2·5 mm., continues to do so with increasing strength as it increases in length up to 20 mm., after which its inhibiting action declines until by the time the leaf has reached a length of about 45 mm. its inhibiting action is negligible. *Vicia faba* and *Phaseolus multiflorus* behave, in general, similarly.

The complete inhibition of axillary bud expansion is only brought about by all the three or four leaves mentioned; removal of one or two of these leaves results in only partial inhibition, so that the buds grow out, but more slowly than would be the case if all the inhibiting leaves were removed. When seedlings are decapitated, leaving only the lowest of the inhibiting leaves attached to the seedling, the inhibition *increases* with distance from the leaf, the growth of the bud being the more retarded the greater the distance of the bud from the inhibiting leaf, the strength of inhibition increasing with increase in length of the intervening portion of the stem. This explains why axillary buds normally grow to a certain length before ceasing to grow. Their growth is only inhibited when, owing to the growth of the main shoot, they attain a certain distance from the inhibiting leaves at which the inhibition is strong enough to stop growth.

The increase of inhibitory activity with distance is itself not easily explained. As Snow pointed out, it makes it difficult to suppose that the transmission of inhibition consists solely in the transport of a soluble substance which induces inhibition of growth. He therefore suggested that such a transmission is one stage only of a complex process, while other stages may be responsible for the increase in inhibition with distance.

In further work Snow experimented with young seedlings which, by decapitation through the epicotyl, had developed two equal shoots growing from the axils of the cotyledons. In such plants the removal of all the leaves longer than 1 mm. from one of the shoots brought about the rapid cessation of its growth, a result attributed to the inhibitory action of the intact shoot. The result was the same whether the plant grew in bright or in very dim light.

At this point reference must be made to the observations of Jost published in 1893 on the influence of leaves on cambial growth, in which it was shown that this growth generally takes place only when leaves are present, and that the effect of the leaves in promoting cambial growth travels downwards. Since this effect is observable, at any rate in *Phaseolus*, even when plants are grown in the dark, it would seem that it cannot be attributed to the production of nutrient material in the leaves and its translocation therefrom. Hence it has been suggested that here again a growth-promoting substance might be formed in the leaves and travel from them to the cambial region; that, indeed, there might be here another example of hormone formation and action. More recently Snow has shown that this influence favouring cambial development can traverse a protoplasmic discontinuity, such as a piece of moist linen, and is therefore in all probability actually a hormone. Among the experiments performed by Jost were some in which the stems of young plants were partially split longitudinally so that attached to the continuous length of shoot was a portion attached only at the top or bottom (cf. Fig. 47). Such strips of tissue attached at the upper end continue to grow in thickness while those attached at the lower end do not grow. This observation was confirmed by Snow in the case of *Vicia faba*. Here, not only does the strip (the ' cut ' half) attached at the upper end (downward pointing strips) grow in thickness, but it grows to practically the same length as the part of the stem remaining in line between stem apex and root (the ' in-line ' half). These results are interpreted by Snow as indicating that there is a downward flow of a substance from the apical region of the shoot furthering the elongation of the short internodes; this substance would be of a similar kind to the growth-promoting substance demonstrated so definitely in *Avena* coleoptiles. There is some evidence that it can travel both upwards and downwards, though not very far.

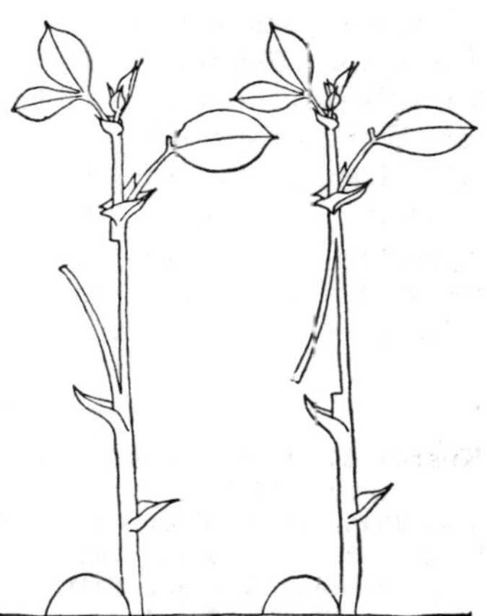

FIG. 47.—For explanation see text (From R. Snow)

There is thus reason to suppose that hormones are concerned in (1) the elongation of the stem, (2) the inhibition of bud development, and (3) the development of the cambium. At one time Snow inclined to the opinion that a separate hormone was concerned in each of these processes, but lately evidence has been accumulating suggesting that the same hormone, auxin, is involved in all of them. Obviously auxin can be held responsible for the elongation of the internodal cells, while Thimann and Skoog have now shown that the application of agar containing auxin to the tops of decapitated stems inhibits the development of the axillary buds. More recently still Snow has found that the application of either auxin-*a* or hetero-auxin to the upper ends of decapitated seedlings of *Helianthus annuus* induces active cambial development. It would thus appear that the same hormone, in travelling downwards through the shoot, induces both cell elongation and cambial development in that shoot, but tends to inhibit the development of the axillary buds.

That the formation of adventitious roots from stem cuttings may depend on the production of a hormone, is indicated by some observations of F. W. Went on *Acalypha* sp. and *Carica papaya*. Detached leaves of these plants were placed with their stalks in water, and the water then subsequently evaporated at low temperature to a small bulk. It was then mixed with warm 3 per cent. agar and the mixture allowed to set. Application of this agar to areas near the top of defoliated stem cuttings of *Acalypha* promoted the development of adventitious roots at the base. A preparation of diastase had the same effect, and as root development was furthered even with diastase preparation which had been boiled, it would appear that the growth promoting substance is not the enzyme itself, which is thermo-labile, but some substance accompanying it in the preparation.

Recent work by Thimann and Went shows that the root-forming hormone (or hormones), to which the name rhizocaline has been given, is widely distributed in the plant kingdom. It has been found in wheat embryos, leaves of *Helianthus annuus, Prunus laurocerasus* and *Malva* sp., etiolated buds and shoots of *Pisum sativum*, the pollen of a number of species, rice polishings and in urine. Chemical tests indicate that rhizocaline is an unsaturated organic acid of about the same acid strength and solubility as auxin-*a* and hetero-auxin, and it is concluded that the root-forming hormone closely resembles the auxins in chemical constitution. In a still later description of the action of the root-forming hormone, F. W. Went refers to a group of these substances of which one is identical with auxin-*a*.

REFERENCES

KÜSTER, E. *Pathologie der Pflanzenzelle.* Teil I. *Pathologie des Protoplasmas.* Berlin, 1929.

—— *Pathologische Pflanzenanatomie.* 3 Aufl. Jena, 1925.

LUNDÊGARDH, H., *Environment and Plant Growth.* (English translation by E. Ashby.) London, 1931.

PFEFFER, W., *The Physiology of Plants.* (English translation by A. J. Ewart.) Vol. II. Oxford, 1903.

CHAPTER XVII

PERIODICITY

ONE of the most evident characteristics of plant activity is that it is rhythmic. To all dwellers in a temperate climate this rhythm is obvious, for it has the profoundest influence on the general aspect of the country, and provides a contrast between summer and winter hardly less striking than the differences in temperature and illumination. For, undoubtedly related to these physical factors of the environment, plant activity in general reaches a maximum in summer and falls to a minimum in winter, to rise to a maximum again with the return of higher temperatures and longer periods of, and more intense, illumination. This rhythm is most impressive perhaps in trees and shrubs, the majority of which lose their leaves in autumn and remain bare until the following spring, giving the whole countryside a completely changed aspect. Herbaceous biennials and perennials for the most part exhibit a similar behaviour, and remain in a dormant condition throughout the winter, in the form of bulbs, rhizomes, or other so-called perennating organs. In annuals the same periodicity occurs, although the rhythmic period corresponds to the whole life-history and the organ of propagation, here the seed, can be regarded as a perennating organ.

In addition to this annual periodicity, most plants exhibit a daily periodicity in their activities. Generally it is obvious that this is directly traceable to daily rhythmic alternations of temperature and illumination. Further reference is made to this daily periodicity in a later chapter.

There are also cases of periodic changes in activity which, being neither daily nor annual, appear to have no direct relation to external conditions. Such, for instance, is the alternation of generations which occurs in the alga *Dictyota*, or the periodic changes in the development of the higher and many of the lower plants in which the period of vegetative development is followed by the development of reproductive organs. Some of these cases are referred to below.

ANNUAL PERIODICITY

The annual rhythm of the higher plants is marked, then, by an alternation of a period of active development with one of rest or comparative inactivity. In some cases it is clear that the rest period is directly due to low temperature or poor illumination which is insufficient to enable the metabolic processes of the plant to proceed. This appears to apply to some of our common weeds such as groundsel (*Senecio vulgaris*) and chickweed (*Stellaria media*), which continue to grow and flower so long as conditions are favourable. But with most plants the rhythm, although corresponding with seasonal periodicity, appears to be something innate or ' aitiogenic ', and not related *directly* to environment, since changes in the periodicity of the environmental factors will not bring about corresponding changes in the rhythm exhibited by the plant. The following considerations will make this clear.

1. Many bulbous plants, such as hyacinths, tulips and narcissi, develop their aerial parts in spring, but when, with the production of flowers, growth ceases, the aerial parts die down, although the conditions may remain suitable for vegetative growth and activity for another 4 or 5 months. The same is the case with many plants in which production of flowers and fruits terminates the activity of the individual either temporarily or permanently. Indeed, we may suppose that the majority of the higher plants exhibit an aitiogenic rhythm of this kind if we are right in assuming that vegetative development will give place sooner or later to the development of flowers and fruits, even if the plants are exposed to the same conditions day after day. Recent work has shown that we have every reason to make this assumption. In such circumstances we must suppose that during vegetative growth reactions proceed which result in the production of substances which lead to the formation of reproductive, instead of vegetative, organs.

2. In the chapter on germination it has been pointed out how the seeds of many species remain dormant even when the conditions of the environment are such as to permit of germination. Some cases, those where germination is delayed on account of insufficient development of the embryo or the need for chemical changes in the seed (' after-ripening '), can be compared with the case of reproduction following vegetative development, inasmuch as the reactions of one period lead up to and determine the next.

3. It can easily be shown that the dormant condition of deciduous trees is not necessarily contemporaneous with low light intensity or low temperature. Molisch quotes the case of the small-leaved lime (*Tilia parviflora*) in this connexion. If the twigs of this tree are placed with their cut ends in a vessel of water in a warm greenhouse immediately after leaf-fall early in October, they remain without any apparent renewal of activity often until the end of February or later. But if twigs of the same plant are removed from the tree in early spring and subjected to the same treatment they sprout at once. Another case quoted by the same authority is that of the cherry and other fruit trees. Twigs taken from the trees in early October and placed with their cut ends in water in a warm room exhibit no renewal of activity by the end of the year, but if removed from the tree early in December, traditionally on St.

Barbara's Day, the 4th of December, and similarly placed in water in a warm room, the flower buds open for Christmas. Such cases can be multiplied almost indefinitely. A rest period appears to be necessary before either leaf or flower buds can continue their development, and this rest is such that the mere provision of suitable external conditions for development is insufficient to reawaken the organs concerned to renewed activity.

IMPOSED AND SPONTANEOUS REST

It will thus be seen that the condition of dormancy of resting plants or organs may be of two kinds, a rest imposed by the external conditions of the environment, such as low temperature and inadequate illumination, a dormant condition which gives place to activity directly the external conditions become favourable, and a rest which cannot be broken merely by exposing the plant to external conditions normal for growth. Various names have been suggested to differentiate these two kinds of inactivity, but here the terms ' imposed rest ' and ' spontaneous rest ' will be used to distinguish between the rest imposed by the conditions of the environment and the rest which appears to develop spontaneously without direct connexion with the external conditions. The duration of the spontaneous rest period varies considerably from species to species. In Northern temperate climates the period of spontaneous rest in winter buds may be over in December, as in the case of the flower buds of *Forsythia*, or it may persist, as in the buds of oak, ash and beech, until well into April. It will be clear that those buds in which the spontaneous rest period terminates early in winter may still remain dormant after the end of the period. Different kinds of buds of the same individual plant may also possess different rest periods. Thus in hazel the spontaneous rest period of the staminate catkins comes to an end at about the end of November, that of the carpellary catkins in January, while that of the leaf buds does not end until March.

PREMATURE RENEWAL OF ACTIVITY

In dealing with the problems of dormancy in seeds it was shown that it is possible by means of various devices to bring resting seeds to germinate before they are capable of doing so under normal conditions of germination. Similarly with other plant organs, such as resting buds, it is possible to bring about premature renewal of activity. Such a premature renewal of activity is sometimes observed in nature. When oak trees are defoliated by attacks of a caterpillar, a fresh crop of leaves may sometimes develop late in the season from buds which would normally rest through the winter and not sprout until the following spring. Such behaviour has been noted with many trees and shrubs and may be produced by artificially removing the leaves from such plants. As well as defoliation many other methods have been devised for breaking down the spontaneous rest period of plants or plant organs. The most important of these are summarized below.

1. *Low Temperature.* Müller-Thurgau in 1885 recorded that potato tubers that normally required a period of rest before sprouting, would sprout at once if kept at a temperature just above 0° C. for 14 days immediately after harvesting. At this low temperature the sugar-starch equilibrium shifts so that more sugar is produced and it is possible that mobilization of the food reserve is the factor responsible for the renewal of activity under these conditions.

2. *Etherization.* Johannsen in 1900 described how lilac and other woody plants might be brought out of their resting condition in autumn and winter by exposure to an atmosphere containing a certain proportion of ether. Thus, when potted plants of lilac were enclosed for 1 to 2 days in the middle of November in a chamber containing 30 to 40 grams of ether per hectolitre of space and kept at about 18° C., the plants came into flower in 3 or 4 weeks. The rapidity of the action depends on the actual time within the rest period when the ether is applied. Application at the beginning of August results in the flowers opening in 5 or 6 weeks, but towards the end of the period of spontaneous rest the buds will open while the plants are still in the ether.

3. *The Warm Water Bath.* In 1909 Molisch published the results of an exhaustive series of experiments on a very simple method of inducing the buds of woody plants and some other perennials to sprout during their period of spontaneous rest. The treatment consists simply in immersing twigs of the plants in a bath of lukewarm water for about 9 hours and then transferring the twigs to a warm greenhouse, where they are kept with their cut ends in water. It is quite easy to immerse the whole of the aerial parts of a plant in the bath. The result of the treatment depends on (1) the duration of the bath, (2) the temperature of the bath, (3) the actual time in the rest period at which the immersion is made, and (4) the species employed.

As regards the first of these factors, submersion for from 9 to 12 hours was found to give the best possible results with most species, although the optimum length of time varies somewhat with different species. Too long a period of immersion leads to injury and death accompanied by a brown coloration, possibly the result of anaerobic respiration at a relatively high temperature.

The optimum temperature also varies over a rather narrow range. For most species 30° C. appears to be the most favourable temperature, as in *Syringa vulgaris* and *Forsythia suspensa,* but with some a higher temperature (35° to 40°) has to be employed, as with *Betula alba, Rhamnus frangula* and *Fraxinus excelsior.*

The position in time in the rest period has a considerable effect on the result of the warm bath treatment. As regards species in which the period of spontaneous rest ends relatively early in winter, it may be stated that on the whole sprouting follows more rapidly on the warm bath treatment up to a certain time, but that, as the spontaneous rest period nears its end the warm bath has less effect and may even have a retarding influence. Thus it was found that the staminate catkins of *Corylus* were not affected by the warm bath treatment in September, but responded well in late October, in November and in December. After this the bath was less effective and in February definitely retarded opening of the catkins.

With different species the effect of the bath may be very different. Among the species which Molisch recorded as responding readily to the treatment were lilac (*Syringa vulgaris* and *S. persica*), *Forsythia suspensa, Cornus mas* and *C. alba, Azalea mollis, Spiraea palmata,* and *S. japonica, Rhamnus frangula,* gooseberry (*Ribes grossularia*), birch (*Betula alba*), sycamore (*Acer pseudoplatanus*), and larch (*Larix decidua*). On the other hand, with beech and lime the warm bath treatment was only effective a short time before the end of the spontaneous rest period.

Two points of particular interest are to be noted with regard to the warm bath. Firstly, the action is quite local, so that if some buds of a twig are submerged in water and others are not, only those buds which are submerged will respond to the treatment. Secondly, the effect of the bath may remain latent. Thus if twigs are bathed and then returned to the ordinary winter conditions which impose inactivity on the buds, they will still respond to the warm bath treatment if they are subsequently transferred to a warm greenhouse.

4. *Injection of Water.* In 1911 Weber recorded that injection of a small quantity of water into the base of resting flower buds of *Syringa* or *Tilia* would break down the condition of spontaneous rest so that the flowering could take place 3 weeks before that of untreated buds of the same plant.

5. *Alcohol and Acid Baths.* Jesenko in 1912 examined the effect of submerging resting twigs in baths of ethyl alcohol ranging in concentration from 1 to 30 per cent., and in baths of hydrochloric and sulphuric acids of concentrations of from 0·5 to 5 per cent., and in solutions of other acids. Moderate temperatures, from 12° to 17° C., and periods of immersion of from 3 to 12 hours were employed. The results obtained by this treatment are very comparable with those given by the warm water bath. In general, 1 per cent. ethyl alcohol and 0·5 to 1 per cent. acid were found to be most effective, but as with the warm bath, the period of immersion and position in time in the rest period were important factors influencing the result. Thus a bath of 1 per cent. alcohol lasting 12 hours was effective with twigs of *Pyrus malus* on the 30th of November, but had a harmful action when applied in January. Carbonic acid (water saturated with carbon dioxide) applied for 12 hours on the 3rd of December induced sprouting of *Larix* buds in 14 days, whereas after a bath lasting 6 hours sprouting did not take place for 18 days, while after a 3-hour bath sprouting did not take place until 19 days after treatment, when the unbathed controls sprouted as well.

Positive results were also obtained with *Populus nigra*, *Carpinus betulus*, *Acer campestris*, and *Sambucus nigra*. On the other hand, no favourable result was obtained with *Salix aurita*.

6. *Injection of Alcohol or Ether*. Jesenko also recorded in the same year the favourable effect on sprouting brought about by injecting into buds a small quantity of a solution of ethyl alcohol or ether. The experiments were carried out in the summer on defoliated trees. A range of alcohol concentrations (1 to 10 per cent.) and ether concentrations (0·1 to 1 per cent.) was employed, while the effects of injecting pure water and merely pricking the buds were also examined. All these treatments were effective in some cases, though to different degrees. Thus in trees of *Tilia grandiflora*, defoliated and treated on the 24th of July, various agents were instrumental in bringing about sprouting in the following order of effectiveness: (1) 1 per cent. alcohol, (2) merely pricking, water and 0·1 per cent. ether, (3) 5 per cent. alcohol. The buds injected with 1 per cent. alcohol began to swell 7 days after the injection, while 16 days after defoliation the untreated control buds were still closed. Injection of dilute alcohol and ether also furthered opening of buds in *Fagus sylvatica*, *Sorbus torminalis*, *Acer platanoides* and *A. pseudoplatanus*, and *Syringa*. The treatment was unsuccessful with *Quercus robur* and very doubtful with *Carpinus betulus*, in both species possibly because the buds are small and pricking brings about relatively extensive injury.

7. *Radium Emanation*. In 1912 Molisch reported that radium, or better, radium emanation, had an effect on resting buds similar to that of ether or the warm water bath. The buds of lilac exposed to radium emanation for 72 hours, commencing on the 27th of November, were opening on the 10th of December, whereas controls in pure air had not opened by the 30th of December. Positive results were also obtained with horse-chestnut, tulip tree (*Liriodendron*) and *Staphylea*, but no response to the treatment resulted in the cases of *Ginkgo*, plane, birch and lime. As with ether and the warm water bath, the treatment must last for a definite time; too short a time has no effect, while too long exposure produces injury. Similarly, with regard to the position in time in the rest period, treatment too early in the rest period has no effect, while late in the rest period the treatment may even retard opening of the buds.

8. *Smoke and Various Gases*. In 1916 Molisch showed that exposure for from 24 to 48 hours to a thick cloud of smoke from burnt paper, sawdust or tobacco, would bring about premature sprouting of the resting buds of many species in much the same way as does ether. It is not clear which constituent of the smoke plays the principal part in this effect; possibly it is acetylene or ethylene. Weber showed in 1916 that exposure of resting buds to air containing acetylene would shorten the rest period, and subsequently that other gases, such as ethylene, coal gas and thymol vapour, have the same effect.

9. *Pressure*. Weber found by experiments made at the end of

1921 that buds of lilac towards the end of the rest period could be brought to open by applying pressure to the bud for a very short time, namely, something less than a minute. The pressure can be applied by squeezing the bud in a metal clip such as is used for closing rubber tubing. Buds of lilac treated in this way on the 22nd of November showed indications of sprouting on the 30th of November, while the leaves began to unfold on the 2nd of December. There was no indication of sprouting in untreated control buds on the 8th of December when the experiment came to an end.

10. *Concentrated Acid and Alkali Baths.* Richter in 1922 recorded premature sprouting of buds of horse-chestnut and lime (*Tilia parviflora*) following immersion for 20 seconds in concentrated sulphuric acid. A treatment lasting half the time was not so effective, while exposure to the acid for 40 seconds resulted in many buds being killed. Premature sprouting could also be induced by treatment for 2 or 3 minutes with concentrated potassium hydroxide solution.

11. *Hydrocyanic Acid.* In 1925 Gassner described experiments in which air containing a small percentage of hydrocyanic acid induced premature sprouting of resting buds. With lilac, at the beginning of December, an exposure of 60 minutes to an atmosphere containing 1 per cent. hydrocyanic acid gave the best results. The same concentration acting for the same time also gave the best results with lily-of-the-valley rhizomes in the middle of December. With these the effect of hydrocyanic acid differed from that of the warm water bath in that the latter chiefly speeded up leaf development, whereas hydrocyanic acid chiefly brought about earlier opening of the flower buds. An exposure of 60 minutes to 1 per cent. hydrocyanic acid was also found the most effective combination for breaking down the rest period of the flower buds of *Prunus avium*. Buds of *Quercus sessiliflora* and *Aesculus hippocastanum* also responded to treatment with hydrocyanic acid, but negative results were obtained with the beech.

12. *Continuous Illumination.* One of the most interesting experiments on breaking down the spontaneous rest period was carried out by Klebs in the years 1913 and 1914. This investigator placed a potted plant of beech at the beginning of its rest period in September, in a chamber subjected to the continuous illumination of an electric lamp. In about a fortnight the winter buds began to unfold, and after the lapse of another fortnight 75 per cent. of the buds on the plant had opened. On removing the plant to an ordinary greenhouse a second set of resting buds was formed. The plant was next, on the 25th of December, transferred again to a continuously lighted chamber, with the result that the second set of resting buds subsequently unfolded. At the nodes of the shoots so produced a third set of buds formed which unfolded in the middle of March. Thus, during the normal resting period, when under ordinary conditions of illumination the plant would have remained dormant even in a warm environment, three successive sets of leaf buds sprouted. This behaviour is the more striking when

it is remembered that the beech is a plant with a long spontaneous rest period, and that methods of breaking down this rest period which are applicable to many species have little effect upon it.

An explanation of the effect of the various agents here mentioned in breaking down the spontaneous rest period is difficult to find, and the very diversity of the methods by which this result is achieved makes the question still more puzzling. We must suppose that renewal of activity after a resting period results from, or is accompanied by, a mobilization of food material as a consequence of renewed enzymatic activity. But whether the various treatments bring about an activation of enzymes previously in an inactive form, whether they have some direct chemical effect on the reserve food material, whether they affect the permeability properties of the cells of the resting organs, or whether such results are secondary effects of some other influence on the protoplasm, we do not know. Certainly in the action of low temperature on the sprouting of potato tubers, as has already been pointed out, the shifting of the starch-sugar equilibrium may be responsible for the renewal of activity, inasmuch as the concentration of sugar is increased. Johannsen considered that the effect of ether might be similar and that a production of sugar from starch and a breaking down of proteins, might result from such treatment. In the case of the warm bath the higher temperature would lead to an increased respiration rate, which under the conditions of immersion in water, we might expect to be largely anaerobic. It may be that in other cases as well, such as those involving wounding and treatment with poisonous substances, the initial result is an increase in respiration rate. There is, however, no evidence that the effect of the warm bath or of other methods of inducing premature sprouting is due to an effect on respiration.

The effects in general appear to be similar to those produced by what are called stimuli, and which are dealt with later, inasmuch as a comparatively slight change in the environment, involving only a small transference of energy, leads to far-reaching results in the organism in which the energy changes are out of all proportion to those involved in the environmental change. It may, or may not, be significant that, as will be seen later, Verworn regarded the first result of all stimulation to be an increase in the respiration rate. We must wait for future work to provide a credible explanation of the methods of breaking down the spontaneous rest period of plants.

LEAF-FALL

The fall of the leaves from the deciduous trees of temperate regions is a periodic phenomenon, and is brought about by definite morphological and physiological changes in the leaf-base. The morphological changes have been the subject of much investigation. The essential feature is the formation of a *separation layer* across the base of the leaf stalk, the separation being effected by the disappearance of the middle lamella from the cell-walls of parenchymatous cells and the rupture of the lignified xylem elements and sieve tubes at the same level. The formation of the separation layer may or may not be preceded by cell division.

Either before or after the formation of the separation layer a protective layer is produced. This, according to E. Lee, may be formed in three different ways: (1) ligno-suberization of existing cells below the separation layer, (2) ligno-suberization of cells after irregular division, and (3) ligno-suberization of cells produced by regular cambial division. At a later date a layer of cork cells is formed below the scar formed by the separation of the leaf.

Leaf-fall in deciduous trees of the temperate zone is obviously related to seasonal change, that is, to external conditions, but there can be no doubt that, as in the case of the resting period, with which it is connected, leaf-fall cannot be controlled by merely controlling the factors of the environment. Moreover, leaf-fall occurs in trees in the Tropics where frequently it appears to bear no relation to changes in the environment, and where, indeed, it may occur at different times in different branches of the same individual. Nevertheless the significance of leaf-fall in the life of woody plants of temperatre regions is considerable. In particular the complete removal of leaves reduces transpiration practically to zero at a time when absorption of water by the roots on account of low soil temperature, may be greatly reduced. It is also generally held that in leaf-fall waste material is removed from the plant, and it is true that fallen leaves frequently contain relatively large quantities of calcium oxalate, calcium carbonate and silicates, substances which may be unwanted by-products of plant metabolism.

The physiological processes in the plant which lead up to leaf-fall are little understood. It is, however, known that it can be induced experimentally in a variety of ways. In particular, a sudden disturbance of water relations of the plant is effective in bringing about leaf-fall. Thus, a great reduction of transpiration, which may be effected by transference of a plant from the open air to a chamber saturated with water vapour, will frequently result in the falling of the leaves, while the same thing may happen through the reverse change of transference from a humid atmosphere to a much drier one. Conditions favouring rapid transpiration combined with low rate of water absorption by the roots, such as are met with in an environment comprising both dry air and dry soil, frequently lead to leaf-fall, as in species of *Fuchsia*, *Impatiens* and *Azalea*.

According to Molisch, plants of many species shed their leaves as a result of subjection to poor illumination, species of *Begonia*, *Fuchsia* and *Coleus* being particularly sensitive in this respect. Woody plants as a rule retain their leaves longer and evergreens longest under such conditions. It is suggested that the effect in this case is related to the reduction in photosynthetic activity, but in what way is not evident.

Exposure to various gases will also induce leaf-fall. Some plants, as, for example, *Mimosa pudica, Robinia pseudacacia* and other members of the Leguminosae, will shed their leaves in from 24 to 48 hours after exposure to air containing a sufficiently high concentration of tobacco smoke, a result most probably attributable to the ethylene and acetylene contained in the smoke. What, however, are the processes involved in this very rapid action are not known.

CHAPTER XVIII

REPRODUCTION

THE formation of reproductive bodies by a plant may be considered a phase in its development. In some species this is very definitely so, where a period of vegetative growth is followed by the production of flowers and fruit, after which the life of the individual ends. This applies to many annuals, such as the sunflower and groundsel, and to some long-lived plants, of which the so-called century plant is a striking example. In this species, *Agave americana,* vegetative growth continues for from 10 to 70 years or more before flowers form and fruit is produced, after which the plant immediately dies. In many bulbous plants, such as tulip, hyacinth and *Crocus*, the vegetative growth of the plant is terminated by the formation of both sexual and vegetative organs of reproduction, flowers and bulbs or corms. In other cases, after a period of vegetative growth, both this and reproductive growth continue together. Such a condition is met with in both annuals and perennials. Yet again, in some perennials there are definite vegetative and reproductive phases in each year's growth, many trees, for instance, first producing flowers before vegetative growth is renewed for the season.

VERNALIZATION

A number of miscellaneous observations have suggested that temperature conditions in an early stage in the development of a flowering plant may influence the formation of reproductive organs. Thus, cabbage plants flower much sooner if exposed to the normal low temperatures of winter than if kept in a greenhouse where exposure to such low temperatures is avoided. Again, winter wheat sown in open ground in autumn flowers early in the following summer, whereas if sown in the spring it tends not to flower at all in the following season. Much systematic investigation of the problems suggested by these observations has been carried out in recent years, particularly in Russia. In work in that country Lysenko and others have shown that if winter wheat is germinated slowly by adding only a limited quantity of water to the grain, and that if, after the appearance of the radicle through the testa, the germinated grain is then exposed to a temperature of from 0° to 5° C. for a few weeks, the seed if sown in spring flowers in the same year, just as if the grain had been sown in autumn and exposed to the normal low temperatures of winter. It is to be noted that the grain need not be sown immediately after treatment; it can be dried and sown later, the treatment remaining effective.

The conclusion drawn by Lysenko is that for normal development, leading to flower formation, the plant has to go through a series of developmental stages, an early one of which is the *vernalization* stage, which requires a definite temperature for its completion. Unless this stage is completed further stages in development cannot proceed. In the case of winter wheat the temperature required for vernalization is low, but in other species or varieties this may not be so. In maize, for example, a temperature of 20° to 30° C. is required for vernalization.

PHOTOPERIODISM

A number of isolated observations during the latter part of the nineteenth century suggested that the length of the daily period of illumination might be an important factor in determining the rate of growth and course of development of various plants, but a systematic investigation of the question was undertaken first in the United States by Garner and Allard, the results of whose earlier researches on the question were published in 1920. Plants of various species and horticultural varieties were grown under conditions which were normal except that the length of day was artificially shortened by darkening the plants for part of the daytime, or lengthened by exposure of the plants to artificial light for part of the night. Garner and Allard found that not only the rate of vegetative growth, but also the time of flowering, could be profoundly affected by the length of day. As an example may be quoted one of their experiments with Soya beans, var. Peking. Seeds were planted on May 8, and the seedlings appeared above ground on May 17. On May 20 the experiment started, one set of seedlings being subjected to daylight for 5 hours a day only (10 a.m. to 3 p.m.), another set exposed to daylight for 7 hours daily (9 a.m. to 4 p.m.), while other seedlings served as controls and were subjected to the full daily period of natural illumination (considerably over 12 hours daily in May, June and July). In Table LXXXVIII are shown the times of flowering and height of the seedlings subjected to different daily periods of illumination.

Table LXXXVIII

Influence of Length of Day on Vegetative Growth and Time of Flowering of Soya Bean, Var. Peking

Daily period of Illumination	Height of Seedling (inches)	Date of Flowering
10 a.m.–3 p.m.	5–6	June 12
9 a.m.–4 p.m.	8	June 10
Full length of day	42–48	July 21

A similar phenomenon was noted with other species, as, for example,

Aster linariifolius, although the behaviour varied not only from one species to another, but also from one variety to another of the same species.

Comparable results were also obtained in the winter, when in some cases a retardation of the time of flowering resulted from artificially increasing the length of day by exposure of the plants to artificial light during hours of normal darkness. An experiment with *Cosmos bipinnata* may be quoted as an example. Two samples of seeds were sown on November 1 and kept in separate greenhouses, one of which received normal illumination, while in the other the period of illumination was continued by means of electric lamps from 4.30 p.m. to 12.30 a.m. each night. The plants exposed to the normal period of illumination all flowered between December 22 and January 2, whereas those subjected to the daily extra 8 hours' illumination showed no sign of flowering even by February 12. The average height of the control plants was, however, only 30 inches, while those subjected to the additional daily period of illumination were, on the average, 60 inches high.

But speeding up the time of flowering may also result from artificially increasing the length of day. Thus in 1912 Klebs had already succeeded in inducing *Sempervivum funkii* to flower in December by exposing the plant to continuous illumination from an electric lamp, a finding which he later extended to a second species, *S. albidum*. From Garner and Allard's extensive series of experiments may be quoted one with *Iris florentina*. Two clumps of this plant were, on October 20, placed respectively in a greenhouse subjected to normal daily illumination and in one subjected to 8 extra hours of artificial light (4.30 p.m. to 12.30 a.m.). The control plants remained practically dormant and showed no sign of flowering by February 12, whereas the experimental plants grew vigorously and flowered between December 24 and December 30. Again, two samples of seeds of spinach (*Spinacea oleracea*) were sown on November 1 and subjected to the same lighting conditions as the two samples of *Cosmos* mentioned above. In this case the control plants grew very slowly and showed no sign of flowering by February 12. The plants subjected to the daily 8 hours of light grew rapidly and all flowered between December 8 and December 23. After this they continued both to elongate and produce flowers.

Not only may the length of day affect the time of flowering, but it may determine whether flowering takes place at all. Thus, as a result of experiments with a wide range of natural species and horticultural varieties, Garner and Allard concluded that flowering and fruiting only take place when the daily period to which the plants are subjected falls within certain limits. It is usual to make an arbitrary distinction between 'short-day' and 'long-day' plants. Short-day plants are those which will flower only if they are subjected to illumination for a period of 12 hours or less each day ; long-day plants, on the other hand, require a daily period of illumination in excess of 12 hours in order to flower. As showing the arbitrary character of this distinction it may be pointed out that there are plants which will only flower when subjected to illumination within a narrow range about 12 hours, and others which are capable of flowering over a range including both long-day and short-day conditions. Garner and Allard introduced the term 'photoperiodism' to express this relation of the time of flowering to the daily length of the period of illumination.

It will be observed that the length of day influences not only the time of flowering or its incidence, but also the extent and kind of vegetative growth. Exposure to a daily period of illumination unfavourable to flower formation, but favourable to vegetative development may lead to a continuance of the latter and the production of gigantism. Further, in later work Garner and Allard concluded that for each species there is an optimal daily light period for maximum increase in length of the shoot. In the temperate zone this optimal light period is furnished by the long days of summer in the case of some species, while for others the shorter days of spring and autumn are optimal. In plants for which the long days of summer are optimal, shortening of the daily period of illumination brings about a number of responses, such as earlier flowering and fruiting, tuber formation, dormancy and branching. The shortening of the days in autumn is regarded as inducing leaf-fall. It is also possible that a plant may be subjected to a daily period of illumination which favours both vegetative development and flowering. In such a case where vegetative growth and flowering and fruiting proceed together, we have what Garner and Allard call the 'ever-blooming' type, or 'ever-bearing type of fruiting.' Such a condition is met with where the daily period of illumination is intermediate between that leading to vegetative development only, and that leading to flowering only. An illustration of this state is afforded by the artificially illuminated spinach plants mentioned earlier.

Garner and Allard's observations were extended by a number of workers, of whom particular mention may be made of Adams, in Canada, and Tincker, in Great Britain. Although Adams felt that not all Garner and Allard's conclusions were justified, he agreed that the experimental data indicate the existence of an upper limit of light duration for certain phases of development in some species, though not in all. Tincker's results confirm completely Garner and Allard's conclusions as regards the influence of length of day on the time of flowering. The runner bean (*Phaseolus multiflorus*), *Chrysanthemum* (var. Mrs. William Buckingham) and Soya bean were found to be short-day plants in Wales, flowering being accelerated by artificially reducing the length of day, while vegetative growth was diminished. On the other hand, a number of grasses (*Anthoxanthum odoratum*, *Alopecurus pratensis*, *Dactylis glomerata*, *Lolium perenne*, *Phleum pratense*, *Avena sativa*) as well as red clover (*Trifolium pratense*) and radish (*Raphanus sativus*) were found to be long-day plants, flowering being retarded by shortening artificially the length of day. *Poa annua* was found to be an indifferent type, flowering and fruiting occurring at the same time in each of four sets of plants exposed respectively to 6 hours, 9 hours, 12 hours and the full normal period of daylight.

The photoperiodic effect is quite local. Thus Garner and Allard exposed one branch of a plant of *Cosmos bipinnata* provided with two branches to natural illumination of winter, the other branch to natural illumination supplemented by a period of artificial light (sunset to midnight). The branch exposed to the short daily period of natural illumination soon flowered, whereas the branch exposed to the longer period of illumination continued to grow vegetatively without producing flowers.

The intensity of illumination appears to be without influence on the time of flowering, although this generalization no doubt only holds within limits. Thus in a series of experiments with Soya beans in which various sets of plants were shaded to different extents, Garner and Allard observed that flowering commenced on the same day in all cases, although with those plants subjected to the greatest shading, the average intensity of direct sunlight reaching the plant was estimated to be scarcely, if at all, greater than 25 per cent. of the normal. Reduction of light intensity during the period of vegetative development tends to bring about a reduction in the yield of seed, whereas a similar reduction in light intensity between flowering-time and maturity of the seed tends to bring about an increase in the yield of seed. From this it may be concluded that seed production is not determined by the total quantity of radiation reaching the plant.

Lysenko's view of photoperiodism is that the vernalization stage of development has to be followed by a *photo* stage for which a certain length of day is necessary. When this stage has been completed the daily period of illumination no longer influences the time of flowering.

It is obvious that photoperiodism may be of much economic importance, inasmuch as, by varying artificially the length of day to which plants are exposed, their vegetative growth and time of flowering and fruiting may, in many cases, be controlled to a considerable extent. For example, some annuals may be made to complete their life-cycle twice in one year, while others can be made to take on characteristics of ever-flowering species and continue to produce flowers and fruits for an indefinite period, while in yet other cases the plant may continue vegetative development indefinitely without flowering. By controlling the length of day varieties which normally flower at different times may be brought to flower at the same time, and so crossing may be effected between varieties which could not otherwise be crossed. It has also been pointed out that photoperiodism may have an important bearing on the question of acclimatization, for the behaviour of plants under different conditions of daily lighting will indicate their behaviour when transferred to other latitudes.

It is also evident that length of day, varying as greatly as it does with latitude, must be an important factor in determining the distribution of plant species.

THE EFFECT OF NUTRITION ON DETERMINING REPRODUCTION

While the work on photoperiodism just described makes it clear that the daily period of illumination may determine whether a plant continues to develop vegetatively or to produce reproductive structures, it offers no explanation of why the length of day should have this effect. Garner and Allard were inclined to think that the degree of hydration of the protoplasm is influenced by the length of daily exposure to illumination, and determines the observed responses of the plant to differences in the daily period of lighting. Moreover, they thought that this regulation of the degree of hydration of the protoplasm is a direct effect of the light period on the internal mechanism of the plant, possibly protoplasmic permeability, rather than an indirect effect of the light period on other external factors which affect water absorption or loss by the plant. They quote observations on *Viola papilionacea* as evidence of the effect of the length of day on the regulation of the degree of hydration of the protoplasm. During the short winter days the leaf buds are dormant. In the spring, with its days of intermediate length, leaves with short petioles, and blue flowers with relatively short stalks, appear. In the long days of summer the leaves have long petioles. Transference of plants with the summer type of leaf to short-day conditions brings about loss of turgidity of the petiole in 2 or 3 days, with consequent partial collapse of the leaf, while blue flowers of the spring type reappear. It will be noted that this behaviour indeed suggests an effect of the daily period of illumination on the amount of water contained in the cell vacuoles, but it affords no indication whatever of the degree of hydration of the protoplasm.

Years before the work of Garner and Allard the problem of the conditions determining vegetative growth and reproduction respectively had been attacked by Klebs. As long ago as 1892 this physiologist published researches on the influence of different environmental conditions in determining the course of development, and in 1896 he published observations on the conditions for reproduction in Algae and Fungi which have become classical. By varying the external conditions of the environment, particularly food supply, humidity and light, the kind of reproduction could be controlled or it might be inhibited altogether. Thus *Vaucheria repens*, in a nutrient solution such as dilute Knop's solution containing 0·2 to 0·5 per cent. salts, exhibits vigorous vegetative development. If the food supply is cut off by transference of the algae to pure water, asexual reproduction is at once initiated and zoospores are produced. Again, when cultivated in damp air, as, for instance, on the surface of damp soil, the plants develop vegetatively, but if transferred to water zoospores develop vigorously, at a room temperature of 15° C., after 2 days. Light also affects reproduction of *Vaucheria repens*. If cultures of the alga growing in water or a weak nutrient solution (0·1–0·2 per cent.) are transferred to the dark, or even to weak light, asexual reproduction follows.

While asexual reproduction results from darkening, it is quite otherwise in regard to the production of sexual organs, which only form if the plants are kept in bright light. While a light of sufficient intensity appears to be essential for the production of sexual organs, the time at which these appear depends on the actual value of the intensity of the light and also on the light conditions to which the plants have been previously subjected. Thus *Vaucheria* that had been standing in a window in a 0·2 per cent. salt solution and had remained sterile under such conditions, formed sexual organs in 4 to 5 days after exposure to a light of a higher intensity, whereas it took 5 to 6 days under the same conditions in the case of plants which had been previously growing in a still weaker light than that of the window. The degree of humidity appears to be without effect on the production of sexual organs, which can form in dry or damp air, as well as in still water. In running water, however, *Vaucheria* species remain sterile. Transference to still water generally leads to active zoospore formation.

Other species of *Vaucheria*, as well as species of *Hydrodictyon*, *Protosiphon*, *Botrydium*, *Spirogyra*, desmids, *Œdogonium*, *Ulothrix*, *Hormidium*, *Conferva*, *Bumilleria*, *Stigeoclonium*, *Draparnaldia*, *Chlamydomonas* and *Hydrurus* among the algae, were experimented upon by Klebs, and also two fungal species, namely, *Eurotium repens* and *Mucor racemosus*. Quite generally Klebs found that the production of reproductive organs or the continuance of vegetative growth could be determined by control of the experimental conditions. Some of the effects, however, observed with one alga do not appear to be applicable in general; thus in *Vaucheria*, as already noted, asexual reproduction is furthered if the filaments are transferred from light to darkness, while it is checked if they are transferred back from darkness to light. The same is the case to a certain extent with *Protosiphon*, *Bumillaria* and *Draparnaldia*. On the other hand, light appears to have no effect on the production of zoopores by *Œdogonium diplandrum*, *Stigeoclonium* and *Hydrurus*. In pure cultures *Hydrodictyon* produces zoospores if it is transferred from a nutrient solution to water and is illuminated. In the dark it does not, as a rule, produce zoospores.

The effect of light on asexual reproduction in the algae thus varies from one species to another. Moreover, the production of zoospores in relation to light often appears to be in the nature of the response to a stimulus, namely, the sudden exposure of the plant to a change in external conditions.

As regards the effect of light on sexual reproduction, the behaviour of the algae is much more regular. In general it may be stated that light furthers sexual reproduction. This has already been stated to be the case in *Vaucheria repens*, and it appears to be a general rule for this genus, in which sexual organs are produced only after exposure to light. The same appears to apply to *Œdogonium*. In *Spirogyra* conjugation can generally be induced in a few days by exposure to bright light. The production of gametes in *Ulothrix* and *Chlamydomonas* is also favoured by light. Other factors, particularly the composition of the medium, may also have an effect on inducing sexual reproduction. Generally, as in *Œdogonium diplandrum*, a comparatively high concentration of inorganic salts in the medium tends to prevent the formation of gametes, while a shortage of such salts in the water furthers it. Thus, plants of the species mentioned, exposed even to sunlight, remain sterile for a long time in a nutrient solution containing 0·1 to 1 per cent. of salt. On the other hand, the presence of sugar in certain concentrations in the nutrient solution furthers the formation of sexual organs. For example, *Vaucheria repens* in bright light forms antheridia and oogonia more rapidly when in a sugar solution than when in water. *Œdogonium diplandrum* also forms sexual organs in a 4 per cent. sucrose solution in 7 days, whereas a similar culture in 0·05 per cent. nutrient solution remains sterile for 22 days.

It is impossible here to go into further details with regard to Klebs's very important work on the algae. It must be pointed out, however, that his researches strongly suggest that active vegetative growth results when the plants are provided with an ample supply of mineral nutrients. As a consequence of exposure to sunlight under such conditions the manufacture of organic food material by photosynthetic activity proceeds vigorously, and it would appear that sexual reproduction results only when this is the case. That this is a correct conception of the chief condition determining the production of sexual organs is supported by the observation that the formation of these is furthered when the medium surrounding the plants contains sugar in solution which, if absorbed by the plant, increases its supply of organic food material. That this is the only factor in determining the formation of sexual organs is scarcely likely, and Klebs's own results suggest otherwise. There does, however, appear every justification for supposing that, broadly speaking, inorganic nutrition favours vegetative growth, while organic nutrition, necessary, of course, for vegetative development, after reaching a certain point determines the oncoming of sexual reproduction.

The complexity of the form and development of the higher plants renders an analysis of the factors influencing this development a much more difficult problem than in the case of Algae and Fungi. But that external conditions can and do profoundly influence this development has already been exemplified by the phenomenon of photoperiodism. Klebs's very considerable contributions to our knowledge of the effect of conditions in determining the course of development of higher plants were published in a series of papers between the years 1900 and 1918. He showed, for example, that a number of species such as *Digitalis*, *Glechoma*, and sugar beet continued their vegetative growth for a year or more without producing flowers if they were maintained in rich soil and exposed to comparatively high temperature and humidity through the winter. Of particular importance and interest were his experiments with *Sempervivum*, to one of which reference has already been made. His general conclusions with regard to the conditions determining respectively vegetative growth and production of flowers were as follows:

1. With vigorous photosynthesis and high light intensity a considerable increase in the supply of water and nutrient salts furthers strong vegetative growth.

2. With vigorous photosynthesis and high light intensity a decrease in the supply of water and nutrient salts leads to production of flowers.

3. If a medium supply of water and mineral nutrients is provided, either vegetative growth or flower production is possible; which of these actually proceeds is determined by the intensity of photosynthesis. By diminishing the latter, as, for example, by reducing the light intensity, vegetative growth continues; on the other hand, exposure to high light intensity leads to the production of flowers.

The formation of reproductive organs in *Sempervivum* thus results when the rate of production of carbohydrates is high compared with the rate of intake of water and mineral nutrients, while vegetative growth is furthered by a lower rate cf production of carbohydrate in proportion to the supply of water and nutrient salts. Such a conclusion that the formation of reproductive structures is determined by the relation of organic to inorganic nutrition, is in agreement with the experience of horticulturists that excessive application of manure to various garden crops may further vegetative growth and retard flowering.

Klebs's ideas have received support from a number of researches during recent years on the nutritional conditions necessary for flower and fruit production in various plants of horticultural importance. The first and best known of these investigations is that by Kraus and Kraybill dealing with the tomato. The general conclusion they drew from their experiments is that the relationship between the amounts of vegetative and of reproductive growth is determined by the ratio of carbohydrate and nitrogenous substances present within the plant and available for its nutrition; by the value, that is, which is now very generally known as the carbohydrate/nitrogen ratio. A good supply of nitrate to the plants with opportunity for photosynthesis and the production of carbohydrate leads to vigorous vegetative growth, but little fruit is produced; here the carbohydrate/nitrogen ratio is relatively low. A moderate supply of nitrate with ample opportunities for photosynthesis leads to an intermediate vegetative growth and maximum fruit production; here the carbohydrate/nitrogen ratio is higher. With poor nitrogen supply both vegetative growth and fruit production are small. Without satisfactory vegetative growth subsequent fruit production cannot be vigorous. Thus fruitfulness is associated with a definite range of values of the carbohydrate/nitrogen ratio, outside of which vegetative growth predominates. Results obtained by Gurjar with the same species confirmed the findings of

Kraus and Kraybill. He determined the carbohydrate and nitrogen contents of plants subjected to different degrees of manuring, and found that plants would develop vegetatively over a range of carbohydrate/nitrogen ratios between 2 and 19, but that fruits formed only so long as the ratio lay within the narrow limits of 4 and 6.

Klebs's views, and the more exact formulation of them conveyed in the idea of the carbohydrate/nitrogen ratio, have been subjected to considerable subsequent testing. Woo found in *Amaranthus* a plant capable of absorbing large quantities of nitrates and still forming reproductive organs readily, and it is evident, if the carbohydrate/nitrogen ratio concept is applicable to this plant, that either the ratio necessary for flowering and fruiting must vary widely with different species, or it may be necessary to limit more exactly the nitrogen compounds to be considered in relation to carbohydrate. Both seem very reasonable suggestions. Much work has also been done on the connexion between the ratio and flowering and fruiting in apple trees. Here the phenomenon of biennial fruiting is frequent, the short shoot or 'spur' which bears the flowers going through a 2-year cycle, in the first year of which growth is vegetative, except that the flower buds are formed at the end of one season, while in the second season these buds develop into flowers. Correlation between the value of the carbohydrate/nitrogen ratio and flowering in the apple has been sought along a number of different lines. Hooker analysed the spurs at intervals throughout the 2-year period and found that at the beginning of the cycle the carbohydrate/nitrogen ratio is relatively low and vegetative growth active. This growth results in the utilization of both carbohydrates and nitrogenous substances, and the amount of both substances decreases until growth in length stops with the production of the terminal bud. The photosynthetic activity of the leaves now brings about an increase in the carbohydrate content of the spur, with the result that the carbohydrate/nitrogen ratio increases, and while this proceeds the flower buds are differentiated. From this time until the cessation of activity at the end of the season both carbohydrate and nitrogen accumulate. In the following spring the enlargement and opening of the flower buds is accompanied by a rapid disappearance of carbohydrate, but not of nitrogen, the latter only disappearing with the development ('setting') of the fruit. A low value of the carbohydrate/nitrogen ratio thus appears to be characteristic of the beginning of the fruiting phase as distinct from flower development. Further evidence of the importance of the carbohydrate/nitrogen ratio in determining the formation of fruits in the apple was obtained by Hooker by manuring experiments, in which more fruit developed when the trees were supplied with nitrogenous manure in the spring so that the nitrogen content of the spurs was increased at the time of fruit-setting.

It was shown by Harvey and Murneek that removal of the leaves from the spurs just before the flower buds are differentiated reduces both the carbohydrate/nitrogen ratio and the formation of flower buds, a result completely in accord with the view already given of the significance of the carbohydrate/nitrogen ratio.

Observations on a number of other plants suggest that the relations between the value of the carbohydrate/nitrogen ratio on the one hand, and reproduction on the other, found to exist in tomatoes and apples, may be more or less general. Nightingale's results with buckwheat, radish, Soya bean and *Salvia*, in which the nitrogen content was varied by manuring, and the carbohydrate content by exposing the plants to different daily periods of illumination, were in general agreement with those already mentioned, as were also those of Gardner with the strawberry.

THE FORMATION OF FRUIT

While the change from purely vegetative growth to reproductive activity may thus in part be influenced by external conditions, it does not necessarily follow that the same conditions which induce the formation of flowers will bring about the ripening of fruit, and it is a matter of common knowledge that the formation of flowers is not necessarily followed by the production of fruit.

In algae with motile antherozoids, Bryophyta and Pteridophyta, fertilization of the ovum can only result if there is a sufficiency of water to enable the motile cell to swim to the oogonium or archegonium. In Bryophyta and Pteridophyta water also plays a part in the preparation of the antheridium and archegonium, both sexual organs absorbing water to a considerable extent. In the case of the antheridium the pressure produced results in the bursting of the organ with the release of the spermatozoids, the cell-walls of which have absorbed water and disappeared as mucilage. An apparently similar absorption of water by the walls of the canal cells of the archegonium results in the disappearance of these walls also, leaving a channel for the passage of the spermatozoid to the ovum.

Some substance which exerts a chemotactic influence (see Chapter XXIV) on the spermatozoids appears to be present in the mucilage formed in the neck canal of the archegonium in many cases. In mosses the substance appears to be sucrose; at any rate, it was shown by Pfeffer that the spermatozoids of mosses move towards higher concentrations of this substance, while the spermatozoids of ferns are attracted by malic acid and its salts, even by a solution as dilute as 0·001 per cent., and to a less extent by maleic acid. Buller showed that a number of other substances, as, for example, potassium nitrate, potassium chloride, potassium bromide, sodium formate and potassium acetate, if present in sufficiently high, but not too high, concentration, would also attract fern spermatozoids. No substance, however, was found to have anything approaching the effect of malic acid. Sugars, alcohols, asparagine and urea have no attractive effect on fern spermatozoids. It is possible that malic acid plays a similar part in bringing about fertilization in *Selaginella* as in the ferns, and it is also possible that the direction of movement of the spermatozoids in other cases is the result of similar chemotactic stimulation by substances so far unidentified.

A very interesting observation by Verkaik calls for mention at this point. It is well known that *Mucor mucedo* comprises two strains, designated by the symbols + and –, and that zygospores normally form only by the union of a hypha of one strain with a hypha of the other. Verkaik found that zygospore formation could be induced in a culture containing the + strain by the addition to it of a piece of agar taken from a culture of the – strain but containing no mycelium. The conclusion to be drawn is that a substance diffuses from the – strain which induces zygospore formation in the hyphae of the + strain; it thus might be regarded as a 'reproduction hormone' analogous to the now well-known growth hormones.

In higher plants the pollen must be carried from the anther to the stigmatic surface, or, in other words, pollination must be effected, as a necessary prelude to fertilization. In some plants self-pollination, in which pollen from the same flower, or from other flowers of the same individual, reaches the stigmatic surface, takes place, and is also effective for subsequent fertilization, but cross-pollination, in which pollen from some other plant is carried to the stigmatic surface, is generally regarded as more usual and also more effective in ensuring fertilization and the production of vigorous plants in the next generation. The part played by various agents, mainly wind and insects, and less frequently water, small birds, snails and so on, in carrying the pollen is extremely well known. Obviously a failure of these agents to function normally may have a very considerable effect on pollination and fertilization. Thus, where wind is the normal agent of pollination, a continuous spell of very still weather during the time of maturation of the pollen, may reduce pollination very considerably. Even more definite in its effects may be the inhibition of the movement of insects that are normal carriers of pollen. Thus among the principal pollinating insects are bees, but these fail to visit flowers in very cold or very windy weather. External conditions may therefore influence pollination and hence fertilization and fruit formation, both directly and indirectly.

In a few species and varieties, it is not necessary for the pollen to germinate after reaching the stigmatic surface, or even for pollination to take place at all, in order that the plant may produce viable seeds. The most notable cases of this kind occur in *Antennaria alpina*, in many species of *Alchemilla* belonging to the subgenus *Eualchemilla*, in *Thalictrum purpurascens*, *Hieracium* spp., *Atamosco* and *Burmannia*. In *Antennaria alpina* few male plants are produced and in those that are formed pollen production is weak or wanting altogether. No reduction division takes place in the formation of megaspores and the egg cell thus contains the sporophytic or diploid number of chromosomes. This latter state of affairs is found in the other species mentioned above, though in some of them, as in *Thalictrum purpurascens*, under natural conditions embryos arise both through fertilization and parthenogenetically. In some cases the embryo may develop as a form of budding from the sporophytic tissue of the nucellus.

Apart from such cases of so-called parthenogenesis or apogamy, development of the pollen tube from the pollen grain is, of course, essential if fertilization is to be effected. But because the pollen reaches the stigmatic surface it does not follow that germination of the former will take place. The stigmatic surface often secretes a fluid in which the pollen grains are held, and when the grain germinates it absorbs water from this fluid and from the surface cells of the stigma, and hence if dry weather, and particularly dry winds, have brought about too great an evaporation of water from the surface or surface cells, absorption of water by the pollen may not be possible. The presence of enough water is clearly, however, only one factor influencing the germination of pollen grains. Thus while the pollen grains of many species will germinate in artificially prepared culture media such as sugar solutions, in other species it has not so far been found possible to germinate the grains away from the stigma or its excretion or in absence of some particular substance. The germination, for instance, of pollen of members of the Ericaceae takes place, according to Molisch, only in presence of malic acid. The pollen may contain, besides sugar, other substances including organic acids, and Branscheidt has recently shown that both the germination of the grain and the growth of the pollen tube are dependent on the hydrogen-ion concentration, although the hydrogen-ion concentration for maximum germination may be affected by the presence of other substances.

In many cases of self-sterility the pollen from a particular plant will not germinate on the stigma of a flower of the same plant, although pollen from other plants of the same variety or species will germinate readily. There are also cases in which a whole variety is self-sterile, inasmuch as the pollen of the variety is incapable of germinating on the stigmas of any plant of that variety. In yet other cases different varieties may be cross-sterile, the pollen of one variety failing to germinate on the stigmas of certain other varieties of the species. As we shall see, failure of pollen to germinate is not the sole cause, but certainly one very definite source of sterility.

Much attention has been given to these questions in recent years, and some advance has been made towards understanding the reasons for sterility of this type in some cases, but the whole matter is still far from clear. It appears that in apples, for example, the stigmatic secretion of some varieties has a definite inhibiting effect on the germination of the pollen of some other varieties. Such inhibition may be related to hydrogen-ion concentration or to the concentration or presence of some other ion or substance, but what this is has not been determined in any example. Actual self-sterility, depending on the failure of the plant's own pollen to germinate on the stigma, must presumably be due to a similar state of affairs. From his researches on self-sterility in *Petunia violacea* Yasuda concluded that the substances which inhibit germination of the plant's own pollen in this species are absent in the unopened flower but develop in the gynaecium before the pollen is shed, and then decrease as the gynaecium grows old. The inhibiting material can be extracted from the gynaecia with water, and it has been found by treating cultures of pollen with a preparation of the substance that not only does it inhibit the germination of the plant's own pollen, but it favours the development of the pollen from other plants.

It has generally been supposed that the substances inhibiting or favouring pollen germination are present in the pistil and its stigmatic excretion, but Branscheidt's work indicates that pollen itself may not be without effect in this matter, for this worker found that the filtrate from a suspension of pollen in a solution of sucrose influenced the germination of pollen of other varieties, although the action of the pollen is generally much less than that of the stigma. Thus, when pollen of a number of different varieties reaches one stigma, it may be that the pollen of one variety may excrete a substance inhibiting the development of the grains of another variety. What has been called the prepotency of alien pollen, that is, the readier germination of alien pollen as compared with pollen of the same individual or variety, when both are present together on one stigma, may be partly due to this influence of the pollen as well as to that of the stigma. Pollen and stigma of the same variety can exert contrary effects, one furthering germination and the other opposing it, which suggests that the differences between the excretions of pollen and stigma may be not only qualitative but also quantitative.

According to Branscheidt, active pollen and stigmatic excretions regularly contain phosphatides and lecithides, reducing and surface-active substances, but not sugar. Proteins and protein decomposition products and diastase are present in the pollen excretion.

Niethammer found that pollen free from sugar would not germinate in hanging-drop cultures, while with grains containing sugar the germination capacity increased with the sugar content. Grains with a higher content of ash constituents and of organic acids also exhibited a greater germination capacity than those in which the content of such substances was low. It is also possible that atmospheric conditions may affect the content of nutrient substances in some pollen grains and so influence their power of development. Nowinski found that starch disappeared from pollen of *Secale cereale* in damp air, while pollen of *Poa annua*, which in damp air was found to be devoid of starch, formed starch when grown on a sucrose solution. Pollen of *Urtica urens* and *Chelidonium majus* also produced carbohydrate reserve ('amyloid') when grown on sucrose solutions. Such a behaviour does not, however, appear to be general, for pollen of a number of species, including *Nymphaea semi-aperta, Nuphar lutea, Sinapis alba* and *Vicia faba*, did not form starch when grown on sugar solutions.

At the time of pollination the respiratory activity of flowers increases considerably. This was shown in 1907 by Jean White for a number of species, the results being summarized in Table LXXXIX. It will be observed that the greatest effects were observed in *Canna indica* and *Pelargonium zonale*, in which the respiration rate increased 6·2 and 5·8 times respectively after pollination. In other cases the effect, although not so marked, was very evident.

Table LXXXIX

Increase in Respiratory Activity of Gynaecium after Pollination

Species	Ratio of Respiration Rates of Pollinated and Unpollinated Gynaecia
Eucalyptus calophylla	1·6
Fuchsia serratifolia	1·6
Pelargonium zonale	5·8
Digitalis purpurea	1·6
Begonia semperflorens	4·0
Tecoma capensis	2·0
Lilium candidum	1·7
Canna indica	6·2
Agapanthus umbellatus	1·5

The germination of the pollen grain results in the development from it of the pollen tube through the tissues of stigma and style to the ovary. Here substances secreted in the tissues of the various parts of the ovary undoubtedly play a part in directing the course of the pollen tube, as first suggested by Molisch in 1889. Certain researches by Miyoshi have become classical. Pollen grains were sown on agar in which various parts of the ovary, including pieces of ovules, styles and stigmas were embedded. The pollen grains germinated and the resulting pollen tubes grew towards the pieces of ovary, the direction of growth being most markedly influenced by fully developed ovules, through the micropyles of which the pollen tubes grew. Among the substances to which the pollen tubes are similarly chemotropically sensitive (cf. Chapter XXIV), sugars (sucrose and glucose) appear particularly active. Miyoshi found no effect produced by ammonium phosphate, peptone and meat extract, whereas Lidforss found in *Narcissus tazetta* that proteins exerted a considerable chemotactic influence. The pollen tubes, according to Miyoshi, are also positively hydrotropic, growing away from a drier region towards a damper; the previous finding of Molisch that the pollen tubes are negatively aerotropic, growing away from air, may possibly be interpreted as a case of positive hydrotropism.

The presence of diastase in pollen grains has been recorded by a number of workers, and this no doubt serves to mobilize to sugar the carbohydrate present in the grains as starch. The importance of sugar for pollen germination and growth has already been mentioned. Reynolds Green also demonstrated the presence in some pollen grains of sucrase, which would serve for the mobilization of sucrose. It seems probable that these enzymes, and perhaps others as well, diffuse through the wall of the pollen tube into the tissues of the gynaecium and there mobilize reserve material which diffuses into the pollen tube. Having regard to the relative sizes of pollen grain and pollen tube by the time the latter has finished its growth it seems fairly obvious that the grain could not provide sufficient material for the full growth of the tube. The tube appears, in fact, to grow as a parasite in the tissues of the ovary.

The pollen tube grows principally through intercellular spaces, and it has been suggested that it forces its way through these spaces, enlarging them where necessary, simply by the hydrostatic pressure exerted by its contents. It is, however, the more general view that the penetration of the pollen tube through the tissues of the gynaecium is assisted by the excretion from the tube of the enzyme pectinase which brings about the dissolution of the pectin of the middle lamellae of the cells.

The rate of growth of the pollen tube is generally remarkably rapid, the embryo-sac being reached and fertilization occurring within 1 or 2 days. In some cases, however, the rate of growth is slower, the time between pollination and fertilization in birch, for example, being normally about a month and in some species of oak as long as a year.

The rate of growth may be expected to depend on the temperature, while there is at least a possibility that variations in the content of the cells of the gynaecium through which the tube passes may involve corresponding variations in the rate of growth of the tubes.

Niethammer has shown that in hanging-drop cultures the rate of growth of pollen tubes can be increased by the addition to the medium of certain substances such as salts of heavy metals, organic acids and alkaloids, while Branscheidt has shown that pollen tube growth can be affected by hydrogen-ion concentration and other factors in the same way as is the germination of the pollen grains. In some varieties of apple, for example, if both the plant's own pollen and that of other varieties are present, the normally more rapid growth of the pollen tube of the other varieties results in cross-fertilization. Such self-sterility probably results from inhibitory action similar to that already considered in relation to pollen germination.

Brink found that pollen of the sweet pea (*Lathyrus odoratus*) is very sensitive to the presence of inorganic salts. Grains grown in a sucrose solution to which had been added 0·0002 M sodium chloride develop tubes only 15 per cent. of the normal length, while a concentration of 0·01 M inhibits the germination of the grain. Magnesium chloride and barium chloride are about 15 times as toxic as sodium chloride. Experiments with pollen of *Nicotiana* showed that this also is highly sensitive to the presence of inorganic salts, although the action of the various salts may not be the same with the two species. Thus the growth of *Nicotiana* pollen tubes is actually increased by magnesium chloride in a concentration of 0·002 M. Calcium chloride in concentrations varying from 0·002 M to 0·02 M furthered the growth of *Lathyrus odoratus* pollen tubes, whereas this salt had a toxic effect on the growth of *Nicotiana* pollen. Brink thought that the favourable effects here mentioned might be due to a lowering of protoplasmic permeability and the consequent prevention of the external diffusion of solutes. He regarded his results as indicating that electrolytes present in the tissues of the style might serve to regulate the diffusion of substances through the plasma membrane of the pollen tube, and so markedly affect its growth.

FERTILIZATION

As regards the physiology of the fertilization process itself, little is known. When the pollen tube reaches the embryo-sac its wall appears to burst and the male cells and some other contents of the apical region of the tube pass into the embryo-sac. This bursting of the thin-walled pollen tube suggests a sudden absorption of water by the tube and stretching of the wall until it can no longer resist the hydrostatic pressure of the contents. The fusion of one of the male nuclei with the egg takes place almost at once, and is followed in many Angiosperms by that of the second male nucleus with one or both of the polar nuclei, although in some cases this fusion appears not to take place. While the fertilized egg cell develops into the embryo the secondary nucleus, either produced by fusion of the polar nuclei with a male nucleus, or by fusion of the two polar nuclei only, gives rise to the endosperm from which the developing embryo later derives its nourishment.

According to the observations made by D. M. Whitaker on *Fucus*, fertilization of the egg brings about a quite definite rise in its respiratory activity. Unfertilized eggs in the dark were found to use about 5·2 c. mm. of oxygen per hour per 10 c. mm. of eggs, while the oxygen consumption of spermatozoids amounted to as much as 25·5 c. mm. per hour per 10 c. mm. of sperms. The magnitude of this respiratory activity will become evident when it is observed that 10 c. mm. of thallus use 2·08 c. mm. of oxygen per hour. After fertilization the rate of oxygen absorption of the egg increased by 90 per cent., this higher rate being maintained for 13 or 14 hours, after which there was a further slight rise. This may correspond to the first division of the egg, which at the temperature employed, namely 18° C., occurred in about half the eggs under observation 15 hours after fertilization. Whether an increase in respiration rate accompanies fertilization in general can at present only be conjectured.

As mentioned above, subsequent to fertilization the fertilized egg and secondary nuclei, by a series of divisions, develop into the embryo and endosperm respectively. Generally the endosperm commences its growth more rapidly than the embryo, which only starts to grow comparatively rapidly after the endosperm has made considerable development. For the development of the endosperm the necessary material is derived from the surrounding nucellus, which in most cases practically disappears, although in a few cases it persists in whole or in part as perisperm, when physiologically it serves the same purpose as endosperm, that is, it forms a reservoir of food material to be utilized by the growing embryo. The latter continues its development up to a certain point, when it enters on a resting stage, the ovule containing it having become the seed. When this stage is reached the whole of the endosperm may have been utilized and the seed is described as exalbuminous, or endosperm may be left surrounding the embryo and the seed is said to be albuminous.

As is well known, the changes in the embryo-sac following on fertilization are accompanied by changes not only in the ovule but also in the outer tissues of the ovary and often in the floral receptacle. These involve, in general, an increase in size, and generally a change in consistency, the ovary-wall either developing sclerenchymatous layers and becoming hard, or else thickening and swelling and becoming succulent. Not only the size, but also the shape, may change, owing to different rates of growth in different parts, while pigments frequently develop and the fruit, instead of remaining green, develops various colours. These changes in external appearance and internal structure are accompanied by changes in chemical composition of which something will be said later.

PARTHENOCARPY

Although the changes in the tissue of the ovary and, sometimes, of the receptacle as well, involved in the formation of the fruit most frequently only take place when fertilization occurs, there are many cases in which they proceed whether fertilization occurs or not. Thus in 1849 Gaertner, in his work on hybridization, in the words of Charles Darwin, ' often insists that the flower of even utterly sterile hybrids, which do not produce any seed, generally yield perfect capsules or fruit, a fact which has likewise been repeatedly observed by Naudin with the Cucurbitaceae '. Darwin himself recorded a number of species which possess varieties forming seedless fruit, such as pears, grapes, figs, pineapples, bananas and so on. The development of fruit without visible seeds may exemplify one of two phenomena described respectively as parthenocarpy and non-parthenocarpic seedlessness. The former term is limited to the development of fruit without fertilization and consequently without development of the embryo (in which latter respect it differs from parthenogenesis), while by non-parthenocarpic seedlessness is meant the development of fruit without viable seeds as the result of fertilization being followed by abortion of the embryo.

Many well-known cases of true parthenocarpy occur among species of fruit cultivated for their edible qualities. In what has been called ' vegetative parthenocarpy ' the fruit develops without any pollination at all. Among such are a number of varieties of orange (seedless oranges), bananas, cucumbers and mulberries. In other cases pollination appears to be necessary for the development of the fruit although fertilization is never effected. Certain varieties of pear, and the Thompson seedless grape, are examples. Moreover, such parthenocarpic development has even been effected with pollen of an entirely different species. Thus Millardet obtained fruits of the grape (*Vitis vinifera*) by pollinating the stigmas with pollen of the Virginia creeper (*Ampelopsis hederacea*). Parthenocarpic development has even been effected in quite a number of cases by applying chemical substances (for example, magnesium sulphate and chloroform vapour), by mechanical irritation and by wounding. All cases of parthenocarpic development resulting from stimulation of any kind, whether by pollen or by other agents, has been called *stimulative parthenocarpy* by Gardner, Bradford and Hooker.

The development of parthenocarpic fruits may include the development of the outer tissues of the ovule so that, although no embryo is produced from the egg cell, the integuments of the ovule develop, with the result that the fruit contains what appear externally to be normal or imperfectly developed seeds. All gradations in these up to apparently normal seed structures have been observed in parthenocarpic fruits, but such seeds are, of course, non-viable.

The conditions favouring the development of parthenocarpic fruits are little understood. In species and varieties where fertilization is essential to the development of the fruit, a frequent event when fertilization fails to occur is the production of an abscission layer at the base of the flower, which in consequence is cut off from the nutrient supplies and falls. In parthenocarpic fruits obviously this abscission layer fails to form and the development of the fruit proceeds normally. In cases of stimulative parthenocarpy, the stimulus, by pollination or otherwise, apparently sets up changes which lead to the inhibition of abscission layer formation ; in vegetative parthenocarpy such inhibition results without the need of previous stimulation. It has been suggested that accumulation of organic nutrient material in the neighbourhood of the flowers may lead to continued growth of the fruiting structures with the inhibition of abscission layer formation.

Non-parthenocarpic seedlessness, due to embryo abortion, can result from the operation of external factors, such as low temperature, which brings about a lowering of the rate, or even complete cessation of embryonic growth. Conditions of food supply can also cause embryo abortion, and while presumably poor nutrition will have this result, curiously enough abortion has been effected in tomatoes by excessive nutrition. It may be that the relation between the available quantities of inorganic and organic nutrients may be important in this connexion, but no definite information on the point has yet been obtained.

In the case of many-seeded fruits there is often a relation between the number and position of viable seeds, and the size and shape of the fruits. In some varieties of apples, pears and strawberries the absence of viable seeds on one side of the fruit involves less development of the fruit on that side, with the consequence that the fruit has a one-sided appearance. In some species of apple there is a greater tendency for premature dropping of the fruit the fewer the number of fertilized ovules. No general rules, however, can be formulated in this connexion, for in some varieties seedless fruits are as large, or even larger than normal fruits of the same variety containing viable seeds.

STERILITY AND FERTILITY

Sterility is the failure to produce viable seeds and is contrasted with fertility, the condition in which such seeds are produced. Much information relating to sterility was collected by Charles Darwin in his work on the *Variation of Animals and Plants under Domestication*, and the conditions producing it he indicated in many cases. Darwin recognized sterility to be due in many instances to evolutionary causes, the self-sterility exhibited by many species being related to the development of devices to further cross-fertilization. The frequency of sterility of hybrids was emphasized. But Darwin also recognized that external conditions might affect sterility and fertility. Among such conditions he mentioned excess of manure, extreme poverty of soil, temperature of soil and time of watering, all of which may influence the formation of viable seeds.

Of later years A. B. Stout has classified the various kinds of sterility into three main types according as it is due to (1) impotence, (2) incompatibility and (3) embryo abortion.

Sterility from impotence results from the failure of either pollen or embryo-sacs to form or develop normally.[1] Impotence is complete when both pollen and embryo-sacs fail to develop, and partial only when one or other of the sex organs are affected. Many cases of complete impotence are recorded by Darwin. Among these are 'double' flowers in which both stamens and carpels are replaced by petals. Doubleness appears to be induced principally by long-continued cultivation in rich soil, although poverty of soil may produce the same result, possibly in different species. Darwin described completely double flowers on plants of *Gentiana amarella* growing on poor chalky soil, and he observed a tendency to doubleness in *Ranunculus repens*, *Aesculus pavia* and *Staphylea* growing under unfavourable conditions. When plants become sterile through hybridization there is, according to both Gaertner and Darwin, a strong tendency for the plants to become double. Luxuriance of vegetative growth, which, as pointed out earlier, may result from excessive inorganic nutrition, may involve absence of flowers altogether. The opinion has also been expressed by horticulturists that long-continued propagation by vegetative means, such as by bulbs, tubers, cuttings and so on, also leads to the production

[1] It is to be observed that Charles Darwin used the term 'self-impotence' as synonymous with self-sterility.

of barren flowers or to a failure to produce flowers at all. According to Darwin, there is no doubt that many plants so propagated are sterile, but whether the mode of propagation is responsible for the sterility he considered there was insufficient evidence to decide.

Partial impotence is met with most commonly in species where the stamens and carpels are borne on separate flowers or on separate plants, the latter being the most effective device possible to prevent self-fertilization. In some species all transitions between perfect flowers and perfect unisexuality are met with. In many hybrids the anthers appear as shrivelled brown, or toughish structures containing no viable pollen ; members of the Caryophyllaceae, Liliaceae and Ericaceae suffer most from this defect. In such 'contabescent' plants, as Gaertner called them, the carpels are not usually affected. The reverse state of affairs, in which the stamens develop normally while the carpels do not, also occurs ; cases of this were described by Gaertner in *Dianthus japonicus*, *Passiflora* and *Nicotiana*. It is also common in a number of plum varieties, a proportion of the carpels developing imperfectly or aborting completely.

Doubling, again, may only affect the stamens and not the carpels, in which case a condition of partial impotence results.

In quite a number of plants the stamens and carpels may appear to be normally formed, but actually the pollen or embryo-sacs fail to develop properly. Here again every transition is found between plants in which only a small proportion of the pollen is non-viable, to plants in which no viable pollen is produced. Plants in which abortion of pollen is common are varieties of grape, blackberry, strawberry and potato. It should be mentioned that a certain proportion of defective pollen may not be sufficient to interfere with fertility, and such plants, although possessing the character of partial sterility to a certain degree, are nevertheless not necessarily sterile. Abortion of the embryo-sac before fertilization, in what appear otherwise to be normal ovules, has been observed in varieties of orange and plum.

Sterility from physiological incompatibility obtains when functional pollen grains and embro-sacs are produced, but nevertheless fertilization cannot be effected. The self-sterility of many plants falls into this category. The case of *Corydalis cava* described by Hildebrand at the International Horticultural Congress in London in 1866 and referred to by Darwin, may be cited as representative of this kind. A number of flowers of this species were pollinated with pollen from other plants of the same species ; out of 63 flowers so treated 58 produced capsules containing on an average 4·5 seeds each. But when 16 flowers in the same inflorescence were inter-pollinated, only three capsules were formed, of which only one contained viable seeds and these were only two in number. Finally, out of 84 flowers self-pollinated no capsules whatever were formed, although the pollen was observed to germinate and the pollen tubes to enter the stigma.

Many observations similar to this have been made with other species.

and particularly with orchids, and in some of these latter plants the pollen and stigma of the same individual appear to exert a toxic action on one another, for both became discoloured and disintegrate.

In a Bohemian variety of apple (' Boemischer Rosenapfel '), which is self-sterile, Osterwalder reported that the sterility resulted from the growth of the pollen tube being insufficient to reach the ovule. In cross-pollinated flowers the pollen tubes reached the ovary about 2 days after pollination, but in self-pollinated flowers the tubes after this time had only penetrated the styles to a depth of 2 to 4 millimetres, and had formed curious club-shaped swellings at the tip, while even after 16 days there was no evidence of fertilization. On the other hand, L. J. Knight found that in the variety Rome Beauty there is a slowing down in the rate of growth of pollen tubes on the stigma of the same plant, but no inhibition of their growth, and he concluded that an important factor in self-sterility of this variety is the relatively slow growth of the pollen tubes. In observations on another self-sterile variety of apple, Yellow Bellflower, Namikawa could not detect even any retardation of growth as a result of self-pollination, and moreover suspected that Osterwalder's conclusions were based on the observation of abnormal cases, for in Yellow Bellflower the pollen tube grows normally outside the stylar tissue, and, exceptionally, when the tube penetrates the style, its growth is very slow and the tip may become swollen. Here, at any rate, the incompatibility is not to be traced to inhibition or retardation of pollen tube growth in the style. Nevertheless, too slow a rate of pollen tube growth would result in sterility, for, unless fertilization occurs within a certain time, an abscission layer forms, sometimes at the base of the style, sometimes at the base of the ovary, or even at the base of the flower, and the flower, or part of it containing the pollen tube, is shed. The slow rate of pollen tube growth in the case of self-pollination as compared with cross-pollination, probably operative in other heterostyled plants, has been observed in heterostyled buckwheat, and is held to be the factor resulting in self-sterility in *Nicotiana*.

What must be regarded as a phenomenon closely allied to incompatibility is the sterility resulting from too early or too late pollination. A number of cases are on record in which penetration of pollen tubes through immature styles results in the falling of the flowers, while the same may happen if pollination is considerably delayed after maturity of the stigma.

Sterility resulting from embryo abortion generally arises from inadequate nutrition of the fertilized egg or of the young embryo arising from it. In dealing with non-parthenocarpic seedlessness it has been mentioned that embryo abortion may give rise to that condition, and that low temperature is an effective factor in producing it, but it probably sets up sterility in some varieties apart from the influence of external conditions. Thus, in one variety of cherry, in spite of apparently favourable external conditions, sometimes 95 per cent. of the fruits may be seedless, owing to abortion of the embryo, while sterility of this kind may reach considerable proportions in other species.

External factors may often induce sterility, and cases in which they have been shown to bring about a condition of impotence by affecting the development of stamens and carpels have already been mentioned. The external factors which may influence development of fruit are, as well as supply of nutrient material, light, temperature, wind, water supply and injury.

The supply of nutrient material, besides affecting fertility through the development of sex organs, may obviously adversely influence the development of fruit if the supply of any particular substance is insufficient. For the rapid increase in size which takes place during the development of the fruit, reserves of material in the plant are no doubt largely drawn on, so that the supply of nutrients, not only during the period of actual fruit development, but prior to this, may be of importance in determining the formation of fruit.

For a similar reason, light, inasmuch as it is a factor necessary for the production of organic food reserves, might be a factor in fruit development. There appears, however, to be no definite evidence that this is so. That light conditions may bring about sterility seems, however, to be true of a number of species, as, for instance, in the rose bay (*Epilobium angustifolium*), where the flower buds are shed unopened when the plants are grown in too weak light. The relation between length of day and failure to produce flowers has already been discussed.

Temperature may be a determining factor in the production of seed. The time taken for the pollen grain to germinate, and the rate of growth of the pollen tube, are both influenced by temperature, and at a temperature below a certain value abscission of the style, ovary or flower, may take place before fertilization is effected.

Temperature may also have an indirect effect on fertilization by influencing the movements of pollinating insects, which are mostly active only above a certain temperature. In some plants normally pollinated by insects, self-pollination is still possible if insect-pollination fails, but in many there are devices, such as the monoecious and dioecious conditions, heterostyly, protandry and protogyny, which prevent it, while incompatibility may also be operative where self-pollination does occur. In varieties of edible fruits, low temperature, through inhibiting the movements of bees or other pollinating insects, may effectively prevent pollination and fertilization.

Wind, as a factor in preventing fertilization, may act in several ways. In the first place, it may bring about evaporation of water from the surface of the stigma, so that this, instead of being covered with a sticky fluid which holds the pollen, becomes dry so that pollen grains no longer adhere to it. In the second place it may damage the flowers by beating them against one another or against the stems of the plant, and thirdly it may affect pollination either favourably or adversely. It, of course, favours pollination of wind pollinated flowers, but has an adverse effect on many insect pollinated flowers, for just as many of these insect visitors of flowers do no work if the temperature is too low, so also they remain inactive if the wind is too high.

The effect of water supply in promoting premature falling of leaves has been dealt with in the previous chapter. The same factors that induce the falling of leaves will also bring about the falling of flowers and fruit. Thus Molisch records that the falling of flower buds of *Camellia* and other plants generally results from shortage of water in the soil or from too hot and dry an atmosphere. According to the same authority, a prolonged dry period of summer weather will induce the falling of immature fruit of many species.

Injury of flowers in a high wind has already been mentioned. Heavy rain may have the same effect. Attacks by insects, fungi and bacteria may also render fertilization impossible by damaging essential parts of the flower or fruit. Smuts, such as *Ustilago antherarum* on *Lychnis*, that attack the stamens of plants, and ergot (*Claviceps purpurea*) that attacks the young ovary of rye, are well-known examples of fungi of this kind. Damage to flowers may also result through the action of chemical substances, which might, for example, have been brought into contact with the flowers in the horticultural practice of spraying, or which might be present in the atmosphere near towns or industrial works.

THE PREMATURE SHEDDING OF FRUIT

The shedding of fruit prematurely must in general be due to one of the factors mentioned in the course of the preceding section of this chapter. Such shedding of fruit is a commercial problem of the fruit-grower, and consequently most information with regard to the conditions inducing it concern edible fruits. Dorsey's investigations on the plum have shown that here premature falling of flowers or fruit occurs at three successive times; the first, soon after flowering, is of flowers containing defective carpels; the second is of flowers with normally developed carpels, in which pollination may have occurred but in which the ovules are unfertilized; the third is of flowers with ovaries containing fertilized ovules in which embryo development has commenced and then stopped prematurely. These three groups of falling flowers and fruits thus correspond respectively to the three types of sterility resulting from impotence, incompatibility and embryo abortion, though the last two may perhaps be traceable to the external factors of temperature and nutrition. In the pecan (*Carya olivaeformis*), a member of the Juglandaceae, more than three-quarters of the total shedding of immature fruit occurs about 4 weeks after the time of pollination. This is the time that elapses between pollination and fertilization, and Adriance, who has examined the factors influencing the setting of fruit in this species, supposes that the shedding of fruit is here due to lack of fertilization following on absence of pollination. It thus corresponds to the second of the falls observed by Dorsey in the plum.

The shedding of immature fruit by apple trees has been examined by Murneek, who concluded that there are four waves of such shedding in this species, occurring at intervals of approximately 12 to 14 days, although the first two waves may overlap. The majority of the ovaries shed in the first ' drop ' appear to be abnormal, and in some cases the gametes of both sexes appear to be non-functional. But some of the fruits contain embryos, so that the shedding of some fruit, even in the first ' drop ' is related to embryo abortion.

The fruits which fall in the second wave show a slight development of the torus, which suggests that they have been stimulated by fertilization or pollination, but it is not clear whether they all contain embryos, although this seems to be the case with the fruit shed in the later drops.

It would seem likely that the shedding of fruit in the later drops, and to a certain extent in the earlier ones, is related to competition among the fruits for food, and is determined in part by food supply.

CHEMICAL CHANGES DURING FRUIT DEVELOPMENT

The morphological changes which are involved in the development of fruit are accompanied by very definite changes in the chemical composition of the tissues. This is, of course, very evident in fruits having an edible value, such as apples, pears, oranges, raspberries, strawberries and so on.

The chief chemical constituents of most fruits are carbohydrates, nitrogenous substances and organic acids, all of which may be present in considerable quantity. Edible fruits also contain small quantities of substances, probably aldehydes and esters, to which their peculiar flavour and aroma are due, and on the presence of which their value as articles of food largely depends. The carbohydrates are those usually found in plants, namely, glucose, fructose and other monosaccharides, sucrose and perhaps maltose and other disaccharides, starch, dextrin, cellulose and possibly other polysaccharides. The chief organic acids are malic, citric and tartaric, the first named being the chief acid in the apple, citric acid that in the orange, lemon and other citrous fruits, while tartaric acid forms the principal acid of the grape.

The chemical changes taking place during the ripening of fruit have been examined in a number of cases. Perhaps the most complete examination of these changes is that made by Miss Archbold on two varieties of apple, namely Worcester Pearmain and Bramley's Seedling, analyses having been made of the fruits taken at intervals beginning from the time when most of the petals had fallen but while the discoloured stamens were still attached. As regards carbohydrates, no starch was found to be present in either variety during the first 3 weeks of development. Starch then begins to appear and increases in amount until a maximum is reached in about 5 to 6 weeks in Worcester Pearmain and 8 to 10 weeks in Bramley's Seedling; the maxima in the two varieties were found to be about 2 per cent. in the former and round about 1·3 per cent. in the latter. During the next stage of development the starch content was found to fall and to reach zero in about another 6 weeks, the time when the fruit is normally gathered in Bramley's Seedling, or a little after this in Worcester Pearmain.

The total sugar content at the commencement of fruit development was found to be about 1 per cent.; the content then rises continuously, reaching a maximum of about 9 to 12 per cent. at about the time the starch disappears and growth ceases. Of the individual sugars, glucose is initially present in greater concentration than the others (in Worcester Pearmain 0·86 per cent. as compared with 0·32 per cent. fructose and 0·48 per cent. sucrose; in Bramley's Seedling 0·81 per cent. as compared with 0·27 per cent. fructose and 0·19 per cent. sucrose). Apart from minor variations the amount of each of these three sugars increases according to an exponential relation until starch appears. During the

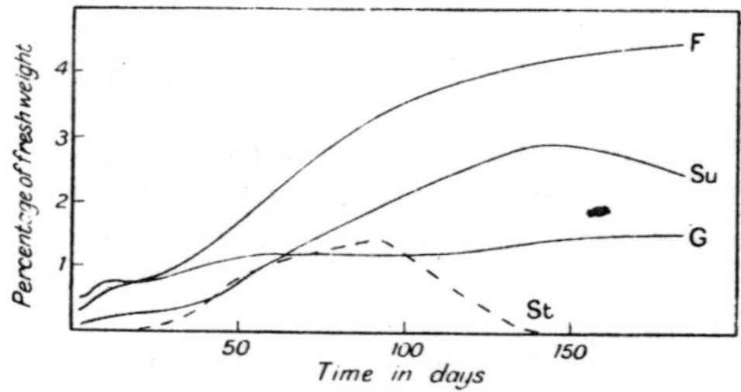

Fig. 48.—Curves to show the changes in sugar and starch content of Bramley Seedling apple during ripening and storage
(After H. K. Archbold)

period of increasing starch concentration the total concentration of sugar increases at an approximately constant rate, but of the separate sugars the concentration of glucose remains practically constant while that of fructose increases rapidly so that in the mature fruit this sugar is much in excess of glucose. Sucrose also increases in concentration, reaching a maximum in Worcester Pearmain some time before the disappearance of starch, and in Bramley's Seedling just before this occurs. The changes in the content of starch and sugars in Bramley's Seedling apples are shown graphically in Fig. 48. It will be observed that hydrolysis of starch in the later stages of development can account for only a small proportion of the sugars formed.

The concentration of acids in the apple rises during the first 3 weeks, reaching a maximum just after the appearance of starch, the values of the maxima in Worcester Pearmain and Bramley's Seedling being respectively about 1·3 and 2 per cent. The acid concentration then slowly falls and by the time the starch has disappeared has only about half the value of the maximum.

During storage after removal from the tree there is a continued loss of dry weight as a result of respiration. Any starch left in the fruit first disappears and with the disappearance of starch the sucrose concentration commences to fall, presumably through inversion into glucose and fructose. At first the concentration of hexose sugars increases, indicating that the inversion of sucrose exceeds the loss of hexose by respiration. As time goes on the rate of inversion of sucrose declines and ultimately produces less hexose than is lost, and consequently the concentration of hexose also falls. The changes in concentration of hexose are chiefly in fructose, for the concentration of glucose remains practically constant throughout the storage period. Acid concentration falls continuously. Whether it is respired directly or whether it is converted first into sugar is not known.

The changes taking place during the ripening of the fruit of the tomato have been examined by Sands. Here throughout the ripening period there is a gradual decrease in the content of starch and pentosans, while sugars and acids increase. Thus in fruit 14 days old the concentration of sugar was found to be 25·7 per cent., while in ripe fruit it was as much as 48 per cent. of the dry matter. In the same period the starch content decreased from 15·8 per cent. to 2·7 per cent.

A slow decrease in acid content of fleshy fruits during ripening has been observed in an American variety of apple by Bigelow, Gore and Howard, in Bartlett pears by Magness, in peaches by Bigelow and Gore, and in grape fruit by Hawkins and Magness. Thus in general fleshy fruit with a definite acid content becomes sweeter as it ripens, both on account of increase in sugar content and of decrease in acid.

A different type of fleshy fruit is the banana, for in this the acid content is presumably negligible. Analyses of this fruit made by Gore show that as it ripens the starch, both of peel and of pulp, decreases, while the sugar content increases. Some of Gore's results are summarized in Table XC.

Table XC

Changes in Starch and Sugar Content of the Pulp of the Banana Fruit during Ripening

Date	State of Fruit	Concentration in per cent. of Fresh Weight of Pulp	
		Sugar (as invert sugars)	Starch
May 2 . .	Green	4·35	19·01
May 3 . .	Slightly yellowing	9·65	13·96
May 4 . .	,, ,,	15·00	8·62
May 6 . .	Yellow	19·65	3·66
May 8 . .	,,	20·08	1·73
May 10 . .	,,	20·61	1·15
May 13 . .	,,	20·05	0·87
May 16 . .	Brown	18·21	1·01

It will be observed that there is a loss in total carbohydrate, attributed to respiration. That only carbohydrate is utilized for this purpose is indicated by Olney's determinations of the respiratory coefficient of bananas during ripening. He found the coefficient remains at unity throughout the ripening period. It will also be noted that the banana is a fruit which, in comparison with the apple, ripens very rapidly.

In many non-fleshy fruits, the seeds occupy a relatively greater proportion of the fruit than is the case with those fleshy fruits discussed above, in which, moreover, plant breeding has been directed to increasing the quantity of flesh in proportion to the quantity of seeds. In non-fleshy fruits the seeds are very frequently characterized by a high content of fat which develops from carbohydrate during the maturation period. One effect of this is that the respiratory quotient of such seeds is greater than unity, owing to the fact that in the formation of fat from carbohydrate oxygen is released ; this is utilized in respiration, and so only a portion of the oxygen required to oxidize carbohydrate is absorbed from outside the plant and determined when measurements are made of the respiratory quotient. As an example of this may be quoted Gerber's observation that the respiratory quotient during the maturation of flax is 1·22.

HYBRID VIGOUR

The increase in vigour of plants produced by crossing different individuals of the same variety, or different varieties of the same species has been long recognized and many instances of it were described by Charles Darwin. Current theories of hybrid vigour are chiefly based on Keeble and Pellew's hypothesis that this vigour is due to the bringing together in the hybrid of complementary dominant factors. A contribution to our knowledge of the physiology of hybrid vigour in maize has been made by Ashby. Maize is a plant in which this phenomenon is strikingly exhibited, and when two strains which had been inbred for years and the first cross between them were grown together, the more vigorous growth of the hybrid was strikingly manifest. Ashby found that the hybrid did not differ from its more vigorous parent in regard to relative growth rate, size of cells, photosynthetic activity of leaves, or the time at which the growth rate begins to lessen. The only physiological differences between the hybrid and its more vigorous parent were an increased percentage of germination in the case of the seeds produced on crossing, and a greater initial weight of embryo. It is apparently this greater initial weight of the embryo which gives an advantage to the plant, an advantage which is maintained throughout life, since the relative growth rate remains the same as that of the more vigorous parent. The increased vigour of the hybrid must then depend on some process involved between fertilization and embryo formation. From observations on reciprocal crosses Ashby suggests that the increased weight of the embryo is largely determined by the influence of the female parent during the development of the embryo. The increased size of the hybrid embryo might be due to earlier division of the zygote, to an increased rate of growth, or to a longer period of growth before the resting period of the seed sets in. At present no information is available to decide which of these causes is operative.

REFERENCES

DARWIN, C., *The Variation of Animals and Plants under Domestication*. London, 1868.

GARDNER, V. R., BRADFORD, F. C., and HOOKER, H. D., *The Fundamentals of Fruit Production*. New York, 1922.

BOOK IV

IRRITABILITY AND MOVEMENT

CHAPTER XIX

THE PRINCIPLES OF IRRITABILITY

In the discussion of Plant Metabolism which formed the subject of Book II it has been pointed out that every plant process depends for its existence and magnitude on a number of factors or conditions, all of which are equally necessary. Now there have for long been known phenomena in which changes in such conditions have produced alterations in the vital processes which appear out of all proportion, or not in any way clearly related, to the change in the conditions. Two well-known cases of such phenomena may be taken as examples. When a seedling broad bean, possessing a main root which has been growing vertically downwards, is displaced through a right angle so that the axis of the main root is now horizontal, the cells on the upper side of the root grow more rapidly than those on the lower side, with the result that the root becomes curved and the tip of the root again comes to point vertically downwards. Here a change in a single condition, namely, the spatial relation of the root axis to the earth's radius, results in a differential rate of growth on the two sides of the root. Again, when a tendril of a sweet pea is brought into contact with a solid support, such as a twig, the side remote from the twig grows very much more rapidly than the side in contact with the twig, with the result that the tendril coils round the twig. Here an alteration in the environment of the tendril produced by the replacement of air on one side of it by a solid body, results also in a differential rate of growth on the two sides of the tendril. To phenomena of this kind, in which an alteration in the external conditions gives rise to changes in the life processes of the plant involving energy transformation out of all proportion to that involved in the changes in the conditions, is applied the term irritability. The change in the external condition of the plant is called the *stimulus*, and the alteration in the life processes of the plant produced by the stimulus is called the *response*.

Verworn, to whom we owe the most reasoned discussion of the subject of irritability of living tissue in general, defined stimulus as 'every alteration in the external vital conditions', and stated that 'the irritability of living substance consists in the capability to respond to stimuli by changes of the vital processes'. This definition of stimulus would include among stimuli a change in carbon dioxide concentration of the atmosphere which, if the carbon dioxide were a limiting factor, might produce a change in the rate of photosynthesis proportional to that in carbon dioxide concentration. Such a change producing such a response would not constitute a stimulus, but it is conceivable that a change in carbon dioxide concentration might produce a response of some kind as a result of the mere change in the condition and not in direct nutritional relationship to the magnitude of the factor.

It may be noted that in this definition the conditions are limited to those external to the plant, and to *vital* conditions, that is, those that affect the life of the plant. There may be external conditions as, for example, the concentration of various inert gases in the atmosphere surrounding a plant, which have no effect on it. Changes in the value of these can have no influence on any plant process and so cannot act as stimuli. With regard to the limitation of stimuli to changes in *external* conditions, we are, however, faced with a difficulty, for there is no logical reason for differentiating between external and internal conditions in this matter. Thus concentration of chlorophyll, an internal factor, may be as important a factor for photosynthesis as intensity of illumination, an external one, and similarly changes in such an internal factor might act as a stimulus just as do changes in an external factor. The difficulty arises from the fact that the organism is in a state of continual change, which is summed up in the word development, and consequently can be regarded as in a state of continued stimulation from within. The limitation of stimuli to changes in external conditions is thus a practical one, but it is possible that in certain cases internal stimuli, as, for example, change in the concentration of a cell constituent, might be recognizable.

Stimuli possess certain characteristics. In the first place, since stimuli consist in changes in vital external conditions, they can, theoretically at any rate, be of as many kinds as there are such conditions. There are thus stimuli due to changes in illumination, temperature, pressure, composition of the external medium, electrical conditions and so on. Secondly, a stimulus may be positive or negative, that is, it may consist of an increase in the value of a condition or a decrease in its value. Thirdly, the stimulus may be directional or equally diffused all round the plant or plant organ. Stimuli may be characterized fourthly by their intensities, and fifthly by the length of time over which they act. Sixthly, there is the question of the nature of the response. Next, the response to the stimulus generally takes some time to become manifest, and hence it can be assumed that the stimulus gives rise to a chain of reactions which follow a definite course. Lastly, the seat of the visible response may not be the same as the place stimulated, and hence a conduction of stimulation, or rather of the excitation, that is, of change in the tissues, through the plant may take place. These various characteristics of stimuli will now be dealt with in more detail.

Fig. 60.—Photographs of a plant of *Oxalis* showing the leaflets in (above) the light position, (below) the dark position

To face page 522

KINDS OF STIMULI

There are a number of conditions of the environment which are necessary to the life of every plant, or to one or other of certain groups of plants. These include carbon dioxide, oxygen, various mineral substances, light, a certain range of temperature, and a certain range of osmotic pressure of the aqueous surroundings. We also know that gravity constitutes a vital factor in the life of the plant. Apart from these essential conditions, others may influence plant processes. Thus many substances without which a plant can grow normally can nevertheless influence its growth. Electrical conditions also can influence the growth of plants. Wounding the tissues of a plant may also act as a stimulus. It is thus possible to distinguish a number of different kinds of stimuli, of which the most important for the student of plant physiology are the following:

1. *Gravitational.* Here the pull of gravity in relation to the plant undergoes a change (produced actually by changes in the position of the plant organ). The response practically always has the effect of bringing the stimulated plant organ into the same relation to the direction of the gravitational pull as it occupied before stimulation took place.

2. *Photic.* Stimulation produced by change in illumination may result from change in the intensity of the light or from changes in its composition. Moreover, a change in the intensity of light of one wave-length may produce a different response from the same intensity change of light with a different wave-length.

3. *Thermic.* Marked response to changes in temperature has been observed with a few plants, but in general temperature appears to play a much smaller part as a stimulus than light and gravity.

4. *Chemical.* Many chemical substances have been observed to act as plant stimuli, and in particular these appear normally to play a part in the life of free-moving cells in lower plants. The movement of the spermatozoids of mosses towards a higher concentration of sugar, and those of ferns towards a higher concentration of malic acid, are the best known of such responses.

5. *Contact.* The response of certain plant organs to contact, as in the case of the tendril that coils round a support, presents perhaps one of the most remarkable types of stimulation. Although it involves a change in the composition of the external medium, the contact stimulus appears to be quite different from chemical stimuli. Stimulation by contact only results when the organ comes into contact with a solid support; it is not effected by contact with liquid or gases and thus is not induced by a jet of water, by rain, or by wind.

6. *Shock.* On the other hand, stimulation can result in a number of cases by the application of sudden pressure, such as that of a blow, and unlike the case of contact irritability, the pressure of liquid or gas is also effective in producing a response. Irritability to such stimulation

is said to be seismonic, and the plant or organ is said to be sensitive to shock.

7. *Electrical.* While animal physiologists have investigated in great detail the stimulus produced by changes in the electrical conditions of the environment, not much is known about the response of plants to such changes. It is highly probable, however, that such effects exist and may be of biological significance.

POSITIVE AND NEGATIVE STIMULATION

In most cases it is usual to think of stimuli as positive, that is, as consisting in an increase in the value of the condition, but a stimulus can equally well be negative. To take a specific example, an increase in temperature can bring about the opening of the flowers of tulip and crocus, but a fall in temperature can bring about the closing of an open flower. Similarly with other stimuli, a decrease in value can bring about a response just as much as can an increase in the value of the condition.

DIRECTIONAL AND DIFFUSE STIMULI

The vast majority of stimuli that have been investigated in plant physiology produce a response that involves movement or, at any rate, curvature of the stimulated organ. Such movements or curvatures are differentiated into *tropic* and *nastic*, according as the stimulus is unilateral, that is, directional, as when a plant or organ is illuminated on one side, or diffuse, as when the light intensity is uniformly increased or diminished all round the plant or organ. Thus a curvature of a shoot towards the more illuminated side is a phototropic curvature, while the opening or closing of flowers with change of light intensity all round is a photonastic one.

INTENSITY OF STIMULATION

Provided the time of action and other conditions are constant, the response depends, within limits, on the intensity of the stimulus. In most cases a measurable response is only observable when the intensity of stimulation reaches a certain value. This value of the stimulus necessary to produce a minimum observable response is termed the *threshold* value of the stimulus. With higher values of the intensity of the stimulus the response increases until a *maximum value* of the stimulus is reached above which no further increase in the response is produced. The terms sub-threshold, sub-maximal and super-maximal are applied to values of the stimulus which are respectively below the threshold value, above the threshold value but below the maximal value, and above the maximal value.

It will be observed that the threshold value must to a certain extent be an arbitrary quantity, since it will depend on the sensitivity of the instrument used to determine the response. Thus, if the response takes the form of an increase in length, a smaller response is measurable with the microscope than with the naked eye.

Many attempts have been made to determine the relationship between the values of the stimulus and the response over the range of sub-maximal values of the stimulus. The law evolved from these attempts is generally spoken of as the Weber-Fechner law, after Weber, who enunciated it first in regard to the sense of touch, and Fechner, who extended it to include all the special senses. Pfeffer appears to have been the first to extend it to plants, which he did in 1888 in regard to the response of bacteria, flagellates and Volvocineae to chemical stimuli.

The Weber-Fechner law may be simply stated. It is, that the response varies with the logarithm of the intensity of the stimulus. While it is no doubt true that this law is approximately correct in some cases, it can at best only hold over a limited range between the threshold and maximum values of the stimulus, while it is equally true that in other cases it is not a correct statement of the relation between intensity of stimulus and response even over this range. According to Verworn, such a law of general application cannot be formulated in mathematical terms, and the law which holds for many living systems cannot be more exactly expressed than by the following statement: 'with increase of the intensity of stimulation the response at first increases rapidly and later more and more slowly'.

There are, moreover, some cases in which, provided a response takes place at all, the latter is constant whatever the value of the stimulus. Or, put in another way, when the threshold value is reached the maximum response takes place. Such cases are said to obey the 'all or none law'.

THE PERIOD OF STIMULATION

With a stimulus of definite intensity, the magnitude of the response is dependent on the length of time during which the stimulus is applied. Just as below the threshold value the stimulus produces no obvious response, so, if the stimulus acts for less than a certain minimum time there is no measurable response. This minimum time which is necessary to produce a perceptible response is termed the *presentation time*. Like the threshold value of the stimulus, and for the same reason, the presentation time must to a certain extent be an arbitrary value depending on the means employed to determine it.

Generally speaking, with increase in the intensity of the stimulus the presentation time is lessened, and also with increase in the period of stimulation the threshold value is lessened. Consequently the question arises whether the amount of stimulation, which would be measured by the product of the intensity of the stimulus and the period of stimulation, does not give a more reasonable quantity to consider in relation to the response to stimulation rather than either the intensity or period of stimulation. In the case of phototropic reaction this is definitely so, and in all recent researches on the question it is the quantity of light rather than either the intensity, or period of stimulation that has been measured.

Since the very essence of stimulation is a *change* in conditions, it follows that a change long continued ceases to be a stimulus. Unless the stimulus is of such a kind as to lead to the death of the tissue if continued for too long a period, prolonged stimulation must simply lead to a fresh position of equilibrium. The consideration of two particular cases will make this point clear. A short exposure of, say, 30 minutes, of resting buds of lilac to hydrocyanic vapour in a certain concentration, acts as a stimulus and leads to their premature unfolding, but prolonged exposure to the same vapour kills the buds. Here we have a case of prolonged stimulation bringing about the death of the tissue. But if a stem growing vertically is exposed to unilateral light the stem may execute a growth curvature by means of which the axis of the stem apex comes to lie in the direction of the incident light. If the unilateral lighting is prolonged no further curvature is executed and the stem apex continues to grow in the same direction, a new state of equilibrium, as regards rate and direction of growth, having been reached. Thus any proportionality between period of stimulation and magnitude of response can clearly hold only within certain limits.

If after stimulation a sufficient time is allowed to elapse, the plant organ may again be stimulated in the same way and the same response result. If, however, the time between successive stimuli is reduced below a certain amount, the response to the second stimulation will be less than that to the first. And if a series of stimulations are given to a plant or plant organ with insufficient time for recovery between the successive stimulations, no response at all may result and the tissues concerned are said to be in a state of *fatigue*.

THE NATURE OF THE RESPONSE

It was realized as long ago as 1826 by Johannes Müller that by whatever kind of stimulus a muscle is stimulated, it reacts in the same way, and later he formulated in regard to sensory structures the so-called law of specific energy according to which the same factor produces different sensations in the different senses, and different factors produce the same sensation in any one sense. This law may be extended to stimuli in general, for the response depends on the nature of the cell or tissue stimulated, and not on the kind of stimulus. Thus both unilateral light and chemical influences bring about the same kind of response, namely movement, in many unicellular organisms, while the stimuli of gravity and unilateral illumination both produce similar curvatures in roots and stems, where the response to stimuli of either kind manifests itself as change in growth rate.

As pointed out in 1858 by Virchow, the effects of stimulation are essentially of three kinds: (1) functional, in which the response simply affects a functional process, as when the sole effect of the stimulus is to increase the respiration rate; (2) nutritive, in which the response to the stimulus takes the form of a changed rate of nutrition; and (3) formative, in which the response is a change in form. These different actions are not necessarily independent, for an increase in respiratory activity may be linked with an increased rate of nutrition, while changes in form are generally dependent on changes in growth rate over the whole or part of an organ, and this change in rate of growth may depend on a changed rate of nutrition and be correlated with an alteration in respiratory activity.

The responses of plants to stimuli that have formed the chief subjects of investigation belong chiefly to the first and third categories. The movement of motile organisms in response to stimuli appear, as far as can be determined at present, to involve merely a change in function, while the same appears to be the case with many of the nastic movements exhibited by multicellular plants as well as many of the responses to seismonic stimuli, such as the movements of leaves of *Mimosa pudica* and those of irritable stamens where the movements are often due to changes in turgor of a cell or group of cells. The tropic curvatures produced in the higher plants by unilateral stimuli are nearly all brought about by changes in growth rate and involve permanent changes in form, and thus belong to the third category.

While it so happens that almost all work on irritability in plants has dealt with movements or curvature, it is not to be assumed that all stimulation of plant material brings about this kind of response. The addition of certain chemical substances to the environment may bring about a change in respiratory activity, while it seems more than likely that the effect of light in bringing about the germination of *Lythrum salicaria* is in the nature of a stimulus. The action of certain of the rays given off by radium preparations may, according to the observations of Molisch, further germination of seeds, the opening of dormant buds and the growth of plants, and the action may be that of a stimulus.

It has already been noted that the responses to unilateral stimuli which result in movement or curvature are spoken of as tropisms or tropic movements and curvatures. When the organisms are free to move, as in the case of unicellular plants such as *Chlamydomonas*, and the spermatozoids of mosses and ferns, the suffixes 'taxis' and 'tactic' are generally used instead of 'tropism' and 'tropic'. Thus the movement of moss spermatozoids towards a more concentrated solution of sugar is a chemotactic movement and an example of chemotaxis.

THE LATENT TIME

The observable response to stimulation rarely takes place at once; usually a measurable time elapses between the time of stimulation and the time of response. This time is called the latent time or reaction time; it may vary from a fraction of a second as in the case of some nastic and shock movements to several hours as in some cases of tropic curvature. The phenomenon indicates that a chain of reactions is involved in stimulation requiring time for their operation. That such must be so is obvious in the case of curvatures depending on different rates of growth on two sides of an organ, where obviously a series of events must be involved between the time when, for instance, light falls on the organ and the time when the resultant differential growth rates begin.

The latent time varies with the intensity of stimulation, the period of stimulation and the external conditions.

The question arises, what do we know of the series of changes taking place between the perception of the stimulus and the response? Here our ignorance of the facts is often almost complete, although certain hypotheses have been put forward. Thus, in the case of stimulation by an electric current, Nernst suggested that the first change produced in the cells is an alteration in ionic concentration at the plasmatic surface. According to Verworn, the first effect of stimulation is very generally to increase the rate of dissimilative processes. In the vast majority of plants, as we have seen, the chief characteristic of dissimilation is the process known as respiration which results, with intake of oxygen, in the formation of carbon dioxide and water. Now this, as we have already seen, releases energy previously existing as chemical energy in the substances, carbohydrates and fats for the most part, from which the carbon of the evolved carbon dioxide is obtained. Hence, we can understand how it is that the energy transformation in the response bears no relation to the energy involved in the stimulation, for the latter is merely a mechanism releasing the energy contained in the substances utilized in the increased respiratory activity which the stimulus induces. Also it can be seen why irritability in aerobic organisms decreases with lessening of the oxygen concentration, for the energy provided by anaerobic respiration is only a small fraction of that provided by the same amount of substrate when respired aerobically.

The way in which energy released as a result of increase in dissimilation brought about by a stimulus is utilized depends on the nature of the particular cell or organ stimulated. It may lead ultimately to altered growth rate or only to turgor changes.

TONUS AND RIGOR

The sensitivity of a plant or plant organ to a stimulus varies with internal conditions in the plant; conditions which are generally undefined. Thus after a plant or organ has been subjected to one set of external conditions its response to a definite light stimulus may be more rapid and greater than that of a similar shoot which has been subjected to a different set of conditions, and which have resulted in different conditions being set up inside the plant. These unknown internal conditions which determine the sensitivity of the plant organ are referred to as *tonus* or *tone*, and as the plant or organ loses tone it responds less readily to stimulation until with the loss of all tone it is in a state of *rigor* in which it does not respond at all.

In some cases it is known that certain external conditions can affect the tone of a plant. Thus the movement exhibited by many leaves in response to stimuli of various kinds depends on the leaves receiving a certain amount of light, for with continued maintenance in darkness the power of response is progressively lessened and finally disappears. Leaves which, owing to their exposure to sufficient light, can respond to stimulation, are said to be in a condition of *phototonus*, while leaves which have lost their power of response owing to absence of light are said to be in a state of *dark-rigor*.

CONDUCTION OF EXCITATION

Frequently, although not invariably, the seat of the response is at a greater or less distance from the part stimulated. The latter has sometimes been called the seat of perception or sensory part in contrast to the seat of response, or, in cases where movement or curvature is involved, the motor part. Thus it has been shown in seedling roots that the seat of perception of the stimulus is chiefly in the apex, while the part that responds with a curvature is naturally the more actively growing region a little distance behind the apex. There must thus be a conduction of the excitation produced by the stimulus at the seat of perception. Various theories have been put forward to account for this conduction of excitation in different cases; reference to them will be made in due course in the following chapters, where the various manifestations of irritability in plants are considered in somewhat more detail.

THE NATURE OF IRRITABILITY

Although it is undoubtedly convenient to consider together the phenomena of irritability, it must be borne in mind that there is no reason to suppose that they involve any special attribute of living material that is not also involved in the phenomena of metabolism and growth. Whether the response of plant organs to stimuli would be explicable in physical and chemical terms if our powers of analysis were capable of dealing with the problem is a question to which at present no answer can be given, but while at present it is convenient and even helpful to retain the conceptions and nomenclature which have been developed by students of this branch of plant physiology, it must be realized that our understanding of the phenomena of irritability is only likely to be advanced by the same methods as have been employed in the study of metabolism and growth, the methods, that is, of physics and chemistry applied to the special material provided by the plant. That considerable advance has been made during recent years by the use of these methods towards a greater understanding of the response of plants to stimuli will be shown in the chapters which follow.

REFERENCES

PFEFFER, W., *The Physiology of Plants.* English Translation. Vol. III. Oxford, 1906.

VERWORN, M., *Irritability.* Oxford, 1913.

CHAPTER XX

MOVEMENT IN PLANTS

FROM what has been stated in the last chapter regarding the general principles of irritability it will be clear that the response to stimulation may take place in various ways. The result may be merely an increase in respiration rate, or it may be increased rate of growth. But very frequently stimulation of a plant or plant organ results in movement, and it will therefore not be out of place to discuss here the phenomena of movement in plants.

In some small aquatic plants the power of movement is obvious. As examples may be cited some unicellular algae such as *Chlamydomonas*, and the zoospores and antherozoids of a large number of species. Here active locomotion takes place owing to the possession by the plant, zoospore or gamete of cilia, by the active movement of which in water the whole cell is propelled. In other plants, notably the Myxomycetes, movement of a different kind, known as 'amoeboid movement', occurs. Here, owing to the absence of a cell-wall, the naked protoplasm can undergo changes of shape, and when such changes proceed continually a kind of creeping movement of the protoplasm results, by means of which the plasmodium may slowly traverse relatively considerable distances.

In rooted plants the most obvious movements are perhaps those performed by the petals in the opening and closing of flowers. Some flowers open and close only once; in other cases the flowers may open every morning and close every evening over a period of several or many days, while in a few cases the opening may be at night and the closure in the morning, or the flowers may open and close many times during the day if external conditions fluctuate. Similar behaviour is observed in the so-called 'sleep' movements of some leaves, such as those of *Trifolium*, *Oxalis* and *Marsilia*. Other well-known cases of movement are those of the leaves of some plants when stimulated by a sudden blow. The best known of these is provided by the sensitive plant *Mimosa*, in which, as a result of a blow or some other stimulus, the petiole, as if hinged at the pulvinus, falls through a considerable angle, and the opposite pairs of leaflets close together. Sensitive stamens, which exhibit a jerking movement on being touched, afford other well-known instances of obvious movements in higher plants. Such movements generally result from changes in turgor, and hence in the size, of certain cells.

Apart from these more obvious types of movement it has been urged that the root and shoot tips of every vascular plant move in space as the plant grows. It is no doubt true that the stem and root apices may be regarded as moving onwards as the plant grows, but even if the material in apical tissues remained the same, which can only partially be the case, the movement of the tip is due to its being pushed forward by increase in the size of the cells behind the tip. Similarly, the curvatures produced in roots and shoots when these are stimulated by exposure to unilateral light or by other means are the results of unequal growth of cells on two sides of the organ, and are thus not very appropriately described as movements. It is, however, often convenient to use one word for all these allied phenomena, and for this reason we use the word movement to include growth curvatures.

It may be true that internal movements take place in all living cells, and in many cases a streaming of the protoplasm is readily observable; in other cases it seems likely that the protoplasts may be capable of executing amoeboid movements similar to those observed in the plasmodia of Myxomycetes.

It will thus depend on the nature of the organ or organism stimulated whether it responds by active locomotory movement, by changes in cell turgor, or by a growth curvature. The free-swimming ciliated zoospore or gamete will respond by active movement, the leaf of *Trifolium* or *Mimosa* by a change in turgor, and the majority of stems and roots by different rates of growth on the two sides of the organ resulting in curvature of the organ. It is possible in some cases that both turgor changes and differential rates of growth may occur in cells of the same organ as a result of stimulation.

The movements met with in plants fall mainly, then, into four types according to the mechanism involved. These are (*a*) active locomotion in water brought about by the action of cilia, (*b*) amoeboid movement, (*c*) movements due to changes in turgor of cells, and (*d*) movements or curvatures connected with the growth of cells. Little is known with regard to the first two types of movement. All plant cells with the power of active movement possess cilia. These are fine hair-like outgrowths of protoplasm which are distributed in very varied ways over the surface of the cell, varying from one in the zoospores of *Synchitrium* to a large number distributed over the whole surface in the zoospores of *Vaucheria*, although a pair inserted at the anterior end constitutes the most usual arrangement. The movement of the organism is effected by the vibratory or twisting motion of these cilia and usually involves the rotation of the organism on an axis in the direction of movement.

Amoeboid movement occurs chiefly in the plasmodia of Myxomycetes, but it is also exhibited by a few algal and fungal zoospores, while changes which take place in the shape of some nuclei, notably those of pollen tubes, appear to be attributable to slow movements of the same kind. Amoeboid movement has been attributed to the contractility of the protoplasm, to local changes in the extent of imbibitional swelling, and to local changes in surface tension, but the actual mechanism is still not clear.

The internal streaming movement of protoplasm is very marked in certain cells, such as the leaf cells of *Elodea*, the internodes of *Chara* and *Nitella* and the staminal hairs of *Tradescantia*, but in many cells no streaming is observable. The streaming may serve the function of transporting material from one part of the cell to another, for various cell inclusions such as starch grains, fat globules, plastids and even the nucleus may be carried in the moving cytoplasm. While a number of attempts have been made at explaining the movement, as, for example, that it is brought about by surface tension changes in the inner plasmatic membrane, there is not at present evidence to justify the acceptance of any theories of protoplasmic movement. The highest recorded rate of protoplasmic streaming appears to be 10 mm. per minute observed by Hofmeister in *Didymium serpula*, one of the Myxomycetes, whereas, according to the same observer, the rate of movement in *Potamogeton crispus* is only 0·009 mm. per minute. The rate of movement is, however, influenced by temperature, various chemical substances and the stage of development of the cell.

A curious type of movement is met with in the Diatomaceae, Oscillariae and Desmidaceae. It is termed *gliding* movement, the organisms moving slowly over a solid surface although they possess the power neither of locomotion by means of cilia nor of amoeboid movement. In the Oscillariae the gliding movement is combined with a rotatory movement around the long axis.

The movement of diatoms is due, according to O. Müller, to the presence outside the cells of a thin layer of cytoplasm continuous through pores in the wall with the internal cytoplasm, the whole being in continuous streaming movement. On account of the friction produced between the external moving protoplasm and the solid surface a movement of the diatom in a direction opposite to that of the external cytoplasm results. No external layer of cytoplasm has been observed in the Oscillariae and it is not known whether the power of movement results from excretion of mucilage with which the filaments are covered, from surface tension changes or from some other cause, but Klebs considered the movements of desmids to be brought about by the excretion of mucilage.

Movements brought about by a change in turgor of certain cells are usually reversible, since the turgor changes themselves are reversible. They depend in part on the elasticity of the cell-wall. When the cell loses water, if the cell-wall is elastic the cell contracts ; if water is again absorbed and the turgor pressure of the cell increases the cell-wall is again stretched and the cell expands. The changes in volume so produced bring about the movement of the type we are considering. In the stamens of plants of the sub-family Cynareae, which are stimulated by contact with a solid body, the effect of stimulation is to produce a sudden fall in turgor of the cells of the filament, water passing from the cells into the intercellular spaces. The filaments in this way may contract to 80 per cent. of their former length.

Perhaps the most frequent examples of movement brought about in this way are provided by leaves and leaflets in which turgor changes take place in the cells of the pulvini. In *Mimosa pudica*, for example, the sudden fall of the leaf is mainly due to loss of water, and consequently turgor, by the cells of the lower parts of the pulvinus. The resulting shrinkage of these cells, combined with the weight of the leaf and the elasticity of the walls of the cells of the upper part of the pulvinus, brings about the fall of the petiole. The water lost by the cells of the lower part of the pulvinus passes into the intercellular spaces and neighbouring cells; on the re-absorption of this water by the cells of the lower part of the pulvinus, their turgidity is gradually regained and the leaf resumes its normal position.

Growth curvatures are very familiar results of stimulation. In this type of so-called movement the rate of growth of the cells in the growing region is affected to a different extent in different parts of the organ. This differential rate of growth on two sides of an organ may be brought about by an increase in the rate of growth on the more rapidly growing side or by a decrease in the growth rate on the less rapidly growing side, or by a combination of both these. In growth curvatures the cells concerned are those undergoing extension and we may therefore suppose that one effect of stimulation is to produce differential turgor changes on the two sides of the organ, so that the cells on the one side extend at a greater rate than those on the other and the cell-walls are more stretched as a result, a considerable proportion of the extension being of the nature of plastic growth and therefore irreversible.

Although it is very general for stimulation of plant organs to result in movement or curvature, such also occur without any apparent change in external conditions. Where movements (or curvatures) take place under apparently constant external conditions it is usual to speak of them as autonomic, autogenic or spontaneous, and to regard them as arising from internal changes, that is, by the action of internal stimuli. They are thus contrasted with the more readily investigated movements which are produced by external stimuli and which are referred to as induced, aitiogenic or paratonic. The most familiar of autonomic movements are the nutation movements executed by the stem apices of plants, which reach their greatest development in climbing plants. The stem apex does not grow vertically upwards, but executes a spiral course in space owing to the position of the most rapidly elongating region moving continuously round the stem.

Other autonomic movements are the so-called variation movements of leaflets which are most pronounced in two leguminous plants, *Desmodium gyrans* and *Eleiotis soraria*, but which occur also in *Oxalis acetosella* and *Trifolium pratense* and some other plants. In these plants some or all of the leaflets execute more or less continuous movements. In *Desmodium* and *Eleiotis* these movements are such that the apex of each of the two basal leaflets moves through an ellipse, while in *Oxalis* and *Trifolium* the leaflets move to and fro through an angle of varying magnitude. Here the movements are brought about by rhythmic changes in turgor of a group of cells at the base of the leaflet stalk.

The opening of both leaf and flower buds involves the parts in movements which are often not brought about by external stimulation. These movements are largely of the nature of curvatures, due to the unequal growth on two sides of the organ. In contrast to the variation movements already mentioned they mostly occur only once and so are termed *ephemeral.* A sort of transition between the two types of movement is met with in the case of flowers which open and close more than once. It will be noticed that such movements are not necessarily autonomic ; changes in light and temperature may in some cases be very effective stimuli in inducing these movements.

The movements of the cilia which bring about the locomotion of free-swimming organs are generally regarded as autonomic, for they continue in apparently unchanged external conditions. As we shall see later, the direction of movement can be determined by external stimuli, just as the direction of movement of a nutating stem apex can be affected by an external stimulus as, for example, the direction of light. Even under constant external conditions the direction of movement may change or movement may even stop.

As regards external stimuli, it has already been pointed out that these may be directional or diffuse, and that the movements resulting from them are described respectively as tropic and nastic. The chief tropic movements are those related to a change in the direction of the action of gravity relative to the plant or organ, and those related to the incident light. The phenomena of movement concerned with these two kinds of stimulation are grouped under the terms geotropism and phototropism respectively. Other tropic stimuli are provided by various chemical agents, by water content, osmotic pressure, contact with a solid body, and electrical conditions, and the corresponding terms chemotropism, hydrotropism, osmotropism, haptotropism (or thigmotropism) and galvanotropism are used to indicate the phenomena displayed by the reacting tissues in response to these various stimuli. As regards nastic phenomena, the most important are photonastic movements, those produced by a general uniform change in light intensity round the plant, but thermonastic and hydronastic curvatures are not uncommon. Mechanical disturbances, such as shaking or blows, often act as stimuli, but it is generally difficult to classify any of the effects as tropic. Where striking the organ on one side results in a curvature, the direction of which depends on the side struck, a tropic phenomenon is undoubtedly involved, but generally the movements resulting from such *seismonic* stimulation, as it is called, are not related to direction, and cannot therefore be regarded as tropic. In tropic stimulation the organ responds by taking up a position related to the direction of the stimulus. With most cases of seismonic irritability, on the other hand, as, for example, when leaves of the sensitive plant (*Mimosa pudica*) are struck, the position taken up depends on the structure of the leaf and is unaffected by the direction of the blow. The phenomenon has thus much in common with nastic stimulation, for nothing directional in the stimulus appears to be transmitted to the responding organ. It would thus appear to be not altogether illogical to classify such cases of seismonic stimulation along with nastic phenomena.

The terms epinasty and hyponasty were introduced by de Vries to express the more rapid growth of the upper or under side respectively of a dorsiventral organ. Thus, in the development of many leaves these, while in the bud, exhibit hyponastic curvature, but as the buds expand the upper side may grow more rapidly than the lower so that the amount of growth on the two sides becomes equal. Not infrequently, and notably in ferns. the upper side may finally attain a greater length than the under side, so that an epinastic curvature results. There is no need to limit the terms to the case of dorsiventral organs, and epinastic and hyponastic curvatures can be observed in hypocotyls and epicotyls, stems, roots, flower stalks and so on. The terms are, perhaps, not too fortunate in suggesting a connexion with nastic curvature, for although this is quite frequently the case, it is not necessarily so, and we find Pfeffer, for example, speaking of 'autogenic epinasty'.

The term *autotropism* requires some explanation. This expression refers to the fact that, apart from the effect of external stimuli, various organs of the plant take up a definite position in relation to the main axis or to one another. Thus, when the action of gravity is eliminated, lateral roots of the first order, that is, those borne on the main root, mostly grow out at right angles to the main root, while lateral roots of a higher order, which are generally without geotropic irritability, grow out at definite angles to the root which bears them. Autotropism is supposed to come into play when, as a result of strong stimulation by light or gravity, so great a curvature takes place that the organ passes its normal position relative to the direction of the light or of gravity, as the case may be. An autotropic curvature then supervenes, bringing the organ back to its normal position. This behaviour is illustrated by the diagrams in Fig. 49, which represent the behaviour of a shoot strongly stimulated geotropically. The shoot, having been placed horizontally, executes a geotropic curvature, but this is so great that the apical part of the shoot passes the vertical position. A growth curvature then takes place in the opposite direction so that finally the shoot apex is directed vertically, sometimes after a number of oscillations. Again, let us suppose a seedling plant is slowly rotated in the dark in a vertical plane so as to eliminate the action of gravity. The shoot then grows in a straight line unrelated to the direction of the gravitational pull. If now the rotation is stopped and the plant is left for a time with its axis making an angle with the vertical, the stem performs a geotropic curvature. If the rotation is now recommenced before the curvature is too pronounced it will be found that the curvature disappears, owing to more rapid growth on the concave side of the shoot. This straightening out of the shoot is attributed to autotropism. This is, perhaps, a better example of autotropism than the former, since there is here no question of gravitational action in the second curvature. The state of affairs was summed up by Pfeffer by saying that 'every disturbance in equilibrium excites reactions which tend to its restoration'. Rectipetality, the resumption of the original direction of an organ after a curvature induced by an external stimulus which is no longer operative, is thus another name for autotropism.

The same directional stimulus that leads to active movement in a definite direction of a free swimming organism will lead to curvature, or sometimes to a twisting or torsion, of an organ of a rooted plant. While all such movements may be termed *tropic*, using the word in its wide sense, the locomotory movements of free swimming organisms, as

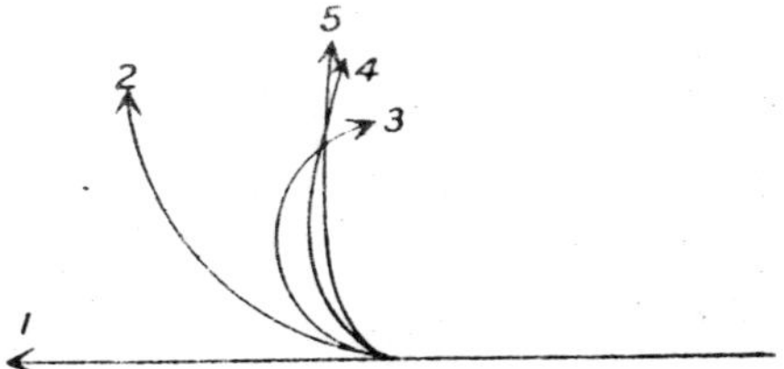

FIG. 49.—Diagrams to represent the course of response of a shoot to geotropic stimulation. Successive stages are numbered 1 to 5
(After J. v. Sachs)

mentioned previously, are often termed tactic, and the special tropism a taxis. Thus the movements of zoospores in relation to the direction of the incident light are described as phototactic, and the phenomenon of such movement as phototaxis. Similarly where a torsion, rather than a curvature, results from, say, the stimulation of gravity, the term geotorsion is sometimes used to denote this special type of geotropism.

It does not follow that a plant movement is due to a single stimulus. A curvature may be tropic or nastic, or it may be the result of both tropic and nastic effects, while, as we have seen, movements may involve both geotropic or phototropic curvatures and autotropic curvature. It is also possible that a curvature may be partly due to growth and partly to turgor changes. It is thus evident that careful analysis may be necessary to determine the actual causes of a particular movement.

In the chapters which follow, the autonomous movements attributed to internal stimulation will first be dealt with and then the effects of the various external stimuli will be considered.

CHAPTER XXI

AUTONOMIC MOVEMENTS

In the investigation of plant movements which cannot be related to changes in the external conditions, and which are therefore considered to be brought about by changes inside the plant, we are at once confronted by a very great difficulty, namely, that of ensuring perfectly uniform conditions of life. And even if we ensure a uniform environment over a certain period of time, there is still the possibility that movements observed during that period may result from external changes which occurred earlier in the plant's history, for, as we have seen, the visible response to a stimulus may not appear until some time after the actual stimulation. There is also the possibility of unsuspected factors being operative. Thus, for example, in what appeared to be a uniform climatic environment electrical conditions, such as conductivity of the atmosphere, might undergo a change, and it cannot be assumed without actual proof that any factor has no influence or that a change in its value cannot act as a stimulus. Thus the movement of the leaves of *Phaseolus vulgaris*, originally supposed to be related to light intensity, inasmuch as the leaves tended to fall at night and rise in the morning, were shown by Rose Stoppel to bear no obvious relation to change in light intensity, and she thought they were brought about by the stimulation of change in the ionization of the air. Subsequent workers have shown this explanation to be unlikely, but the case indicates the possibility of overlooking a change in an environmental factor.

The chief cases of autonomic movements in plants have been mentioned in the previous chapter. They are nutation, or circumnutation, of stems, the variation movements of certain leaves and the general movement of free swimming forms. With the last it is not proposed to deal further, but the phenomena of nutation and autonomic variation movements call for further treatment.

NUTATION

The most important series of observations on nutation were carried out by Charles Darwin and published in his books on *Climbing Plants* and *The Power of Movement in Plants*. While the phenomenon is most pronounced in stems, and especially in those of climbing plants, it is also met with in other organs, such as leaves, particularly tendrils, roots, stolons, flower stalks, filamentous algae and fungal sporangiophores, although often only to a slight extent. Indeed, Darwin affirmed that all the organs of every plant while they continue to grow are continually nutating. He made many observations on nutation in seedlings, using species from 29 families, and from these observations it appears that nutation commences very early in the life of the plant. Nutation of the radicle commences as soon as it protrudes through the seed coat, while the hypocotyl or epicotyl nutates even before it has completely penetrated through the soil and while still arched. The course taken by the growing apex may at first be very irregular, but in the shoot this becomes more regular as the seedling develops, while the amplitude of the movement generally increases. Thus in the case of the hop (*Humulus lupulus*) Darwin in his earlier work on *Climbing Plants* describes the first two or three internodes as remaining straight, although from his later work on seedlings we must interpret this as meaning that their nutatory movement is slight. The next internode, however, commences to nutate very definitely and, moving slowly at first, increases its speed, and the normal full rate of movement is soon reached. As long as growth continues so does the nutation, although after a time each internode in turn becomes still.

It is very general for the nutation in all members of one particular species to take place in the same direction. Thus, the hop, black bryony (*Tamus communis*) and *Scyphanthus elegans* move in the same direction as the sun, whereas the majority of twining plants move in the opposite direction. Among such are the ferns *Lygodium scandens* and *L. articulatum*, the monocotyledons *Ruscus androgynus*, *Asparagus* sp. and *Roxburghia viridifolia*, and the dicotyledons *Wistaria chinensis*, *Phaseolus vulgaris*, *Jasminum pauciflorum*, *Thunbergia alata* and *Convolvulus sepium*. In a few species the movements are variable; thus, of 17 plants of *Loasa aurantiaca* Darwin found that five followed the sun, eight revolved against the sun, while in the remaining four the course of revolution was first in one direction and then in another. As a general rule, however, all the plants in one family revolve in the same direction, although there are exceptions to this.

Both the amplitude of the nutating movement and its rapidity vary greatly in different species, the amplitude being much greater in twining than in non-twining plants. Darwin records a case of *Ceropegia gardnerii*, a member of the Asclepiadaceae, in which the extreme tip performed a circle 62 inches (1·575 metres) in diameter. At the other extreme is the case of the cactus *Cereus speciocissimus*, in which the extreme amplitude of the movement was less than 0·05 inch. (1·27 mm.). The rapidity of movement, on the other hand, may be as great in non-twiners as in twiners, but in all cases it is influenced by external conditions and may display considerable variations even when external conditions appear to be constant. Average times taken by plants of different species to make a single complete revolution, as observed by Darwin, are shown in Table XCI. The most rapidly nutating stem observed by Darwin was that of a plant of *Scyphanthus elegans*, which made one revolution in 1 hour 17 minutes.

Table XCI

Approximate Average Time taken by Nutating Plants of Various Species to make one Complete Revolution

Species	Time in hours
Lygodium scandens . . .	6
Ruscus androgynus . . .	3
Tamus communis . . .	3
Lapagera rosea . . .	14
Roxburghia viridifolia . .	24
Humulus lupulus . . .	2
Wistaria chinensis . . .	3
Phaseolus vulgaris . . .	2
Jasminum pauciflorum . .	7

The length of stem and number of internodes taking part in the nutating movement also varies from one species to another. In seedlings and small plants only a centimetre or so of the stem may nutate, involving only a single internode, whereas in *Hoya carnosa* Darwin observed a length of shoot equal to 32 inches (81 cm.) and involving seven internodes, nutating.

According to Darwin's observations, nutation is also executed by tendrils ; indeed, in *Passiflora gracilis* the average rate of six revolutions was 1 hour 1 minute a revolution, a rate higher than that observed in any stem.

The nutating movement of stems, as we have seen, is due to the continuous shifting of the position of most rapid growth round the stem. Why this shifting should occur is not known. Darwin thought it indicated that the growth changes in the cells required periods of rest, and it certainly means that the growth in any one part of the stem apex is rhythmic.

The nutating movement of growing tendrils is presumably also related to the shifting regularly round the tendril of the position of most active growth, but in many of the circumnutating leaves observed by Darwin, and certainly in the case of older leaves, it would seem that the movements are not due to differential growth rates but to reversible changes in turgor in the cells of the leaf-base, so that the movements are of the ' variation ' type described below.

No carefully controlled observations on the effect of external conditions appear to have been made on nutation, but Darwin's observations indicate that, within limits at any rate, increase in temperature leads to increase in the rate of movement.

As long ago as 1844 Dutrochet recorded the effect of temperature on the rate of nutation of tendrils of *Pisum sativum*, a revolution being accomplished in about 10 hours at 5° C. and in 80 minutes at 24° C. These numbers give a temperature coefficient (Q_{10}) of 2·89, and hence of the same order as that found in many metabolic processes.

As regards other external factors, light is not as a rule an essential condition for nutation, and many etiolated plants nutate as actively as normal plants grown in light. In some cases, however, nutation becomes very feeble if plants are kept in prolonged darkness.

EPHEMERAL MOVEMENTS

A second group of autonomic movements caused by differential rates of growth on two sides of an organ are characterized by the movement taking place once only, for which reason they are called ephemeral. Such movements are the bending of hypocotyls and epicotyls, and the expanding of leaf and flower buds. The twisting or torsion observed in a number of stems, tendrils and flower and fruit stalks are similar movements, as are the movements performed in some flowers by stamens, style or stigma in bringing about pollination.

VARIATION MOVEMENTS

Variation movements, as we have seen, are caused by reversible changes in turgidity of a group of cells. It is thus understandable that the same sequence of movements may periodically recur. While many variation movements are brought about by external stimuli, a number of them fall into the class of autonomic movements, since they continue under constant external conditions.

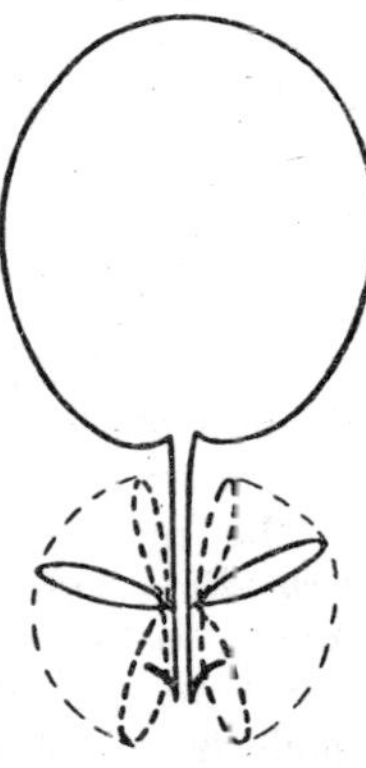

FIG. 50. — Diagrammatic representation of a leaf of *Eleiotis soraria* to indicate the autonomous movements of the basal leaflets

(After W. T. Saxton)

One of the most striking instances of variation movement is that which occurs in the small lateral leaflets of *Desmodium gyrans* and *Eleiotis soraria*. In both these plants the leaf comprises a large terminal leaflet and two much smaller opposite lateral leaflets (cf. Fig. 50). It is these lateral leaflets which display the variation movements. Fixed, of course, at the base, they move so that the tip of each leaflet describes an ellipse in a plane parallel to the petiole of the leaf. The movement upwards and downwards is comparatively rapid, but at the top and bottom positions the movement is too slow to be followed by the eye. The time taken for a complete revolution of a leaflet at about 28° C. was found to be about 4 minutes in each case, that of *Desmodium gyrans* being determined by Kabsch and that for *Eleiotis soraria* by Saxton. The movements of the two leaflets rarely synchronize, the most common conditions being those in which one leaflet in its movement lags only a few seconds behind the other, or one leaflet is almost exactly a half-revolution behind the other.

In *Desmodium* the movements are brought about by changes in turgidity of the pulvinus of the petiole of the leaflets, but in *Eleiotis*, according to Saxton, there is very little change in shape at the basal end of the petiole, the movement being chiefly due to changes at the other end of the petiole at its point of junction with the lamina.

The autogenic variation movements of the leaflets of *Oxalis acetosella*, *O. hedysaroides* and *Trifolium pratense*, in which the leaflets move backwards and forwards through an angle, are, like those of *Desmodium*, due to periodic changes in the turgidity of cells in the pulvinus. The movements are, however, much slower than those of *Desmodium gyrans* and *Eleiotis soraria*, for a complete movement in *Oxalis acetosella* takes from 45 minutes to 2 hours, while that of *Trifolium pratense* takes about twice this time.

In *Averrhoa* as well, a genus of Oxalidaceae, the leaflets execute spontaneous movements. Here the leaf is a pinnate one with about 10 pairs of leaflets as well as a terminal one. In this plant, during a warm sunny day the leaflets slowly rise one after the other, the rise of each leaflet being followed by its more rapid fall. At night the movement ceases.

Variation movements comparable to those met with in leaves are also encountered in the flowers of *Stylidium adnatum*, where the column oscillates. Whether an apparently similar movement of part of the labellum of the orchid *Megaclinium falcatum* is due to periodic turgor changes or to changes in growth rate is not known.

The movements are much affected by temperature. Those of *Desmodium gyrans* only occur within certain limits of temperature, namely, between 16° and 42° C. Within this range the rapidity of movement increases with rise of temperature up to about 35°, above which the rate falls off until at the upper limit of the temperature range movement no longer occurs.

As regards the effect of light, while variation movements continue if the plants are transferred from light to darkness, after a prolonged period in the dark the movements gradually become more feeble and finally cease. A certain exposure to light is evidently required as a condition for the movements to proceed. As mentioned in Chapter XIX, this fact is expressed by describing the organs capable of executing the movements as a result of exposure to a sufficient amount of light as being in a condition of *phototonus*, while those in which the movement had ceased as a result of long exposure to darkness are described as being in a condition of *dark rigor*. The mode of action of light in conditioning the movements is obscure.

EPINASTY AND HYPONASTY

As already mentioned, the terms epinasty and hyponasty were introduced by De Vries to indicate the more rapid growth of the upper and lower sides respectively of an organ, which thus results in a curvature convex on the upper side in the case of epinasty and convex on the lower side in the case of hyponasty. Such movements are conspicuous in opening leaf and flower buds, but are also present in other organs, as, for example, many side stems, branch inflorescence axes and the thalli of liverworts.

While such curvatures are often traceable, at any rate in part, to the operation of external stimuli, sometimes the movements appear to be of an autonomic character. Such include the unfolding of fern fronds, which takes place in the dark and when the frond is slowly rotated on a klinostat with a horizontal axis. Very frequently autonomic curvatures of this kind are combined with others. Thus, in fern fronds Miss Prankerd distinguishes six types of movement: nutation, rectipetality (autotropism), sagging (due to the weight of the developing pinnae), epinasty, phototropism and geotropism. Frequently the position assumed by leaves appears to be due to the combined action of epinasty and either heliotropism and geotropism, or both.

De Vries attempted to show epinasty present in such cases by comparing the curvature of leaves placed horizontally with the morphologically upper surface uppermost with that exhibited by similar leaves placed with the morphologically under surface uppermost. The greater curvature of the leaf in the latter position he explained by the combined action of epinasty and geotropism in this position, whereas they were opposed in the more normal position. He also showed that curvatures were executed by leaves placed in a vertical plane with the midrib horizontal. Here it was assumed that geotropic curvature was impossible and that the greater curvature of the morphologically upper side (so that the midrib curves in a horizontal plane) must be due to epinasty. Again, if a plant, for example, of *Dahlia* or *Coleus*, is rotated on the klinostat with horizontal axis, the axis of the plant being parallel or in line with the axis of the klinostat, the leaves develop a marked curvature attributed to epinasty, that is, the convex curvature of the morphologically upper side becomes accentuated so that the leaves curve towards the base of the plant.

CHAPTER XXII

GEOTROPISM

That, as a general rule, the main roots of vascular plants grow vertically downwards and the main shoots vertically upwards are among the most obvious facts of plant growth, while the fact that this orientation persists in plants maintained in darkness at least suggests that gravitation may be concerned in determining the direction of growth of these organs. The first definite attempt to prove this influence of gravity was made in 1806 by T. A. Knight, who attached young seedlings to a wheel which was rotated in a vertical plane. By this device the action of gravity in any direction relative to the seedling was exactly compensated by its action for the same length of time in the opposite direction. On the other hand, although the gravitational effect was thus eliminated, the wheel was rotated rapidly enough (150 revolutions a minute) to produce a centrifugal force of roughly 3·5 times the force of gravity, so that, in effect, in Knight's experiment, as far as the seedlings were concerned, the force of gravity was replaced by a force 3·5 times as great directed radially outwards from the centre of the wheel. If the force of gravity determines wholly or in part the direction of growth of the shoot and root, it would then be expected that in Knight's experiment the roots of the seedlings would grow outwards from the centre of the wheel and the shoots towards the middle. This is, indeed, what Knight found to be the case.

In the investigation of geotropism it is necessary to ensure the absence, as far as possible, of stimuli other than the gravitational one. Thus, light often produces responses similar to those produced by gravity, and therefore the experimental investigation of geotropism has very generally been carried out in the dark. The instrument known as the klinostat has also proved of the greatest value in the investigation of geotropic and other phenomena of irritability. This instrument is simply a revolving wheel or disk to which the experimental material is attached, so that Knight's wheel can really be regarded as the first klinostat. When the klinostat is revolved in a vertical plane at a slow speed the gravitational action of the plant is eliminated, while the centrifugal force developed is negligible. Thus at a speed of two revolutions an hour and with the experimental material 10 cm. from the centre of the disk, the centrifugal force acting on the material is little more than 10^{-8} times the force of gravity. If, however, the rate of rotation is much slower, say one revolution in several hours, the material may remain in approximately the same position relative to gravity long enough for geotropic curvature to take place, or rather curvatures shifting in position regularly round the organ may result, so that a movement resembling nutation is produced.

In the *intermittent* klinostat the disk remains stationary for a time and then, by a clockwork or electrical device, is suddenly shifted through an angle so that the experimental material remains in a series of desired positions relative to the direction of the gravitational force. For some purposes the horizontal klinostat is used, in which the disk is horizontal and revolves about a vertical axis. With this type of instrument it is possible to examine the combined action of gravity and centrifugal force.

When an organ becomes orientated so that it grows towards the centre of the earth it is positively orthogeotropic, when it grows vertically away from the earth it is negatively orthogeotropic. When the axis of the organ tends to lie at right angles to the direction of the gravitational force it is said to be diageotropic, and when it takes up an angle with the vertical other than a right angle it is plagiogeotropic. Most main roots are positively geotropic, while a few rhizomes, as, for instance, those of *Yucca*, also appear to be positively geotropic. In a few monocotyledons positive geotropism is exhibited by the cotyledon and serves to bring the main axis of the seedling into the substratum. Among lower plants rhizoids are sometimes geotropic. Thus the rhizoids of *Marchantia* are geotropic and so, according to Noll, are the root-like appendages to some marine algae, although he remarks that geotropism plays a very minor part in the life of these forms. In non-vascular plants it would seem that geotropic irritability is not generally developed to the extent that it is in vascular plants.

Negative geotropism is strongly developed in the aerial main stems of most flowering plants, while occasionally organs of other morphological value exhibit negative geotropism. Among such are the pneumatophores of mangroves, which are morphologically roots. The foliage shoots of mosses are generally negatively geotropic, while in the sporangiphores and stalks of the fruit bodies of many fungi negative geotropism is strongly developed; this is the case both with smaller fungi such as *Mucor* and *Phycomyces*, and the larger Ascomycetes and Basidiomycetes.

While many rhizomes and runners are diageotropic, side stems and roots of the first order, and foliage leaves, are commonly plagiogeotropic, but side stems and roots of a higher order generally possess very little geotropic irritability.

In an earlier chapter it has been pointed out that the manner of response to a stimulus must depend on the material stimulated. Thus motile organisms respond by active movements and so exhibit geotaxis, while growing stems and roots respond by executing a geotropic curvature or a torsion, the latter particularly in dorsiventral organs, or a combination of the two. Variation movements resulting from geotropic stimulation have also been observed. Thus, in *Phaseolus vulgaris* and *P. multiflorus*, inversion of the plant brings about a comparatively rapid movement of the petiole due to turgidity changes in the cells of the pulvinus, so that the laminae tend to take up a position assumed by the leaves normally in the dark.

THE COURSE OF THE GEOTROPIC REACTION

The positively geotropic reaction is most readily observed in primary roots, and the negatively geotropic reaction in primary stems. The course of the visible reaction was described many years ago by Sachs for the case of *Vicia faba*, seedlings of which, because of ease of handling, are still a favourite material for observing the reaction. When such a seedling with its root growing normally downwards is transposed so that its root is horizontal, the change of the position of the root in relation to the direction of gravity constitutes a stimulus to which the root replies by executing a curvature, the result of which is that the root tip is again directed vertically downwards. As the curvature results from the unequal growth of the two sides of the root it is obvious that it must take place in the growing region; this, as we have seen (cf. Chapter XVI), is a very short distance behind the root apex. The method of demonstrating this is well known. By means of Indian ink a series of equidistant marks, say 1 or 2 mm. apart, are made on the unstimulated root. After stimulation it will be seen that the curvature occurs in the region of most active growth at the time (cf. Fig. 51). At ordinary laboratory temperature, that is, about 16 to 18° C., curvature is noticeable after 1 or 2 hours from the commencement of stimulation, but as the growing zone is short, only a length of a few millimetres of root is involved in the curvature. In the stem the growing zone is generally very much longer, and the length of stem involved in an apogeotropic curvature is also very much longer and may extend to many decimetres. Commonly, in the shoot, as in the root, the curvature commences a short distance behind the apex, but the region of most active curvature shifts down the stem away from the apex so that at the conclusion of the reaction the region of greatest curvature may be some considerable distance from the stem apex. Quite frequently, both in root and shoot, the curvature does not stop when the apex is directed vertically downwards or upwards, but continues further. This curvature is followed by one in the reverse direction tending to orientate

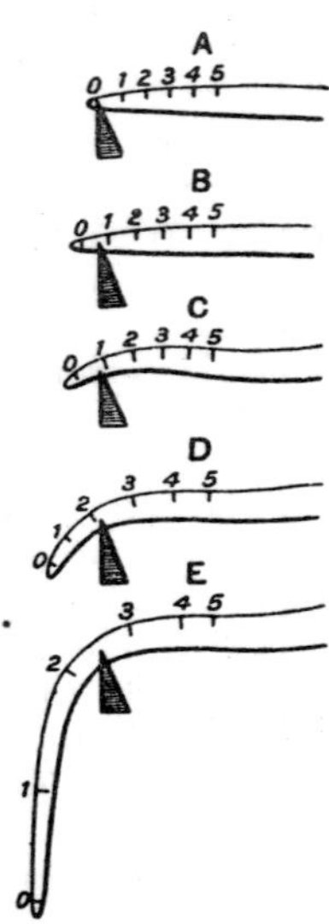

Fig. 51.—Diagram to indicate the region of curvature and growth in a geotropically stimulated root.
(From J. v. Sachs)

the apical part of the organ vertically, a movement which, as we have seen, is generally attributed to autotropism, but which may alternatively or in part be due to a second geotropic stimulation produced by the displacement of the apical part of the organ by the growth curvature below it. However, plants exhibiting the 'overcurvature' straighten themselves on the klinostat, and in this case the curvature involved must be attributed to autotropism, since geotropic stimulation is eliminated. It has been stated that this secondary movement may itself carry the root or shoot past the vertical position, when geotropic stimulation again comes into play and induces a second curvature in the original direction. The attainment of the equilibrium position would thus be preceded by a number of oscillations of decreasing amplitude. Recent attempts to confirm the existence of this oscillatory movement have not, however, succeeded in doing so.

It has been pointed out earlier that a growth curvature must result from the more active rate of growth of the organ on the convex as compared with the concave side. The phase of growth concerned is that of cell stretching, so that the result of geotropic stimulation is the increased turgor of the cells on one side of the organ as compared with those on the other side, the consequence of which is the greater stretching of the cell-walls on the former side, a stretching which is for the most part ultimately fixed and hence inelastic. But in the autotropic counter-reaction in roots the straightening may be brought about in part by a contraction of the cells on the longer side. When this is the case we must suppose that the autotropic counter-reaction must occur while the stretched cell-walls are still capable of a considerable elastic contraction.

Seedlings of the Gramineae have provided a favourite material for experiments on irritability both to gravity and to light, and none more than those of the oat (*Avena sativa*). On germination of a cereal grain such as that of wheat, barley, oat, rye, or maize, there develops from the seed in an upward direction a hollow cylindrical organ, which according to some authorities is to be regarded as the cotyledon, or part of it, and which, owing to the claims of other structures to be regarded as the cotyledon, is now generally known by the non-committal term of the coleoptile. Within the coleoptile there forms the first foliage leaf which ultimately breaks through it. It is with the behaviour of the coleoptile that practically all investigations on the irritability of seedlings of this type have been concerned.

The coleoptile is connected below with the scutellum by an axis of varying length, short in *Avena* and *Hordeum* but longer in *Setaria* and *Sorghum*. This axis can be regarded as hypocotyl or epicotyl, according as the coleoptile or the scutellum is considered to be the cotyledon. In most researches with seedlings in which this axis is significant the coleoptile is referred to as the cotyledon and the axis as the hypocotyl, but Mrs. Arber, in her work on *Monocotyledons*, regards the coleoptile and scutellum as together forming the cotyledon, and the axis as formed by fusion of part of the cotyledon with the hypocotyl. Since it is convenient to have a name for this axis, that used for it by Mrs. Arber will be employed here, namely, *mesocotyl*, but this is not intended to imply the acceptance of any particular theory with regard to the morphological interpretation of the structures present in the grass seedling.

According to the observations of Sachs, the geotropic curvature in roots of *Vicia faba* involves a decrease in the mean rate of growth, the rate of growth of the convex side being only slightly more than that of the unstimulated root, while that on the concave side is decidedly less. While the growth of the unstimulated root was 10·5 mm., that of the axis (middle line) of the stimulated root in the same time was only 8·4 mm., that of the convex side 10·8 mm., and of the concave side only 6·1 mm. Recent measurements with refined methods made by Cholodny, Dolk and U. Weber on coleoptiles and shoots indicate, however, that in these cases the rate of growth of the axis is unchanged by stimulation.

The nodes of grasses respond to geotropic stimulation with a growth curvature, and it must be supposed that the cells in these tissues retain or recover their power of growth. Here, since stimulation brings about renewed growth, it is obvious that the stimulated node must grow larger than the unstimulated one. Indeed, after curvature the convex side may be as much as 5 times as long as before, while the concave side may suffer a definite contraction ; actually shortenings of 10 to 40 per cent. of the original length have been observed. Comparable geotropic curvatures also occur in some pulvini such as those of *Dianthus* spp. and *Tradescantia fluminensis*.

INTENSITY OF GEOTROPIC STIMULATION

Since the force of gravity is approximately constant over the surface of the earth, variations in its intensity cannot be obtained directly as in the case of a factor such as light. There are, however, two ways in which the intensity of the geotropic stimulus can be varied experimentally. The first method consists in varying the angle made by the experimental material with the vertical. If a main root is kept in the vertical position the intensity of the stimulus is of course zero, while if it is displaced to lie horizontally the intensity of the stimulus is the total force of gravity, g. If the angle made with the vertical is less than 90° the intensity of the stimulus is something less than g, and the less the angle the less the intensity of the stimulus. Actually, if α is the angle made by the stimulated organ with the vertical the intensity of the stimulus is $g \sin \alpha$. This relation, proposed by Sachs, was shown experimentally to hold good by Fitting and others, since it is found that the presentation time, that is, the time required to give the smallest observable response (see p. 425) increases in proportion as $\sin \alpha$ decreases ; that is, the presentation time is inversely proportional to the angle made by the stimulated organ with the vertical.

In Table XCII are given some of the values obtained by Rutten-Pekelharing with regard to the relation of the angle made by oat coleoptiles with the vertical and the presentation time for geotropic reaction at a temperature of from 23° to 24° C.

Table XCII

Relation between Angle of Inclination of *Avena* Coleoptiles to the Vertical and the Presentation Time for Geotropic Reaction

Angle (α)	sin α	Presentation Time in Seconds (t)	t sin α
90° . . .	1·00	269	269
120° . . .	0·866	332	288
60° . . .	0·866	326	282
135° . . .	0·707	340	240
45° . . .	0·707	366	259
150° . . .	0·50	538	269
30° . . .	0·50	540	270
15° . . .	0·259	871	225
10° . . .	0·174	1415	246

They show that the product of the presentation time and the sine of the angle made by the coleoptile with the vertical is approximately constant over a wide range of angles.

The second method of varying experimentally the intensity of geotropic stimulation is to substitute centrifugal force for gravity by the use of a vertically rotating wheel or disk in the manner first employed by Knight. In this way centrifugal forces varying in value from a fraction of the force of gravity to many times as great can be obtained. It may be noted in passing that in experiments of this kind the seedlings have to be enclosed in order to prevent the excessive evaporation in the wind to which they would otherwise be exposed when the rate of rotation is high. Rutten-Pekelharing, using this method of investigation with oat coleoptiles and roots of *Lepidium sativum*, found that the presentation time was again inversely proportional to the intensity of the stimulus as the numbers shown in Table XCIII indicate.

Table XCIII

Relation between Intensity of Stimulation of *Avena* coleoptiles by Centrifugal Force and the Presentation Time at 16°–17° C.

Centrifugal Force (C)	Presentation Time in Seconds (t)	Product (Ct)
0·08	3900	312
0·14	2230	312
0·25	1300	325
0·67	441	296
1·04	310	322
3·00	100	300
5·76	53	305
10·08	31	312
23·86	13	310
41·76	7	292
58·43	5	292

It will be noticed from Table XCII that the intensity of stimulation is the same whether the stimulated organ makes an angle of $\alpha°$ or $180-\alpha°$ with the vertical, that is, whether the organ is in the position A or B in Fig. 52. It follows that a completely inverted organ, as, for example, a root in the position C in the diagram, is not subjected to stimulation. The position is not, however, stable in this respect, for a slight deviation from this position, which might result from other stimuli, internal or external, will result in its making an angle with the vertical, and response to stimulation will be a sharp curvature as shown at D, whereas in the normal position of non-stimulation any slight disturbance from the vertical results in a response which brings the root directly into its original alignment. The equilibrium positions E and C are therefore distinguished as stable and labile positions respectively.

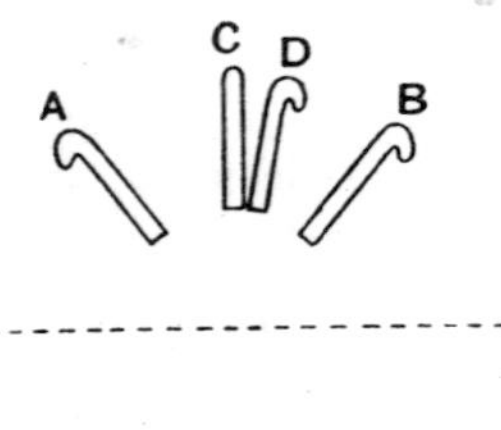

Fig. 52.—For explanation see text

QUANTITATIVE RELATIONS BETWEEN GEOTROPIC STIMULUS AND RESPONSE

When a geotropic organ is displaced from its stable position of equilibrium it is subjected to the influence of the geotropic stimulus. If the stimulus is continued for only a very short time and the organ is then returned to its original position, no curvature may result as the threshold value may not have been reached. With longer periods of stimulation a curvature may result, even when the return to the original position has been made before the beginning of the curvature. Generally a curvature so produced is not permanent as an autotropic curvature will follow. Since the essence of a stimulus is a *change* in a condition, if the new position is maintained the stimulus dies out. It has been pointed out when dealing with the principles of irritability that prolonged stimulation leads either to death of the tissue or to the acquisition of a new position of equilibrium. With geotropic stimulation the latter is the case, the new position of equilibrium involving a permanent curvature and the adoption by the apical part of the organ of a new direction in relation to the other parts of the plant.

The values for the presentation time given in Tables XCII and XCIII indicate that for stimulation by gravity itself (and not by centrifugal force many times greater) the presentation time may be quite considerable; hence, when a klinostat is used to eliminate the action of gravity the slow continuous sequence of stimulations to which the tissue is subjected by the continuous change in its positions are, unless the movement is very slow, sub-threshold in value and produce no response.

It is to be noted, however, that a number of sub-threshold stimulations add together to produce a response if the time elapsing between the stimulations is not too great. This was shown by Fitting for the epicotyl of the broad bean by using a klinostat in which periods of slow rotation alternated with short periods of rest during which the organ was stimulated. When the ratio of the period of stimulation to the period of non-stimulation was 1 : 5 the presentation time (the sum of periods of stimulation only) was the same as with continuous stimulation, but if the period of rest relative to the period of stimulation was increased the presentation time also had to be increased, thus indicating that the stimulations commenced to die out.

In measuring the amount of response to geotropic stimulation the most correct criterion to employ would appear to be the difference in the amount of growth on the two sides of the organ. This is approximately proportional both to the radius of curvature, and the deviation of the apex from its original position. Generally speaking, it is the last-named quantity which is most readily measured.

The numbers given in Tables XCII and XCIII indicate that whatever the intensity of stimulus and period of stimulation within the limits employed, the same (minimum) response is produced by the same value of the product of intensity of stimulus and its time of action. Now this quantity, the product of intensity of stimulation and its duration can be regarded as the quantity of stimulus. It is at least reasonable to expect that if the minimum response depends on the quantity of stimulus, that any definite response is also determined by the quantity of stimulus. We are thus led to the *stimulus-quantity law* according to which the amount of response is determined by the quantity of stimulus, that is, the product of intensity of stimulus and its duration. The law will obviously only hold within limits, for, as we have already seen, the condition of stimulation gradually passes away as a fresh position of equilibrium is reached. There appears, however, to be very little information regarding the quantitative relationship between stimulation and response; and whether, within limits, the amount of response is proportional to the quantity of stimulation, or some other relation holds, is not at present clear. If the Weber-Fechner law (cf. p. 425) holds, then with constant duration of stimulation the response will be proportional to the logarithm of the intensity of the stimulus, and this would mean that as the quantity of stimulus increased the amount of response would increase, but not at the same rate. Data on the matter are, however, still wanting.

It will be observed that the first visible response to geotropic stimulation takes place some time after stimulation, and there is thus a latent time (cf. p. 427), or, as it is usually called by continental workers, a reaction time. The existence of a latent time indicates that between the perception of the stimulus and the response a series of reactions occur which require time. Moreover, as will be shown later, the seat of visible response to the stimulus by the plant is not necessarily the seat of perception, thus indicating a chain of reactions between perception and visible response, not only in time, but in space.

The relation between the intensity of the stimulus and the length of the latent time was investigated by Bach. Seedlings of *Vicia faba* were rotated on a wheel in a vertical plane at different speeds, so that they were subjected to centrifugal forces of various intensities. The stimulation appears to have been continued until the first curvature was observed, so that the duration of stimulation has the same value as the reaction time, but it is to be expected that the quantity of stimulus involved in each case is actually less than the product of intensity and duration since, as we have seen, the stimulus fades out with prolonged duration and, even if this were not so, from the very existence of a latent time it seems clear that the stimulus at the end of this time cannot play a part in producing the curvature. Indeed, Bach found with gravitational stimulation no significant difference in the latent time in plants stimulated for the presentation time only and those stimulated until curvature commenced. However this may be, Bach's data indicate that with increasing stimulation the reaction time diminishes up to a point, but that beyond this point increase in stimulation produces no further effect in speeding up the chain of reactions. There appears thus to be a minimum latent time.

Table XCIV

Relation between Intensity of Stimulus and Latent Time for Shoots of *Vicia faba* seedlings

Intensity of Stimulation in Gravitational Units	Temperature in Centigrade degrees	Mean Latent Time in Minutes
0·014	17–21·5	277
0·056		206
0·099		180
0·20		125
0·31		115
0·80		113
1·25–2·19		87
6·5–10·0	16–19	72·5
10–20		64·8
20–30		61·5
30–40		68·2
40–50		70·4
50–60		63·0
60–70		80·0
70–80		62·5
80–90		76·0
100–100		68·7

The minimum latent time for seedling shoots of *Vicia faba* thus appears to be about 67 minutes when the experiments are carried out at about 18° C. This is only slightly less than that observed with control plants placed horizontally and so exposed to gravitation stimulus, in which case the average latent time was 72 minutes. It may thus be concluded that with an intensity of stimulation equal to that of gravity the minimum latent time is almost reached, and that, provided duration of the stimulus is sufficient to produce a curvature at all, the period of stimulation is without influence on the latent time. In earlier observations on roots of *Vicia faba* and *Lupinus albus* Czapek also found a minimum latent time, but in his experiments carried out at about 17° C. this was only reached when the intensity of stimulation was 40 times the force of gravity. The minimum latent time in these experiments was found to be about 45 minutes.

THE INFLUENCE OF EXTERNAL CONDITIONS ON GEOTROPIC IRRITABILITY

Various external conditions may influence geotropic irritability, notably temperature, light and various chemical substances.

Temperature. The influence of temperature on the presentation time and latent time for geotropic reaction was examined by Czapek in 1898 in seeding roots of *Lupinus albus* and *Zea mais* and hypocotyls of *Helianthus annuus.* The values he found for the presentation time and latent time at different temperatures in the first of these is shown in Table XCV. These results suggest that temperature has little effect on the response to geotropic stimulation until the extremes of the temperature range under which the plant lives are approached. Bach's experiments with shoots of *Vicia faba,* however, indicated a much greater dependence of reaction to geotropic stimulation on temperature.

Table XCV

Effect of Temperature on the Presentation Time and Latent Time for Geotropic Reaction in Seedling Roots of *Lupinus albus*

Temperature in Centigrade degrees	Presentation Time in Minutes	Latent Time in Minutes
5	45	360
10	30	120
15	20	80
20	20	80
25	20	80
30	20	70
39	25	120

His results are summarized in Table XCVI. They show that the presentation time is much shorter for this organ than for roots of *Lupinus albus.* Both sets of data agree, however, in that towards the lower and higher limits of the normal temperature range for growth, both presentation time and latent time increase. The results of a recent careful investigation by Miss Hawker of the influence of temperature on the geotropism of the epicotyls of seedlings of *Lathyrus odoratus,* agree with those of Bach rather than with the earlier ones of Czapek. Miss Hawker found the sensitivity to geotropism in her material to reach a maximum when the epicotyls were 6 to 10 cm. long. The values

Table XCVI

Effect of Temperature on the Presentation Time and Latent Time for Geotropic Reaction in Seedling Shoots of *Vicia faba*

Mean Temperature in Centigrade degrees	Presentation Time in Minutes	Latent Time in Minutes
14	14	122·8
17	11	115·4
20	7·5	97·9
25	3	64·8
30	2	48·2
35	4	80·8

for the presentation time and latent time for epicotyls at this stage of development are given in Table XCVII. In the epicotyls of the sweet pea, as in the seedling shoots of the broad bean, maximum sensitivity to gravitational stimulation occurs at about 30° C., while the presentation time is very nearly the same in the two species examined. The latent time, however, is decidedly shorter in the sweet pea than in the broad bean.

Table XCVII

Effect of Temperature on the Presentation Time and Latent Time for Geotropic Reaction in Epicotyls of *Lathyrus odoratus*

Temperature in Centigrade degrees	Presentation Time in Minutes	Latent Time in Minutes
5	60	180
10	26	100
15	16	55
20	8	35
25	4	31
30	3	30
35	5	32
40	15	50

Miss Prankerd's observations on the effect of temperature on the latent time for geotropic response of fronds of *Asplenium bulbiferum* show the same general relationship as in the species already discussed. Both the pr esentation time and latent time are reduced with increase in temperature between 8° and 20°. As the numbers in Table XCVIII show, the presentation time at 10° is about double that at 20°, while the latent time at 10° is about 2·36 times that at 20° C. The results are thus in agreement with the Van't Hoff rule. A temperature of 30° is progressively injurious to the graviperceptional sensitivity of the frond, exposure

Table XCVIII

Effect of Temperature on the Presentation Time and Latent Time for Geotropic Reaction in Fronds of *Asplenium bulbiferum*

Temperature in Centigrade degrees	Presentation Time (That at 20° = 1)	Latent Time in Hours
8	3	19
10	2	13
13	1·7	11
16·5	1·3	8
20	1	5·5

to this temperature increasing the presentation time at once by one-third, and by two-thirds after exposure to the temperature for a day. A time factor thus appears to operate as regards the power of perception, but the latent time is the same as that at 20° and does not appear to be affected by the time factor introduced by the high temperature. That this latter affects adversely the initial part of the chain of actions (that is, the perception) and not the subsequent stages (the reaction) is also indicated by the results of experiments in which the fronds were stimulated at one temperature and allowed to respond at another. Thus, stimulation at 20° followed by transference to a temperature of 30° produces the same response as when the temperature is allowed to remain at 20°, whereas stimulation at 30° followed by response at 20° results in a decrease of about 56 per cent. in the number of fronds responding.

That the temperature to which the material is subjected before and during stimulation may have a considerable effect on sensitivity to the gravitational stimulus had been shown earlier by Rutgers, who stimulated coleoptiles of oat at various temperatures to which they had been subjected for 1 hour or longer prior to a stimulation. After stimulation they were allowed to respond at room temperature (18° to 21° C.). As the numbers given in Table XCIX show, the presentation time is greatly affected by temperature. At lower temperatures the presentation time was the same whether subjection to the experimental temperature was continued for 1 or 4 hours. Continued exposure to a temperature of 30° C. increased the sensitivity but, as the numbers in the table show, continued exposure to higher temperatures reduced the sensitivity. We again have an example of the time factor introduced by high

Table XCIX

Effect of Exposure to Various Temperatures previous to, and during stimulation, on the Presentation Time for Geotropic Reaction of Coleoptiles of *Avena sativa*

Temperature in Centigrade degrees	Presentation Time in Minutes after these different Periods of Exposure to the Experimental Temperature					
	1 hour	2 hours	4 hours	6 hours	12 hours	24 hours
0	72	—	72	—	—	—
5	16	—	16	—	—	—
10	10·7	—	10·7	—	—	—
15	6	—	6	—	—	—
20	4·3	—	—	—	—	—
25	2·3	2·3	2·3	2·3	—	—
30	3·5	3·2	2·2	1·8	1·7	1·7
35	2·5	3·5	4	4	—	5
37	9·3	16	—	21·7	—	21·7
38	11·5	19·2	38	53	75	347
39	23	40	—	—	—	—
40	260	—	—	—	—	—

temperature. The latent time was found to be very little affected by the temperature.

Results in agreement with these findings were obtained by Miss Hawker with sweet pea seedlings. Seeds were germinated at 20° C. and then subjected to various treatments before stimulation at 20° C. and 30° C. The presentation times and latent times at 20° C., at the most sensitive stage of the epicotyl, after the various treatments are shown in Table C.

Table C

Effect of Various Treatments before Stimulation on the Presentation Time and Latent Time at 20° C. for Geotropic Reaction of Epicotyls of *Lathyrus odoratus*

Treatment	Presentation Time in Minutes	Latent Time in Minutes
Grown at 5° C. all the time	25	60
Grown at 5° C.; then exposed to 20° C. for 24 hours before stimulation	12	
Grown at 5° C.; then exposed to 20° C. for 48 hours before stimulation	8	35
Grown at 20° C. all the time	8	35
Grown at 20° C.; then exposed to 5° C. for 24 hours before stimulation	16	

The bearing of the information summarized above on the theory of the mechanism of the geotropic reaction will be dealt with later in this chapter; it may, however, be noted here that the effect of temperature on the geotropic reaction is similar to that of temperature on a number of metabolic processes, in that there is an increase in the rate of the process with increase in temperature up to about 30° C. above which there is a falling off in rate. It would appear that here also a time factor operates as in the case of metabolic processes, but the methods of experimentation so far evolved to examine the effect of temperature on geotropic irritability do not permit a very exact examination of this question.

Light. The few recorded observations that bear on the question of the influence of light on irritability to gravity suggest that exposure to light brings about a decrease in graviperception. In 1895 Czapek had already found that illumination of *Avena* and *Phycomyces* for 10 minutes delayed the appearance of geotropic curvature. A more systematic investigation of the question was made by Krones in 1914. Seedlings of *Avena* were illuminated for various periods (0·5, 1 and 2 hours) with different light intensities (125, 250 and 500 metre-candles) and then stimulated in a horizontal position for 3 minutes, experiments being conducted at about 21° C. The higher the light intensity and longer the exposure, the smaller the percentage of seedlings which executed geotropic curvatures as a result of this stimulation. These results are summarized in Table CI. For comparison it should be observed that the percentage of curvatures of seedlings kept in the dark, was, on the average, 80·5 per cent. The numbers given in the table suggest that, within limits, the degree of lowering of the geotropic sensitivity is determined by the quantity of light falling on the organ. It will be recalled that in this material light brings about a lowering of the rate of growth, but according to Pringsheim and Krones, differences in growth rate affect the time relations in the geotropic reaction very little. The explanation of the effect of light in lowering the sensitivity to graviperception must therefore be sought elsewhere than in the light-growth reaction.

Table CI

Effect of Pre-illumination on the Sensitivity to Gravity of Seedlings of *Avena sativa*

Intensity of Illumination in MK.	Percentage of Seedlings executing Geotropic Curvature after Pre-illumination for		
	0·5 hour	1 hour	2 hours
125	60·6	51·7	49·9
250	52·3	43·1	38·6
500	43·5	35·1	30·5

Pressure. The effect of reduced atmospheric pressure on geotropic irritability of the radicles of *Phaseolus vulgaris* was recorded in 1912 by Paál. To determine the influence of pressure on the presentation time and latent time the radicles were stimulated in air of known pressure for various times and then the seedlings were rotated on the klinostat, the presentation time being taken as the period of stimulation necessary for 50 per cent. of the roots subsequently to exhibit a curvature. An attempt was also made to distinguish between the effects of reduced pressure on presentation time and on latent time. To this end the radicles were stimulated in normal air and then transferred to air under reduced pressure.

Table CII

Influence of Reduced Atmospheric Pressure on the Presentation Time and Latent Time for Reaction to Gravity of Radicles of *Phaseolus vulgaris*

Pressure in Atmospheres	Presentation Time in Minutes	Latent Time (that under atmospheric pressure = 1)
1	6	1·0
0·74	20	1·09
0·61	25	1·15
0·34	30	1·39
0·21	35	1·60
0·14	50	1·80
0·08	70	2·20

Paál's values for the latent time given in the above table were obtained when the seedlings were maintained under the reduced pressure throughout the periods both of perception and of response. When the perception took place in air at normal pressure it was found that the response was indeed slower in air under reduced pressure than under normal pressure, though the lengthening of the latent time was not very significant except in quite low pressures (cf. Table CIII).

Table CIII

Influence of Reduced Pressure on the Latent Time for Reaction to Gravity of Radicles of *Phaseolus vulgaris* stimulated in Normal Air.

Pressure in Atmospheres	Latent time under reduced pressure / Latent time under normal atmospheric pressure
0·34	1·19
0·28	1·40
0·21	1·42
0·14	1·53

The not improbable suggestion has been made that the slowing down of both the perception and response phases of the geotropic reaction in reduced pressure is related to respiration.

Oxygen. Since in aerobic organisms growth only takes place with a sufficient supply of oxygen, it follows that in continued absence of oxygen, curvatures dependent on differential rates of growth cannot take place. The actual concentration of oxygen necessary for geotropic curvature to take place may be very low. Wortmann mentions that seedling roots of *Vicia faba*, *Phaseolus vulgaris* and *Ph. multiflorus* and stems of *Paeonia peregrina* do not respond to geotropic stimulation in an atmosphere devoid of oxygen. He noted that on addition of oxygen to the environment the power of geotropic response returns even in the case of plants which have been deprived of oxygen for weeks. Later he extended his observations to *Helianthus annuus* and *Lepidium sativum*. Kraus also observed that flowering shoots of *Taraxacum*, *Ranunculus* and *Anthriscus* fail to respond to gravitational stimulation in an atmosphere of carbon dioxide or of hydrogen, but regain the power of response when again brought into contact with oxygen.

The most critical examination of this question appears to be that made in 1917 by van Ameijden who found that when coleoptiles of *Avena sativa* and hypocotyls of *Sinapis alba* after being maintained in an atmosphere of nitrogen for 6 hours, are stimulated in nitrogen and then replaced in air, no curvature takes place. If the pre-treatment in nitrogen is for a shorter period, say 1 to 3 hours, a curvature still occurs, while if after the longer period in nitrogen stimulation is carried out in air and the latter atmosphere maintained, a reaction takes place, although the curvature is less than that which occurs under normal conditions. In *Sinapis alba*, according to Correns, a concentration of from 3 to 4 per cent. oxygen is necessary for geotropic reaction, but in *Helianthus* curvatures were observed in much lower oxygen concentrations. The observations of van Ameijden on geotropic response of *Avena sativa* subjected to an atmosphere with a low oxygen content

(4·3 per cent.) show that the coleoptile of this plant under such a condition remains for many hours capable of reacting normally to the gravitational stimulus, but that after 24 hours there seems to be a definite reduction in sensitivity to gravity.

Various Gases. Molisch found that small quantities of tobacco smoke have the effect of greatly lessening the sensitivity to graviperception of seedlings of a number of species including *Vicia sativa*, *Pisum sativum* and *Ervum lens*. Smoke from other sources has a like effect and so have small amounts of coal gas. According to Knight and Crocker, the ethylene and carbon monoxide present are chiefly responsible for, while acetylene probably contributes to, the adverse influence of smoke and coal gas.

Other Poisonous Substances. Chloroform and ether also lower the sensitivity to graviperception. According to Czapek, chloroform in certain concentrations, in addition to lowering the power of roots of *Vicia faba* and *Lupinus albus* to perceive a gravitational stimulus, also inhibits all expression of response, but that while no reaction is visible during exposure to such chloroform solutions, curvature can follow removal of these plants into normal conditions. Several workers have recorded a lengthening of the latent time for the geotropic reaction when the material has been subjected to treatment with a solution of eosin before stimulation, and Neukirchen has confirmed this with seedlings of cereals and also obtained the same effect with formaldehyde and copper sulphate. According to the same observer, pre-treatment with arsenious acid accelerated the action.

Metallic Ions. Some divergence appears in the recorded results on the effect of ions of the alkali and alkaline earth metals. While Cholodny found that K, Na and NH_4 ions retarded the geotropic reaction in seedling roots and that the alkaline earth ions Ca and Mg had little effect, Warner found that only K and Na exerted a retarding effect, as compared with controls in distilled water, while Li and NH_4, as well as the bi-valent ions Ca, Mg, Sr and Ba, accelerated the reaction, the order of the ions in bringing about the accelerating action being the same as their order of absorption by plant tissue (cf. p. 74). Cholodny has, however, objected that distilled water, as compared with balanced nutrient solutions, itself retards the geotropic reaction, so that although acceleration of the reactions in comparison with that obtained with distilled water may occur, there may be not merely no furthering of the reaction, but even retardation in comparison with that which takes place in seedlings growing on a substratum containing nutrient salts in a balanced mixture.

VARIATIONS IN GEOTROPIC IRRITABILITY DURING DEVELOPMENT

Already in 1905 Fitting had found that the presentation time and latent time for the geotropic reaction were greater in *Vicia faba* epicotyls

1–2 cm. long than in epicotyls 3–5 cm. long, and Bach also showed that the presentation time and latent time decreased in this species with development of the seedling.

That considerable variation in geotropic irritability occurs during development has been made plain by the careful investigations of Miss T. L. Prankerd and other workers of her school. It has been shown that fern fronds, as they develop, exhibit a 'grand period' in respect of geotropic irritability. Thus, in fronds of the fern *Asplenium bulbiferum* geotropic irritability is present from the first appearance of the frond until it is quite uncurled. Over this period the irritability, as indicated both by presentation time and latent time, rises to a maximum as development proceeds, and then falls, until at the end of the period the frond is no longer sensitive to the gravitational stimulus. To express the different stages of development in fern fronds certain terms were applied, namely, infant, when the leaflets are contained in the apical coil, adolescent when the leaflets are unfolding, and mature when the frond has uncurled. By the time the last stage is reached the irritability to gravity, as we have just seen, has disappeared. The earlier stages can be subdivided into early infant (fronds 0·8 to 3·0 cm. long), middle infant (fronds 3·0 to 5·5 cm. long) and late infant (fronds over 5·5 cm. long with no leaflets unfolded), and adolescent 1, adolescent 2 and so on, according to the number of pairs of leaflets unfolded. Each of these stages can be subdivided if necessary. From observations made by Miss Waight it appears that at 20° C. and 85 per cent. humidity the presentation time decreases from 8 hours at a very early stage of development to a minimum of 0·5 hour when the fifth to the seventh pair of leaflets are unfolding. It then rises to about 6 hours just before the fronds are mature and graviperception is no longer observable. The latent time follows a similar course, being about 16 hours at the very early stage of development, falling to a minimum of 5 hours and then rising. Actual data obtained by Miss Waight are shown in Table CIV.

Table CIV

Variation in Presentation Time and Latent Time of Fronds of *Asplenium bulbiferum* in Relation to Graviperception during Development

Stage of Development	Presentation Time in Hours	Latent Time in Hours
Early infant 1st stage	8	16
,, ,, 2nd ,,	4	10
Middle infant	3	8
Late infant	2	6
Adolescent 1	1·5	5·5
,, 2	1	5·5
,, 3–4	0·75	5·5
,, 5–7	0·5	5
,, 8	1	5·75

A similar course in the development of geotropic irritability in fronds of another fern, *Osmunda regalis*, was observed by Miss Prankerd, but in this species the power of response to the gravitational stimulus persists to a later stage than in *Asplenium bulbiferum*, while the sensitivity is greater. There also appears to be a definite increase in sensitivity to gravity in the fronds as the plant grows older. The results are shown in Table CV. The sensitivity values given in the last column are obtained by assuming that the sensitivity is inversely proportional to the presentation time.

Table CV

Variation in Presentation Time and Latent Time of Fronds of *Osmunda regalis* in Relation to Graviperception during Development

Stage of Plant	Stage of Frond	Length of Frond in cm.	Presentation Time in Minutes	Latent Time in Hours	Relative Sensitivity
1st year . .	Early infant	0·46	300	7·5	2
	Middle infant	0·8	120	3·5	5
	Late infant	1·6	25	2·8	24
	Adolescent 1	2·6	25	2·7	24
	Mature	2·8	20	2·0	30
	Mature 1	3·3	40	3·0	15
2nd year. .	Early infant	0·63	180	6·6	1
	Middle infant	1·2	45	2·9	4
	Late infant	2·8	18	2·25	10
	Adolescent 1	5·5	15	2·7	12
	,, 2	7·5	12	2·0	15
	Mature	8·4	3	2·0	60
Adult. . .	Early infant	2·6	150	6·15	1
	Late infant	13·3	7	1·75	21
	Adolescent 1	14·8	6	1·38	24
	,, 2	15·9	5	1·5	30
	,, 3	21·3	3	1·5	50
	,, 4–6	33·2	1	1·38	150
	Sub-mature	38·2	1 or ½	1·25	150 or 300
	Mature	44·4	1	1·63	150

Similar changes in geotropic sensitivity during development are probably usual throughout vascular plants. At any rate they have also been observed in the seedlings of a number of flowering plants examined by Miss Hawker. Here again, with development of the organ the sensitivity increases, reaches a maximum and then declines. The values for the presentation time obtained by Miss Hawker in experiments carried out at 20° C. are summarized in Tables CVI and CVII.

Table CVI

Variations in Sensitivity to Stimulation by Gravity of Epicotyls of Seedlings of *Lathyrus odoratus* during Development

Length in cm.	Presentation Time in Minutes	Latent Time in Minutes
0–1	20	45
1–2	16	42
2–4	12	39
4–6	9	35
6–10	8	35
10 + ...	12	40

Table CVII

Variations in Sensitivity to the Gravitational Stimulus of Hypocotyls of Seedlings during Germination

Length of Organ in cm.	*Ricinus communis*		*Cheiranthus allionii*		*Calendula officinalis*		*Nigella damascena*	
	Presentation Time in Minutes	Latent Time in Minutes	Presentation Time in Minutes	Latent Time in Minutes	Presentation Time in Minutes	Latent Time in Minutes	Presentation Time in Minutes	Latent Time in Minutes
0–1 .	—	—	} 20	} 60	} 15	} 70	18	55
1–2 .	—	—					11	50
2–3 .	—	—	7	40	12	60	12	57
3–4 .	}	}	6	40	11	60	—	—
4–5 .	} 20	} 45	6	45	12	60 +	—	—
5–6 .	}	}	13	45 +	—	—	—	—
6–9 .	10	40	—	—	—	—	—	—
9–12 .	5	35	—	—	—	—	—	—
12–15 .	6	35	—	—	—	—	—	—
15–18 .	20	45	—	—	—	—	—	—
18 + .	20 +	45 +	—	—	—	—	—	—

The general conclusion to be drawn from Miss Hawker's results is that in organs of, theoretically, unlimited growth above ground, sensitivity to the gravitational stimulus increases until a maximum is reached, after which the sensitivity falls off somewhat to a value which remains constant. In organs of limited growth, such as hypocotyls, the sensitivity increases to a maximum and then falls off until the organ loses its sensitivity with cessation in elongation. Exceptions to this rule were, however, observed by Miss Brain in three species of *Lupinus* (*L. albus*, *L. polyphyllus* and *L. arboreus*), in which the sensitivity of the hypocotyl to gravity increases to a maximum value which is maintained until growth in length ceases. In roots also, according to Miss Hawker, the sensitivity remains constant, at any rate for some time, after the maximum value has been reached. (Table CVIII.)

Table CVIII

Variation in Sensitivity to the Gravitational Stimulus of Primary Roots of Seedlings during Development

Length of Root in cm.	Presentation Time in Minutes		
	Lathyrus odoratus	*Helianthus annuus*	*Cucurbita pepo*
0–2 . . .	60	40	30
2–4 . . .	30	25	20
4–6 . . .	25	20	18
6 + . . .	25	20	18

Variations in sensitivity to stimulation by gravity were also observed by Miss Brain in the development of the inflorescence in *Lupinus albus*, and *L. arboreus*, the presentation time of the inflorescence axis of the former species falling from 20 minutes to 2·5 minutes when sensitivity was greatest, and then rising to 10 minutes. In the inflorescence axis of *L. arboreus* there was observed a fall in presentation time from 120 minutes to 20 minutes, without any subsequent decrease in sensitivity.

DIFFERENCES IN SENSITIVITY TO THE GRAVITATIONAL STIMULUS IN DIFFERENT SPECIES

From the values for presentation time and latent time already given it will be evident that all tissues, even in the same stage of development and under the same external conditions, probably do not react with the same rapidity to stimulation by gravity. As long ago as 1898 Czapek determined the presentation time at 25° C. for reaction to the gravitational stimulus of a number of species, and obtained values varying from 15 minutes for sporangiophores of *Phycomyces blakesleeanus*, coleoptiles of *Avena sativa* and *Phalaris canariensis* and hypocotyls of *Beta vulgaris* to 50 minutes for the radicles of a large-seeded variety of *Vicia faba* and the epicotyls of *Phaseolus vulgaris*. These values, compared with ones determined later, all appear high, a result later attributed by Czapek to light shortage and impurities in the air of his laboratory. Some later determinations by Fitting and Bach are summarized in Table CIX. The temperature in Fit-

Table CIX

The Presentation Time for Geotropic Stimulation in Various Plant Organs

Species	Organ	Presentation Time in Minutes	Observer
Vicia faba	Epicotyl	6–7	Fitting
,, ,, ...	,,	5	Bach
,, ,, ...	Radicle	6	,,
Helianthus annuus . . .	Hypocotyl	5–6	Fitting
,, ,, ...	,,	< 3	Bach
Phaseolus multiflorus . .	Epicotyl	6–7	Fitting
,, ,, ...	,,	3–4	Bach
,, ,, ...	Radicle	7–8	,,
Sinapis arvensis . . .	,,	20–25	Fitting
Sinapsis alba	,,	20–25	,,
Lens sp.	,,	20–25	,,
Cucurbita pepo	Hypocotyl	6	Bach
Tropaeolum sp. . . .	Epicotyl	8–9	,,
Panicum sanguinale . .	Coleoptile	10	,,
Setaria alopecuroides . .	,,	12	,,
Lupinus albus	Hypocotyl	20–25	,,
Capsella	Flowering shoot	< 2	,,
Sisymbrium officinale . .	,, ,,	3	,,
Plantago lanceolata . .	,, ,,	3	,,
Plantago media . . .	,, ,,	3	,,

ting's experiments was 22° C., in Bach's between 20° and a few degrees under 30° C.

But since, as we have seen, the sensitivity varies with development, a better criterion of the sensitivity in different species is provided by the values for the presentation time when the organs are in their most sensitive condition. Such values, found by Miss Hawker, at 20° C., are summarized in Table CX. The minimum values for the ferns examined by Miss Prankerd are included for the sake of comparison. From the data given in this table it will be seen that the most sensitive material to gravitational stimulus so far examined is the mature frond of the adult plant of *Osmunda regalis,* a striking fact in view of Darwin's doubt regarding the possession by fern fronds of the power of response to the gravitational stimulus. On the whole dicotyledons are more sensitive than monocotyledons and conifers.

Table CX

Minimum Presentation Time and Latent Time for Various Plant Organs

Species	Organ	Minimum Presentation Time in Minutes	Minimum Latent Time in Minutes
Ricinus communis . .	Hypocotyl	5	35
Cheiranthus allionii . .	,,	6	40
Calendula officinalis . .	,,	11	60
Nigella damascena . .	,,	11	50
Helianthus annuus . .	,,	5	52
,, ,, . .	Root	20	—
Beta vulgaris	Hypocotyl	10	46
Lathyrus odoratus . .	Epicotyl	8	35
,, ,, . .	Root	25	43
Cucurbita pepo . . .	Hypocotyl	5	43
,, ,, . . .	Root	18	—
Allium cepa	Cotyledon	6	40
Canna indica	Seedling	27	120
Commelina coelestris . .	,,	18	62
Asparagus officinalis . .	Epicotyl	3	46
Larix europea	Hypocotyl	25	75
Picea pungens, var. *glauca*	,,	30	80
Cupressus arizonica . .	,,	40	90
Asplenium bulbiferum .	Adolescent frond	20	300
Osmunda regalis . . .	Mature frond	0·5–1	75

It is perhaps advisable to mention that the determinations of presentation time by different investigators are not strictly comparable unless all internal and external conditions are the same, and the same criterion is used to determine the first indication of curvature. Miss Prankerd and other workers of her school take as the presentation time the period which will generally produce a movement of about 5°, and only very rarely exceed 10°, in 75 per cent. of the individuals used. In Bach's experiments the time required to induce curvature in only 50 per cent. of the individuals was taken as the presentation time. Having regard to these considerations the degree of agreement between Bach's results and Miss Hawker's is all that can be expected and the numbers given in Table CX can be confidently accepted as affording a satisfactory indication of the variation in geotropic sensitivity in different species.

BILATERAL SYMMETRY IN GEOTROPISM

In Table CX are recorded the minimum presentation times for a number of seedling organs. It has, however, been found by Miss Brain and confirmed by Miss Hawker, that some seedlings exhibit a physiological zygomorphy in regard to response to the gravitational stimulus, in that the response is more rapid when stimulation is effected in one plane than when it is effected in another. Thus, in seedlings of *Lupinus polyphyllus* stimulated at 20° C. in the cotyledonary plane, that is, with the plane of the cotyledons horizontal, the minimum presentation time of the hypocotyl is 20 minutes, whereas when stimulated in the intercotyledonary plane, that is, with the plane of the cotyledons vertical, the minimum presentation time is 4 times as great. A similar behaviour was observed in *Lupinus arboreus* where the presentation time for stimulation of the hypocotyl in the cotyledonary plane was found to be 20 minutes, as compared with 60 minutes in the intercotyledonary plane. In *Lupinus albus*, on the other hand, the seedling exhibits radial symmetry in respect to geotropic stimulation, the minimum presentation time being 20 minutes whatever the plane of stimulation. Other species in which this bilateral symmetry in regard to geotropic stimulation has been observed are *Cucurbita pepo*, *Helianthus annuus* and *Aquilegia vulgaris*.

DIAGEOTROPISM

Diageotropic organs are those which normally orientate themselves with their axis or surface at right angles to the direction of the gravitational force. They can conveniently be divided into two groups, those with a morphologically radial organization such as horizontally growing rhizomes, and those with a dorsiventral structure, such as many leaves. Among the former may be cited the rhizomes of *Eleocharis palustris*, *Sparganium ramosum*, *Scirpus maritimus*, *Adoxa moschatellina* and *Circaea lutetiana*. When such a rhizome is tilted out of the horizontal it executes a curvature so that the apical region again comes to lie horizontally. If the organ is physiologically, as well as morphologically radial, it will be observed by reference to Fig. 53 that its stable and labile positions of equilibrium are quite different from those of an orthogeotropic organ. In the transversely geotropic, radially organized structure both A and C are stable positions of equilibrium, for, owing to its radial organization, there is no alteration in the relation of the rhizome to gravity brought about by rotating it on its own axis. Furthermore, the positions B and D are both labile positions of equilibrium. If diagetropic orientation is the result of simple geotropic stimulation, and the growing part is free to curve in any direction, a curvature should result in bringing the apical part of the organ to lie parallel with the horizontal axis of the klinostat. If the diageotropism is accompanied by orthogeotropism, the latter can be eliminated on the klinostat and the diageotropism can still be apparent. But a true diageotropism has never been demonstrated in this way and it is in doubt whether it actually exists. Rawitscher's experiments with *Asparagus plumosus* in which the side shoots exhibit diageotropism certainly indicate that in this plant the conditions are complex. Thus, if a primary vertically directed shoot with its diageotropically directed side shoots is displaced through a right angle, the growing tips of the side shoots all curve so that the tips come again to lie horizontally, but they do not curve half in one direction and half in the other, but are all directed towards the morphological base of the plant. If after some days the shoot system is replaced in its original position with the main axis again directed vertically, the side shoots which were on the physically upper side during the displacement now again curve so that the tips take up a horizontal position, but they are now directed upwards towards the main axis. The side shoots which were on the under side of the main axis during the displacement also curve, but they are directed away from the main axis. Again, if such a shoot system of *Asparagus plumosus* is displaced so that the main shoot points obliquely downwards, the tips of all side shoots at first curve in the upward direction. This curvature brings the tips of the side shoots on the under side to their stable horizontal position, but takes the tips of the shoots on the upper side farther away from it. But these latter now curve in the reverse direction so that they also come to be orientated horizontally.

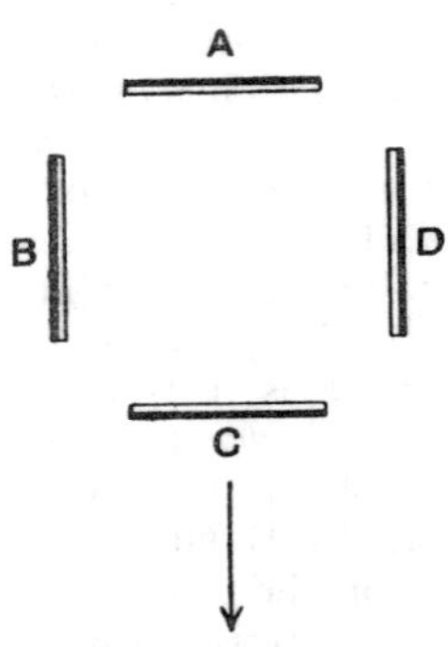

FIG. 53.—Scheme to illustrate stable and labile positions of equilibrium of a diageotropic radially symmetrical organ

(To distinguish the two sides of the organ, one is shown black, the other white; this is not intended to indicate any kind of dorsiventrality. The arrow indicates the direction of the gravitational pull.)

Among dorsiventral organs exhibiting diageotropism are many leaves, for example, those of *Fragaria vesca, Viola odorata* and *Malva neglecta.* Here also it is extremely doubtful whether a simple geotropic action will explain the normal orientation of these leaves, but no complete analysis of the phenomena involved has so far been achieved. Even the different parts played by lamina and petiole in bringing about orientation movements of such leaves are for the most part not recognized without doubt.

In some shoots with a morphologically radial organization there appears to be physiological dorsiventrality in regard to geotropism. Thus diageotropic side stems of *Atropa belladonna*, when twisted through 180° C. so that the formerly upper surface now faces downwards, execute a curvature in the growing region so that the apex is directed horizontally inwards. If side shoots of *Taxus* and *Picea* are similarly twisted, the original position is regained by the apex owing to a torsion of the stem. In the case of these conifers, although the stem itself is organized radially, the shoot as a whole exhibits a certain degree of dorsiventrality since the needles are a different size and differently orientated, on the upper and lower sides, respectively, of the shoot.

PLAGIOGEOTROPISM

Plagiogeotropism, using the word in its narrow sense so as to exclude diageotropism, refers to the special type of geotropism in which the rest position of the organ is such that the axis makes an angle other than a right angle with the vertical. The most familiar examples of plagiogeotropic organs are side roots of the first order which generally make an acute angle with the vertical. When these are displaced from their normal position they respond to the stimulus by a curvature which tends to re-orientate the growing apices in their original direction in relation to gravity. It is to be noticed that the angle made by side roots with the vertical is not constant from one lateral root to another, and in general the angle made by the root with the vertical increases with distance from the apex of the main root. Some side shoots and leaves also exhibit plagiogeotropism.

As with diageotropism, so with plagiogeotropism, the question arises whether a true simple plagiogeotropism, in which the position taken up by the organ depends on gravity only, really exists. It may be noted firstly, that a plagiogeotropic organ can become orthogeotropic, as, for example, when the apical bud of a stem or the apical part of a root is removed and its place taken by a lateral branch. When this takes place the formerly plagiogeotropic organ becomes orthogeotropic. It has therefore been suggested that plagiogeotropism is not inherent in the lateral branch, but that the branch is really orthogeotropic and that the actual position taken up by it is the resultant of orthogeotropism and some internal influence such as epinasty, but other explanations have been offered. Thus Lundegårdh has supposed that the rest position of side roots is determined by the combined action of normal geotropism, lengthening force and negative geotropism. However, the analysis of the problem is extremely difficult and a discussion of it would be out of place here.

THE SEAT OF GRAVIPERCEPTION

In our consideration of the principles of irritability it has been mentioned that the seat of perception of the stimulus is not necessarily the place of response. In the case of growth curvatures the response must obviously occur in the region of growth, and direct observations show that the curvatures in such cases take place in cells undergoing elongation. That the seat of perception of the gravitational stimulus was not that of the response was suggested by Ciesielski's experiments described in 1871 in which he found that decapitated radicles of seedlings of *Pisum*, *Ervum* and *Vicia*, when placed horizontally, did not execute a geotropic curvature. If seedlings with intact radicles were placed with these latter horizontal and were decapitated after a short time, the geotropic curvature took place. These results which were confirmed and extended a few years later by Charles and Francis Darwin, strongly suggest that the seat of graviperception is in the root tip. It has been shown that decapitation or other injury to the root tip does not materially lower the growth rate of the root for a time, so that failure of the geotropic reaction in decapitated roots cannot be accounted for by suppression of growth. There is, however, still the possibility that the damage to cells in the root may constitute a 'shock' stimulus which, by conduction of the stimulation so produced, has the effect of suppressing the geotropic curvature. To overcome possible complexities introduced by decapitation Czapek devised his famous experiment in which seedling beans are rotated on the klinostat while the radicle is made to grow into a little glass boot so shaped that the apical part of the radicle comes to make a right angle with the rest of that organ, including the growing zones. It is then possible to subject the apex, but not the growing region, to gravitational stimulation, and vice versa. Czapek thought by this means to have shown that when the apex was directed horizontally the growing region executed a curvature, while when the growing region was horizontal and the tip directed vertically downwards no curvature took place, thus indicating that the root tip was the seat of perception of the stimulus, the stimulation being subsequently directed to the seat of curvature, or motor part. However, repetition of Czapek's experiments by several observers has in every case given negative results. Further, it has been pointed out that even if Czapek's interpretation of his experimental results were correct it would not follow that the growing region behind the apex is insensible to gravity. If it were stimulated, only to a less degree than the root tip, the result recorded by Czapek would still be obtained, since in the case of the plant with horizontal growing region and vertical root tip any slight curvature induced by stimulation of the growing region would bring the tip into a position in which it would be stimulated and so bring it back to the vertical. That this may indeed be the case is indicated

FIG. 54.—The arrangements of a seedling on the klinostat in Piccard's method

To face page 469

by the observations of Newcombe that curvature of decapitated roots can be induced by the action of centrifugal force if this is great enough, namely, of a magnitude 8 to 40 times the force of gravity.

Further confirmation of the view that perception of the gravitational stimulus is not confined to the root apex is obtained by the use of a method introduced by Piccard in 1904. In this a disk is rotated in a vertical plane about a horizontal axis and the seedling attached excentrically with the root directed at an angle over the axial line of the disk (cf. Fig. 54). When the disk is rotated at a speed sufficiently high to produce a reaction to centrifugal force, if both the apical part of the root and the more basal part are sensitive organs for graviperception the reaction of the apical and basal parts will then tend to be in opposite directions. Working with this method Haberlandt found that with seedlings of *Vicia faba*, *Phaseolus* and *Lupinus*, if the apex projects to the extent of 1·5 to 2 mm. beyond the axis of rotation the root curves in the direction to be expected if only the tip were perceptive; if on the other hand only 1 mm. of the root projects beyond the axis of rotation the root curves in the direction to be expected if only the part of the root behind the apex were perceptive. This result is evidence that not only the apex, but the part of the root behind the apex including the growing zone, also perceives the stimulus. But since, owing to the shorter distance from the axis of rotation, the average intensity of stimulus is less in the apical region than towards the base, it follows that the perception or some early stage in the response, is much stronger in the *apical* region than in the region behind the apex.

As regards the perception of the gravitational stimulus by negatively geotropic organs, Francis Darwin in 1899 found that when the coleoptile of certain grass seedlings, namely, those of *Setaria* and *Sorghum*, is pushed into a horizontal glass tube, the mesocotyl (see p. 448) first curves upwards, and the curvature continues so that the mesocotyl curves into a spiral. This was held by Darwin to indicate that the coleoptile is the seat of geotropic perception in these plants, it being assumed that while the horizontal position is maintained the stimulation continues with the result that the curvature constituting the response continues as well.

Further investigation on the localization of graviperception in Gramineae seedlings has been made by Piccard's method. Francis Darwin in this way decided that the coleoptile must be the chief seat of perception in *Sorghum*, and the same result has subsequently been obtained in other grasses, as, for example, *Setaria*. In seedlings of *Avena*, *Hordeum* and *Phalaris*, Von Guttenberg found that the apical part of the coleoptile is more sensitive than the basal part, and he estimated that in his experiments with *Avena* the apical region was about 6 times as sensitive as the base. It is likely that this ratio depends on the intensity of stimulation used and possibly on other internal and external conditions.

It may be noted that the coleoptile is not only the perceptive organ, but that so long as it is capable of growth it is capable of reacting to the geotropic stimulus. But when the power of growth has ceased, obviously its capacity to react by means of a growth curvature also ends.

The behaviour of a number of dicotyledonous seedlings, *Brassica napus*, *Lepidium sativum*, *Linum usitatissimum* and *Vicia sativa* was examined by Herzog who concluded that over an apical region from about 11 to 18 mm. long the sensitivity was about the same, but the stem below this appears not to be perceptive although curvature reaction is possible in the region.

From what has been written above, and from a number of scattered observations on geotropic curvatures resulting from stimulation of adult organs, such as branches of trees, it is clear that the older view that graviperception is localized in a very small region at the growing apex of young organs is no longer tenable, although it is true that these regions are much the most sensitive to geotropic stimulation. It may even be that every living cell is capable of geotropic perception, although often the sensitivity may be very low. Whether this is a correct view or not, it is clear that a geotropic curvature can only take place in a region growing in length, while geotropic perception is generally greatest in the apices where, indeed, the first phase of growth, cell division, is active, but in which the phase of extension has not set in. Consequently there must very generally be a conduction of the excitation, and this, along with the fact that some appreciable time must elapse between perception of the stimulus and the observed response, and also the lack of any obvious and simple connexion between change in the relations of an organ to gravity and the resulting response, all indicate without a doubt that stimulation must set up a chain of reactions of which the observed curvature is the last link. Little enough is known of the nature of this chain of reactions, but a summary of our knowledge is given below.

THE MECHANISM OF PERCEPTION OF THE GRAVITATIONAL STIMULUS

The perception of the gravitational stimulus constitutes the first of that chain of actions which leads to the observed response. We must therefore inquire into what effect a displacement from the normal position in relation to gravity can have on the cells of the perceptive region. It is conceivable, as pointed out by Francis Darwin, that strains will result from this displacement which were not present in the cells previously, but as the same author remarked, it is clear that this cannot be the only means of graviperception as plants are still geotropic when supported throughout their length.

Pfeffer considered that changes in pressure, or changes in position, of different parts of the cell, resulted from displacement of the organ stimulated, and constituted the actual stimulus. He regarded it as uncertain whether the liquid contents of the cell or more solid inclusions were responsible for the stimulation, but a theory that has obtained a considerable measure of support attributes to solid inclusions in the cell the part of perceiving the external stimulus.

This theory is known as the statolith theory, and in its present form is largely due to Haberlandt and Němec who put forward their

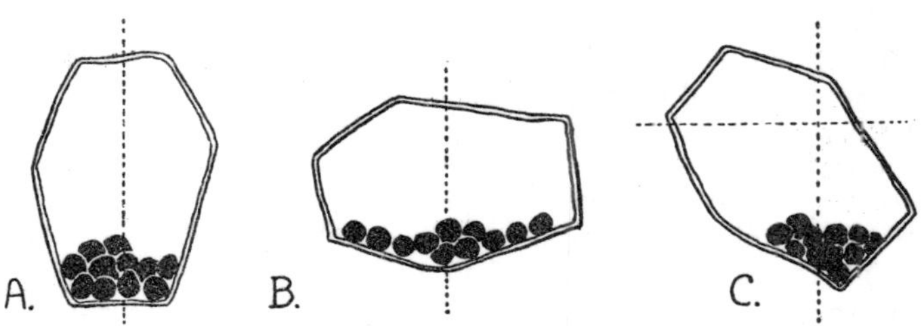

FIG. 55.—A statocyte viewed in section after geotropic stimulation in various positions (From L. E. Hawker)

opinions independently in 1900. According to them, the gravitational stimulus is perceived owing to comparatively heavy starch grains, free to move in the cell, falling, after displacement of the cell, to the lowest possible position, and there coming into contact with the protoplasm or internal plasmatic membrane (cf. Fig. 55). That starch grains do move in the manner described there is no possible doubt. Thus, in

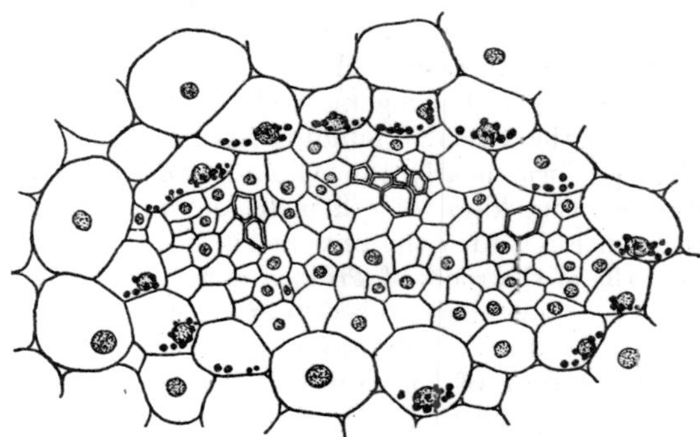

FIG. 56.—Vascular strand in cotyledon of *Allium cepa* showing statolith starch grains (From L. E. Hawker)

Fig. 56 is shown the distribution of starch in the endodermal cells of *Allium cepa* after displacement of the plant through a right angle. The moving starch grains are termed statoliths and the cells containing them statocytes.

According to Miss Prankerd, throughout the plant kingdom other bodies, besides starch grains, are met with which are free to move under the influence of gravity, and which can therefore be described as statoliths. Thus Miss Prankerd distinguishes the following types of statocyte :

1. In one kind of statocyte the statoliths are starch grains which may be simple or compound, larger, the same size or smaller than the fixed grains in other tissues. This is no doubt the commonest type.

2. In some statocytes characterized by the presence of these *amylo-statoliths* the nucleus is differentiated both in size and staining capacity from those of neighbouring cells.

3. In yet others the nucleus occupies a definite position in the cell, and though apparently free from the moving starch grains is itself free to move and so moves with them.

4. The statoliths are starch-containing chloroplasts (chlorostatoliths).

5. The nucleus is connected in some way with starch grains or starch-containing chloroplasts and the whole system appears to form a unit which moves under the influence of gravity.

6. The statocyte is a cell elongated in the vertical plane and containing a thick strand of protoplasm with starch grains and a nucleus embedded in it and fixed to each end of the cell. On displacement from its normal orientation the whole strand of protoplasm with its inclusions swings to the lower side of the cell. This is probably a rare type.

7. The statolith is a crystal of calcium oxalate usually occurring singly in the cell. This is probably also a rare type but has been described by Miss Prankerd as occurring in the nodes of the wheat plant. Here the statocytes occupy more than half the total volume of the node and form a definite tissue. These statocytes are of two kinds; the more usual type, containing movable starch grains, form groups of cells adaxial to the vascular bundles, but the crystal statocytes of this seventh type form nearly the whole of the ground tissue.

Probably yet other kinds of cell inclusions may function as statoliths. Thus Němec recorded the presence of heavy bodies of unknown nature in the cells of the roots of *Trianea* and in the stems of members of the Characeae. Giesenhagen also observed in the rhizoids of *Chara*, bodies of unknown nature which moved under the influence of gravity.

Evidence in support of the statolith theory has been sought (1) in the general correspondence between the capacity for graviperception and the presence of a statolith apparatus, and (2) by experiments in which the effect on graviperception of removal of the statolith apparatus is determined.

With regard to the first line of approach there exists now a vast amount of evidence that in general where statoliths are present the organ has the property of graviperception. Thus movable starch is present in such definitely graviperceptive organs as root tips and grass coleoptiles, but absent, as Haberlandt showed, from non-geotropic roots of *Hedera*, *Hoya* and *Ficus*. Similarly, Tischler found that aerial roots of orchids are free from statolith starch and are non-geotropic. In the liverworts Němec found general agreement between the possession of graviperception and the presence of statoliths. Thus in *Pellia epiphylla* the sporangiophore is very sensitive to the gravitational stimulus and statoliths are well developed; in *P. calycina* graviperception is feeble and the development of a statolith apparatus is poor, while in the sporangiophore of *Aneura pinguis* both statolith starch and the power of geotropic perception appear to be absent.

In the Gramineae it has been shown by von Guttenberg that sensitivity to gravity runs parallel with the development of statolith starch. In particular Miss Hawker has examined the statolith apparatus in seedlings of about 80 species including dicotyledons, monocotyledons and conifers. Over this very wide range of species she found a very close correlation between the development of statoliths and sensitivity to gravity. It may be noted that in roots the statoliths first appear in the root cap of the radicle, while in dicotyledonous seedlings the statoliths in hypocotyls and epicotyls are usually limited to the cells of the endodermis.

With regard to the statoliths of the endodermis of stems, it is stated that the moving starch grains disappear as the cells grow older and the region of stem containing them loses its power of growing in length. But in nodes of grasses the statolith apparatus persists, and as we have already seen for the case of wheat, may be very strongly developed. The retention of the power of geotropic response in this tissue has already been noted.

Reference has already been made to the bilateral symmetry in regard to geotropic sensitivity of certain seedlings, notably *Lupinus* spp. and *Cucurbita pepo*. Miss Hawker found in these seedlings that the statocytes are not isodiametric in transverse section, so that when stimulated in one plane the statoliths will fall on a smaller area of protoplasm than they would if stimulated in a plane at right angles, while they will on the average travel through a longer distance and so take longer to reach the layer of protoplasm at the base of the cell. In all such cases examined it was found that the more rapid fall of the statoliths on to the greater area of protoplasm takes place when the seedlings are stimulated in the plane wherein sensitivity is greatest.

A few cases are on record in which there is apparently no correspondence between graviperception and statolith development. Tischler, for example, found that non-geotropic roots of *Pyrola* and *Pistia* contain movable starch grains. Also statoliths appear to be absent from most Thallophyta, and although it is true that geotropic sensitivity is, on the whole, poorly developed in the Algae, many Fungi exhibit strong geotropism. But that in vascular plants there is a close correspondence between sensitivity to gravitational stimulation and the development of a statolith apparatus seems to be well established.

The second line of evidence in support of the statolith theory consists in experimentally altering the extent of statolith development and observing the effect on graviperception. The decapitation of the root apex, or the removal of the cortex of the stem, including the statolith-containing endodermis, while a very efficacious method of statolith removal, may in itself involve a 'shock' stimulus which renders the interpretation of any effects on graviperception difficult. The removal of statolith starch by inducing metabolic changes has also been attempted. Thus Haberlandt found that prolonged exposure of plants of *Capsella*, *Linum* and *Ruta* to low temperatures brings about the disappearance of starch from the endodermis. This is accompanied by a loss in the power of geotropic perception. Francis Darwin similarly found that exposure of seedlings of *Setaria* to high temperatures caused a disappearance of starch and a loss of geotropic irritability. But he found that phototropic irritability was lost at the same time, so that there is no evidence that the loss of geotropic irritability in this case is to be related to the disappearance of starch. Similarly, it is possible that the disappearance of starch and of geotropic irritability by exposure to low temperatures may also not be directly related.

It may be significant that when plants are starved the moving starch in the endodermis is persistent, thus suggesting that it may play some part in the life of the plant other than that of a storage material. Fräulein Zollikofer succeeded in bringing about the removal of this starch, without unduly starving the plants, by continuous exposure of the latter to light for 2 to 4 days, followed by removal of them to the dark. In plants of *Tagetes* and *Dimorphotheca* removal of the statolith starch in this way was accompanied by loss of the power of geotropic reaction, and it is important to note that the power to react to the stimulus of unilateral light was not lost except in cases where growth had stopped.

Another type of experiment in which evidence in regard to the statolith theory has been sought consists in observing the behaviour of statolith starch in plants on a klinostat. Such experiments have led to very divergent conclusions. Thus Jost found that with rotation at low speeds just sufficient to produce a curvature in response to the stimulus of centrifugal force there was no movement of starch. An experiment by Buder, on the other hand, suggests a different conclusion. A *Lepidium* seedling is placed with its root horizontal for 12 minutes so that the statoliths reach the lower side of the cells. The root is then rotated through 180° so that the former under side is now uppermost. After another 12 minutes the statoliths will have travelled to the other side. It might now be assumed that the two oppositely directed stimuli cancel one another. But the statoliths are now collected on one side of the cells and if the root is now alternately stimulated in these same two positions, but for so short a time that the starch has no time to move (10 seconds), and if this treatment is continued, the effect will be that the statoliths will remain on one side of the cells, and on the statolith theory we should therefore expect a curvature concave on this side, since the root will only be stimulated when this side is the lower. In Buder's experiments this was actually the case, and his results therefore support the statolith theory, but other workers have not confirmed his finding. The most recent experiments of this kind appear to be those of Fräulein v. Ubisch who, working with seedlings of *Nasturtium*, could find no general parallelism between the position of the statolith starch grains and geotropic perception. Taking all the evidence into consideration it seems definitely to be the case that a wide parallelism exists between the possession of graviperception by an organ and the presence of statoliths. But there are on record a number of cases in which no such parallelism is observable. Whether in such cases a statolith apparatus is indeed present, but possibly owing to the nature and size of the particles has so far escaped observation, or whether some entirely different mechanism is present, or whether the relationship between moving starch grains and graviperception is quite different from that usually supposed by upholders of the statolith theory, it is at present impossible to say.

THE FIRST STAGE IN THE STIMULATIVE PROCESS

If the statolith theory is correct it means that the actual stimulation of the protoplasm is effected by the pressure of the statoliths on it, and the falling of the statoliths is therefore to be regarded not as the first stage in the processes between stimulus and response, but preliminary to them. Whether the statolith theory is correct, or whether the stimulation is effected by a general displacement of cell contents consequent on turning it through an angle, we have still to find what the first consequence of stimulation is. It has been stated earlier that Verworn suggested that the immediate response to stimulation is always a change in respiration rate. There is very little information available on this point. From qualitative experiments with *Vicia faba* Miss Schley in 1930 concluded that the respiration of a geotropically stimulated root is greater than that of an unstimulated one, while she also concluded that the rate of the respiration on the convex side is greater than that on the concave side throughout the whole period of perception and response. Later, Navez did indeed find that when a seedling of *Vicia faba* is geotropically stimulated by being displaced through 90° there is a very considerable rise in respiration rate which in his experiment reached a maximum of approximately 30 per cent. above the normal rate about 2·5 hours after the commencement of stimulation, after which it slowly and regularly fell to the normal value (cf. Fig. 57). It is very desirable that this observation should be extended to other material in order that it may be determined if this increased respiration rate is a general phenomenon of geotropic stimulation.

From measurements of the rate of movement of statolith starch grains in the endodermis of seedlings of *Phaseolus multiflorus* under different amounts of stimulation, G. and F. Weber decided that geotropic stimulation brought about a decrease in the viscosity of the protoplasm.[1] Fräulein Zollikofer was, however, unable to confirm this result.

Some writers have thought that geotropic stimulation brings about electrical changes in the protoplasm. Having regard to the colloidal character of the protoplasm, it is not improbable that disturbances in the relative position of particles in the cell consequent on its displacement may result in alteration of the charges on the various phases of the protoplasm, but we have not, as yet, any definite evidence as to what these changes may be.

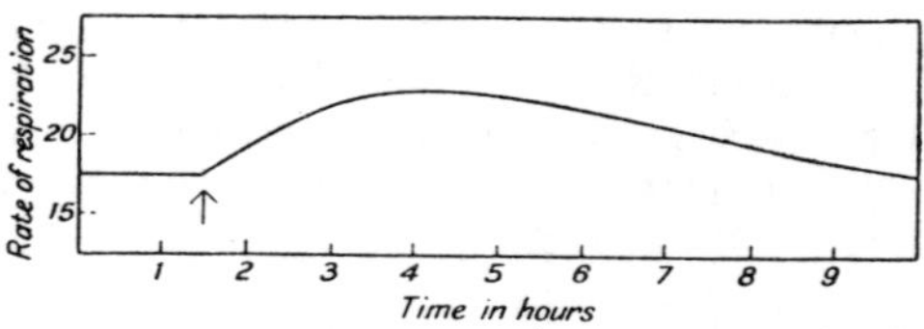

Fig. 57.—Effect of geotropic stimulation on a geotropically stimulated root of *Vicia faba* (After A. E. Navez)

INTERMEDIATE STAGES IN GEOTROPIC REACTION

We have already seen that whether or not the stimulus of a changed orientation to gravity is transmitted to the protoplasm by means of statoliths or by other changes produced in the cell by its displacement, and whether or not the first result of stimulation is a change in respiration rate or in electrical conditions in the protoplasm, there is often a conduction of excitation, since the region in which the stimulus is most readily perceived is frequently not that in which the observed geotropic response takes place.

There appear to be three ways in which conceivably the excitation might be conducted, by pressure, by an electric current or by the transference of some chemical substance. All the available evidence indicates that the last of these possibilities furnishes the actual mechanism for the conduction of excitation, and that, indeed, growth regulators or hormones, similar to those which are active in various phenomena of growth (cf. Chapter XVI), function similarly in conducting gravitational excitation.

As we shall see in the next chapter, the action of such substances in bringing about the conduction of phototropic excitation was indicated by Boysen Jensen in 1910, and confirmed by subsequent workers. That

[1] It will, however, be observed that this method of estimating protoplasmic viscosity involves the assumption that the starch grains fall *through* the protoplasm, whereas the statolith theory assumes that they fall *on* the protoplasm.

a similar mechanism might be present in the case of geotropism was indeed suggested by this work on phototropism, and also indicated by experiments by Boysen Jensen in which strong geotropic curvatures were produced in coleoptiles of *Avena* from which the apex had been removed and replaced by another apex fixed on with gelatin, and then stimulated, whereas only a feeble curvature was produced in similar decapitated coleoptiles left without any apex. It was further shown by Brauner in 1923 that when an unstimulated tip of a coleoptile of *Avena* was fixed with a drop of water to the top of a geotropically stimulated but decapitated coleoptile, a curvature resulted. The reverse experiment was performed in 1924 by Stark, who showed that a geotropically stimulated tip can bring about curvature in an unstimulated stump. Keeble, Nelson and Snow have reported similar results with roots of *Zea mais*. A decapitated root, as indicated earlier in this chapter, is practically insensitive to the gravitational stimulus. Now Snow in 1923 showed that the sensitivity to geotropic stimulation can be largely restored if the tip is replaced on the decapitated stump by means of a drop of water. Cholodny in the following year showed that restoration of the geotropic sensitivity to decapitated roots of *Zea mais* was even better effected by coleoptile tips than by root tips. Keeble, Nelson and Snow found that if intact roots are stimulated, then decapitated, and the apices placed on the stumps of unstimulated decapitated roots, in a proportion of the roots a curvature is produced in the direction of the side of the tip which had been lowest during its stimulation. The same result was obtained if, instead of stimulated root tips, stimulated coleoptile tips were fixed to the decapitated roots. The curvature was again towards the side of the tip which had been lowest when stimulated. Again, when decapitated roots were stimulated and then unstimulated tips of either roots or coleoptiles placed on the stimulated stumps, a proportion of the roots curved in the direction of the side which had been lowest during stimulation.

The hormone theory put forward by Cholodny and Went, and accepted by Keeble, Nelson and Snow, explains these results by supposing that in the tips of the root and coleoptile there is produced a growth-promoting substance, which under geotropic stimulation undergoes redistribution so that the upper and under sides of the tip during stimulation contain different concentrations of the substance. The latter thus diffuses to the growing region in different concentrations on the two sides and therefore sets up unequal rates of growth on the two sides so that a curvature results. But since curvature will also take place in a stimulated stump attached to an unstimulated tip it would also appear that in such experiments the growth-promoting substance must diffuse in equal concentration on all sides of the root tip and then in the stimulated stump become unequally distributed on the two sides.

Cholodny found that the rate of growth of decapitated coleoptiles is accelerated by the addition to them of tips of either coleoptiles or roots; conversely the rate of growth of decapitated roots is retarded when either coleoptile tips or root tips are placed on them. This will explain the fact that roots and coleoptiles curve in opposite directions. In the root the excess of growth substance on the lower side thus leads to less rapid growth on that side and a positively geotropic curvature; in the coleoptile, on the other hand, it leads to more rapid growth on that side and a negatively geotropic curvature. The observations of Keeble, Nelson and Snow that the placing of a coleoptile tip, root tip or half root tip excentrically upon the stump of a decapitated maize root leads to a curvature towards the side covered by the tip, are also in agreement with the hormone theory.

Quantitative determinations of the amount of growth hormone in the tips of unstimulated and geotropically stimulated *Avena* coleoptiles and in the upper and lower halves of the tips of stimulated coleoptiles were made by Dolk in 1929 by the method indicated in Chapter XVI, that is, by allowing the growth substance to diffuse out from the tips into agar plates for a definite time and then observing the curvature produced by a unit quantity of the agar when placed over one side of the decapitated coleoptile stump. Dolk showed in this way that the total amount of growth substance in stimulated and unstimulated coleoptile tips is the same, but that in stimulated tips there is more of the growth substance in the lower half than the upper half.

Similar experiments were made by Miss Hawker in 1932 with roots of *Vicia faba*, with similar results. Here also gravitational stimulation brings about a greater concentration of growth substance in the lower part of the tip than in the upper.

Finally, in support of the hormone theory of geotropism may be cited Boysen Jensen's recent observation that roots of *Vicia faba* produced from seeds treated with erythrosin contain no growth substance and also have no power to react to gravitational stimulation. Later, formation of growth substance begins, and the roots then acquire the power to react to the stimulus of gravity. In roots of *Pisum sativum* seedlings developed from seeds similarly treated, the development of the growth substance is depressed and the roots fail to react to gravity, recovery of the power to react developing slowly with the formation of the growth substance.

According to Miss Schley's investigation on geotropically stimulated roots of *Vicia faba*, the sequence of changes which take place in a geotropically stimulated root are firstly increased respiration, secondly increased acidity, thirdly increased turgor and next increased production of hydrolysable sugars with a corresponding decrease in polysaccharides on the more actively growing side of the organ. It may be noted that Phillips found in stimulated nodes of *Zea mais* a greater percentage of water on the concave side during the early stages of curvature, followed by a higher percentage of water on the convex side. An investigation has been made more recently by Warner on the changes in sugar and acid content which occur in geotropically stimulated shoots of *Helianthus annuus* and *Silphium perfoliatum* and in etiolated shoots of *Dahlia variabilis* and *Solanum tuberosum*. According to Warner's observations, geotropic stimulation of these shoots results in a higher content of reducing sugars in the convex under side than in the concave upper side. This difference in sugar content is not a consequence of mere curvature, for it does not occur in shoots passively curved without stimulation. The content of free acid was found to be greater in the upper side than in the under side of geotropically stimulated shoots of *Dahlia*.

While we can understand the changes in sugar content in stimulated organs observed by Schley and Warner in relation to increased growth on the convex side of the curving organ, the relation between these changes and the growth hormone are not clear. It has already been pointed out in Chapter XVI that the action of the growth substance appears to be on the cell-walls in the stretching zone. To what extent the changes in the content of sugars in stimulated organs are induced by the growth substance, or what degree of independence there is between the two sets of changes is at present quite unclear.

It will be observed that the work so far considered on the changes taking place as a result of geotropic stimulation, all relates to growth curvatures. But it has already been noted that sometimes geotropic response takes the form of movement due to change in turgor. As we shall see in dealing with seismonic irritability there are cases known where stimulation results in the production of a hormone and in which the observed response is not a growth curvature but a variation movement due to changes in turgor of cells remote from the place of stimulation. It is therefore at least credible that in those cases where geotropic stimulation brings about variation movements that the conduction of the excitation is here also effected by a hormone which becomes unequally distributed at the seat of perception of the stimulus. Further, from the work of Phillips and Miss Schley already quoted, and from some much earlier work of Kraus, a change in water content of the two sides of the stimulated organ appears to be an accompaniment of geotropic curvature, so that there is the possibility that the hormone may in all cases lead to turgor changes, but in most cases to the further change of differential growth rates on the two sides of the organ.

REFERENCE

RAWITSCHER, F., *Der Geotropismus der Pflanzen*. Jena, 1932.

CHAPTER XXIII

PHOTOTROPISM

THE curvature of shoots towards unilateral light is a well-known phenomenon and is exemplified by almost any plant growing in a room lit by a window on one side only. The response to the directional stimulus of light was formerly more generally known as heliotropism, but the term phototropism which is now usually employed is preferable, since the source of the light constituting the stimulus need not be the sun. The stems of higher plants are generally positively orthophototropic, that is, the curvature resulting from the stimulus tends to direct the apex of the responding organ towards the source of light. Particularly sensitive phototropic organs are the coleoptiles of grasses. Among lower plants the sporangiophores of the Mucorineae, and particularly those of *Phycomyces blakesleeanus*,[1] are positively phototropic, and so are the stalks of the fructifications of many of the larger fungi.

Many leaves orientate themselves so that they lie with their surface in a plane at right angles to the direction of the incident light; they are thus diaphototropic. Negative phototropism, on the other hand, appears to be comparatively rare under natural conditions. No doubt some subterranean roots are negatively orthophototropic, and among such are those of white mustard (*Sinapis alba*). Contrary, however, to the general opinion, most roots are devoid of phototropic irritability. Negative phototropism is exhibited by the rhizoids of fern prothallia and of liverworts. As will be shown later, the same organ may in many cases exhibit either positive or negative phototropism according to the intensity of the light.

As in the case of geotropism, so with phototropic irritability, the kind of response depends on the nature of the stimulated material. Organs with the power to grow in length respond by executing growth curvatures; in other cases the response may take the form of a variation movement produced by turgor changes. Most motile organisms respond by active movements which are termed phototactic, and are said to exhibit phototaxis.

As noted in the last chapter, very many plant organs respond both to the stimulus of unilateral light and to gravity. So, for example, if a shoot growing vertically upwards is subjected to the stimulus of unilateral light a curvature will result so that the apex of the shoot tends to lie in the direction of the incident light. But inasmuch as this in-

[1] The species usually described as *Phycomyces nitens*.

volves a deviation from the vertical it becomes subjected to the action of gravity and its final position will depend on the operation of the two stimuli. In the investigation of phototropic irritability it may therefore be necessary to eliminate the action of gravity, which, as we have seen, may be accomplished by the use of the klinostat. Thus, if unilateral light is allowed to fall for a definite period, on a negatively orthogeotropic organ growing vertically, it will be subjected to the phototropic stimulus but not to the gravitational stimulus during that period. If, after stimulation, it is slowly rotated on a klinostat with horizontal axis, the gravitational stimulus will not be applied to the organ when it executes a curvature in response to the stimulation by unilateral light.

THE COURSE OF THE PHOTOTROPIC REACTION

The phototropic reaction, although practically universal in the shoots of higher plants, is very readily observed in the coleoptiles of members of the Gramineae, and these, particularly those of the oat, *Avena sativa*, have formed the favourite material for observations on phototropism. If such a coleoptile, growing vertically upwards, is exposed for a time to unilateral light and then rotated in the dark on a klinostat with horizontal axis, a curvature results on account of unequal growth on the two sides of the organ. There is a latent time, the first visible curvature occurring some time after the beginning of stimulation. This curvature commences at the apex of the coleoptile and extends gradually towards the base, so that the whole of the organ may be involved and the curvature may not be complete in 24 hours, although generally it has reached its maximum and an autotropic straightening of the organ has set in before this time has elapsed; the autotropic reaction is, however, said to be generally weaker after phototropic than after geotropic reactions.

The course of phototropic curvature is not a constant one, but varies with the quantity of light which falls on the stimulated organ, and not only the amount of curvature but also the latent time and the time taken for the maximum curvature to be reached depend on the quantity of light used in stimulating the organ. Further reference to this question will be made later.

It has already been noted that geotropic curvatures may take place in the nodes of grasses, a renewal of growth occurring in the nodal tissues, when subjected to the geotropic stimulus. A phototropic stimulus does not have this effect, but if the renewal of growth is first brought about by the geotropic stimulus, then a phototropic curvature can take place in those tissues that have regained the power of growth. On the other hand, a phototropic curvature will take place in the nodes of members of the Commelineae as a result of the stimulus of unilateral light alone. This again must be due to a renewal of growth in tissues in which it had previously come to a standstill.

THE STIMULUS QUANTITY LAW

With other conditions constant, the amount of response to unilateral light depends on the intensity and duration of the stimulus. The researches of Fröschel and of Blaauw in 1909 show that, as in the case of geotropic stimulus, the amount of response depends not on the intensity alone nor on the duration of stimulation alone, but on the quantity of light falling on the organ, the quantity being the product of the intensity of the light and its duration. Thus, as shown by Rutten-Pekelharing for graviperception, the presentation time for the phototropic reaction of the coleoptile of *Avena sativa* varies inversely with the intensity of stimulation. This is indicated by some numbers obtained by Blaauw and summarized in Table CXI. From this it would appear that over a very wide range of light intensities the minimum observable response is effected by approximately the same quantity of light.

Table CXI

Relation between Intensity of Phototropic Stimulation and the Presentation Time for Coleoptiles of *Avena sativa*

Intensity of Stimulation in Metre-candles (MK)	Presentation Time in Seconds	Quantity of Stimulus in Metre-Candle seconds (MKS)
0·00017	154800	26·3
0·00085	21600	18·6
0·00477	3600	17·2
0·0898	240	21·6
5·456	4	21·8
1902	0·01	19·0
26520	0·001	26·5

This relation only holds if all other conditions, both internal and external, are constant. In this connexion it is significant to note that whereas Blaauw in 1909 found the quantity of light necessary to produce an observable response in oat coleoptiles was about 21 MKS, Konrad Noack in 1914 found a value for the same quantity of 12 MKS, while in 1922 von Guttenberg obtained a value as low as 2·3 MKS. These very great differences in observed sensitivity may be partly accounted for by possible differences in the composition of the source of light, for, as we shall see, the effectiveness of light is a function of the wave-length, but probably differences in external and internal factors have much to do with it. As regards external factors, temperature and the presence of injurious gases in the atmosphere may be operative. Internal factors are not so readily defined, but it may well be imagined that the seedlings used by different workers differed in constitution. It may further be noted that the coleoptile is not circular, but elliptical, in cross-section, and that, as shown by Du Buy and Neuernbergk, the magnitude of the reaction is different according to whether the light falls at right angles to the long or to the short axis of the cross-section. There is thus a dorsiventrality in regard to response to phototropic stimulation in this organ.

Although, at least within the limits indicated in Table CXI, the same quantity of stimulus produces the same response under otherwise constant internal and external conditions, it does not follow that the amount of response is proportional to the quantity of stimulation, and indeed, this is far from being the case. Actually, it was found by Arisz in 1915 that with increase in the quantity of stimulus there was an increase in the response until the latter reached a maximum, which was found to occur in Arisz's experiments when the quantity of stimulus was about 200 MKS, after which, with increasing quantity of stimulus, the response fell off until with a stimulus of 10,000 MKS an initial positively phototropic curvature soon became reversed and a permanent negative curvature resulted. With a somewhat greater stimulus of 13,600 MKS the curvature was a negatively phototropic one from the commencement. With still greater quantities of stimulus the curvature was again positive. The values obtained by Arisz for the response to stimuli up to 2800 MKS are shown in Table CXII.

Table CXII

Influence of Quantity of Light used in Stimulation of Coleoptiles of *Avena sativa* on the Amount of Phototropic Curvature

Quantity of Light in Metre-candle seconds (MKS)	Maximum Deviation of Coleoptile Apex in mm.
7·6	0·7
12·4	1
18·1	1·6
26·4	2·3
45	3
65	3·3
75	4
100	5
140	4·7
237	5·4
560	4
1500	3
2800	1·2

That the same organ might execute either a positively or negatively phototropic curvature according to the amount of stimulus was recorded as long ago as 1872 by N. J. C. Müller for seedling stems of cress (*Lepidium sativum*), where, as in the oat coleoptile, the curvature was found to be positive in moderate values of the stimulus but negative when high values of the stimulus were employed. Later Stahl found a similar reversal in filaments of *Vaucheria*, while Oltmanns found that the sporangiophores of *Phycomyces blakesleeanus* exhibited the same phenomenon, and indeed that the reversal occurred with considerably lower stimulation than in *Lepidium* and barley (*Hordeum*). Indeed it may be, as Jost has supposed, that every organ which is sensitive to the stimulus of unilateral light will respond with a positive or a negative curvature, and which of the two results depends on the quantity of stimulus and other conditions.

It seems probable that the stimulus-quantity law only holds within limits. If the intensity of the stimulus is very low, prolonged duration of its action does not, according to Arisz, bring about the negative curvature in *Avena* even when the quantity of stimulus exceeds 10,000 MKS. It may be noted that with as low a light intensity as $1{\cdot}7 \times 10^{-7}$ MK Richter observed a phototropic reaction in *Vicia villosa*, but the actual quantity of stimulus employed is not on record.

According to Castle, the reversal of the direction of curvature when sporangiophores of *Phycomyces* are subjected to high light intensities is due to infra-red radiation from the source of light and can therefore be regarded as negative thermotropism. Castle was able to show that when this infra-red radiation is removed by absorption with a copper sulphate filter, light of high intensity, devoid of the infra-red rays, does not produce the negative phototropic curvature, the sporangiophores remaining uncurved. The explanation offered by Castle of this 'indifference' will be given later in this chapter.

Summation of Sub-threshold Stimulations. It has already been noted in the case of geotropism that a series of sub-threshold stimulations will bring about a reaction if the time allowed to elapse between consecutive stimulations is not too great. The same has been shown to be true of phototropism. By placing between the source of light and the experimental organ opaque sectors which could be rotated at varying speeds (300 to 27,000 revolutions a minute), and which could be arranged so that from about 0·5 to about 0·03 of the incident light was cut off, Nathansohn and Pringsheim showed that, within the range of conditions in their experiments, the phototropic effect was always the same for the same quantity of light. Actually they used seedlings of *Brassica*, which were placed between a constant source of light on one side and a source partly obscured by the rotating sectors on the other. The position between the two lamps was found in which there was no phototropic response, indicating that the stimulus from the two sides must have been the same. It was found that always complete compensation resulted when the quantity of light falling on the two sides of the seedlings was the same, no matter what the speed of rotation of the sectors nor the amount of light cut off by the sectors, or in other words, the relative durations of the periods of stimulation and rest. We have seen that in the case of gravitational stimulation, Fitting found that if the period of rest is more than 5 times as long as the period of stimulation, the stimulations begin to die out before the rest periods are over, and the presentation time has to be increased. In the experiments of Nathansohn and Pringsheim, at any rate, the rest period can apparently be 32 times as long as the period of phototropic stimulation without the stimulus beginning to fade out.

Sine Law. As with geotropic stimulation, so with phototropic stimulation, the sine law holds; that is, the intensity of the stimulus is proportional to the sine of the angle made by the direction of the incident light with the illuminated surface.

Resultant Law. Similarly, just as when a geotropically irritable organ is stimulated by both gravity and centrifugal force the organ responds as if stimulated by a force equal to the resultant of the two, as given by the parallelogram of forces, so with phototropic stimulation by light impinging on the organ from two different directions the same law holds.

Surface Area Effect. It was shown by Von Guttenberg that the screening from light of a portion of an *Avena* coleoptile on the illuminated side has the effect of increasing the presentation time in the same proportion; that is, the presentation time is inversely proportional to the area of equally sensitive tissue illuminated.

Latent Time. The latent time for phototropic response varies greatly with different material. The phototactic response of actively moving zoospores and other bodies occurs in general a negligible time after stimulation and a measurable latent time is non-existent. The phototropic variation movements of some pulvini may occur within a minute or two of unilateral light falling on them. A longer latent time is the rule with growth curvatures, but in the case of sporangiophores of *Phycomyces*, it may, according to Oltmanns, be no more than from 1 to 3 minutes. From Arisz's observations on *Avena* coleoptiles it would appear that with a stimulus quantity of only 5 MKS the latent time may be as long as 100 minutes, but is reduced to about 70 minutes with a stimulus of 20 MKS and to about 10 minutes with 112 MKS. Although with a stimulus of 700 MKS a greater response resulted, the latent time was longer, namely, about 30 minutes. While the latent time thus varies with the value of the stimulus and possibly with other external conditions, it would appear that on the whole the response to phototropic stimulation takes place considerably more rapidly than does the response to geotropic stimulation, where, as the numbers given in the previous chapter show, the latent time is rarely less than 30 minutes.

THE INFLUENCE OF WAVE-LENGTH OF THE LIGHT ON PHOTOTROPIC STIMULATION

A considerable number of observations were made from the years 1842 onwards into the twentieth century with a view to determining the relative activities of the different wave-lengths of light in inducing phototropic response. Different observers were generally agreed that response was brought about by the use of light from the blue end of the spectrum, but the greatest divergence was displayed in the conclusions drawn with regard to the action of the longer rays of the visible spectrum. Thus Dutrochet in 1844 concluded that cress seedlings were insensitive to unilateral red light, but that other seedlings responded to its stimulus.

While Sachs in 1864 decided that phototropic response could only be induced by blue light, Wiesner in 1879 decided that *Vicia* seedlings reacted to light of all colours except yellow.

Much of this earlier work is of no more than historical interest, for two reasons. In the first place, the coloured lights were rarely pure in the particular colour used, and secondly, no account was taken of the varying intensity of the light used in comparative experiments. In more recent years the question was reopened by Blaauw in 1909, who attempted to show that the divergence in the results of previous workers was due to neglect of the intensity factor. In his own experiments with *Avena* coleoptiles he found that with short periods of illumination response only took place with blue light, but that with prolonged illumination the reaction with the blue might be negligible, while curvatures were induced by ultra-violet and red radiation. It may perhaps be supposed that with prolonged action of blue light the stimulus is approaching the value where the primary positive curvature gives place to the negative, while with ultra-violet and red rays the action is much feebler and only with prolonged irradiation is the first positive curvature induced. Blaauw's conclusions with regard to the relationship between wave-length and phototropic reaction of *Avena* coleoptiles are that with decreasing wave-length from the red to the green the action is feeble, but that from 500 μ it increases very rapidly with decreasing wave-length, a maximum being reached in the indigo at 465 μ, below which the sensitivity decreases slowly. With sporangiophores of *Phycomyces* Blaauw obtained a generally similar result.

A careful examination of the reaction of *Pilobolus* sporangiophores to light of different wave-lengths was made by Miss Parr in 1918. Miss Parr experimented with the sporangiophore in its most sensitive condition, which is reached just before the tips begin to swell to produce the sporangia. She obtained light of different wave-lengths by splitting up white light from an electric lamp (Nernst or 200 Watt nitrogen-filled tungsten filament Mazda). The intensity of radiation was measured by means of a thermopile. The period of stimulation required to produce a curvature in half the sporangiophores of a sample within 1 hour was measured for various ranges of wave-length. The curvatures observed were probably the second positive curvature. The results with the lights from the Nernst lamp are shown in Table CXIII. Similar results were obtained with the Mazda lamp.

These results show that although with decreasing wave-length of light the intensity of the incident light regularly decreased, the period of stimulation required to produce the same response also decreased. Hence with decreasing wave-length, light becomes regularly more effective in inducing phototropic curvature. Judging from the numbers given in Table CXIII, blue light is roughly 15 to 20 times as effective as red light, and violet light about 50 times as effective as red light.

The phototropic sensitivity of the *Phycomyces* sporangiophore to light of different wave-lengths has been recently investigated by Castle.

Table CXIII

Response of Sporangiophores of *Pilobus* to Light of Different Wave-lengths

Range of Wave-length in μ	Mean Wave-length in μ	Colour	Intensity of Light in ergs per sq. cm. per sec.	Period of Stimulation
712–704 . . .	708	Red	1·616	78
695–638 . . .	667	,,	1·097	76
647–616 . . .	631	Orange	0·647	75
660–564 . . .	612	,,	0·344	73
625–553 . . .	589	Yellow	0·305	72
619–551 . . .	585	,,	0·295	72
570–510 . . .	540	Green	0·149	69
543–507 . . .	523	,,	0·128	67·5
510–482 . . .	496	Blue	0·097	65
479–461 . . .	470	,,	0·077	63
472–458 . . .	464	Indigo	0·063	62
444–432 . . .	437	,,	0·055	60
430–408 . . .	414	Violet	0·038	56
410–385 . . .	398	,,	—	55

His results agree on the whole with those of Miss Parr, for he concluded that in the visible spectrum the sensitivity decreased with increase in wave-length, light at the red end of the spectrum having only 0·12 per cent. of the activity of that of violet light in inducing phototropic curvature. He found, however, maximum sensitivity with violet light, for sensitivity in the ultra-violet was somewhat less than in the violet. Results confirmatory of Castle's work were obtained by Wiechulla.

INFLUENCE OF EXTERNAL CONDITIONS ON PHOTOTROPIC IRRITABILITY

As with the reaction to the stimulus of gravity, so with phototropism, the response is much influenced by external conditions, including temperature, light and oxygen concentration.

Temperature. The influence of temperature on the reaction of coleoptiles of *Avena sativa* to the stimulus of unilateral light was investigated by Miss de Vries in 1914. The seedlings were subjected to particular temperatures for different lengths of time, and the quantity of light necessary to produce a standard curvature (a deviation of the coleoptile apex of 1·5 mm. from the vertical) was determined. The results obtained are summarized in Table CXIV. It will thus be seen that the length of time for which the seedlings are maintained at the experimental temperature does not influence the sensitivity of the coleoptile to the phototropic stimulus between the temperatures of 0° and 25°. From 27·5° to 30° the sensitivity is considerably increased by maintenance at these temperatures. At 31° again the sensitivity

Table CXIV

Influence of Temperature on the Quantity of Light necessary to produce a standard Phototropic Curvature of the Coleoptiles of *Avena sativa*

Temperature in Centigrade degrees	Quantity of Light in MKS necessary to produce a Standard Curvature after these periods at the Respective Temperatures							
	1 hour	2 hours	4 hours	6 hours	12 hours	18 hours	24 hours	48 hours
—2 . .	200	200	—	—	—	—	—	—
0 . .	160	160	160	160	—	—	—	—
5 . .	70·5	70·5	70·5	70·5	—	—	—	—
10 . .	52·5	52·5	52·5	52·5	—	—	—	—
15 . .	24·5	24·5	24·5	24·5	—	24·5	—	24·5
20 . .	20	20	20	20	—	20	—	—
25 . .	9·5	9·5	9·5	9·5	—	9·5	—	—
27·5 . .	9·2	7·2	5·6	4·8	4	4	4	—
30 . .	8	6	4	3	2	2	2	2
31 . .	8	8	8	8	—	8	—	—
32·5 . .	9·2	12	13·6	14·4	—	14·4	—	—
35 . .	10	15	20	22	25	26	26	26
37 . .	40	64	80	88	—	92	92	—
37·5 . .	48	72	104	120	—	176	184	184
38 . .	56	84	128	160	—	272	320	—
39 . .	120	176	240	280	—	400	—	—
40 . .	1680	—	—	—	—	—	—	—

is uninfluenced by the length of time for which the seedlings are maintained at this temperature, but at still higher temperatures the sensitivity falls off with continued duration of the temperature. These results are thus similar to those recorded for geotropic stimulation (cf. p. 455) and can be regarded as due to the operation of a 'time factor' similar or analogous to that operating in metabolic processes. Up to 30° C. the results, according to Miss de Vries, suggest a mean temperature coefficient (Q_{10}) of about 2·6.

The latent time is also affected by temperature, the time decreasing with temperature from 5° to 30°, above which it rapidly increases with rise of temperature. Miss de Vries's results, in which perception and reaction took place at the same temperatures are summarized in Table CXV. It will be observed that length of time to which the seedlings were subjected to the various temperatures before stimulation had little effect on the latent time with temperatures below 35°, but at this and higher temperatures the time taken for the curvature to be brought about was increased with increase of pre-heating at the higher temperature.

Light. It has been found that the response of a plant to unilateral light is greater if it has been previously kept in complete darkness than if it has been exposed to (uniform) light previous to stimulation. Now if we have a fixed source of light and a plant is rotated near it on a klinostat with a vertical axis in such a way that the plant remains in the same horizontal plane with the source of light, the plant will be

Table CXV

Influence of Temperature on the Latent Time for Phototropic Reaction of Coleoptiles of *Avena sativa*

Temperature in Centigrade degrees	Latent Time in Minutes after these periods at the Respective Temperatures before Stimulation			
	1 hour	2 hours	6 hours	18 hours
5 . .	300	300	—	—
10 . .	180	—	180	—
15 . .	120	—	120	—
20 . .	90	90	90	90
25 . .	80	—	80	—
30 . .	70	60	60	60
35 . .	100	—	120	150
37 . .	360	540	—	720
38 . .	∞	—	—	—

subjected to illumination by the same quantity of light on all sides. By stopping the klinostat in a desired position, the plant can then be stimulated by unilateral light. In this way Arisz exposed *Avena* coleoptiles to the equivalent of uniform illumination of 30 MK for various periods before stimulating them with unilateral light. With pre-illumination up to 7200 MKS it was found that the sensitivity to the action of unilateral light declined so that the stimulus necessary to produce the same response had to be increased. With further increase in the quantity of pre-illumination the sensitivity again increased. Arisz's experimental findings are summarised in Table CXVI.

Table CXVI

Effect of Pre-illumination of Coleoptiles of *Avena sativa* on Sensitivity to Unilateral Light

Amount of Pre-illumination in MKS	Amount of Light required to produce Standard Response in MKS
0	20
600	200
900	365
1800	800
7200	2500
27,000	1200
54,000	800
432,000	600

The same result is obtained if the organ is stimulated by different quantities of light on two opposite sides. It may be noted that the effect of pre-illumination is not permanent. Arisz found that even 1 minute after illumination with 2500 MKS the effect was lessened, while after 1 hour it had disappeared.

Oxygen. In 1878 Wiesner found that seedlings of *Lepidium sativum*, *Sinapis alba*, *Phaseolus multiflorus* and *Vicia sativa* were unable to execute phototropic curvatures in an atmosphere devoid of oxygen.

Correns in 1892 recorded that for seedlings of *Helianthus* to respond to phototropic stimulation a concentration of at least 1 per cent. oxygen was necessary, while response in seedlings of *Sinapis alba* only took place if as much as 6 per cent. oxygen was present.

In her experiments on the influence of oxygen on phototropic irritability, Miss van Ameijden obtained similar results to those she obtained with regard to geotropic stimulation. When seedlings of *Avena sativa* or *Sinapis alba* are kept in an atmosphere of nitrogen, they are unable to execute a phototropic curvature. If, however, the seedlings are kept for a comparatively short period, say 3 hours, in nitrogen, then stimulated in nitrogen and air re-admitted, a curvature results. If the seedlings after 8 hours in nitrogen are stimulated in air a phototropic curvature results, but the response is less than that of seedlings not subjected to the pre-treatment with nitrogen. If, however, the stimulus is given in nitrogen there is no response, even if the seedlings are transferred to air immediately.

It would thus seem that the power of perceiving the phototropic stimulus is not immediately lost when oxygen is withdrawn, but gradually disappears with continued absence of this gas. The presence of oxygen is necessary for the reaction to take place.

Various Gases. The gases that reduce sensitivity to the gravitational stimulus probably affect phototropic irritability in the same way. Among such substances are ethylene, acetylene and perhaps other constituents of coal gas.

CHANGES IN PHOTOTROPIC IRRITABILITY DURING DEVELOPMENT

No quantitative estimations of the changes in sensitivity to the stimulus of unilateral light have been made comparable with those obtained in regard to graviperception. Nevertheless, there is a certain amount of information available indicating that similar changes in sensitivity to light take place during development. Thus Miss Parr mentions that the sensitivity of *Pilobolus* sporangiophores to the phototropic stimulus decreases as the tips of these organs swell. The sensitivity of *Phycomyces* sporangiophores also decreases with age.

In quite a number of cases the direction of curvature changes as the organ develops. Thus the flower stalks of *Linaria cymbalaria* are positively phototropic under their natural conditions of growth, but by the time the fruiting stage is reached the same stalks are negatively phototropic. A somewhat similar change occurs in the behaviour of the flower stalk of *Tropaeolum*.

DIAPHOTOTROPISM AND PLAGIOPHOTOTROPISM

A number of radial plant organs orientate themselves with their axes at right angles to the direction of the incident light and are thus diaphototropic. Many leaves and other dorsiventral organs take up a position determined by the direction of the light, and very frequently they can be described as diaphototropic inasmuch as they come to lie with their surfaces in a plane at right angles to the direction of the incident light.

Among radial organs which appear to take up a position at right angles, or at some other angle, to the incident light are the shoots of the ivy (*Hedera helix*), *Atropa belladonna, Pellionia, Goldfussia anisophylla* and *Selaginella,* but no analysis, approaching completeness, of the factors determining the orientation of these shoots has been made, so that the parts played respectively by gravity and light in determining the direction of growth of these organs is generally not clear. In *Hedera* and some other plants, such as *Cucurbita pepo* and *Linaria cymbalaria,* unilateral illumination induces a dorsiventral structure in the shoot.

The orientation of some leaves, such as those of *Erica,* appears to be independent of external conditions and determined solely by the relation of the leaves to the stem which bears them. Other leaves take up a position determined by the direction of the incident light, some with the surface of the leaf in a plane at right angles to the direction of the incident light, others with their edges directed to the source of light, so that they lie in a plane parallel to the direction of the light, while yet others are orientated with their surface at right angles to light of lower intensities, but come to direct their surface in a plane parallel to the direction of strong light. In the so-called 'compass plants' such as *Lactuca scariola* investigated by Stahl and more recently by Karsten, the leaves lie in the north to south plane, and as in other cases of orientation to light, the arrangement appears to be of ecological significance. In these plants the surface of the leaf is exposed to approximately the full light intensity during the morning and evening hours, while the edge of the leaf is exposed to the direct rays during the hours of most intense illumination, when the full strength of light might adversely influence the leaf tissues.

In open sunny habitats sheltered from the wind the leaves of *Aster linosyris* were found by Huber to be directed towards the south, and he suggested the term 'gnomon plants' for such cases.

In many rosette leaves the surface takes up a position at right angles to the direction of the light; as, for example, in species of *Plantago, Rumex* and *Primula.* Such leaves in the dark, according to Frank, exhibit orthogeotropism, and he supposed that their normal position in the light was due to the combined action of diaphototropism and negative geotropism, the action of the former predominating. However, more recent investigations have shown that light may affect geotropic reaction, and it is possible that in some cases of apparent diaphototropism, actually an organ which is orthogeotropic in the dark becomes through the action of light diageotropic, without the direction of illumination having any direct effect in determining the orientation of the organ.

The thallus of *Marchantia* apparently provides an example of this kind. Here the dorsiventrality itself is first induced by light, but when permanently established the thallus in higher light intensities orientates itself at right angles to the direction of the incident light and thus takes up a diaphototropic position. In darkness the thallus exhibits negative orthogeotropism, but when subjected to the action of unilateral light the thallus approaches the horizontal position more and more with increasing light intensity. The simplest explanation of this behaviour is that of Frank, according to which with increasing light intensity the diaphototropic action predominates more and more over the geotropic action. But Sachs found that when the thallus was rotated in diffuse light on a disk so that it was subjected to a centrifugal force 3·2 times that of gravity, the growing region became orientated at right angles to the direction of the force and so exhibited diageotropism. But whether the thallus is also truly diaphototropic is not clear.

Charles Darwin emphasized the diaphototropic behaviour of the cotyledons of many plants. Thus he pointed out that if seedlings are exposed at a window the hypocotyls, being phototropic, bend towards it, and the upper surface of each cotyledon is thereby maintained at right angles to the direction of the light. But if the hypocotyl is fixed, the cotyledons still orientate themselves at right angles to the direction of the light by means of a torsion executed by the petioles. The same behaviour was noted by Frank for many foliage leaves.

The movements of leaves in response to the stimulus of unilateral light are chiefly brought about by changes in the petiole (cf. Fig. 58, Plate III). Thus in *Sparmannia africana*, a plant in which the leaf laminae orientate themselves at right angles to the direction of the incident light, the region of greatest curvature is at about 1 cm. from the basal end of the petiole, which is about 12 cm. long. Sometimes the lamina itself may take part in the curvature, but as the curvatures are generally due to unequal growth on the two sides of the organ they are naturally limited to regions in which growth is still possible. Where the orientation of leaves in relation to the direction of the stimulus is such that the petiole must twist itself in order that the lamina may take up the appropriate position, a torsion is executed. The mechanics of such torsions are far from being understood at present, but they also are due to differential changes in growth rate. Turgor changes are also responsible, wholly or in part, for some phototropic leaf movements. For example, in leaves of Malvaceae, investigated in 1888 by Vöchting, movements are due partly to turgor changes in the cells of the pulvinus and partly to differential rates of growth in the petiole.

THE SEAT OF PERCEPTION OF THE PHOTOTROPIC STIMULUS

As so much of our knowledge of phototropism is based on experiments carried out on seedlings of Gramineae, it is not surprising that it is on this material also that much of the work on the localization of perception of the stimulus of unilateral light has been carried out. The first experiments to this end appear to be those made by Charles and Francis Darwin on *Phalaris canariensis* and *Avena sativa.* Opaque caps of tin-foil or blackened glass were placed over the tips of the coleoptiles, which then failed to execute a curvature in the direction of the incident light. But Rothert later showed that although the tip is the most perceptive part of the coleoptile, the base is not completely insensitive. A more complete examination has been made more recently by Sierp and Seybold in 1926 and by Lange in 1927. These workers determined the presentation time when different zones of the coleoptile of *Avena* seedlings were illuminated with the same light intensity. The results of Sierp and Seybold are shown in Table CXVII, where sensitivity is taken as inversely proportional to the presentation time. It will be observed that the apical $\frac{1}{4}$ mm. of the coleoptile is 3600 times as sensitive as the basal region 2 mm. below the apex, while if we further take into consideration the estimated number of illuminated cells in each zone, a cell at the apex must be approximately 36,000 times as sensitive as a cell towards the base.

Table CXVII

Distribution of Sensitivity to Unilateral Light of Coleoptiles of *Avena sativa*

Distance of Zone from Apex in mm.	Relative Sensitivity	Number of Illuminated Cells	Relative Sensitivity of Individual Cells
0–0·25 . .	3600	100	36,000
0·25–0·50 . .	300	300	900
0·50–0·75 . .	70	500	140
0·75–1·00 . .	18	1000	18
1·00–1·25 . .	7	1000	7
1·25–1·50 . .	4	1000	4
1·50–1·75 . .	3	1000	3
1·75–2·00 . .	2–1	1000	2–1
2·00→ . .	1	1000	1

Lange, who claimed to have used an experimental arrangement and a method of calculation giving greater exactitude, also found a considerable fall in sensitivity to light with increasing distance from the apex, and the fall was of the same order as that found by Sierp and Seybold. His results are summarized in Table CXVIII.

Rothert also showed that in *Setaria*, *Sorghum* and *Panicum* the power of perception of the stimulus is confined to the coleoptile (called by him the cotyledon) while the curvature in response to the stimulus is mainly, though not entirely, executed by the mesocotyl (the hypocotyl in Rothert's terminology). But more recently it has been recorded by F. A. F. C. Went that the mesocotyl here also has the

Table CXVIII

Distribution of Sensitivity to Unilateral Light of Coleoptiles of *Avena sativa*

Distance of Zone from Apex in mm.	Relative Sensitivity
0–0·1	6475
0·1–0·2	5665
0·2–0·3	4106
0·3–0·4	3046
0·4–0·5	1749
0·5–0·6	734
0·6–0·7	382
0·7–0·8	179
0·8–0·9	118
0·9–1·0	79·1
0·0–0·5	3887
0·5–1·0	287
1·0–1·5	25·3
1·5–2·0	2·47
0·0–1·0	2304
1·0–2·0	14·3
2·0–3·0	1·25

capacity to perceive the stimulus, but is much less sensitive than the coleoptile so that a much greater stimulus than has to be applied to the coleoptile must be applied to the mesocotyl to produce a response. This has lately been confirmed by Buch for members of the Paniceae.

In the hypocotyl of dicotyledons the region most highly perceptive of the phototropic stimulus is the upper part of the organ, but the rest of it is not insensitive. Rothert further showed that in shoots of *Dahlia* covering the stem apex with an opaque cap reduced the sensitivity to unilateral light but did not eliminate it, so that here again it may be concluded that while the stem apex is the most sensitive region, the stem below has also some power of perception.

Experiments have been carried out by many workers to determine whether the seat of perception of the stimulus in diaphototropic leaves is in the petiole or the lamina. Thus Charles Darwin covered the laminae of leaves of *Tropaeolum majus* and *Ranunculus ficaria* with black paper and concluded that the diaphototropic movements were then carried out to the same extent as by uncovered leaves. These observations were confirmed for *Tropaeolum* by Rothert and were extended by Krabbe to *Fuchsia* and *Phaseolus*, the same behaviour in the latter species being also observed by Haberlandt. According to the last-named worker, however, the diaphototropic reaction in *Tropaeolum* is not complete in leaves protected from the light by black paper, while in *Begonia discolor* and *Monstera deliciosa* the leaves can orientate themselves normally when the petiole is covered with tin-foil and the lamina exposed. The same was found by Vöchting to apply to *Malva verticillata*. The more recent work of Ball indicates that in *Oxalis macra* the petiole is strongly phototropic and will execute a strong curvature after the lamina has been removed. In *Sparmannia africana*, on the other hand, both the lamina and the petiole perceive the stimulus of unilateral light. It must therefore be concluded that in some cases both petiole and lamina have the power of perceiving the stimulus of unilateral light, while in other cases either the petiole or the lamina is the predominatingly perceptive organ. Where the stimulus is perceived by the lamina there must be a conduction of excitation as the movement is brought about by changes in the petiole which executes a curvature or a torsion or a combination of the two.

Again, in the leaves of Malvaceae referred to earlier, in which movement is brought about partly by turgor changes in the pulvinus and partly by growth curvature of the petiole, variation movements of the pulvinus can be induced not only by unilateral light falling on the pulvinus itself, but also by light falling on the leaf lamina. Here also there must be a conduction of excitation from lamina to pulvinus.

THE MECHANISM OF PERCEPTION OF THE PHOTOTROPIC STIMULUS

As long ago as 1832 de Candolle suggested that the curvature of stems towards a source of light could be explained as due to the more rapid growth through etiolation of the side remote from the light. But, as Sachs pointed out, this theory will not explain the negative phototropism of roots which grow more rapidly in the dark than in the light; further, phototropic response takes place in translucent organs such as fungal hyphae, in which the difference in light intensity on the two sides is too slight for the curvature to be brought into line with etiolation in this way. Sachs himself thought, from the similarity of geotropic and phototropic behaviour, that as with the former curvatures, so with phototropic ones, the important point was the direction in which the light passed through the plant, and not that one side was illuminated more strongly than the other. While in any case it is difficult to understand how light can maintain a direction in a complex medium such as a cellular plant organ, experiments made by Charles and Francis Darwin and later by von Guttenberg effectively disposed of the light direction theory. If one half of a phototropically sensitive organ such as an oat coleoptile is shaded from light while it is illuminated equally from opposite sides, so that the direction of the light is at right angles to the plane joining the middle of the shaded half and the middle of the unshaded half of the organ (cf. Fig. 59) the organ curves in this plane directly away from the shaded part, and so at right angles to the direction of the incident light. The reaction thus appears to be determined by the different quantities of light falling on the two sides of the organ. The results of experiments described by Buder in 1920 confirm this conclusion. When a longitudinal half of an oat coleoptile is shaded from the light and is then illuminated from above the organ curves away from the shaded side. In another, and ingenious, experiment made by the same worker the coleoptile was illuminated from within. In a detached coleoptile a fine cylindrical glass rod, silvered on the outside, was inserted longitudinally through the greater length of the coleoptile. Through this rod by means of mirrors a narrow beam of light was thrown. At its upper end, in the apex of the coleoptile, the rod was bent so that the light entered the coleoptile tissues from within and at right angles to the surface. When subjected to the action of unilateral light in this way a curvature resulted so that the illuminated side was concave. The curvature is thus away from the source of light, but the illuminated side, as when the coleoptile is illuminated from without, grows less rapidly than the non-illuminated side.

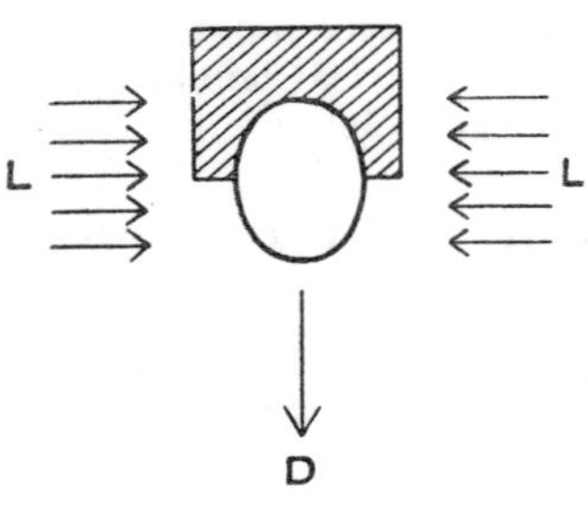

FIG. 59.—Diagram to illustrate the direction of response of the coleoptile of *Avena* when this is partly shaded and illuminated from two opposite sides

L indicates the direction of incident light, D the direction of curvature

The most convincing evidence that phototropic curvature results from the direct effect of impingement of light of different intensities (or rather of different quantities of light) on the two sides of the organ is provided by the work of Blaauw, who showed how the curvature was the result of the light-growth reaction. It has already been observed (p. 352) that when the sporangiophore of *Phycomyces* is illuminated equally from four sides for a short time, the rate of growth of the sporangiophore increases after a short time, reaches a maximum and then slows down to the normal value. When illuminated on one side only the two sides will receive different quantities of light and the light-growth reactions on the two sides of the sporangiophore will be different and a curvature will result. Blaauw held that owing to the cylindrical form of the sporangiophore and the refractive index of the contents the rays of light are refracted so that actually the side of the sporangiophore remote from the source of light is more strongly illuminated than the side nearer the source of light. Consequently the farther side grows more rapidly than the near side, since in *Phycomyces* the light-growth reaction is positive, and a positively phototropic curvature results. Confirmation of this view is provided by the observation of Buder that when the sporangiophores are placed in liquid paraffin (' paraffin oil '), owing to the refractive index of the latter, unilateral light no longer tends to be focussed on the side remote from the light and the side towards the source of light is now the more strongly illuminated. And corresponding to this, the near side now grows more rapidly than the remote side and the curvature is phototropically negative.

Further, it has been noted earlier that whereas with low light intensities the sporangiophore of *Phycomyces* is positively phototropic, with higher light intensities the organ executes a negatively phototropic curvature. This change in the direction of curvature corresponds with changes in the light-growth reaction with increasing light quantity. Whereas with comparatively low light intensities the reaction is positive, in higher intensities the reaction is negative, the exposure to the light resulting in a lowering of the growth rate.

More recently, in 1930, Castle has described experiments in which sporangiophores of *Phycomyces* illuminated from above were compared with others illuminated laterally. It was found that with equal quantities of light the growth reaction and the phototropic curvature always synchronized, and the conclusion is drawn that the tropic curvature and the growth reaction are induced by the same light-sensitive system. In the next year he showed that phototropic indifference, that is, failure to produce a curvature, occurred over a wide range of exposures to light, namely, from 104·6 foot-candle seconds upwards. But over this range exposure to unilateral illumination produced a distinct growth reaction. Indifference to phototropic stimulation is therefore not due to the absence of a light-growth effect but is characterized by a failure of the light to produce differential rates of growth on the two sides of the sporangiophore. With exposures below the lower limit mentioned above, the latent time is longer the shorter the exposure. The conclusion is therefore drawn that with these shorter exposures insufficient chemical change takes place to promote the maximum rate of the processes involved during the latent period, but that with an exposure of 104·6 foot-candle seconds or longer these processes are enabled to attain their maximum rate. Thus with shorter periods the rate is sub-maximal and the less illumination on the darker side of the sporangiophore results in a slower rate of the processes involved on that side as compared with the rate on the more illuminated side; hence a differential rate of growth producing a curvature results. With higher light intensities the maximum rate is reached on the darker as well as the lighter side, with the consequence that the growth rates on the two sides are the same and a curvature does not occur.

A comparison of the light-growth reaction observed in stems of higher plants with the phototropic reaction of these organs affords further evidence of the connexion between the two phenomena. Thus, in *Helianthus globosus* over a wide range of light quantities the light-growth reaction is negative, the retardation of the growth rate resulting from exposure to light being greater the larger the quantity of light. When illuminated on one side only, the near side, that is, the one more strongly illuminated grows less rapidly than the other so that a positive curvature results.

(As regards roots, these are mostly without phototropic sensitivity, and in correspondence with this exhibit no light-growth reaction. An exception is provided by *Sinapis alba*, the roots of which are negatively phototropic and possess a negative light-growth reaction. It must be supposed that the light is here refracted in the same way as in the *Phycomyces* sporangiophore.

More recently a comparison of the phototropic curvature of *Avena* coleoptiles with the light-growth reaction of this organ has been made by van Dillewijn, and here again the curvature resulting from the action of unilateral light over a considerable light range is that to be expected from the light-growth reaction.

Some objections have been put forward in recent years to the theory that phototropic curvature results from a quantitative difference in the light-growth reaction on the two sides of the stimulated organ. F. W. Went in 1927 calculated that although the phototropic curvature of the *Avena* coleoptile is in the direction to be expected on the theory, the actual rates of growth of the two sides are not those which would take place if the light-growth reactions determined them. A similar conclusion was drawn from the rate of growth of seedlings illuminated uniformly and unilaterally respectively. Cholodny in 1930 also found that no considerable alteration in mean growth rate occurred in coleoptiles stimulated sufficiently to produce a phototropic curvature. These results with the *Avena* coleoptile are in marked contrast to those of Castle on the *Phycomyces* sporangiophore.

In leaves Haberlandt considered that epidermal cells constituted the seat of light perception. He found that in some leaves the epidermal cells have dome-shaped outer walls which act like lenses and more or less focus the light on to a spot in the middle of the basal wall or rather the protoplasm lining of the wall. If the leaf is displaced through an angle relative to the direction of the incident light the spot of light will shift to some other place on the wall. The movement of this spot of light from one part of the protoplasmic layer to another is supposed, in Haberlandt's theory, to constitute the actual stimulus, and the response to result in the leaf orientating itself so that the spot of light is once more central. In some plants, as for example, *Fittonia*, the epidermal cells have a plane upper surface and in these the light cannot be focussed on the basal wall. There are, however, among the epidermal cells two-celled hairs, in which the upper cell has the form of a biconvex lens which focusses the light on the protoplasmic layer of the cell below. In support of Haberlandt's theory are his own observations on leaves submerged in water. When so treated, owing to the different refractive index of water as compared with air, the light is no longer focussed on the basal wall of the epidermal cells, and the leaves are found no longer to orientate themselves in relation to the light. An exception to this rule is *Tropaeolum*, the leaves of which, under water, can take up their normal position in regard to light. This is explained on Haberlandt's theory on the ground that owing to the waxy particles (' bloom ') on the surface, the latter remains dry and in contact with air held between the wax particles. But if the bloom is removed the surface of the leaf becomes wet on submersion and the power of orientation is lost.

Evidence against the theory is provided by Kniep's experiments in which the leaf was covered with a layer of liquid paraffin ('paraffin oil'). Under this condition also the light is not focussed on the middle of the basal wall; indeed this spot is less brightly illuminated than the rest of the wall, yet the leaf reacts to light almost as readily as an untreated leaf. The same was found to be the case by Nordhausen with leaves covered with gelatin, and later in leaves from which the lens character of the epidermal cells had been removed by rubbing. Haberlandt's theory of light perception in leaves can thus by no means be regarded as proved.

THE CONDUCTION OF PHOTOTROPIC EXCITATION

From what has been stated earlier it is clear that in phototropic stimulation the seat of most acute perception is very generally not that of the greatest response, and there must therefore be a conduction of excitation. The mechanism for the conduction of phototropic excitation is now generally regarded as the same as that which brings about the conduction of the excitation in the case of stimulation by gravity, that is, a substance furthering or retarding growth is conveyed from the seat of perception, and, reaching the seat of reaction in different quantities on the two sides of the organ, brings about different rates of growth on the two sides.

We can suppose that light might act in three ways. In the first place the first stage in the phototropic action might be a photochemical reaction leading to the production of a growth-promoting or growth-retarding substance. This would be produced in different quantities on the two sides in relation to the different quantities of light, and being conducted mainly in a longitudinal direction would reach the seat of reaction in unequal concentration on the two sides.

Alternatively the action of unilateral light might in some way result in a redistribution of growth substance already present with again the result that the two sides of the organ at the place of response would receive different quantities of the substance.

Thirdly, as suggested by Brauner in 1922, light might increase the permeability of the cell membranes, so that the substance would be transported to the place of response more rapidly on the illuminated side than on the dark side.

Evidence that the conduction of phototropic stimulation in the coleoptile of *Avena* is brought about by the movement of a substance was produced by Boysen Jensen in 1910 and 1913. He showed that insertion of a small mica disk on the illuminated side of the coleoptile just below the apex did not prevent the response, while insertion of the disk on the darker side did prevent it, a result which could be explained on the ground that a growth-promoting substance travelling on the darker side brought about an increased rate of growth on that side of the coleoptile. The insertion of a disk of mica on the illuminated side did not stop this movement of growth substance, whereas the insertion of the disk on the darker side did prevent it and so the curvature was inhibited. But Boysen Jensen further showed that the coleoptile apex could be removed and then re-fixed on the stump with gelatin without the coleoptile losing the power to react phototropically. Hence the stimulation could be conducted through a layer of gelatin, again suggesting the passage of an actual substance. Support for such a view came from the experiments of Paal, described in 1919, on the coleoptile of the grass *Coix lacrima*. It was found that when the apex of the coleoptile was removed and then placed excentrically on the stump a curvature resulted similar to a tropic one.

The establishment of the hormone theory of phototropic conduction is largely due to F. W. Went, who suggested that the effect of unilateral illumination of the coleoptile is to bring about a movement of growth substance, which we now know as auxin, in the apex to the less illuminated side, so that there thus results an unequal distribution of the growth substance. This, it will be observed, is the second of the possible actions of light mentioned above.

F. W. Went placed a light-stimulated coleoptile apex on two small agar blocks so that the half-apex on the illuminated side was on one block and the darker half on the other block. The growth substance from the two sides of the apex then diffused into the respective agar blocks. When these were applied excentrically to decapitated coleoptile stumps in the manner employed by Paál with actual tips, it was possible from the curvatures produced to obtain a measure of the amount of growth substance in the two halves of the coleoptile apex. The results showed that there was always more growth substance in the shaded than in the illuminated side. It is not yet clear whether exposure to light reduces the quantity of growth substance present in the coleoptile apex, but there is no doubt that exposure of the apex to unilateral light results in an unequal distribution of the auxin in the coleoptile apex.

Experiments on seedlings of *Raphanus sativus*, described by van Overbeek in 1933, show that here also unilateral illumination results in an uneven distribution of the growth substance so that it becomes more highly concentrated on the shaded side. We have noted in an earlier chapter that van Overbeek found that in *Raphanus* the same quantity of auxin produced more growth in the dark than in the light. Hence with unilateral illumination there are actually two effects; in the first place there results a movement of the auxin to the shaded side, in the second place the auxin left on the illuminated side is less sensitive in furthering growth than is that on the shaded side. Both effects tend to bring about positive phototropic curvature. More recently Thimann and Skoog have observed that in *Vicia faba* the effect on the stem of growth substance is greater in the dark than in the light, although the production of growth substance by the plant takes place only in the light.

The course of events in phototropic stimulation thus appears to be very similar to that in geotropic stimulation. In both, stimulation leads to a redistribution of the growth substance in the perceptive region, and the growth substance, travelling longitudinally to the elongating region, there sets up differential rates of growth. In the case of phototropic stimulation there is the additional factor of the effect of light on the activity of the growth substance, a factor which affords an explanation of the light-growth reaction.

As in geotropic stimulation, so with phototropic phenomena, the analysis of the actions involved by means of the methods evolved in the Dutch school at Utrecht has only been proceeding during the last few years, and it cannot yet be assumed that all phototropic phenomena, including those exhibited by diaphototropic leaves, can be explained on the same grounds. But there is at least the probability that the explanations of phototropism in terms of redistribution of growth substance and the influence of light on the activity of the latter may prove a general one.

CHAPTER XXIV

CHEMOTROPISM AND OTHER TROPIC PHENOMENA

Our information regarding the responses of plants to the stimuli of gravity and unilateral light is much greater than that concerning the action of any other stimuli. Nevertheless, it is known that tropic response to changes in other conditions of the environment takes place. Among such stimuli are those brought about by various chemical substances, by changes in the osmotic pressure of the environment, by changes in temperature and electrical conditions, and by contact of the organ with some solid object. Inasmuch as these stimuli are directional, we meet with the phenomena of chemotropism, osmotropism, thermotropism, galvanotropism (or electrotropism) and thigmotropism (or haptotropism) respectively.

CHEMOTROPISM

The movement of spermatozoids of mosses and ferns in response to changes in the concentration of certain chemical substances is a well-known phenomenon. Such chemotactic responses and many others were investigated by Pfeffer and described by him in 1884. His method of investigation was to place a capillary tube, open at one end and containing a solution of the substance, in a drop of liquid containing the spermatozoids or other motile organisms. As the substance diffused from the open end of the capillary tube a concentration gradient would be set up, and organisms exhibiting positive chemotaxis would swim along the concentration gradient to the most concentrated place and so reach and enter the capillary tube. On the other hand, negative chemotaxis would be displayed by the movement of organisms away from the open end of the capillary.

In this way Pfeffer showed, as mentioned in Chapter XVIII, that fern antherozoids are strongly chemotactic towards malic acid and malates and move rapidly towards a capillary tube containing 0·01 per cent. of sodium malate, while they still move, although more slowly, towards a tube containing this salt in one-tenth this concentration. According to Buller, in high concentrations of malic acid, from about 0·2 per cent. upwards, the antherozoids of *Gymnogramme martensii* exhibit negative chemotropism, but this is not observed with the salts of malic acid. A considerable number of other substances also exert a chemotactic influence on antherozoids of this fern and probably of others; among such substances are a number of potassium salts, including the nitrate, sulphate, chloride, bromide, iodide, formate, acetate, phosphate and tartrate, and also a number of salts of sodium (although neither chloride nor nitrate), ammonium, rubidium, magnesium and caesium. While the fern antherozoids were found to respond also to the stimulus of maleic acid, though not as readily as to that of malic acid, a number of organic substances, including ethyl alcohol, glycerol, glucose, sucrose, lactose, urea and asparagine, were found to be without effect. According to Pfeffer, the chemotactic movements of the antherozoids of both ferns and *Selaginella* are brought about in nature by malic acid secreted by the archegonia.

Pfeffer also found that in certain mosses, *Funaria hygrometrica*, *Leptobryum pyriforme* and *Brachythecium rivulare*, the antherozoids exhibit positive chemotropism towards sucrose, but no chemotactic response of moss antherozoids could be obtained with a number of other carbohydrates and glycosides.

Among Thallophyta the zoospores of *Saprolegnia* are definitely known to respond chemotactically to phosphates, but there is little information regarding the existence of chemotaxis in the Algae, although members of the Volvocineae are said to respond to the stimulus of a number of substances. The Bacteria, on the other hand, appear to be generally sensitive to the stimulus of a variety of chemical substances, particularly peptone and salts of potassium.

The chemotropic response in non-motile forms has been examined particularly in fungal hyphae, and in the pollen tubes of higher plants. Some information is also available regarding chemotropism of the roots and shoots of seedlings.

Our knowledge of the chemotropism of fungi is largely due to Miyoshi, an account of whose researches was published in 1894. The species he worked with were *Saprolegnia ferox*, *Mucor mucedo*, *Penicillium glaucum*, and *Aspergillus niger*. His method of investigation consisted in injecting leaves of *Tradescantia discolor* with a solution of the substance under investigation, and then sowing the spores of the fungus on the under surface of the leaf. Where positive chemotropism was present the hyphae resulting from the germination of the spores ultimately grew towards, and into, the stomata. Similar experiments were also carried out with pieces of onion scale epidermis and thin sheets of mica which were perforated with small holes and placed on plates of gelatin containing the substances to be examined. By these experiments Miyoshi showed that ammonium salts and phosphates, as well as glucose and sucrose, have a definite chemotropic action. Thus a positive reaction of the hyphae of *Mucor mucedo* was obtained with a concentration of ammonium nitrate of 0·05 per cent. and of glucose of 0·01 per cent. Glycerol and quinic acid did not appear to act as chemotropic stimuli in these experiments.

As regards chemotropism in higher plants the behaviour of pollen tubes was also examined by Miyoshi. His experiments, in which he placed ovules and pieces of stigma on plates of 10 per cent. gelatin and sowed pollen grains on the gelatin, have already been mentioned. The growth of the resulting pollen tubes towards the ovules and pieces of stigma he attributed to chemotropism, and by experiments carried out in a similar fashion to those described above with fungi, he was able to show that here sucrose in particular, but also glucose and dextrin, induced a chemotropic response, whereas ammonium salts and phosphates were without effect.

In 1904 Newcombe and Rhodes recorded the existence of a chemotropic reaction in the roots of *Lupinus albus* when these were placed in contact with certain salts contained in blocks of gelatin. A positive curvature was observed with sodium monohydrogen phosphate while negative curvatures occurred with ammonium nitrate, calcium nitrate, potassium nitrate and magnesium sulphate. In the following year Sammet recorded experiments on the chemotropism of the roots of 14 plant species, the roots in this case being grown in water. Roots of *Lupinus albus*, *Vicia sativa* and *Sinapis alba* were found to be particularly sensitive. In general they showed a positive reaction to sodium chloride, potassium nitrate, acetic acid, glycerol, ethyl alcohol, ether, sucrose and camphor when these were present in low concentrations, but with higher concentrations the reaction was reversed. In the same year Lilienfeld also published observations on the chemotropism of roots of *Lupinus albus* and a number of other species, the chemical substance being held in gelatin. Lilienfeld employed a large number of substances and concluded that they could be classified into five groups in relation to their power of inducing chemotropic curvature. These groups comprised : (1) substances which exert no chemotropic influence (calcium carbonate, magnesium carbonate, glucose, sucrose, lactose) ; (2) substances which in all concentrations employed induce only positive curvatures ; these include ammonium sulphate, potassium sulphate, and carbonates of lithium, ammonium, potassium and sodium ; (3) substances which in all concentrations examined induce only negative curvatures, such substances including nearly all chlorides, sulphates and nitrates (but not potassium sulphate and potassium nitrate), and many acids ; (4) substances which induce a positive or negative curvature according to the concentration, and including sodium tartrate, potassium nitrate and potassium ureate ; (5) substances giving indefinite results, such as ammonium oxalate, urea, asparagine and some dyes.

It may be noted that Lilienfeld found some roots much more sensitive than others ; thus the roots of *Ervum lens* appeared to be practically insensitive.

Certain criticisms of this earlier work have been advanced by Porodko, whose own work must be regarded as the most critical we have on the subject of chemotropism of roots. His experiments were carried out chiefly with *Lupinus albus*, the roots being grown mainly in agar. Among the substances employed were boric acid, ethyl alcohol, glycerol, chloral hydrate, urea, glucose, phenol, aniline and a number of inorganic salts, acids and alkalies. These and the earlier experiments leave no doubt that chemotropic curvatures are executed by roots, and that roots of *Lupinus albus* are particularly favourable objects for observing the chemotropic reaction. Porodko concluded that the course of such a chemotropic reaction depends on the concentration of the stimulant; in very low concentrations the reaction is positive, but as the concentration is increased the reaction becomes negative. In higher concentrations still the reaction is again positive, but this is related to injury of the tissues.

But all these three phases of chemotropic reaction were not always observed. Sometimes one was missing, sometimes another, and, indeed, Porodko found that the chemotropic reaction of roots depended markedly on the substance. In general the non-electrolytes examined, namely, ethyl alcohol, sucrose, urea, phenol and boric acid, were found to be inactive. A positive reaction was never observed with these substances and a weak negative reaction only occasionally in high concentration of the substance.

With acids and bases the positive chemotropic action is very weak or wanting altogether, but the negative reaction is quite definite. Arranged in order of increasing negative chemotropic activity the acids examined were citric, tartaric, trichloracetic, sulphuric, butyric, acetic, hydrochloric and formic. With salts both the positive and negative chemotropic curvatures were observed. From a consideration of the magnitude of the positive and negative curvatures produced by the different salts Porodko concluded that chemotropic action is chiefly determined by ions and not by undissociated molecules and that, moreover, the positive chemotropic curvature is determined by the anion, the negative by the kation. The actual curvature produced by a salt in any particular concentration is the algebraic sum of the action of the two ions in the solution at that concentration, and hence may be positive or negative according to the particular ions concerned and their concentration. According to Porodko, the chemotropic activity of the different kations varies inversely with their electrolytic solution pressure, while the chemotropic activity of the different anions varies with their lyotropic action.

Again according to Porodko, the response is not due to the intensity of the stimulus alone, nor to the time of action alone, but to the product $C^n t$, where C is the concentration of the substance, t is the time of action and n has a value of about 2·5 to 3.

Although both Lilienfeld and Sammet recorded that decapitated roots were capable of executing chemotropic curvatures, these may have resulted from shock or injury, and the more critical experiments of Porodko with intact roots stimulated over limited regions indicate that the chemotropic stimulus is perceived chiefly, if not exclusively, by the root tip. There must thus be a conduction of chemotropic, as of phototropic and geotropic, stimulation. Whether the stimulation and conduction here also involves a redistribution and longitudinal transport of growth substance has not yet been investigated, but it seems not improbable that such may be the case. Porodko himself attributed chemotropic stimulation to differential changes produced in the emulsoids of the protoplasm on the two sides of the stimulated root. Thus in the cells which perceive the stimulus the degree of hydration of the disperse phase of the plasmatic emulsoids changes, positive chemotropism involving an increase, and negative chemotropism a decrease, in the hydration potential of the disperse phase. The equilibrium, in respect of this potential, between the cells of the root apex and the growing zone, is thus disturbed, with the result that water passes from the neighbouring cells into the stimulated ones, or vice versa. In this way a conduction of stimulation to the growing zone takes place and the change in hydration of the colloids of the cells of the growing zone leads to changes in the growth rate in this region. The changes being different on the two sides of the roots, a curvature results in consequence.

AEROTROPISM

A particular case of chemotropism, in which the organ is sensitive to changes in concentration of a gas, particularly oxygen, in a definite direction, was termed aerotropism by Molisch. According to him, roots are positively aerotropic, a root executing a curvature tending to bring the apex in the direction of the highest oxygen concentration, while pollen tubes growing in water are in this sense negatively aerotropic. Pfeffer recorded that Celakovsky found hyphae of *Dictyuchus monosporus* to be positively aerotropic to oxygen.

Also according to Molisch, roots are negatively aerotropic to gases other than oxygen, including carbon dioxide, ammonia, coal gas, hydrochloric acid and chloroform vapour. Sammet also found that roots of *Lupinus albus* and other species growing in a saturated atmosphere were positively aerotropic towards oxygen, carbon dioxide, ammonia and the vapour of a number of organic substances, including methyl alcohol, acetone and ether.

With increasing concentration, however, the aerotropic curvature may be negative. Negative aerotropic curvatures were also observed by Sammet with shoots of a number of species, including those of *Brassica napus*, *Triticum vulgare* and *Avena sativa*. No reaction could be observed with the sporangiophores of *Phycomyces*.

It is agreed by Molisch and Sammet that the aerotropic stimulus is perceived by the growing zone in which the curvature takes place, for the curvatures are still executed in roots stimulated after removal of the apex. In this respect the perception of the aerotropic stimulus stands in marked contrast to those of gravity and light where, as we have seen, the stimulus is most acutely perceived in the root apex. However, having regard to Porodko's observations on the chemotropism of roots, it would seem desirable to have further evidence relating to the perception of the aerotropic stimulus.

HYDROTROPISM

What may be regarded as another special case of chemotropism, and indeed, of aerotropism, has been designated by the term hydrotropism. This refers to the reaction of many plant organs in response to unilateral change in the concentration of water vapour. Hydrotropism appears to have been demonstrated first by Johnson in 1829, but it was brought into prominence by Sachs in 1872 by means of an experiment which is familiar in most botanical laboratories. Seeds of *Pisum sativum* or *Vicia faba* are sown in damp sawdust contained in a vessel the lower side of which consists of a perforated zinc plate which is not horizontal but which makes an angle of, say, 45° with the horizontal. The radicles produced by the germinating seeds tend to grow down vertically on account of their geotropism, and on reaching the sloping floor of the vessel, grow out through the perforations. If the outer air is saturated with water the roots continue to grow downwards, but if not the roots may execute a curvature so that the apex grows towards the damp sawdust in the vessel. A root may, indeed, even enter the vessel again through one of the perforations.

While roots are for the most part positively hydrotropic, some seedling shoots have been stated to exhibit negative hydrotropism. This has been recorded for *Linum* by Molisch and for *Solanum tuberosum* by Vöchting, but there is no indication that such a reaction is by any means general.

Among lower plants Molisch found the rhizoids of some liverworts, including *Marchantia polymorpha*, were positively hydrotropic.

Negative hydrotropism is exhibited by many fungi. Molisch found the sporangiophores of *Phycomyces* and other Mucorineae to react in this way, and so do the stipes of the fructifications of some of the larger fungi such as *Coprinus*. The sporangiophore of *Phycomyces* is particularly sensitive to the hydrotropic stimulus, but Pfeffer quoted Steyer as describing this sporangiophore as positively hydrotropic when the general humidity is low while it appears to be diahydrotropic under conditions of intermediate humidity.

According to Walter, the sporangiophores of *Phycomyces* exhibit a humidity-growth reaction in response to a change in humidity of the environment similar to the light-growth reaction, and he would relate hydrotropic curvature with the humidity-growth reaction in the same way as phototropic curvature can be related with the light-growth reaction. How far such a connexion is general remains to be seen.

In 1915 Hooker published an account of an investigation of hydrotropism in the roots of *Lupinus albus*. He concluded that these are positively hydrotropic, but that a hydrotropic curvature only takes place if the relative humidity is between 80 and 100 per cent. Further, the roots only react over a certain range of moisture differences. Thus at 20° C. the minimum moisture gradient for hydrotropic reaction of the roots examined is 0·2 per cent. per cm. and the maximum gradient one of 0·5 per cent. per cm. The optimum gradient has a value of 0·4 per cent. per cm.

The perception of the hydrotropic stimulus by roots was shown by Hooker to be most acute at the root apex, while the region behind the apex, although less sensitive, is not without some power of perception. The condition thus appears to be similar to that in the cases of geotropism and chemotropism, and, aerotropic phenomena possibly excepted, it would seem extremely probable that roots may exhibit a similarity in this respect towards the perception of all tropic stimuli.

THERMOTROPISM

A change in temperature on one side only of an organ may constitute a stimulus of a directional character, and a number of observations of thermotropism are on record. Most of these rest on dubious interpretation of experimental findings, and the conclusions of different observers are often contradictory. It seems certain, however, that peduncles of *Anemone stellata, A. nemorosa* and *Tulipa sylvestris*, which, when these plants are growing in the open, curve towards the sun and so follow it throughout the day, do so on account of temperature changes, at any rate in part, for when covered *in situ* with a dark vessel so that the plants are in the dark the peduncles still execute the same movements and follow the sun. It has already been noted that Castle attributed the negative phototropic curvature of sporangiophores of *Phycomyces* to infra-red radiation, that is, to heat, but although Wortmann had much earlier recorded such a curvature when the sporangiophores were subjected to the radiation of heat from a dark hot plate, subsequent experiments by Steyer, quoted by Pfeffer, led the latter to conclude that *Phycomyces* sporangiophores exhibit no response to unilateral change in temperature. Further, Hooker has concluded that what has been generally known as thermotropism in roots is chiefly a demonstration of hydrotropism.

Actually thermotropism can be regarded as a special case of phototropism, the stimulus in the former case being due to the longer waves of infra-red radiation, as compared with the shorter waves of visible light. It has been shown in the last chapter that as regards the stimulus of unilateral light the sensitivity to its perception rapidly declines with increasing wave-length throughout the visible spectrum. If this rule applies also in passing from the visible spectrum to the longer-waved radiation of the infra-red, then we should expect there to be a much lower sensitivity to heat than to light of the same quantity of energy.

GALVANOTROPISM

When an electric current is passed through liquid containing certain motile organisms, in some cases a movement of the organisms takes place towards either the positive or negative pole. Where the movement is not induced in dead organisms it cannot be attributed to cataphoresis, and is regarded as a response to the stimulus of the electric current and is termed galvanotaxis. Such galvanotactic response has been observed in bacteria and in some members of the Volvocaceae.

Galvanotropic curvatures in roots have been observed by Elfving and others, but the interpretation of their observations is difficult. It has been suggested that the observed curvatures resulting from the action of an electric current may be chemotropic in character, arising from the electrolysis of substances in the medium, or may be the result of injury, in which case they belong to the category of traumatotropic curvatures dealt with below. Thus in 1907, Miss Bayliss (Dr. Bayliss-Elliott) found that when non-polarizable electrodes were used with a current of sufficient magnitude to produce a curvature when ordinary electrodes were used, no curvature took place. Twenty years later Navez made a similar observation, for he found that when care was taken to eliminate as far as possible products of polarization and to prevent the migration of any such products when formed, there resulted no curvature from attempted galvanotropic stimulation. In cases of observed galvanotropic curvature of roots Navez supposed that the surface cells of the root are injured by the action of products of polarization. The injured cells then act as electrodes applied to the internal tissues which are thus subjected to electrolysis. The alterations thus produced in the internal tissues lead to differences of growth of the cells on the two sides of the root so that a curvature results. If this interpretation is correct it would seem that so-called galvanotropic curvatures are really curvatures due to injury and therefore belong to the next category to be considered, a view put forward by Gassner in 1906. However, Dr. Bayliss-Elliott, in the work referred to above, concluded definitely that galvanotropic curvatures could be produced without injuring the root. Her experiments were carried out with seedling roots of a number of species : *Vicia faba*, *Phaseolus multiflorus*, *Zea mais*, *Pisum sativum*, *Helianthus annuus*, *Cucurbita pepo* and *Ricinus zanzibarensis*, and she showed that the so-called galvanotropic curvatures not involving injury were actually chemotropic in character, the actual stimulus being the ions produced by electrolysis of the medium. With higher strengths of current Dr. Bayliss-Elliott agreed with Gassner that the curvature might be due to injury.

Thus, although one cannot regard the problem of galvanotropism as solved, the evidence at present available suggests that in low strengths of current the response may be due to a chemical stimulus, in higher strengths of current to injury.

TRAUMATOTROPISM

Charles and Francis Darwin, in their work on the *Power of Movement in Plants*, recorded a number of cases of roots executing growth curvatures as a result of wounding the apex by treatment of one side of the latter with silver nitrate. In various roots examined, including those of *Vicia faba*, *Pisum sativum*, *Phaseolus multiflorus*, *Gossypium herbaceum* and *Cucurbita ovifera*, the curvature was away from the injured side. Such a response to one-sided injury is now generally known as traumatotropism. Darwin's observations have now been confirmed and extended by several workers. The most important work on traumatotropism is that of Stark, published in 1917 and 1921. This worker found that in seedlings of a number of species the hypocotyl or epicotyl would execute either positive or negative curvatures as a result of amputation of a cotyledon. The species in which this was observed were *Agrostemma githago*, *Cucurbito pepo*, *Helianthus annuus*, *Lupinus albus*, *Ricinus communis*, *Sinapis alba* and *Silybum marianum*, and this work confirmed earlier observations by Sperlich and by Heidmann. It is curious to note that whereas a traumatotropic reaction was observed with amputation of a cotyledon of *Phaseolus multiflorus* no such reaction was observed in *Phaseolus vulgaris* similarly treated. It may be noted that partial, instead of complete, amputation of the cotyledon either reduces the amount of the reaction or the latter may not occur at all.

Traumatotropic curvatures can also be induced in more adult plants by a similar operation. Thus removal of part of a leaf lamina will induce curvature in the petiole and the midrib towards the damaged region. Such a response was observed in 16 species out of 24 examined. A similar effect is produced in compound leaves by removal of the leaflets on one side of the main axis. Here out of 91 species examined 58 showed a response, members of Rosaceae, Leguminosae and Ranunculaceae being particularly sensitive. Similarly, complete removal of leaves can lead to traumatotropic curvature of stems. Also removal of flowers from one side of an inflorescence axis can induce curvature of the axis. Curvatures can also be produced, as shown by Darwin, by other methods of injury, such as by burning and by local treatment with silver nitrate. Stark found that there can be a conduction of the excitation over a distance of 10 cm. or more, and that the curvature can extend over a similar length of stem and is not limited to the internode in which it starts.

Stark also investigated traumatotropism in seedlings of Gramineae. In seedlings of the *Avena* type in which the mesocotyl (cf. p. 448) is poorly developed, stimulation of the base of the coleoptile by local application of silver nitrate has a greater effect than stimulation of the apex The reaction generally appears in the region of maximum growth near the apex. In grass seedlings of the *Panicum* type in which the mesocotyl is well developed, the sensitivity is less in the coleoptile than in the mesocotyl. In the latter organ sensitivity is greatest at the upper end and decreases towards the base. The seat of the reaction is here chiefly in the upper part of the mesocotyl.

Stark also showed that if a coleoptile tip was cut off and replaced on its own stump, or on the stump of another coleoptile of the same or of a different species, traumatic stimulation by means of silver nitrate was conducted across the cut surface.

It has been suggested that traumatic stimulation can be regarded as a special case of chemotropism. Wounding results in certain chemical changes in the injured cells, with the production of substances which bring about a chemotropic curvature. In recent years several workers have shown that traumatotropic curvatures in *Avena* coleoptiles can be explained in terms of a redistribution of growth substance in the stimulated organ. Thus when one side of the coleoptile is wounded the growth substance on that side is reduced and so its transport is retarded on that side, but not on the uninjured side. This will reduce the rate of growth on the stimulated side and the organ will thus perform a positive traumatotropic curvature. However, after 2·5 to 3 hours from the time of wounding the growth substance is again produced on the wounded side and the organ straightens itself or even produces a negative curvature. How far this view offers explanation of all traumatotropic curvatures remains to be seen, but it seems scarcely to fall into line with Stark's conclusion that weaker traumatic stimulation often leads to an increase in growth rate, while stronger stimulation leads to a decrease in the rate of growth.

CONTACT IRRITABILITY (HAPTOTROPISM OR THIGMOTROPISM)

One of the most striking examples of curvature displayed by a number of flowering plants is that resulting from the stimulus of contact with a solid object. This is most familiar in plants which climb by means of tendrils, but the power of response to the stimulus is not confined to these plants, although most highly developed in them.

Tendrils were defined by Charles Darwin as filamentary organs sensitive to contact and used exclusively for climbing. Morphologically they are to be regarded as formed by the modification of leaves or part of a leaf, peduncles and branches. Frequently the leaflets of a compound leaf may be modified into tendrils, in other cases the leaf tip, and in others the petiole, is of tendril character. In some cases the morphological value of the tendril is difficult to determine.

Certain families of plants contain many tendril-bearing species, and among such are Cucurbitaceae, Bignoniaceae and Leguminosae, but many other families include plants possessing tendrils, as, for example, *Smilax* (Smilaceae), *Corydalis* and *Dicentra* (Fumariaceae), *Passiflora* (Passifloraceae), and *Ampelopsis* (Vitaceae). The tendril may be simple as in *Bryonia dioica* or branched as in *Pisum* and *Lathyrus*. In these the tendrils are morphologically to be regarded as derived from parts of the leaf, while in *Lathyrus aphaca* the whole of the leaf is reduced to a compound tendril. Leaf tendrils are also met with in *Cobaea scandens* and in species of *Bignonia*. Leaf tips prolonged into tendrils are presented by *Gloriosa superba* and *Corydalis claviculata.* Petioles irritable to contact and functioning as tendrils occur in *Clematis vitalba, Solanum jasminoides* and various species of *Tropaeolum.*

The usual response of a tendril to the stimulus of contact with a solid body is, of course, the promotion of differential rates of growth of the two sides of the tendril, the side remote from the seat of stimulation growing more rapidly than the near side. Response to contact stimulation also occurs in the stems of the various species of the dodder (*Cuscuta*) and in the aerial roots of the orchid *Vanilla.*

Stark and others have shown that many organs apart from tendrils exhibit a response to the stimulus of contact. As examples may be cited all etiolated seedlings examined as well as older green organs, including stems, petioles and inflorescence axes of a number of plants. In such cases contact, or stroking with a solid object, leads to a curvature concave on the stimulated side. While the sensitivity to contact stimulation varies greatly and is most highly developed in tendrils, it must be concluded that it is a widely distributed property of plant tissue, at any rate among vascular plants.

Among Thallophyta it has been found that the branches of some marine algae and the hyphae of some fungi are similarly sensitive to contact.

In some cases the response to stimulation does not end with differential rates of growth on the two sides of the organ. In various species of *Bignonia* and *Ampelopsis*, and in some other plants, sucker-like disks are produced on the tendrils as a result of contact stimulation. These disks are usually, but not always, produced at the ends of the tendril and constitute organs of attachment of the tendril to a supporting surface, in many cases the attachment being aided by the secretion of a sticky fluid from the disk. In *Cuscuta* the final response to contact stimulation is the production of haustoria which penetrate the tissues of a suitable host providing the stimulating .terial.

Contact and Shock Stimulation. Pfeffer laid great stress on the difference between the stimulus of contact as exemplified in the behaviour of tendrils, and the stimulus of shock. In the case of tendrils the stimulus consists of contact with a solid body only; there is no response in these organs to contact with, or pressure of, water, mercury or any other liquid, nor to stroking with a wet rod coated with gelatin gel. In shock stimulation any sort of mechanical disturbance, whether caused by striking with a solid rod, or by a jet or current of water or by wind, can act as a stimulus.

It is true that Van der Wolk found that curvatures were produced in coleoptiles of *Avena sativa* stimulated by contact with a rod coated with gelatin, a result confirmed by Stark and extended by him to seedlings of a number of other species. Stark further showed that in some seedlings a jet of water would produce a curvature, as, for example, in *Cannabis sativa* and *Agrostemma githago*. In this respect seedlings thus exhibit a marked difference in behaviour from tendrils which react to contact with a solid body only. It may be that with seedlings stimulated by contact with water or a gelatin gel we are dealing with a different phenomenon from that of haptotropism, namely, the type of irritable behaviour to which the name rheotropism has been applied and which is discussed at the end of this chapter.

Intensity of Contact Stimulus. A method employed by Charles Darwin for examining the sensitivity of tendrils to contact was to place a small loop of cotton or platinum wire on a tendril and observe whether a curvature resulted or not. Thus he found that a curvature was produced in tendrils of *Cissus discolor* by a loop of soft thread weighing 9·25 mg., while a curvature was produced in *Corydalis claviculata* by a weight of 4·05 mg. and in *Passiflora gracilis* by a loop of platinum wire weighing only 1·23 mg. The tendrils of *Sicyos angulata* are very much more sensitive, even than those of *Passiflora,* for a perceptible curvature was observed by Pfeffer to result here from stimulation with a thread of cotton weighing only 0·00025 mg. On the other hand, weights of 0·1 gm. or more may be necessary to produce curvatures in less sensitive tendrils.

For this method to be made quantitative it would be necessary to ensure that the tendrils were always orientated in a definite direction, a condition which does not appear to have been realized in Darwin's experiments. Also if the stimulus-quantity law holds in the case of contact stimulation it would also be necessary to measure the time of action of the stimulus. It was, however, the view of Pfeffer that the stimulus in such cases is due not to the pressure of the weight on the tendril, but to the contact of the weight, slight as it is, falling on the tendril.

This view finds support from the observation made by Darwin of cases in which small weights did not produce a curvature, even when left on the tendril. Further, application of a small weight may produce a curvature which is not permanent even when the stimulus is maintained. Such is the case already mentioned of stimulation of the tendril of *Passiflora gracilis* by placing on it a weight of 1·23 mg. Although, as already mentioned, a curvature was thus produced, it was not permanent even when the weight was left suspended on the tendril. It would appear that the falling of the weight on to the tendril effects the stimulus, the response to which is the curvature, but that the continued pressure of the weight on the tendril is not a stimulus, and the original curvature subsequently disappears owing to autotropism.

The method adopted by Stark to obtain different quantities of stimulus was that of stroking one side of the stimulated organ with a small rod of cork having a square cross-section, the stroking being effected by the sides, not the corners, and the number of strokes giving a value for the quantity of stimulation. In this way Stark was able to obtain a variety of quantitative data relating to contact stimulation, reference to which will be made later.

The Course of the Haptotropic Reaction. The result of stimulation by contact is, as we have seen, usually a growth curvature with the stimulated side forming the concave surface of the curvature. With weak stimulation of short duration a resultant curvature does not necessarily persist, for it may be followed by a straightening due to autotropism. But with continued stimulation the differential growth on the two sides continues so that the tendril forms a series of coils. The chances of a tendril coming into contact with a support are increased by its circumnutatory movement as it grows. Whether it actually coils round an object with which it comes into contact depends on the dimensions of the object. Since the radius of curvature of the organ after response depends on the different rates of growth on the two sides it is clear that an object may be too wide or too narrow for the tendril to coil round it. Thus Sachs pointed out that while the fine tendrils of *Passiflora gracilis* can coil round threads of silk, the thick tendrils of the vine (*Vitis*) require a support of at least 2 to 3 mm. diameter. Sachs further observed that a tendril, in order to become firmly attached to a support, must be firmly pressed to it. This means that the tendril tends to form a coil the radius of curvature of which is generally less than the radius of the support. Hence it follows that fresh parts of the tendril are brought into contact with the support and so fresh parts of the tendril are stimulated and the tendril grows round the support as a spiral coil.

The reaction of tendrils and other organs to contact, as distinct from shock, being the production of a growth curvature, it follows that the reaction can only take place so long as the organ has the power of growth. In most cases tendrils are of limited growth and can react to the stimulus of contact from the time when they have reached about 25 per cent. of their full length until fully grown. Growth may be very rapid, in some stages the length being increased by 50 per cent. or more in a day. Frequently one side of an unstimulated tendril grows a little faster than the other so that a weak curvature results. When the tendril approaches the horizontal the under side is generally somewhat concave. The curvature in tendrils resulting from contact stimulation generally commences at the place of application of the stimulus and extends a little on either side of it.

The course of the haptotropic reaction in seedlings has been carefully examined by Stark. The extent of the reaction naturally depends on the growth activity of the organ, the reaction being greatest at that stage of development of the organ when it is growing most rapidly. In the hypocotyls of dicotyledons the curvature commences in the zone of most active growth, and spreads from there, the maximum curvature being generally at the apex and the weakest at the base of the hypocotyl. In *Avena* coleoptiles the curvature appears to start about 1 cm. below the apex, while in *Panicum*, if the whole seedling is stimulated, the curvature commences at the top of the mesocotyl and is either limited to that region or spreads downwards; generally the coleoptile takes no part in the curvature. In *Zea mais* the curvature commences at the apex of the mesocotyl, but spreads from there in both directions so that both coleoptile and mesocotyl are involved in the curvature which is strongest in the region where coleoptile and mesocotyl join.

Fitting found that in tendrils of *Passiflora gracilis* and *Pilogyne suavis* contact stimulation involved an increased rate of growth not merely on the side of the tendril remote from the place of contact, but also in the medium zone. The increase in the rate of growth may be very considerable; thus the convex side may grow from 40 to 200 times as rapidly as before stimulation, while no appreciable decrease in growth rate may occur on the stimulated side.

A counter reaction sets in at different times in different seedlings. Most frequently it commences at the apex while the basal part of the hypocotyl is still executing the haptotropic curvature, with the result that the organ may become S-shaped. In a number of cases, including coleoptiles of *Avena sativa*, the counter reaction may be great enough to produce a negative curvature, which is then followed by another positive curvature so that a sort of pendulum movement results. The counter reaction appears to be autotropic in character and not to be accounted for by geotropism, nutation or in terms of negative haptotropism.

The Influence of the Strength of the Stimulus on the Course of the Reaction. It has already been noted that Stark obtained different quantities of stimulation by varying the number of times the organ was stroked with a cork rod. He found that the stronger the stimulus, the greater the proportion of seedlings which react. This, as in the case of other tropic phenomena, only holds within limits. The lower limit is fixed by the threshold value of the stimulus, but in most cases a single stroke constituted a stimulus above the threshold value. Of 28 species examined, only in 4 cases was a single stroke insufficient to induce a curvature. The upper limit is, of course, reached when all the individuals in a sample respond. Some of Stark's results are tabulated in Table CXIX.

With increasing stimulation the mean curvature of a population of stimulated seedlings increases, at first rapidly with increase in the quantity of stimulus and then more slowly. Here Stark has sought to show that the relation between quantity of stimulus and mean response obeys the Weber-Fechner law.

Latent Time. Stark also found that the stronger the stimulus, the sooner the curvature starts; that is, the shorter is the latent time. The latent time of the reaction to contact stimulation is generally quite short, notably shorter than in the case of geotropic or phototropic reaction. Darwin observed that tendrils of *Bignonia capreolata* became slightly curved 10 minutes after coming in contact with a stick, while

Table CXIX

Relation between Quantity of Stimulus and Response in the Haptotropic Reaction of Seedlings

Species	Percentage of Individuals responding with Different Quantities of Stimulus					
	1 stroke	5 strokes	10 strokes	20 strokes	50 strokes	100 strokes
Agrostemma githago	70	76	97	100	100	100
Avena sativa	44	100	—	100	—	—
Cannabis sativa	58	—	100	100	—	—
Panicum nutiaceum	78	92	90	100	90	—
Phaseolus vulgaris	0	0	17	41	69	89
Ricinus communis	29	—	53	100	93	100
Silybum marianum	0	41	—	79	92	100
Sinapis alba	70	95	94	100	100	—
Zea mais	33	—	—	93	100	—

within 30 minutes they had completely curved round it. A tendril of ***Passiflora gracilis*** after a single delicate touch on its concave surface at the tip produced a definite curvature in 2 minutes, while in ***Passiflora punctata*** a definite curvature took place 5 to 10 minutes after stimulation. Among organs other than tendrils Van der Wolk found a latent time for the coleoptile of ***Avena sativa*** of about 9 minutes. Stark found the latent time for contact stimulation in a number of seedlings was related to the thickness of the organ. He stimulated equally seedlings of a number of species. In the thin seedlings of ***Agrostemma githago, Phalaris canariensis*** and ***Panicum miliaceum*** more than 50 per cent. of the individuals examined had reacted in less than 20 minutes; indeed, in ***Agrostemma*** about 50 per cent. had reacted in about 8 minutes, and in ***Brassica napus*** the same proportion in about 12 minutes. On the other hand, in the stouter seedlings of ***Ricinus communis*** and ***Silybum marianum*** about 50 minutes elapsed before 50 per cent. of the seedlings stimulated showed visible curvatures, while in ***Lupinus albus***, the thickest species examined, about 70 minutes were necessary for this.

In addition to the diameter, the rate of growth and the length of the growing zone affect the latent time. Obviously the more rapid the growth and the longer the growing region, the sooner will the differential rate of growth of the two sides result in a visible curvature. Since the rate of growth depends on the stage of development of an organ, it is clear that the latent time will vary during the development of the organ, and this holds equally for tendrils and other organs.

It has already been mentioned that Stark found that the latent time of haptotropic reaction was reduced with increase in the quantity of the stimulus. Stark's numbers illustrating this in the case of seedlings of ***Agrostemma githago*** are shown in Table CXX.

Table CXX

Relation between Quantity of Stimulus and Latent Time for Haptotropic Reaction in Seedlings of *Agrostemma githago*

Quantity of Stimulus (Number of Strokes)	Percentage of Seedlings exhibiting Curvature after Various Times				
	20 min.	40 min.	60 min.	80 min.	100 min.
1 . .	25	60	70	70	70
10 . .	36	79	93	97	97
20 . .	70	93	93	100	100
50 . .	74	89	93	100	100
100 . .	73	89	95	100	100

Localization of Perception of the Stimulus of Contact and the Conduction of Excitation. It is very easy to limit the stimulus of contact to a very small area and so to determine how far the perception of the stimulus is limited to certain regions. In tendrils it has been shown by Darwin and others that all parts of the organ are not equally sensitive to the stimulus. Frequently the basal part of the organ, although capable of growth, has not the power of perception. In many cases, including members of the Cucurbitaceae, only the under, or concave, side of a tendril is sensitive, while in other species, including *Cobaea scandens*, *Smilax aspera* and *Cissus discolor*, the tendrils are sensitive to contact on all sides. Tendril-like petioles and the stem of *Cuscuta* are similarly sensitive all round. In *Ampelopsis veitchii* sensitivity to contact is limited to a small region at the tip of the tendril.

With sharply localized stimulation of tendrils the curvature commences at the place of stimulation and only extends about 1 cm. from this place. There is thus only slight conduction of excitation.

In seedlings Stark found a definite difference between dicotyledons and Gramineae as regards localization of perception. In dicotyledons, such as *Agrostemma*, the whole of the hypocotyl is sensitive to contact, but the zone of greatest sensitivity coincides with the most actively growing region. When stimulated in this region the resulting curvature commences in the stimulated place and spreads towards the base, if growth is still possible there. If stimulated in the basal region the curvature sometimes commences there and spreads upwards. In other cases a slight curvature results over the whole hypocotyl, the maximum curvature finally occurring a little below the apex. In most cases, however, of local stimulation at the base of the hypocotyl, the curvature commences at the apex and travels towards the base. In spite of the differences in individual behaviour, presumably depending on differences in the extent of conduction of excitation in different individuals, it is clear that such conduction is a more marked feature of haptotropic response in seedlings than in tendrils. Thus in *Agrostemma githago* the excitation can be conducted for a distance of 3 to 4 cm., in *Helianthus annuus* for about 7 cm. and in *Ricinus communis* for 5 to 10 cm. On the other hand, in hypocotyls of *Brassica napus* the reaction, as in tendrils, scarcely extends beyond the place of perception, so that the curvature is practically limited to the place at which the stimulus is applied.

In the Gramineae two types, as exemplified by *Avena* and *Panicum* respectively, can be differentiated in regard to localization of perception to the stimulus of contact. In the *Avena* type the apical region of the coleoptile, that is, the uppermost 3 or 4 mm., is without sensitivity to contact, but the remainder of the coleoptile possesses about the same sensitivity throughout. When the mesocotyl is developed at all it is without contact irritability. In seedlings of the *Panicum* type, on the other hand, the mesocotyl is the more sensitive organ, the coleoptile possessing only weak sensitivity to contact. In seedlings of Gramineae a conduction of haptotropic excitation can also occur to a considerable extent, while in older stems, petioles and inflorescence axes of a number of vascular plants a conduction of the excitation over a distance of 5 cm. has also been observed, while in *Clematis* the excitation can be conducted for a distance of 10 cm.

It may be noted that the sensitivity to contact may not only vary in different regions of the same organ, but also at different times during the development of the organ. Thus, Darwin mentioned that a half-grown tendril of *Passiflora gracilis* is not sensitive, but that when fully grown the tendril is very sensitive.

We have little information at present regarding the mechanism of perception and conduction in the case of irritability to contact and whether here also perception of the stimulus involves an uneven distribution of growth substance has not yet been determined.

Haptotropic Stamens. Before leaving the subject of haptotropism reference may be made to the sensitive stamens of certain Cactaceae (*Opuntia* and *Cereus*) which when touched move towards the stimulated side. It thus appears legitimate to describe these movements as haptotropic. The phenomena of irritability in these stamens nevertheless appear very similar to those in other irritable stamens in which the response to stimulation is independent of the direction of the stimulus, and which are in consequence considered as nastic phenomena. Stamens exhibiting a less degree of haptotropic sensitivity are met with in *Sparmannia* and some species of Tiliaceae and Portulacaceae. The mechanism of movement in these stamens is not known.

RHEOTROPISM

It was recorded in 1883 by Jönsson and later confirmed by Newcombe and by Juel that the roots of many, but not all species, when grown in a stream of running water, execute a curvature so that the root apex comes to be directed against the direction of the current. This phenomenon is described as positive rheotropism, since the roots curve in the direction from which the water comes. The sensitivity to the stimulus of running water varies with the stage of development of the root, and the species. Generally speaking, seedling roots are more sensitive than those of more adult plants. As regards specific differences, the roots of some species are quite insensitive, while those of the Gramineae and Cruciferae are most sensitive. According to Jönsson, the hyphae of *Botrytis cinerea* are positively rheotropic and those of *Mucor* and *Phycomyces* negatively rheotropic. Among roots Newcombe found that with *Raphanus sativus* a water velocity of 2 cm. a minute produced no response, while a maximum positive response was reached with a water velocity of 100 to 500 cm. a minute. With much higher velocities (e.g. 1000 cm. a minute) curvature was observed to result but was attributed by Newcombe to mechanical conditions unconnected with stimulation. A zone of about 15 mm. was found to be perceptive in roots of *Raphanus sativus*, whereas the growing region is only about 6 mm. long.

Several interpretations have been given of rheotropism. Pfeffer regarded the phenomenon as a case of hydrotropism, but against this view it has been pointed out that only the root tip is sensitive to hydrotropic stimulation whereas, as we have just seen, a much greater length of root is perceptive of the rheotropic stimulus. Newcombe also found that rheotropic curvatures occurred in roots covered with a collodion membrane through which water penetrated very slowly.

More recently Sen-Gupta has reported that rheotropic curvatures are not produced by a stream of pure distilled water, but that if very small quantities of certain salts are present in the water the roots react rheotropically. If this should prove to be general it would appear that rheotropism is a special case of chemotropism.

A third view of rheotropism is that the curvatures are a response to the stimulus of the pressure of water. In this case the question arises whether rheotropism is to be regarded as a special case of contact stimulation. We have already noted that tendrils are not stimulated by the pressure of water or wind, a fact which led Pfeffer to lay emphasis on the necessity for contact with a solid body to effect haptotropic stimulation. But here Stark's work has shed a different light on this question, for as mentioned earlier he found that curvatures resembling those produced by stroking with a solid body could also be produced by a jet of water impinging on the sensitive organs.

CHAPTER XXV

NASTIC PHENOMENA

It has been pointed out earlier that the essential difference between a tropic and a nastic movement or curvature is that in the former the stimulus is directional and the response has a relation to this direction, whereas in nastic phenomena the response has no relation to the direction of the stimulus and can be brought about generally by a change in a condition all round the organ, so that the stimulus can be described as diffuse. In some cases of nastic irritability indeed, the stimulus may be applied unilaterally, but inasmuch as the response is independent of the direction of the stimulus, the phenomenon is not a tropic, but a nastic one. The position may be made clear by citing two well-known examples. When flowers of *Crocus* or *Tulipa* which have been kept at, say, a temperature of 10° C. are brought into a higher temperature of, say, 20° to 25° C., the flowers open, that is, growth curvatures occur at the base of the perianth leaves. The movements of the perianth leaves involved in the opening of the flowers are thus thermonastic movements, brought about by the change in temperature all round the flower. The stimulus is thus a diffuse one. The second example is that of the irritable stamens of *Helianthemum* or *Mesembryanthemum*. When touched the stamens of these plants execute a movement which is in the same direction wherever the stamens may be touched. The response to the stimulus being thus independent of the place receiving the stimulus, that is, to its direction, the movement is a nastic one, and the phenomenon can be referred to as seismonasty or haptonasty (thigmonasty) according as one regards the stimulus as one of shock or one of contact.

Reference was made in the last chapter to irritable stamens which exhibit haptotropism, and it seems likely that whether the response in irritable stamens is tropic or nastic depends on the anatomical structure of the stamens. And in general it must be supposed that where a diffuse stimulus produces a curvature, the organ concerned must possess a certain degree of dorsiventrality or asymmetry which is most likely to be morphological, but which may be only physiological; otherwise the change in environment constituting the stimulus should produce equal changes in growth rate on all sides of the organ. A unilateral stimulus, as we have seen, gives rise in a previously radial organ to a physiological dorsiventrality which may lead to a curvature. The incidence of tropic or nastic curvatures thus depends in great degree on whether an organ is radially organized or not, but as we have seen, tropic stimulation can occur in a dorsiventral organ. Thus while a completely radial organ cannot perform a nastic curvature, a dorsiventral or asymmetrical organ, even if only physiologically so, might respond by curvature to either a unilateral or a diffuse stimulus or both. Where a unilateral stimulus produces a nastic curvature the response being independent of the direction of the stimulus, we must suppose that either only certain cells are capable of both perception and response, or that conduction of excitation is always to the same place wherever the stimulus is perceived. In any case there is no reason to suppose that the mechanism of perception, the conduction of excitation and the mode of reaction are essentially different in tropic and nastic phenomena.)

Most of the conditions which can provide tropic stimulation can also give rise to nastic stimuli. Thus we have the phenomena of photonasty, thermonasty, hydronasty, haptonasty, seismonasty and chemonasty, according as the stimulus is produced by light, temperature, water, contact, shock or chemical substances. As with tropic phenomena, the response generally takes the form of movement, using the term in its wide sense to include curvature, and the mechanisms of response are also the same, involving growth curvatures, torsions or variation movements attributable to turgor changes in the cells of certain tissues, or a combination of these. Not infrequently plant movements taking place under natural conditions result from a combination of the reaction to two or more nastic stimuli or from a combination of the effects of both tropic and nastic stimuli.

PHOTONASTY

The daily periodic movements which occur in the opening and closing of many flowers, and the so-called 'sleep' movements of some foliage leaves, appear at first sight to bear an obvious relationship to the periodic alternation of day and night, and because of their connexion with the oncoming of night they have been described as *nyctinastic*. The most obvious change in conditions as day gives place to night is the alteration in light intensity, but this is also accompanied very generally by a change in temperature, so that nyctinastic movements might be either photonastic or thermonastic or both. Further, associated with the change from day to night are changes in the electrical conditions, a fact now familiar to every user of a wireless receiver, and it cannot be assumed that such electrical changes are without effect on plants, although at present little is known of any such effect.

Whether nyctinastic movements are due chiefly to light or to temperature change is a question which can be answered in a few, but in by no means all, cases.

The nyctinastic movements of leaves are for the most part attributed to photonasty. The movements may be brought about by differential growth on two sides or by turgor changes in the cells of the leaf base. A type of photonastic growth movement of leaves is exhibited by *Impatiens* and other members of the Balsamineae in which the leaves droop, that is, hang vertically downwards, at night, whereas during the day they are expanded more or less horizontally. The reverse movement, also brought about by growth, occurs in other cases, the leaves pointing more or less vertically upwards at night. This type of nyctinastic movement is observed in, among other plants, *Amaranthus* spp., *Nicotiana glauca* and *Chenopodium album*. It may be noted that these two types of movement may occur in the same genus, for in *Polygonum convolvulus*, the leaves droop at night, while in *Polygonum aviculare* the leaves stand erect at night. In these plants it can easily be shown that the movements are related to light by darkening the plants during the hours of daylight when the nyctinastic movement follows.

Since these movements result from differential rates of growth on two sides, it follows that the movements only occur while the leaves retain the power of growth. As these become fully grown the movements become less and when a leaf is mature no longer take place.

Variation photonastic movements of leaves, resulting from changes in turgor of cells in the leaf-bases, or so-called motile pulvini, occur in many members of the Oxalidaceae and Leguminosae and in some other plants, including *Marsilia*. The actual movements show a certain amount of variation among themselves. In many cases the leaves are expanded more or less horizontally during the day and droop at night. This behaviour is exemplified by the leaves of *Desmodium gyrans*, where the drooping of the leaf lamina is accompanied by an upward movement of the petiole. In *Trifolium* the three leaflets are expanded horizontally during the day, but at night the two lateral leaflets bend downwards and their stalks twist so that the laminae move towards one another until their upper surfaces are close together with their midribs making an angle of about 45° with the petiole of the leaf. At the same time the terminal leaflet bends right over and folds about the midrib until it comes to lie over the lateral leaflets. It may thus move through an angle of 90° to 180° and comes to rest in its night position with its morphologically under surface uppermost. Darwin found these nyctinastic movements essentially the same in all species of clover. The day and night positions of the leaves of *Oxalis* are shown in Fig. 60.

In *Robinia* and *Amorpha* the leaves are compound and the leaflets pinnately arranged in pairs with a terminal leaflet. During the day the leaflets are expanded more or less horizontally, but at night they all hang vertically downwards, the under surface of each pair being virtually in contact. The sleep movements of the leaves of *Averrhoa* were recorded by Bruce in 1785. Here also the leaf is a compound one with about 10 pairs of leaflets pinnately arranged and a terminal leaflet. As noted earlier, during the day the leaflets execute spontaneous movements, at night they all hang vertically downwards, as in *Robinia*.

The type of nyctinastic movement in which the leaflets sink at night is fairly common. Occasionally the movement of the leaflets is accompanied by movement of the leaf petiole which rises in *Centrosema* and *Phaseolus*, and sinks in *Amphicarpaea*.

The reverse type of nyctinastic movement in which the leaflets rise at night is also met with. In ten species of *Lotus* Darwin observed that the leaf petiole rises at night while the three leaflets rise until they are vertical. In *Coronilla rosea, Mimosa pudica* and *Mimosa marginata* the leaves are compound with a number of pairs of opposite leaflets as in *Robinia*; here the leaflets rise at night until the upper surfaces of the two leaflets of each opposite pair approach one another. There is, however, this difference between *Coronilla rosea* and *Mimosa pudica*: in the former species the leaflets, as well as rising at night, are twisted so that their apices are directly towards the base of the petiole, whereas in *Mimosa* the apices of the leaflets in their night position are directed towards the base of the leaf. In *Mimosa pudica* the petiole of the leaf slowly sinks during the day and rises during the night, while in *Coronilla rosea* the petiole is curved slightly downwards during the day and rises and straightens itself at night.

Enough has been stated to indicate that nyctinastic movements in leaves are frequent and varied. Such movements are not limited to adult leaves but also occur in cotyledons; indeed the movements according to Darwin are more widely distributed in cotyledons than in foliage leaves. In the former, as in the latter, they include both growth movements and variation movements due to changes in cell turgor. As is to be expected, the latter type of movement continues longer than growth movements, which naturally cease when the cotyledons cease growth. The same applies to foliage leaves, in which nyctinastic growth movements can continue only as long as the leaf is capable of growth, whereas nyctinastic variation movements can continue as long as the leaf remains alive.

In *Mimulus tilingii* whole foliage shoots, which are more or less horizontally expanded during the day, rise at night, a movement shown by Vöchting to be photonastic.

As is well known, many flowers close at night, but not in all cases is the closure related to change in light intensity. The closing of flowers of *Oxalis* and allied genera, among others, are, however, photonastic, while the same is true of most flowers which open at night and close in the daytime, such as the evening primrose (*Oenothera*), tobacco (*Nicotiana*) and certain cacti, for example, *Cereus grandiflorus*. In many Compositae the florets execute photonastic movements which bring about the closure of the whole inflorescence. In the daisy (*Bellis*) this is effected by curling of the ligulate ray florets inwards over the disk; in *Hieracium* all the ligulate florets bend upwards and inwards.

In *Tragopogon brevirostris* the inflorescence closes in this way when the light intensity is reduced, but also closes when the intensity of illumination is increased beyond a certain value. Whether such a reversal of movement would occur in other cases with sufficiently high light intensity is not known.

The photonastic movements of flowers are in all probability growth movements, since they closely resemble the movements, to be described later, induced by change in temperature and which have been shown to be due to different rates of growth on the two sides of the perianth leaves.

The photonastic movements described above take place periodically with plants growing under natural conditions, the periodicity in movement corresponding with the periodicity in the stimulus of change in light intensity. In very many cases this periodicity appears to become fixed to some extent, so that when a plant which has been performing these movements under natural conditions is transferred to conditions involving continuous illumination or continuous darkness the movements continue for a time, although with regularly decreasing intensity. The time for which they continue varies with the species. Thus in darkness the movements of the inflorescence of *Bellis perennis* cease after 1 or 2 days, those of the flowers of *Oxalis rosea* after 3 or 4 days, while those of the leaves of *Impatiens noli-me-tangere* and *Mimosa pudica* may continue for from 4 to 8 days. It would appear that the normal periodicity resulting from periodic stimulation induces an autonomic movement which sooner or later dies out under constant illumination or darkness. If after this condition has been reached, the plant is then stimulated by transference to darkness or light as the case may be, the photonastic response again occurs, although not always to the same extent as before. It may, perhaps, be supposed that the greater response under continued natural alterations of light intensity is due to the induced autonomic movement supplementing that of the purely photonastic curvature. The point may be illustrated by reference to Pfeffer's researches on the leaf movements of a variety of *Sigesbeckia orientalis*. If the periodicity in the sleep movements of the leaves of this plant was brought to an end by subjecting the plant to continuous illumination, the photonastic movement of the leaves after a single night's exposure to darkness was slight, namely, movement through an angle of 10° to 30°. If the plant was then subjected to daily alternations of light and darkness, the movement resulting from each subsequent exposure to darkness was greater, until after the fifth stimulation in this way the movement was through an angle of 70° to 100°, the normal movement under the normal alternations of night and day. Obviously, where a single stimulation results in the full amplitude of movement, subsequent stimulations cannot produce any increase in the movement. This is the case in the leaves of *Acacia* (*Albizzia*) *lophantha*. Here, after a period of continuous illumination sufficiently long to bring about cessation of the periodic movement, a single night's exposure to darkness causes the leaflets to rise to their highest position.

Even with a single stimulation there is usually more than a single movement, the first movement being followed by a few others of smaller amplitude.

THERMONASTY

The best-known thermonastic curvatures are those which occur in the flowers of *Crocus* and *Tulipa*, where an increase in temperature brings about opening of the flowers, and a fall in temperature closure of the flowers. In *Crocus* a change in temperature of no more than 0·5° C. may bring about a movement of the perianth leaves. Some other flowers also show a thermonastic reaction, including those of *Ornithogalum umbellatum, Colchicum autumnale, Anemone nemorosa* and *Ranunculus ficaria,* although in these cases a change in temperature of 5° to 10° C. may be required to induce a response. In *Crocus* and *Tulipa* the latent time may be only a few minutes, but is longer in the other cases mentioned above.

Thermonastic movements have also been recorded as occurring in leaves with motile pulvini, including those of *Oxalis acetosella, Mimosa pudica* and *Desmodium gyrans* and in some shoots and peduncles.

While the thermonastic movements of leaves with motile pulvini are due to changes in turgor of the cells, the thermonastic movements of flowers are brought about by changes in the growth rate of the perianth leaves. With increase in temperature there results in the perianth leaves of *Tulipa*, for example, marked increase in the rate of growth in the upper side of the perianth leaf, which is followed later by an increased growth of the lower side. Growth takes place chiefly towards the base of the perianth leaf. With a fall in temperature growth is first more rapid on the lower side, and later on the upper side. According to the researches of Bünning, published in 1929, increase in temperature brings about an increase in the modulus of elasticity of the walls of the cells of the upper side of the perianth leaves of *Tulipa*.

Measurements of the rate of growth of the two sides of the perianth leaves of *Crocus* and *Tulipa* after thermonastic stimulation were published in 1904 by Wiedersheim. The results of some of his observations on *Crocus* are shown graphically in Fig. 61. It will be observed that after a change in temperature from 9·3° to 20·8° C. the upper side immediately begins to grow rapidly. The lower side at first shows only negligible growth, but after about 2 hours the lower side begins to grow. Similar relations were observed in the case of *Tulipa*.

In a number of plants possessing thermonastic flowers, maintenance at a comparatively low temperature results in the flowers remaining closed throughout their whole development. This was observed by Pfeffer to be the case with flowers of *Crocus* and *Oxalis rosea*, and the inflorescences of *Hieracium* and *Leontodon*.

It appears possible that a reversal of direction of thermonastic movement may occur with high temperatures. Thus closure of the flowers of *Crocus* has been observed under such conditions, and it has been suggested that such a reversal of thermonastic response may explain the incidence of the 'sleep' position of leaves of many tropical plants in the hottest hours of the day.

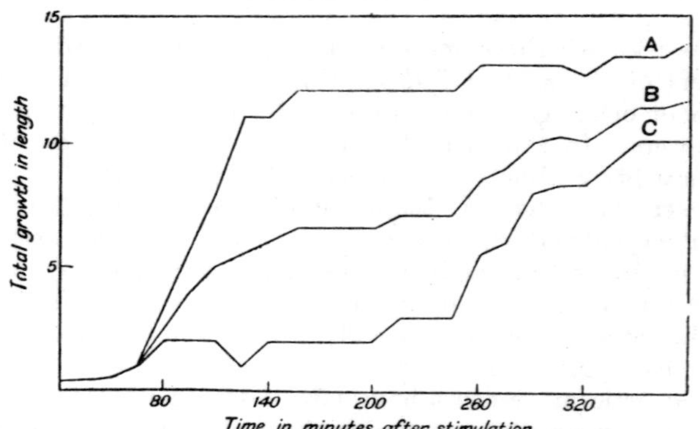

FIG. 61.—Curves to illustrate the course of growth on the two sides of a perianth member of *Crocus* as a result of thermonastic stimulation

HYDRONASTY

Changes in the relative humidity of the atmosphere may induce movements which can be regarded as hydronastic. Indeed, Jost expressed the opinion that in all nyctinastic leaves, the night and day positions are partly determined by changes in humidity as well as by changes in light intensity and temperature. Among leaves which have been recorded as taking up the sleep position as a result of increased humidity of the air may be mentioned those of *Oxalis* and *Myriophyllum proserpinacoides*.

CHEMONASTY

The production by chemical stimuli of curvatures which are independent of the direction of the stimulus, and which are therefore to be regarded as chemonastic, occur notably in some insectivorous plants. Of these, *Drosera rotundifolia* in particular formed the subject of detailed observations by Charles Darwin, described in his book on *Insectivorous Plants*, in which are also recorded data on other species of *Drosera*, and on *Dionaea muscipula*, *Aldrovanda vesiculosa* and species of *Pinguicula* and *Utricularia* among others.

It is well known that the upper surface of a leaf of *Drosera* bears, in addition to small non-secreting hairs, a large number of glandular hairs or tentacles. In *D. rotundifolia* Darwin found these numbered from 130 to 260 on a single leaf. They are short and erect on the middle of the leaf and longer and more inclined outwards the nearer they are to the margin. On the edge of the leaf they either extend outwards in the plane of the leaf or are actually curved downwards. Each tentacle consists of a pedicel terminating in a knob-like ellipsoidal gland which secretes a viscous fluid containing proteolytic enzymes which digest the bodies of insects which come into contact with the tentacles (Chapter XIII).

Darwin found that whereas a drop of distilled water [1] placed on a leaf of *Drosera* has no effect on the tentacles, placing on a leaf a drop of liquid containing one of a number of nitrogenous substances caused the tentacles to curve over towards the middle of the leaf. Among substances found by Darwin to bring about curvature of the *Drosera* tentacles in this way were milk, urine, egg albumin, meat infusion, saliva, isinglass and a number of ammonium salts. Most alkaloids, on the other hand, had no effect. The non-nitrogenous organic substances tested by Darwin brought about no response; such substances were sucrose, starch, ethyl alcohol and olive oil. Various salts, however, particularly those of sodium, brought about the curvature of the tentacles, while others, notably those of potassium, were without effect or doubtful in their action. Most acids, also, induced a chemonastic curvature.

Of the substances tested by Darwin ammonium phosphate proved the most active. A drop of solution containing little more than 0·000423 mg. of the crystallized salt placed on the glandular head of an outer tentacle was sufficient in some cases to induce curvature of the tentacle within 6·25 hours, while when comparatively large quantities of phosphate were placed on the leaf, the tentacles in one case moved within 15 minutes. Ammonium carbonate and nitrate were found to be less effective than the phosphate. Relative data found by Darwin for the three salts are summarized in Table CXXI.

Table CXXI

Minimum Quantities of Ammonium Salts required to induce Chemonastic Reaction of the Tentacles of *Drosera rotundifolia*

Salt	Minimum Quantity in mg. required to induce Curvature of Outer Tentacles	
	Solution placed on Glands of Middle of Leaf	Solution placed directly on Gland of Outer Tentacle
Ammonium carbonate . . .	0·0675	0·00445
Ammonium nitrate . . .	0·027	0·0025
Ammonium phosphate . . .	0·0169	0·000423

[1] It was later reported by Correns that distilled water induced a feeble response.

It is to be observed that stimulation not only induces movement of the tentacles but also results in an increase of the secretion of the digestive fluid by the glands. The chemical nature of the secretion itself also appears to change, since Darwin states that as a result of stimulation the secretion becomes acid.

The course of the chemonastic curvature in *Drosera* tentacles has been examined in more recent times by H. D. Hooker. The curvatures are due to unequal growth on the two sides. The unequal growth commences towards the base of the tentacle and extends upwards, the side nearer the margin of the leaf growing more rapidly. In Hooker's experiments the curvature reached its greatest extent in 3·25 hours, after which more rapid growth on the concave side led to a straightening of the tentacle, which was complete in about 24 hours. As a permanent result of such behaviour the tentacle increases in length. In three successive stimulations Hooker found that a tentacle increased in length by 0·43 mm., 0·37 mm. and 0·24 mm. after which the tentacle was no longer capable of growth and was incapable of responding to further stimulation. With successive stimulations the curvature is more and more limited to the basal part of the tentacle.

In the chemonastic reaction of *Drosera* tentacles there is a very definite conduction of the excitation. The seat of perception appears to be the glandular tip of the tentacle and application of the stimulus at this place results in curvature of the stalk. Moreover, stimulation of tentacles in the middle of the leaf induces curvature of the tentacles as far away as the margin, while Darwin showed that a decapitated tentacle still possessed the power of response by curving when a neighbouring tentacle was stimulated.

Darwin showed that certain visible changes in cells of the tentacles are associated with stimulation; changes which he referred to as 'aggregation', and which have also been studied by a number of subsequent observers. Before stimulation the cells contain a layer of protoplasm within the wall and surrounding the vacuole, the latter appearing as a homogeneous purple liquid. After stimulation the cell contents appear to be aggregated into purple masses contained in a colourless or almost colourless liquid. According to later observers, these visible changes are due to a change in the relative quantities of protoplasm and vacuole in the cells, the volume of the protoplasm increasing at the expense of the vacuole which becomes broken up into a number of smaller vacuoles. At the same time the protoplasm begins actively streaming. Next a precipitate appears in the cell-sap of the vacuole, the precipitated substance being frequently, though not always, tannin. The precipitated particles are often stained red owing to the adsorption of the pigment present in the vacuole.

These changes in the cells of the tentacles consequent on stimulation are first observed in the cells of the glands and then extend down the tentacles. Conversely, as the tentacles recover their original direction the state of aggregation gives place to the normal non-aggregated appearance, and just as the condition of aggregation travels down the tentacle after stimulation, so the disaggregation travels upwards in the reverse direction.

It has been supposed that in the progress of aggregation through the tentacle we see the transmission of the excitation. There are, however, serious arguments against this view. When a marginal tentacle responds as a result of stimulation of another tentacle (so-called secondary stimulation) the progress of aggregation in the former still travels from the upper to the lower part of the tentacle, and not in the reverse direction as one would expect to be the case if the transmission of the excitation were brought about through the aggregation. Again, in such cases of secondary stimulation the aggregation does not set in with the curvature of the stalk but only with increased secretion by the gland. The aggregation thus appears to be related to the secretory function and not to the conduction of the excitation.

Nevertheless, it is highly probable that the conduction of the excitation is brought about by the transport of some substance, and since the chemonastic response of the tentacles is a growth curvature it may be that auxin is again the substance involved. The excitation can be transmitted through parenchymatous cells, for Darwin showed that such transmission still took place when the vascular bundles of the tentacle were cut through. On the other hand, such treatment has been shown to reduce the rate of conduction of the excitation and it may be supposed that the hormone travels more rapidly through the elongated cells of the vascular bundles than through the shorter parenchymatous cells of the outer tissues.

Dionaea muscipula is a North American species of the Droseraceae. In this plant the leaf consists of a somewhat flattened leaf stalk terminating in a bilobed lamina. The two lobes of the lamina are not in the same plane but, in their normal position, form an angle of something more than a right angle with one another. The margin of each lobe bears a row of rigid spike-like projections and the upper surface of each lobe has three fine spines projecting at right angles to the leaf surface. In addition to these the leaves bear three other kinds of emergences. On the upper surface of the leaf lobes there are a large number of short glandular papillae, while octofid papillae occur mainly on the petiole and under side of the lamina and simple hairs on the under surface of the lamina. The short glandular papillae are secretory, but only produce their secretion when stimulated by a nitrogenous substance, such as a piece of damp meat or egg albumin. The stimulus also induces a slow movement of the leaf lobes which close together, the midrib acting as a hinge and the margins moving upwards and approaching one another. The movement is, however, very much slower than the haptonastic movement of the same leaves which will be described later. In Darwin's experiments it required about 3 or 4 hours for secretion to commence when the leaves were stimulated chemically and a period of the order of 24 hours for the leaves to close. Inorganic non-nitrogenous substances were reported by Darwin to have no effect.

In *Pinguicula vulgaris* it was observed by Darwin that placing insects on a leaf at a point near the edge induced the margin to curve over and inwards towards the midrib. The same result was obtained by stimulating the leaf at the margin with an infusion of raw meat or a weak solution of ammonium carbonate. Here also, as well as inducement of a chemonastic movement, there is a stimulation of secretion by the glands which are borne on the surface of the leaf. The chemonastic response takes place much more rapidly in *Pinguicula* than in *Dionaea*, for the shortest time in which a definite movement of the *Pinguicula* leaves was observed after placing a nitrogen-containing liquid on the leaf was 137 minutes. Other species of *Pinguicula* exhibit the same general behaviour, which is also observed in species of *Drosera* when strongly stimulated.

A chemonastic movement is induced in stamens of *Berberis* by ammonia. The movement in this case takes the form of a sudden contraction of the filament. Ammonia also brings about a chemonastic movement of the stigmas of the same species which thereby close together.

Chemonastic movement of leaves has been observed in *Mimosa pudica*, in which chloroform vapour induces movement of the pulvinus, and in *Callisia repens* in which coal gas induces a curvature of the leaves. While the chemonastic reaction in *Drosera* tentacles is brought about by unequal rates of growth on the two sides of the organ, the movement in the case of *Mimosa* leaves and probably of *Berberis* stamens, is due to turgor changes.

In some insectivorous plants chemical stimulation produces a response in the form of increased secretion of the digestive fluid, but no chemonastic curvature results. Such is the case in the leaves of *Drosophyllum lusitanicum* and the pitchers of *Nepenthes*.

HAPTONASTY

Reference has been made earlier to the difference between the stimulus of contact with a solid body and that due to shock, in which the exciting factor may be solid, liquid or gaseous. The most conspicuous examples of response to contact are tropic in character, inasmuch as the direction of curvature resulting from stimulation is related to the direction, or place of application of the stimulus. There are also, however, examples of contact stimulation in which the direction of the curvature or movement resulting from stimulation is unrelated to the direction of the stimulus and which are therefore to be regarded as cases of haptonasty. Chief among materials exhibiting haptonastic curvatures are some of the insectivorous plants considered in the preceding section of this chapter, notably *Drosera*.

The tentacles of *Drosera rotundifolia* respond to the stimulus of contact in the same way as they respond to chemical stimulants, and the stimulus of contact induces an increase in the secretion of the digestive fluid by the glands, as well as the curvature of the filaments of the tentacles. Again, as with chemonasty, there can be both direct and indirect stimulation. The former is produced when the glandular head of an outer tentacle is touched; the response then consists in the bending of the tentacle filament towards the middle of the leaf. Indirect stimulation is effected by touching the tentacles of the middle part of the leaf when the outer tentacles also respond by bending over towards the middle of the leaf.

In both direct and indirect stimulation there is a conduction of the excitation just as with stimulation by chemical substances; in direct stimulation the excitation is conducted from the seat of perception, the glandular head, to the base of the filament where curvature begins, while with indirect stimulation the conduction takes place from the glands of the tentacles of the middle of the leaf, through cells of the leaf lamina to the base of the outer tentacles.

In the case of direct stimulation the latent time may be quite short. Thus Darwin observed in one instance a distinct curvature of a tentacle 10 seconds after a small piece of raw meat had been placed on the gland, while the tentacle had curved right over so that its head reached the middle of the leaf in less than 17·5 minutes after the first application of the stimulus. In this case response appeared to be exceptionally rapid, but it seems to be quite usual for a considerable curvature to result within 5 or 6 minutes from the first application of the stimulus. It may be noted that contact with a solid body here constitutes the stimulus, for a response in the form of a curvature of the tentacles results when any solid material, such as glass, paper, cork, gold leaf, thread or hair is used to effect contact, whereas neither drops of water nor the pressure of a wind have any effect.

With indirect stimulation the latent time is longer, as is only to be expected in view of the greater distance through which the excitation is conducted. For example, Darwin recorded that after irritating the central glands of a leaf with a camel-hair brush several of the outer tentacles exhibited curvature in 70 minutes, while within 5 hours all the sub-marginal tentacles were curved. It may be noted that all had straightened after 22 hours. In another case of similar stimulation the response was more rapid, some tentacles showing curvature in 20 minutes, while all the sub-marginal and some of the extreme marginal tentacles were curved within 4 hours; the margin of the leaf had also curved over in this time. Recovery of the original position occurred within 17 hours from stimulation. In a number of experiments in which stimulation was effected by placing various objects on the leaf, the time for maximum curvature varied from one to 24 hours, and the time for recovery from 1 to 10 days. The nature and size of the object affect these times, but this influence may be due not only to differences in the magnitude of the stimulus, but also to the fact that some objects may provide a chemical stimulus in addition to the one of contact. Such is probably the case in the natural stimulation of the tentacles by insects. Indeed, as we have already seen, there is complete recovery from simple contact stimulation with non-digestible materials within a day, but after stimulation with the body of an insect the tentacles may remain curved for a week or more, a result attributed to the continued chemical stimulation as the proteins of the insect body are digested.

The leaves of *Pinguicula* also respond to the stimulus of contact as well as to that of certain chemical substances. Thus when small pieces of glass are placed on the margins of the leaves the curving over of the margins is induced, recovery subsequently taking place.

The leaves of *Dionaea* also respond by movement when stimulated by contact, but the more general view at present is that this is a case of seismonic irritability rather than of haptonasty. The question is therefore considered in the next section of this chapter.

SEISMONASTY

By seismonasty, or seismonic irritability, is meant the phenomenon met with in a number of plants in which response is made to what are often termed mechanical 'shocks', such as blows, shaking or pressure. It is immaterial by what agency the mechanical disturbance is produced, whether by contact with a solid body or by pressure of liquid or gas, and in this respect, as already pointed out, seismonic irritability is differentiated from contact irritability in which the stimulus is contact with a solid body only. Whether there is any fundamental difference between the two cases is perhaps doubtful, but as pointed out by Pfeffer, to distinguish between them would still be justified even if it should transpire that they are closely related, since our classification of tropisms and nasties is based on the kind of exciting agency or stimulus.

Generally speaking, the same response as that made to seismonic stimulation is also made to stimulation by electrical means, as for example, when an electric current is passed through the organ. The usual method employed for stimulating in this way is to subject the plant to an electric shock by means of an induction coil; that is, a high voltage alternating current is passed through the organ. The stimulation phenomena involved can be included under the term *electronasty,* and since the response is generally similar to that induced by mechanical shock it is convenient to consider cases of electronasty along with those of seismonasty.

The most familiar examples of 'shock' stimulation are provided by a number of plants possessing motile pulvini. The best known, and most spectacular of these plants, are the sensitive plants *Mimosa pudica* and *M. spegazzinii.* There are, however, a number of other species, particularly in the Leguminosae and Oxalidaceae, which also

A

Fig. 62.—Photographs of a plant of *Mimosa* showing the leaves in A, the unstimulated position, B the stimulated position

To face page 533

show a definite response to shock stimulation, although not so noticeable a one as that given by *Mimosa pudica*. Among such are *Desmanthus plenus*, a species nearly related to *Mimosa*, and *Biophytum sensitivum*, a member of the Oxalidaceae. Other plants giving a definite response by variation movements of the pulvini are *Oxalis acetosella, Robinia pseudacacia* and *Acacia lophantha.*

The leaves of two species of Droseraceae, *Dionaea muscipula* and *Aldrovanda vesciculosa*, respond to seismonic stimulation by movement, but the mechanism of response is different from that of *Mimosa pudica* and other species with motile pulvini as the movement is brought about by differential rates of growth on the two sides of the leaf.

Seismonic irritability is also met with in the stamens of a number of species. Notable among these are members of the sub-family Cynareae of the Compositae, especially *Cynara scolymus, Centaurea americana* and *Centaurea jacea.* Similarly, irritable stamens are also found in Berberidaceae, Tiliaceae, Cistaceae and Cactaceae. The movements of these stamens in response to mechanical stimulation appear to be brought about, as in the case of leaves with motile pulvini, by changes in turgor of certain cells.

Stigmas sensitive to the stimulus of shock are known to occur in several families, including Scrophulariaceae and Bignoniaceae. The best-known cases are provided by species of *Mimulus* and *Bignonia.* In two species of a genus of Compositae, *Arctotis*, namely *A. aspera* and *A. calendulacea*, according to von Minden the style is sensitive to the shock stimulus, while seismonasty is exhibited by the labellum of the orchid *Masdevallia muscosa.*

A considerable number of investigations have been made on the seismonic irritability of *Mimosa pudica.* When a leaf of this plant is stimulated by a blow or by shaking, or electrically, the leaf and its parts take up new positions which give the organ an altogether different appearance. The leaf of *Mimosa pudica* is a compound one comprising usually a main petiole bearing near the tip two pairs of segments each of which is composed of a secondary petiole bearing a number of pinnately arranged pairs of leaflets (cf. Fig. 62). On stimulation the main petiole falls, the secondary petioles close together somewhat so that the leaf is less spreading, while the leaflets, which were expanded in one plane fold towards one another in pairs, so that the upper surfaces of each pair are practically in contact. There are thus three movements involved depending on changes in the leaf pulvinus, the pulvini of the sub-petioles and the leaflet bases respectively. The extent of the movement depends on the strength of the stimulus. With a slight stimulus applied to the main pulvinus the response may be complete with the falling of the main petiole, while with a greater stimulus the movement of the sub-petioles is involved and the leaflets close up progressively from the base towards the tip. A slight stimulus applied to the tip of one of the leaflets may bring about no more than the closure of one or a few pairs of leaflets. With stronger stimulation the excitation proceeds farther until the sub-pulvini and main pulvinus of the leaf are involved. With still stronger stimulation the excitation extends still further and travels up and down the main stem from the base of the leaf first affected and successive leaves in both directions respond, the main petiole first moving, then the sub-petioles and the leaflets in successive pairs, first those at the base, then the next pair and so on towards the tip. In this way every leaf of the plant may eventually respond to the original stimulus and adopt the drooping position. The comparatively rapid conduction of the stimulus through the whole of the aerial part of the plant is one of the most striking features of the seismonic irritability of *Mimosa pudica*.

After stimulation there is a slow recovery to the normal unstimulated condition. During the recovery movement the plant is insensitive to shock stimulation, but after recovery is complete the maximum sensitivity is slowly regained. Thus immediately after recovery the response to a stimulus of a particular value is less than at some time later when the sensitivity has been fully restored. With continued stimulation, as, for example, by shaking, the plant may regain its original position. Exact measurement of the intensity or quantity of a shock stimulus is not, generally speaking, possible, and so quantitative data regarding the relation between stimulus and response are for the most part lacking in the case of stimulation of *Mimosa pudica* by shock. It is nevertheless clear that external conditions play an important part in determining the sensitivity of the plant. Thus after maintenance for a time at a temperature of 10° C. or lower the power of response may be lost for a considerable period or even permanently, while at temperatures below 15° there is generally no response to mechanical stimulation or only a partial one. At temperatures from 25° to 30° the leaves respond most readily, while at temperatures above 40° there is again an absence of response.

Although purely mechanical shock by means of blows is difficult to regulate, the quantity of electrical stimulation can be varied at will by the use of an induction coil. This method has been employed by Bose to investigate the effects of external conditions on the response of *Mimosa pudica* to electrical stimulation. The intensity of stimulation could be varied by altering the relative positions of primary and secondary coils while the duration of the stimulus in Bose's experiments was controlled by means of a clockwork device. The amount of the response was determined by automatically recording the movement of the primary petiole. According to Bose's records, a diminution of light intensity results in a lessening of the response to the same quantity of electric stimulation, but with continued darkness the response rises and then falls again, remaining at a lower value so long as darkness continues. With return to light there is a momentary slight depression in the amount of response, but almost immediately the response increases.

As regards the effect of temperature Bose found the greatest response at 32° C. Thus in a given specimen the measured amplitude of movement at this temperature was 52 mm., at 27° it was 22 mm., while at 22° C. it was only 2·5 mm. There was practically no response below 19° C., but Bose obtained some response at a temperature as high as 42° C.

According to Bose's observations, various gases, including carbon dioxide, hydrogen sulphide, nitrogen peroxide, ammonia and the vapours of ether and chloroform, all bring about a diminution in the response of *Mimosa* to electrical stimulation. In small concentrations, however, acting for a short time, the response may be increased in some cases. This was observed with carbon dioxide and chloroform vapour. An increase in sensitivity was also observed in the presence of ozone.

The latent time for the seismonastic response in *Mimosa* is usually very short, under favourable conditions being less than a second, while the reaction may be complete in 5 or 6 seconds. Under electrical stimulation Bose found in a highly sensitive specimen a latent time of only 0·075 second, while in less vigorous specimens the time was still only 0·12 second. The reaction was complete, that is, the maximum fall of the petiole occurred, in about 1·5 seconds, after which recovery commenced. Recovery was found to be fully accomplished in from 8 to 20 minutes, the time depending on external conditions and on the strength of the stimulus, recovery being slower after stronger stimulation.

Bose also showed that a succession of sub-threshold electrical stimulations can bring about response in *Mimosa pudica*, their effect in this, as in other cases of irritability, being additive if the period between successive stimulations is not too long.

Considerable light has been thrown in comparatively recent years on the mechanism of the conduction of excitation in *Mimosa pudica* and *M. spegazzinii*. As long ago as 1824 Dutrochet showed that the excitation in *Mimosa pudica* was still conducted through a ' ringed ' stem in which all tissues outside the xylem had been removed. As shown by McDougal in 1896, the excitation can also be conducted through a region of the stem in which the living cells have been killed by hot water. Further, in 1916 Ricca showed that if a branch of *M. spegazzinii* is cut completely across and the two parts of the branch are then re-connected by a tube of hot water, conduction of excitation will take place through the latter. The conclusion is therefore drawn that the excitation must be conducted through the water in the xylem.

That the conduction of excitation is not brought about simply by a change in hydrostatic pressure of the water in the xylem is evident from the slow rate at which the conduction takes place under unfavourable conditions, and direct measurements of the pressure in a tube of water intercalated between the lower and upper parts of a shoot and conducting the excitation, show that variations in the resistance of water play no part in the transference of the excitation.

Now *Mimosa* can be stimulated by burning the tip of a leaflet. Ricca stimulated *Mimosa* plants in this way and then cut a number of pieces from the stem and allowed the contents to diffuse into a small quantity of water. Cut branches of a plant of *Mimosa* were then allowed to absorb this liquid with the result that the characteristic response to stimulation resulted. The conclusion drawn by Ricca was, therefore, that the excitation is conveyed by a substance produced in the protoplasm as a result of stimulation, and that this substance diffuses into the xylem, through which it is carried in the water, and on reaching a pulvinus there acts as a stimulus and induces the typical response. Ricca's theory of the conduction of excitation in *Mimosa pudica* and *M. spegazzinii*, like the current theories of conduction of excitation in geotropic and phototropic stimulation, is thus a hormone theory.

Some eight years later the question was further examined by Snow, who repeated and confirmed with *Mimosa pudica* Ricca's observation on the conduction of excitation through a shoot which had been cut across and then joined with an intercalated tube of water. The rate of conduction of the excitation in cut shoots was found to be approximately the same as the rate of movement of the transpiration stream as indicated by the movement of solutions of eosin and methylene blue. Thus Snow found that the average rates of conduction of the excitation to the second, third and fourth leaf from the base were respectively 10·7, 10·1 and 7·9 cm. a minute, while the rates of movement of the transpiration stream varied from 7 to 18·5 cm. a minute. Further, in damp weather, when the transpiration stream would be expected to be slow, conduction of excitation upwards was slower than usual, while conduction downwards did not take place at all. It may be noted, however, that when an intact plant is stimulated the excitation travels much more readily upwards than downwards. While this behaviour may be regarded as affording further support for the view that the excitation is conducted in the transpiration stream, the fact that there is any downward conduction of excitation at all may be accounted a difficulty. Snow himself considered that further investigation is necessary to explain this conduction of excitation in a basipetal direction.

In the *leaves* of both *Mimosa pudica* and *M. spegazzinii*, however, Snow has shown that the conduction both in the acropetal and basipetal directions, does not depend on the movement of the transpiration stream. Thus rapid conduction was observed in leaves of *Mimosa pudica* that had been kept for some hours in air saturated with water vapour. Also, in very damp weather, leaves conducted just as well as in dry weather, thus exhibiting a marked difference in behaviour from the stem. Conduction was also observed in the leaves of a shoot that was stimulated under water and which had been kept in this environment for the previous 3 hours. In leaves of *Mimosa spegazzinii* stimulated by a cut, Snow found more rapid conduction of the excitation in both acropetal and basipetal directions the damper the conditions, the conduction being much more rapid in leaves of a submerged shoot, less rapid in leaves in damp air, and least rapid in leaves in dry air.

In the leaf the excitation is conducted at least as rapidly in a basipetal as in an acropetal direction. In detached leaves of *Mimosa pudica* stimulated near the tip Snow found that frequently the main pulvinus fell after an interval of 2 seconds, and he justly points out that it cannot be supposed that a water current could pass from the tip of a pinna to the main pulvinus of the leaf in so short a time. As this observation would also suggest, conduction in the leaf can be much more rapid than in the stem.

For these and other reasons Snow came to the conclusion that there are two types of conduction of excitation in *Mimosa*, the 'normal' conduction in the xylem of the stem and a ' high-speed ' conduction in the leaf which has nothing to do with the transpiration current. Snow confirmed an observation made by Herbert in 1922 to the effect that if petioles of *Mimosa pudica* are cut through in various ways the excitation can be conducted across the cut so long as some of the phloem is left intact even if all the xylem is cut through, whereas if the phloem is all cut through and some of the xylem left the conduction characteristic of the leaf no longer takes place. When, however, the more rapid conduction in the phloem is eliminated it is possible to demonstrate the existence in the leaf of a slower type of conduction in an acropetal direction depending on the movement of the transpiration stream.

Snow found that in and near the main pulvinus of the leaf methylene blue can pass very rapidly from xylem to phloem. This suggests the possibility that the connexion between the conduction of the excitation in stem and in leaf may be brought about by the passage of the stimulating hormone from the xylem to the phloem in the main pulvinus, but how the conduction takes place in the phloem is not explained. It may, however, be noted that the exudate from the phloem of a cut stem or petiole when applied to the cut distal end of another petiole sets up excitation in the latter.

A phenomenon called by Snow *multiple conduction* has been observed in *Mimosa spegazzinii*. When stimulated by a cut in the petiole the excitation appears to be conducted in successive waves. When the first wave of excitation reaches them, the leaflets do not close together completely. After a few seconds a second wave appears to reach them and the leaflets rise higher and sometimes several such waves may travel up the pinnae before the pairs of leaflets quite close together.

Other investigators, namely Seidel in 1923 and Ball in 1927, have come to the same conclusion as Snow that there are two methods of conduction of excitation in *Mimosa*. Umrath also found different rates of conduction in each of the aerial organs, and spoke, for example, of a ' slow ' and a ' rapid ' conduction in the leaf. Both Seidel and Ball, like Snow, agreed with Ricca that normal transmission in the stem is effected by means of a hormone. With regard to the second method of conduction, Seidel supposed this to be brought about by changes of pressure occurring in certain cells of the phloem, known as ' tube cells ' or ' tannin tubes ', a view of conduction of excitation in *Mimosa* originally hypothesized by Haberlandt. Snow, however, does not favour this view, for which there is no experimental evidence, and he thinks the ' high-speed ' conduction of excitation is by way of cambiform cells which occur not only in the phloem but also at the inner side of the xylem in the petiole and between the vessels in the leaflets.

Ball found that submersion of shoots of *Mimosa pudica* under water led to a very rapid conduction of excitation, the velocity of conduction being on the average about 2 to 2·5 cm. per second at about 30° C. Removal of all tissues external to the wood over lengths of 1 cm. to 6 cm. had practically no effect on the rate of conduction of excitation, so the phloem cannot be concerned in it, but on the other hand, it only occurs if the pith is wholly or partially present. Further characteristics of this high-speed conduction in submerged stems are that only the primary pulvini are affected, the excitation not being conducted through the leaves, while the rate of conduction is not related to the rate of water movement in the xylem. Ball suggests that in rapid conduction the hormone produced at the place of stimulation causes contraction of neighbouring cells, probably those in the pith, which in turn produce more hormone as a result, so that a ' perfect relay mechanism ' is present by means of which the excitation can be conducted in either direction independently of any water movement. Ball further suggests that the ' high-speed ' conduction observed by Snow is of similar character to that observed by him in submerged shoots, being due, in the leaves examined by Snow, to the contraction of highly turgid cells in the phloem. Ball's conclusion with regard to the conduction of excitation in *Mimosa* is thus that there are two modes of conduction, that in the xylem, generally called ' normal ' conduction, which takes place by means of a hormone carried in the transpiration stream, and ' rapid ' conduction, which is independent of the movement of water and which is effected by the hormone inducing changes in turgidity in living cells, either of the pith, as in the case of submerged stems, or in the phloem, as in the cases of high-speed conduction in the leaf investigated by Snow. The normal conduction operates best under drier conditions, when turgor in living cells is low and the movement of water in the tracheae likely to be most rapid, whereas rapid conduction is most effective under damp conditions where turgidity of the cells is greatest and the movement of water in the xylem slow, conditions which reach their greatest development in submerged shoots.

The most recent work on the conduction of excitation in *Mimosa pudica* is that of Houwink. To follow the conduction of excitation this worker has made use of the discovery of Bose, that when an electrode is inserted into the stem or petiole of a *Mimosa* plant the potential of the electrode changes with respect to that of the earth when the excitation reaches the tissue at the electrode. By inserting electrodes at a certain distance apart and connecting them with recording apparatus it was possible to obtain information with regard to the rate of conduction of the excitation under different conditions.

Houwink concluded from his researches that there are three mechanisms of conduction of the excitation in *Mimosa*, but whether one or other of these, or more than one, operates depends on the kind of stimulus. When stimulated by a local lowering of temperature, effected by the application of a drop of water at 0° C., an excitation results which cannot pass through a killed region of the plant nor through a zone cooled to 5° C. It is therefore concluded that this type of conduction is through living cells. It depends on the condition of the cells and is most rapid in a young shoot in damp air. The excitation can be propagated in this way through stems, petioles, pinnae and pulvini, but is stopped at the base of the pinnae. This type of conduction is regarded as identical with Ball's 'rapid' conduction in the stem and with Umrath's 'slow' conduction in the leaf.

The second type of conduction is met with when stimulation is effected by wounding. The excitation can be conducted through a killed zone, and it seems clear that the mechanism of conduction in this case is that demonstrated by Ricca as due to a hormone.

A third kind of conduction was observed by Houwink when stimulation was effected by cutting a pinna. The excitation thus set up can be conducted by the two methods already mentioned, but also by a third method which only affects the main pulvinus. This third type of conduction was observed to continue through a cooled zone, but not through a killed region of the petiole. It occurs in young leaves, particularly when in damp air, and is regarded by Houwink as identical with Snow's 'high-speed' conduction and Umrath's 'fast' conduction. Houwink appears to incline to the view that this type of conduction is due to the transmission of a pressure change in the phloem, as supposed by Haberlandt.

It has already been mentioned that the movements of the leaves and leaflet of *Mimosa* are brought about by changes in turgor of the cells of the pulvini, which result in a change in the shape of the pulvinus. The turgor change is principally associated with a loss of water from the cells of the lower half of the pulvinus, the water passing partly into cells of the neighbouring stem and petiole but chiefly perhaps into the intercellular spaces of the pulvinus, which exhibits a darker colour as the result of the displacement of the air in the intercellular spaces by water. How these turgor changes are brought about is not understood. The two possible explanations are that the osmotic pressure of the cell-sap is suddenly reduced by internal chemical changes, an effect which would result in the immediate passage of water out from the cells owing to the disturbance of osmotic equilibrium between the cells concerned and neighbouring unaffected cells, or that the permeability of the cells affected is suddenly increased as a result of stimulation, so that both osmotically active substances and water pass out through the cell membranes. The question was investigated by Blackman and Paine, who showed in 1918 that when an excised pulvinus of *Mimosa pudica* is placed in a small quantity of water and stimulated there is, indeed, an increase in the rate of exosmosis of electrolytes from the pulvinus, but the amount of the increase is quite insufficient to account for the loss in turgor of the cells. It would thus appear that when the pulvinus is stimulated, the hormone which is produced or conducted to it, according as stimulation is direct or indirect, induces the sudden disappearance of a considerable proportion of the osmotically active contents of the cells, presumably either by some chemical change or by adsorption. Whatever the mechanism of this inactivation of osmotically active substances may be, the change is a reversible one, for as the effects of excitation die out the substances concerned must be re-formed and water re-absorbed by the cells involved as the osmotic pressure rises.

Other Sensitive Plants in Leguminosae and Oxalidaceae. As already mentioned, sensitiveness to the stimulus of shock comparable to that exhibited by *Mimosa* is found in other members of the Leguminosae and in a number of species of Oxalidaceae. In many cases the sensitivity is low, as in *Oxalis acetosella* and other species of *Oxalis*, in which complete falling of the leaflets only results after repeated shaking of the plant, while in other cases, as, for example, in *Robinia pseudacacia*, even this treatment only results in a slight movement of the leaflets. However, in *Desmanthus plenus* and *Neptunia oleracea*, among Leguminosae, and in species of *Biophytum* (*B. sensitivum*, *B. reinwardtii* and *B. apodiscias*), among the Oxalidaceae, a degree of sensitivity approaching that of *Mimosa* is developed, at any rate in plants growing in the tropics.

In *Biophytum sensitivum* and *B. reinwardtii* only the leaflets move in response to a shock or wound stimulation, but in *B. apodiscias* the petiole moves as well. The movements take place in the reverse direction to those executed by *Mimosa*, for the leaflets fall while the petiole rises. The mechanism of the movement is, however, probably similar to that of *Mimosa*, depending on turgor changes in the main pulvinus and sub-pulvini.

The rate of conduction of the excitation was measured by von Faber, who found velocities of 17 to 20 mm. per second in *B. sensitivum* and 20 to 25 mm. per second in *B. apodiscias*. In the latter plant slower rates of conduction were found in the floral axis and in the roots, namely, 5 to 7 mm. per second and 5 to 8 cm. per minute (0·8 to 1·3 mm. per second) respectively. Whether these different rates are related to different types of conduction, as in *Mimosa*, has not been determined. That a hormone is involved, as in *Mimosa*, is indicated by the fact that, as shown by MacDougal for *B. sensitivum*, and by von Faber for all three species mentioned above, the excitation is conducted over a zone of killed tissue. But von Faber's observation that the excitation travels at least twice as rapidly in the basipetal direction as in the basifugal direction does not support the view that the hormone is carried in the transpiration stream, and the necessity for further investigation is indicated.

Dionaea. The leaves of *Dionaea* respond by movement of the lobes not only to chemical stimulation, but also to the stimulus of contact. Darwin's observations indicated that the irritability in this case was to be regarded as a true case of haptonasty, but later work suggests that it is an example of seismonic irritability. Darwin found that drops of water or a thin broken stream of water falling on the leaves from above did not induce movement, nor did blowing strongly on the leaf produce any effect, and he concluded that, as in the case of *Drosera*, the plant is stimulated neither by heavy rain nor by gales of wind, and he stated definitely that the sensitiveness is 'related to a momentary touch rather than to prolonged pressure; and the touch must not be from fluids, such as air or water, but from some solid object'; that, in fact, the irritability is a case of haptonasty. On the other hand, Pfeffer thought it possible that seismonic and contact irritability might both be developed in the leaf of *Dionaea*. Macfarlane in 1902 reported that a steady stream of water directed against the hairs on the upper surface of the leaf induced closure, while Brown and Sharp in 1910 found that closure takes place if water is dropped on these hairs so as to bend them. Jost states that stimulation is effected by a water jet or by stroking with damp gelatin, as well as by contact with a solid object.

The principal seat of perception of the mechanical stimulus, whether it is one of contact or shock, is found in the fine spines which occur, three on each lobe, on the upper surface of the leaf (see p. 529). The sensitivity of the filaments increases towards the base. When these filaments are touched the two lobes of the leaf close together. Darwin found that there were two distinct types of movement of the leaves of *Dionaea*. When the filaments are touched lightly and quickly with a solid object, such as a piece of thread, the lobes close rapidly, the marginal spikes of the two lobes meet and interlock, and the lobes of the leaf bulge out so that the lower, and now outer, surface of each lobe is convex. Recovery of the original orientation of the lobes takes place slowly and may require several hours for completion. Stimulation in this way by contact with a solid object, such as a piece of wood, paper, stone or glass, does not induce secretion by the glands of the leaf.

The second type of movement is produced when stimulation is effected by an insect, a piece of meat or albumin. In this type of movement the lower surfaces of the leaf lobes may become concave, which has the effect of bringing the upper surfaces of the two lobes in close contact with the insect or other body. At the same time the glands are stimulated to secrete their digestive fluid. The leaves may remain closed for many days, and in some cases observed by Darwin never reopen. Even when they reopen they are incapable of response to similar stimulation for many days if ever. From what has been said in the previous section of this chapter, it would appear likely that in this second type of movement we have to do with the combined effects of seismonastic and chemonastic action.

The first type of movement, that induced by brief contact with a solid object, was the subject of investigations by Brown and Sharp and later by Brown alone, the results being published in 1910 and 1916 respectively.

It is difficult to determine the quantity of stimulation in the case of a contact stimulus. At temperatures of 2° C. and 15° C. Brown and Sharp found that usually two successive stimulations were necessary to bring about closure of the leaf, and it was of little importance whether the sensitive bristles were bent moderately or much. With very slight bending, however, three to five applications of the stimulus might be necessary. At higher temperatures (35° C.) one application of the stimulus was usually sufficient to induce closure of the leaf. At 15° C. the interval elapsing between successive stimulations was found to be important; leaves stimulated at intervals of 20 seconds responded after the second stimulus, those stimulated at intervals of 1 minute responded after the second to fifth stimulation, when the interval was increased to 2 minutes from five to seven stimulations were necessary, and with an interval of 3 minutes between successive stimuli six to nine of these latter were necessary to induce closure of the leaves. With fewer stimulations than those necessary to bring about complete closure a partial closure of the leaves might result.

The experiments of Brown and Sharp in which leaves of *Dionaea* were stimulated by an electric current produced by means of an induction coil, without providing exact quantitative data, suggested that the response is determined by the quantity of stimulus rather than by either intensity or duration of stimulation solely, for these workers found that when a succession of such stimuli was applied, the number necessary to induce response varied inversely as the intensity of a single stimulus.

The latent time of the response of *Dionaea* leaves may be very short. Burdon-Sanderson obtained a latent time of about 1 second at 20° C., complete closure of the leaflets taking place within 5 or 6 seconds. Brown later showed that as the time between successive mechanical stimulations increased so did the latent time. Thus when the interval between two stimulations was 20 seconds rapid closure took place immediately after the second stimulation. When the interval between successive stimuli was 4 minutes the average time between the first stimulus and the beginning of movement was found to be 22 minutes, while closure itself, instead of being almost instantaneous, took 7·33 minutes. With 20 minutes elapsing between each stimulation the average time between the first stimulus and the beginning of movement was as much as 268 minutes, while the time occupied in the closing of the leaf was 94 minutes.

As regards the seat of perception of the mechanical stimulus, it has already been noted that this is chiefly in the bristle-like hairs borne on the upper surface of the leaf lobes. Scraping the surface of the lobes or squeezing the leaf with forceps will induce closure of the leaf, so it may be concluded that perception is actually not limited to the hairs.

According to Brown and Sharp, stimulation results in cells of the leaf becoming compressed. According to Haberlandt, there is, near the base of each bristle, a layer of cells readily bent and therefore easily compressed by the bending of the bristle. Brown and Sharp found that closure results so long as the bristle is bent; and if release from the bent position is prevented the response continues. Response is always to the downward movement of the bristle and never to its upward movement. Further evidence that compression of the cells at the base of the bristle is essential for response is provided by the following observations. If a bristle is cut off near the base, and after a time the stump is bent by pressing it with a needle, a normal response is induced. Also if the bristle is bent by directing a jet of air against it, or by dropping water on it, the leaves respond. The response to mechanical stimulation is thus due to compression of cells, and the cells concerned are not limited to the region at the base of the bristles, but, as mentioned above, all cells of the leaf, or at least cells in all parts of the leaf, are sensitive in this way.

Stimulation of the leaf brings about a considerable increase in the rate of growth of the under side of the leaf lobes. The first stage in the promotion of growth of these cells appears to be a decrease in the osmotic pressure of the cells on the upper side of the leaf, and this, owing to the resultant disturbance of equilibrium, leads to the passage of water from the cells affected to those of the lower side. The concomitant stretching of the walls then becomes fixed. The subsequent recovery of the leaf when it reopens is brought about by the slow enlargement by growth of the cells of the morphologically upper side of the leaf lobes.

The movement of *Dionaea* leaves is thus due to unequal growth on the two sides of an organ, as in many other nastic, as well as tropic, curvatures. Whether the conduction of the excitation from the seat of perception to the place of accelerated growth on the lower side of the leaf lobes is brought about by means of a growth promoting hormone is not known, but must be considered, having regard to the known facts, as quite likely. Even in 1916, before the discovery of the part played by hormones in the conduction of geotropic excitation, Brown was impressed with the similarity of the mechanism of movement in leaves of *Dionaea* to that of geotropic curvature. It is to be observed that the mechanism of the seismonastic response in *Dionaea* is thus definitely different from that in *Mimosa*.

The peculiar freely swimming aquatic plant *Aldrovanda vesiculosa*, a species of the Droseraceae nearly allied to *Dionaea*, appears to exhibit the same mechanism of response to seismonic stimulation as does the latter plant.

Sensitive Stamens. The best-known cases of stamens which respond to mechanical stimulation are those of the Cynareae, including species of *Cynara*, *Centaurea* and some related Compositae. The irritability of these stamens was investigated in particular by Pfeffer. The stamens in these plants are, of course, syngenesious, and when one filament is touched a contraction of all the filaments occurs, bringing about a decrease in their length which may amount to 10 to 30 per cent. in *Centaurea jacea* and 8 to 20 per cent. in *Cynara scolymus*. There is no change in the area of the cross-section of the filaments, but all parts of the filaments are equally concerned in the shortening except the ends, where the percentage shortening is less than elsewhere.

As in the *Mimosa* pulvinus, the movement of these stamens is brought about by a fall in turgor of the living cells, water passing out into the intercellular spaces and displacing the air therein which escapes. If cut filaments are first injected with water and then stimulated there results an escape of water from the cut end of the filament. The non-living xylem elements, owing to the elasticity of their walls, contract along with the living cells attached to them.

According to Linsbauer, the latent time of response in stimulated stamens of *Centaurea americana* can be less than 1 second, while the contraction of the filaments can be completed in 7 to 13 seconds. Recovery of the original length may take from 50 to 60 seconds.

An isolated stamen responds to stimulation in the same way as one attached to others in the flower, and it is held that there is not here a conduction of excitation from one stamen to another, but that the contraction of one stamen exerts a pull on neighbouring ones, which acts as a stimulus.

The stamens of *Berberis* respond to mechanical stimulation by executing a sudden curvature, which results in the anther, instead of being directed outwards, curving inwards and coming in contact with the stigma. Here also, under certain conditions, water may be observed to escape from the stamens on stimulation, and hence it was concluded by Pfeffer that here, as in the Cynareae, the movement is brought about by turgor changes in the living cells. In other cases of seismonastic irritability of stamens the evidence is at present lacking to decide whether the movements are due to growth or to turgor changes.

These staminal movements are, of course, related to pollination and form part of a pollination mechanism. Thus the shortening of the filaments in *Cynara* or *Centaurea* causes the style and stigma to push out pollen, while the sudden inward curvature of the stamens in *Berberis* may result in jerking pollen from the anther on to an insect which by its touch may have stimulated the stamen to movement. But a biological significance cannot be found for all cases of seismonic irritability, and in particular it is not evident what significance, if any, the response of *Mimosa* and *Biophytum* can have in the life of the plant.

REFERENCES TO LITERATURE IN TEXT

ADAMS, J., The effect on certain plants of altering the daily period of light. *Ann. Bot.*, **37**, 75–94, 1923.

—— Duration of light and growth. *Ann. Bot.*, **38**, 509–23, 1924.

ADRIANCE, G. W., Factors influencing fruit setting in the Pecan. *Bot. Gaz.*, **91**, 144–66, 1931.

AMEIJDEN, U. P. VAN, Geotropism and phototropism in the absence of free oxygen. *Rec. trav. bot. Néerlandais*, **14**, 149–217, 1917.

ANDERSON, D. B., The structure of the walls of the higher plants. *Bot. Rev.*, **1**, 52–76, 1935.

ARBER, A., *Monocotyledons.* Cambridge, 1925.

ARCHBOLD, H. K., Chemical studies in the physiology of apples, XII. Ripening processes in the apple and the relation of time of gathering to the chemical changes in cold storage. *Ann. Bot.*, **46**, 407–59, 1932.

ARISZ, W. H., Untersuchungen über den Phototropismus. *Rec. trav. bot. Néerlandais*, **12**, 44–216, 1915.

ARMSTRONG, E. F., Enzymes. *Chem. World*, 117–18, 1913.

ARTHUS, M., *Nature des Enzymes.* Thèse, Paris, 1896 (Quoted by W. M. Bayliss in 'The Nature of Enzyme Action', Third Edition, London, 1914.)

ASHBY, E., Studies in the inheritance of physiological characters. I. A physiological investigation of the nature of hybrid vigour in maize. *Ann. Bot.*, **44**, 457–67, 1930.

—— Studies in the inheritance of physiological characters. II. Further experiments upon the basis of hybrid vigour and upon the inheritance of efficiency index and respiration rate in maize. *Ann. Bot.*, **46**, 1007–32, 1932.

ASPREY, G. F., Studies on antagonism. I. The effect of the presence of salts of monovalent, divalent and trivalent kations on the intake of calcium and ammonium ions by potato tuber tissue. *Proc. Roy. Soc.*, B, **112**, 451–72, 1933.

—— Studies on antagonism. II. The effect of previous immersion of potato tuber tissue in solutions of monovalent, divalent and trivalent kations on the subsequent absorption of the ammonium ion. *Proc. Roy. Soc.*, B, **113**, 71–82, 1933.

ATKINS, W. R. G., *Some Recent Researches in Plant Physiology.* London, 1916.

AUBERT, E., Recherches sur la respiration et l'assimilation des plantes grasses. *Rev. gén. Bot.*, **4**, 203–82, 321–31, 337–53, 421–41, 497–502, 558–68, 1892.

BAAS BECKING, L. G. M., The state of chlorophyll in the plastids. *Zesde Internat. Bot. Congres Amsterdam. Proc.* Vol. II, 265–6. Leiden, 1935.

BACH, A., Eine Methode zur schnellen Verarbeitung von Pflanzenextrakten auf Oxydations fermente. *Ber. deut. chem. Ges.*, **43**, 362–3, 1910.

—— Zut Theorie der Oxydasewirkung. I. Mangan- und eisenfreie Oxydasen. *Ber. deut. chem. Ges.*, **43**, 364–6, 1910.

—— Zur Theorie der Oxydasewirkung. II. Einfluss der Metallsalze auf die weitere Umwandlung der Produkte der Oxydasewirkung. *Ber. deut. chem. Ges.*, **43**, 366–70, 1910.

—— Les ferments oxydants et réducteurs et leur rôle dans le processus de respiration. *Arch. sci. phys. Geneve*, Sér. 4, **35**, 240–62, 1913.

Bach, A., Über den Mechanismus der Oxydations-Vorgänge. *Ber. deut. bot. Ges.*, **46**, 3864–8, 1913.

Bach, H., Über die Abhangigkeit der geotropischen Präsentations und Reaktionszeit von verschiedenen Aussenbedingungen. *Jahrb. f. wiss. Bot.*, **44**, 57–123, 1907.

Baeyer, A., Ueber die Wasserentziehung und ihr Bedeutung für das Pflanzenleben und die Gärung. *Ber. deut. chem. Ges.*, **3**, 63–78, 1870.

Ball, N. G., Phototropic movements of leaves—The functions of the lamina and the petiole with regard to the perception of the stimulus. *Sci. Proc. Roy. Dublin Soc.*, **17** (N.S.), 281–6, 1923.

—— Rapid conduction of stimuli in *Mimosa pudica*. *New Phytologist*, **26**, 148–70, 1927.

Barendrecht, H. P., Enzymwirkung. *Zeitschr. f. physik. Chem.*, **49**, 456–82, 1904.

—— L'uréase et la theorie de l'action des enzymes par rayonnement. *Rev. trav. chim.*, **39**, 2–87, 1920.

Barker, J., Analytic studies in plant respiration. IV. The relation of the respiration of potatoes to the concentration of sugars and to the accumulation of a depressant at low temperatures. Part I. The effect of temperature-history on the respiration/sugar relation. *Proc. Roy. Soc.*, B, **112**, 316–35, 1933.

—— Analytic studies in plant respiration. V. The relation of the respiration of potatoes to the concentration of sugars and to the accumulation of a depressant at low temperatures. Part II. The form of the normal respiration/sugar relation and the mechanism of depression. *Proc. Roy. Soc.*, B, **112**, 336–58, 1933.

Bärlund, H., Permeabilitätsstudien an Epidermiszellen von *Rhoeo discolor*. *Acta Bot. Fennica*, **5**, 117 pp., 1929.

Bayliss, J. S., On the galvanotropism of roots. *Ann. Bot.*, **21**, 387–405, 1907.

Beal, W. J., The vitality of seeds. *Bot. Gaz.*, **40**, 140–3, 1905.

—— The vitality of seeds buried in the soil. *Proc. Soc. Prom. Hort. Sci.*, **31**; 21–3, 1911.

Becquerel, P., Recherches sur la vie latente des graines. *Ann. Sci. Nat., Bot.*, Sér. 9, **5**, 193–311, 1907.

—— La vie latente des graines et des spores. *Zesde Internat. Bot. Congres Amsterdam. Proc.* Vol. II, 279–81, Leiden, 1935.

Benecke, W., Über die Giftwirkung verschiedener Salze auf Spirogyra und ihre Entgiftung durch Calciumsalze. *Ber. deut. bot. Ges.*, **25**, 322–37, 1907.

Bennet-Clark, T. A., The respiratory quotients of succulent plants. *Sci. Proc. Roy. Dublin Soc.*, **20**, 293–9, 1932.

Berthelot, Sur la fermentation glucosique du sucre de canne. *Comp. rend. acad. sci.*, **50**, 980–4, 1860.

Bertrand, G., Sur le latex de l'arbre à laque et sur une nouvelle diastase contenue dans ce latex. *Comp. rend. soc. biol.*, Sér. 10, **1**, 478–80, 1894.

—— Sur la laccase et sur le pouvoir oxydant de cette diastase. *Comp. rend. acad. sci.*, **120**, 266–9, 1895.

—— Sur la recherche et la présence de la laccase dans les végétaux. *Comp. rend. acad. sci.*, **121**, 116–68, 1895.

—— Sur la présence simultanée de la laccase et de la tyrosinase dans le suc de quelques champignons. *Comp. rend. acad. sci.*, **123**; 463–5, 1896.

—— Sur l'intervention du manganese dans les oxidations provoquées par la laccase. *Comp. rend. acad. sci.*, **124**, 1032–5, 1355–8, 1897.

—— Sur l'emploi favorable de manganese comme engrais. *Comp. rend. acad. sci.*, **141**, 1255–7, 1905.

Beyer, A., Ueber die Keimung der gelben Lupine. Landw. Versuchs-Stationen, **9**, 168–78, 1867.

Beyerinck, M. W., Die Bakterien der Papilionaceenknöllchen. *Bot. Zeit.*, **46,** 725–35, 741–50, 757–71, 781–90, 797–802, 1888.

Bigelow, W. D., and Gore, H. C., Studies on peaches. *U.S. Dept. Agric. Bur. Chem., Bull.* **97,** 32 pp., 1905.

Bigelow, W. D., Gore, H. C., and Howard, B. J., Studies on apples. *U.S. Dept. Agric. Bur. Chem., Bull.* **94,** 100 pp., 1905.

Blaauw, A. H., Die Perzeption des Lichtes. *Réc. trav. bot. Néerlandais,* **5,** 209–372, 1908.

—— Das Wachstum der Luftwurzeln einer *Cissus*-Art. *Ann. Jard. Bot. Buitenzorg.* **26,** 266–93, 1912.

—— Licht und Wachstum I. *Zeitschr. f. Bot.,* **6,** 641–703, 1914.

—— Licht und Wachstum II. *Zeitschr. f. Bot.,* **7,** 465–532, 1915.

—— Licht und Wachstum III. (Die Erklarüng des Phototropismus.) *Meded. Landbouwhoogeschool Daaraan Verbouden Inst.,* **15,** 89–204, 1919.

Blackman, F. F., Experimental researches on vegetable assimilation and respiration. II. On the path of gaseous exchange between aerial leaves and the atmosphere. *Phil. Trans. Roy. Soc.,* B, **186,** 503–62, 1895.

—— Optima and limiting factors. *Ann. Bot.,* **19,** 281–95, 1905.

—— Analytic studies in plant respiration III.—Formulation of a catalytic system for the respiration of apples and its relation to oxygen. *Proc. Roy., Soc.* B., **103,** 491–523, 1928.

—— Respiration and oxygen concentration. Fifth Internat. Bot. Congress, Cambridge, 1930. *Report of Proceedings,* 423–4, Publ. Cambridge, 1931.

Blackman, F. F., and Matthaei, G. L. C., Experimental researches on vegetable assimilation and respiration. IV. A quantitative study of carbon dioxide assimilation and leaf temperature in natural illumination. *Proc. Roy. Soc.,* B, **76,** 402–60, 1905.

Blackman, F. F., and Parija, P., Analytic studies in plant respiration I. The respiration of a population of senescent apples. *Proc. Roy. Soc.,* B, **103,** 412–45, 1928.

Blackman, F. F., and Smith, A. M., Experimental researches on vegetable assimilation and respiration. IX. On assimilation in submerged water-plants and its relation to the concentration of carbon dioxide and other factors. *Proc. Roy. Soc.,* B, **83,** 389–412, 1911.

Blackman, V. H., The compound interest law and plant growth. *Ann. Bot.,* **33,** 353–60, 1919.

Blackman, V. H., and Paine, S. G., Studies in the permeability of the pulvinus of *Mimosa pudica. Ann. Bot.,* **32,** 69–85, 1918.

Bode, H. H., Beiträge zur Dynamik der Wasserbewegung in den Gefässpflanzen. *Jahrb. f. wiss. Bot.,* **62,** 92–127, 1923.

Bodnár, J., Über die Zymase und Carboxylase der Kartoffel und Zuckerrube. *Biochem. Zeitschr.,* **73,** 193–210, 1916.

Böhm, J., Ueber die Respiration der Kartoffel. *Bot. Zeitschr.,* **45,** 671–5, 681–92, 1887.

Bolam, T. R., *The Donnan Equilibria,* London, 1932.

Bonnier, G., Recherches sur la chaleur végétale. *Ann. sci. nat., Bot.,* Sér. 7, **18,** 1–34, 1893.

Bonnier, G., and Mangin, L., Respiration des tissus sans chlorophylle. *Ann. sci. nat.,* Sér. 6, **18,** 293–379, 1884.

Bonte, Helene, Vergleichende Permeabilitätsstudien an Pflanzenzellen. *Protoplasma,* **22,** 209–42, 1934.

Bose, J. C., On diurnal variation of moto-excitability in *Mimosa. Ann. Bot.,* **27,** 759–79, 1913.

—— *The Motor Mechanism of Plants,* London, 1928.

Boucherie, Rapport sur un mémoire de M. le docteur Boucherie relatif à la conservation des bois. *Comp. rend.,* **11,** 894–8, 1840.

BOURQUELOT, E., La synthèse biochimique des glucosides d'alcools, IV. Galactosides d'alcools. *Ann. de chim.*, Sér. 9, 7, 153–226, 1917.

BOURQUELOT, E., and BRIDEL, M., Synthèse des glucosides d'alcools à l'aide de l'emulsine et reversibilité des actions fermentaires. *Ann. chim. et phys.*, Sér. 8, **29**, 145–218, 1913.

BOUSSINGAULT, J. B., Recherches chimiques sur la végétation, enterprises dans le but d'examiner si les plantes prennent de l'azote à l'atmosphère, *Ann. chim. phys.*, **67**, 5–54, 1838.

—— De la végétation dans l'obscurité. *Ann. sci. nat., Bot.*, Sér. 5, **1**, 314–24. 1864.

—— Sur les fonctions des feuilles. *Comp. rend. acad. sci.*, **61**, 493–506, 605–13, 657–65, 1865.

—— *Agronomie, Chimie agricole et Physiologie.* Vol. 4, Paris, 1868.

—— Sur les fonctions des feuilles. *Ann. sci. nat., Bot.*, Sér. 5, **10**, 331–43, 1869.

—— *Agronomie, Chimie agricole et Physiologie.* Vol. 5, Paris, 1874.

BRAIN, E. D., Bilateral symmetry in the geotropism of certain seedlings. *Ann. Bot.*, **40**, 651–64, 1926.

—— A comparative study of geotropism in three species of *Lupinus*. *Journ. Linn. Soc., Bot.*, **49**, 375–89, 1933.

BRANSCHEIDT, P., Zur Physiologie der Pollenkeimung und ihrer experimentellen Beeinflussung. *Planta*, **11**, 368–456, 1930.

BRAUNER, L., Lichtkrümmung und Lichtwachstumreaktion. *Zeitschr. f. Bot.* **14**, 497–547, 1922.

—— Über den Einfluss der Koleoptilspitze auf die geotropische Reaction der Avenakeimlinge. *Ber. deut. bot. Ges.*, **41**, 208–11, 1923.

BRENCHLEY, W. E., MASKELL, E. J., and WARINGTON, K., The inter-relation between silicon and other elements in plant nutrition. *Ann. App. Biol.*, **14**, 45–82, 1927.

BRENCHLEY, W. E., and THORNTON, H. G., The relation between the development, structure and functioning of the nodules on *Vicia faba* as influenced by the presence or absence of boron in the nutrient medium. *Proc. Roy. Soc.*, B, **98**, 373–99, 1925.

BRIGGS, G. E., Experimental researches on vegetable assimilation and respiration XIII.—The development of photosynthetic activity during germination. *Proc. Roy. Soc.*, B, **91**, 249–68, 1920.

—— Experimental researches on vegetable assimilation and respiration XVI. The characteristics of subnormal photosynthetic activity resulting from deficiency of nutrient salts. *Proc. Roy. Soc.*, B, **94**, 20–35, 1922.

—— A consideration of some attempts to analyse growth curves. *Proc. Roy. Soc.*, B, **102**, 280–5, 1928.

—— The accumulation of electrolytes in plant cells—a suggested mechanism. *Proc. Roy. Soc.*, B, **107**, 248–69, 1930.

—— The absorption of salts by plant tissues, considered as ionic interchange. *Ann. Bot.*, **46**, 301–22, 1932.

—— Experimental researches on vegetable assimilation and respiration XXI.—Induction phases in photosynthesis and their bearing on the mechanism of the process. *Proc. Roy. Soc.*, B, **113**, 1–41, 1933.

BRIGGS, G. E., KIDD, F., and WEST, C., A quantitative analysis of plant growth. Part I. *Ann. App. Biol.*, **7**, 103–123, 1920.

BRIGGS, G. E., and PETRIE, A. H. K., On the application of the Donnan equilibrium to the ionic reactions of plant tissues. *Biochem. Journ.*, **22**, 1071–82, 1928.

—— Respiration as a factor in the ionic equilibria between plant tissues and external solutions. *Proc. Roy. Soc.*, B, **108**, 317–26, 1931.

BRIGGS, L. J., and SHANTZ, H. L., Hourly transpiration rate on clear days as determined by cyclic environmental factors. *Journ. Agric. Res.*, **5**, 583–649, 1916.

BRINK, R. A., The physiology of pollen. *Amer. Journ. Bot.*, **11**, 218–28, 283–94, 351–64, 417–36, 1924.

—— The influence of hydrogen-ion concentration on the development of the pollen tube of the sweet pea (*Lathyrus odoratus*). *Amer. Journ. Bot.*, **12**, 149–62, 1925.

BROOKS, S. C., The accumulation of ions in living cells—a non-equilibrium condition. *Protoplasma*, **8**, 389–412, 1929.

BROWN, A. J., and WORLEY, F. P., The influence of temperature on the absorption of water by seeds of *Hordeum vulgare* in relation to the temperature coefficient of chemical change. *Proc. Roy. Soc.*, B, **85**, 546–53, 1912.

BROWN, H. T., and ESCOMBE, F., Static diffusion of gases and liquids in relation to the assimilation of carbon and translocation in plants. *Phil. Trans. Roy. Soc.*, B, **193**, 223–91, 1900.

—— The influence of varying amounts of carbon dioxide in the air on the photosynthetic process of leaves and on the mode of growth of plants. *Proc. Roy. Soc.*, B, **70**, 397–413, 1902.

BROWN, H. T., and MORRIS, G. H., A contribution to the chemistry and physiology of foliage leaves. *Journ. Chem. Soc., Trans.*, **53**, 604–83, 1893.

BROWN, W., On the preparation of collodion membranes of differential permeability. *Biochem. Journ.*, **11**, 40–57, 1917.

BROWN, W. H., The mechanism of movement and the duration of the effect of stimulation in the leaves of *Dionaea*. *Amer. Journ. Bot.*, **3**, 68–90, 1916.

BROWN, W. H., and SHARP, L. W., The closing response in *Dionaea*. *Bot. Gaz.*, **49**, 290–302, 1910.

BUCH, H., Über den Phototropismus der Panizeen. *Mem. Soc. Fauna Flora Fenn.*, **6**, 167–73, 1929–31.

BUCHNER, E., Alkoholische Gärung ohne Heferzellen. *Ber. deut. Ges.*, **30**, 117–24, 1110–13, 1897.

BUDER, J., Untersuchungen zur Statolithenhypothese. *Ber. deut. Ges.*, **26**, 162–93, 1908.

—— Neue phototropische Fundamentalversuche. *Ber. deut. bot. Ges.*, **38**, 10–19, 1920.

BULLER, A. H. R., Contributions to our knowledge of the spermatozoa of ferns. *Ann. Bot.*, **14**, 543–82, 1900.

BÜNNING, E., Über die thermonastischen und thigmonastischen Blutenbewegungen. *Planta*, **8**, 698–716, 1929.

BURDON-SANDERSON, J. S., On the electromotive properties of the leaf of Dionaea in the excited and unexcited states. *Phil. Trans. Roy. Soc.*, **173**, 1–55, 1882.

—— On the electromotive properties of the leaf of *Dionaea* in the excited and unexcited states. II. *Phil. Trans. Roy. Soc.*, **179**, 417–49, 1888.

BÜRGERSTEIN, A., Beobachtungen über die Keimkraftdauer von ein bis zehnjährigem Getreidesamen. *Verhandl. k. k. zool.-bot. Ges. Wien.* 1895. (Cited by H. Molisch in ' Pflanzenphysiologie als Theorie der Gartnerei '. 2 Aufl. Jena, 1918.)

BUTKEWITSCH, W., Umwandlung der Eiweisstoffe durch die niederen Pilze im Zusammenhange mit einigen Bedingungen ihrer Entwickelung. *Jahrb. f. wiss. Bot.*, **38**, 147–240, 1903.

CANDOLLE, A. P. de, Physiologie végétale. Paris, 1832.

CASTLE, E. S., Phototropism and light-sensitive system of *Phycomyces*. *Journ. Gen. Physiol.*, **13**, 421–35, 1930.

—— The phototropic sensitivity of *Phycomyces* as related to wave-length. *Journ. Gen. Physiol.*, **14**, 701–11, 1931.

—— Phototropic ' indifference ' and the light-sensitive system of *Phycomyces* *Bot. Gaz.*, **91**, 206–11, 1931.

Castle, E. S., On 'reversal' of phototropism in *Phycomyces*. *Journ. Gen. Physiol.*, **15**, 487–9, 1932.

Chambers, R., Microdissection studies. I. The visible structure of cell protoplasm and death changes. *Amer. Journ. Physiol.*, **43**, 1–12, 1917.

Channon, H. J., and Chibnall, A. C., The ether-soluble substances of cabbage leaf cytoplasm. IV. Further observation on diglyceridephosphoric acid. *Biochem. Journ.*, **21**, 1112–20, 1927.

—— The ether-soluble substances of cabbage leaf cytoplasm, V. The isolation of n-nonacosane and di-n-tetradecyl ketone. *Biochem. Journ.*, **23**, 168–75, 1929.

Chibnall, A. C., Investigations on the nitrogenous metabolism of the higher plants. Part II. The distribution of nitrogen in the leaves of the runner bean. *Biochem. Journ.*, **16**, 344–62, 1922.

—— Investigations on the nitrogenous metabolism of the higher plants. Part III. The effect of low temperature drying on the distribution of nitrogen in the leaves of the runner bean. *Biochem. Journ.*, **16**, 599–607, 1922.

—— Investigations on the metabolism of the higher plants. Part V. Diurnal variations in the protein nitrogen of runner bean leaves. *Biochem. Journ.*, **18**, 387–94, 1924.

—— Investigations on the metabolism of the higher plants. Part VI. The role of asparagine in the metabolism of the mature plant. *Biochem. Journ.*, **18**, 395–404, 1924.

Chibnall, A. C., and Grover, C. E., A chemical study of leaf cell cytoplasm. I. The soluble proteins. *Biochem. Journ.*, **20**, 108–18, 1926.

Chibnall, A. C., Miller, E. J., Hall, D. H., and Westall, R. G., The proteins of grasses. II. A new method of preparation. *Biochem. Journ.*, **27**, 1874–84, 1933.

Chibnall, A. C., Piper, S. H., Pollard, A., Smith, J. A. B., and Williams, E. F., The wax constituents of the apple cuticle. *Biochem. Journ.*, **25**, 2095–110, 1931.

Chibnall, A. C., Williams, E. F., Latner, A. L., and Piper, S. H., The isolation of n-triacontanol from lucerne wax. *Biochem. Journ.*, **27**, 1885–93, 1933.

Chien, S. S., Peculiar effects of barium, strontium and cerium on Spirogyra. *Bot. Gaz.*, **63**, 406–9, 1917.

Chodat, R., *Monographies d'Algues en Culture Pure*, Berne, 1913.

—— Darstellung und Nachweis von Oxydasen und Katalasen pflanzlicher und tierischer Herkunft. Methoden ihrer Anwendung. *Handbuch biol. Arbeitsmethoden*, Abt. IV, Teil 1, 319–410, 1925.

Chodat, R., and Bach, A., Untersuchungen über die Rolle der Peroxyde in der Chemie der lebenden Zelle. V. Zerlegung der sogenannten Oxydasen in Oxygenasen und Peroxydasen. *Ber. deut. chem. Ges.*, **36**, 606–8, 1903.

Cholodny, N., Zur Frage nach der Rolle der Ionen bei geotropischen Bewegungen. *Ber. deut. bot. Ges.*, **41**, 300–11, 1923.

—— Über die hormonale Wirkung der Organspitze bei den geotropischen Krümmung. *Ber. deft. bot., Ges.*, **42**, 356–62, 1924.

—— Mikropotometrische Untersuchungen über das Wachstum und die Tropismen der Koleoptile von *Avena sativa*. *Jahrb. f. wiss. Bot.*, **73**, 720–58, 1930.

—— Lichtwachstumsreaktion und Phototropismus. *Ber. deut. bot. Ges.*, **49**, 243–7, 1931.

Chudiakow, N. V., Beiträge zur Kenntnis der intramolekularen Athmung. *Landw. Jahrb.*, **23**, 333–89, 1894.

Ciesielski, T., Untersuchungen über die Abwärtskrümmung der Wurzel. *Beitr. z. Biol. Pflanzen*, 1, 1–30, 1872.

Collander, R., Über die Permeabilität pflanzlicher Protoplasten fur Sulphosäurefarbstoffe. *Jahrb. f. wiss. Bot.*, **60**, 354–410, 1921.

COLLANDER, R., Über die Durchlässigkeit der Kupferferrozyanidniederschlagsmembran für Nichtelektrolyte. *Kolloidchem. Beihefte*, **19**, 72–105, 1923.

—— Über die Durchlässigkeit der Kupferferrozyanidmembran für Säuren nebst Bemerkungen zur Ultrafilterfunktion des Protoplasmas. *Kolloidchem. beihefte*, **20**, 273–87, 1924.

—— Über die Permeabilität von Kollodiummembranen. *Soc. Sci. Fennica. Commentationes Biologicae*, **2**, No. 6, 48 pp., 1926.

—— Permeabilitätsstudien an *Chara ceratophylla*, I. Die normale Zusammensetzung des Zellsaftes. *Acta Bot. Fennica*, **6**, 20 pp., 1930.

COLLANDER, R., and BÄRLUND, H., Permeabilitätasstudien an *Chara ceratophylla*, II. Die Permeabilität für Nichtelektrolyte. *Acta Bot. Fennica*, **11**, 114 pp., 1933.

COLLINGS, G. H., The influence of boron on the growth of the soybean plant. *Soil Sci.*, **23**, 83–104, 1927.

COPELAND, E. B., The relation of nutrient salts to turgor. *Bot. Gaz.*, **24**, 399–415, 1897.

CORRENS, C., Ueber die Abhängigkeit der Reizerscheinungen höherer Pflanzen von den Gegenwart freien Sauerstoffes. *Flora*, **75**, 87–151, 1892.

—— Zur Physiologie der Ranken. *Bot. Zeit.*, Abt. 1, **54**, 1–20, 1896.

—— Zur Physiologie von *Drosera rotundifolia*. *Bot. Zeit.*, Abt. 1, **54**, 21–31, 1896.

COWARD, KATHERINE H., Some observations on the extraction and estimation of lipochromes from animal and plant tissues. *Biochem. Journ.*, **18**, 1123–6, 1924.

CRAFTS, A. S., Movements of organic materials in plants. *Plant Physiol.*, **6**, 1–38, 1931.

CROCKER, W., Mechanics of dormancy in seeds. *Amer. Journ. Bot.*, **3**, 99–120, 1916.

CROCKER, W., and DAVIS, W. E., Delayed germination in seed of *Alisma plantago*. *Bot. Gaz.*, **58**, 285–321, 1914.

CROCKER, W., and GROVES, A method of prophesying the life duration of seeds. *Proc. Nat. Acad. Sci.*, **1**, 152–5, 1915.

CURTIS, O. F., The upward translocation of foods in woody plants. I. Tissues concerned in translocation. *Amer. Journ. Bot.*, **7**, 101–24, 1920.

—— The upward translocation of foods in woody plants. II. Is there normally an upward transfer of storage from the roots or trunk to the growing shoots? *Amer. Journ. Bot.*, **7**, 286–95, 1920.

—— The effect of ringing a stem on the upward transfer of nitrogen and ash constituents. *Amer. Journ. Bot.*, **10**, 361–82, 1923.

—— Studies on solute translocation in plants. Experiments indicating that translocation is dependent on the activity of living cells. *Amer. Journ. Bot.*, **16**, 154–68, 1929.

CURTIS, O. F., and SCOFIELD, H. T., A comparison of osmotic concentrations of supplying and receiving tissues and its bearing on the Münch hypothesis of the translocation mechanism. *Amer. Journ. Bot.*, **20**, 502–12, 1933.

CZAPEK, F., Untersuchungen über Geotropismus. *Jahrb. f. wiss. Bot.*, **27**, 243–339, 1895.

—— Zur Physiologie des Leptoms der Angiospermen. *Ber. deut. bot. Ges.*, **15**, 124–31, 1897.

—— Weitere Beiträge zur Kenntniss der geotropischen Reizbewegungen. *Jahrb. f. wiss. Bot.*, **32**, 175–308, 1898.

DACHNOWSKI, A., The toxic property of bog water and bog soils. *Bot. Gaz.*, **46**, 130–43, 1908.

DAKIN, W. J., The West Australian pitcher plant. *Journ. Roy. Soc. Western Australia*, **4**, 37–53, 1918.

DARLINGTON, H. T., Dr. Beal's seed vitality experiment. *Rep. Michigan Acad. Sci.*, **17**, 164–5, 1915.

—— Dr. W. J. Beal's seed viability experiment. *Amer. Journ. Bot.*, **9**, 266–9, 1922.

—— The 50-year period for Dr. Beal's seed viability experiment. *Amer. Journ. Bot.*, **18**, 262–5, 1931.

DARWIN, F., Observations on stomata. *Phil. Trans. Roy. Soc.*, B, **190**, 531–621, 1898.

—— *Lectures on the Physiology of Movement in Plants*, London, 1907.

—— On a method of studying transpiration. *Proc. Roy. Soc.*, B, **87**, 269–80, 1914.

—— The effect of light on the transpiration of leaves. *Proc. Roy. Soc.*, B, **87**, 281–99, 1914.

DARWIN, F., and PERTZ, D. F. M., On a new method of estimating the aperture of stomata. *Proc. Roy. Soc.*, B, **84**, 136–54, 1911.

DASTUR, R. H., Water content, a factor in photosynthesis. *Ann. Bot.*, **38**, 779–88, 1924.

—— The relation between water content and photosynthesis. *Ann. Bot.*, **39**, 769–86, 1925.

DASTUR, R. H., and DESAI, B. L., The relation of water content, chlorophyll content, and the rate of photosynthesis in some tropical plants at different temperatures. *Ann. Bot.*, **47**, 69–88, 1933.

DAVIS, W. A., and SAWYER, G. C., The estimation of carbohydrates. IV. The presence of free pentoses in plant extracts, and the influence of other sugars on their estimation. *Journ. Agric. Sci.*, **6**, 406–12, 1914.

DAVIS, W. E., and ROSE, R. C., The effect of external conditions upon the after-ripening of the seeds of *Crataegus mollis*. *Bot. Gaz.*, **54**, 49–62, 1912.

DAY, D., Some effects on *Pisum sativum* of a lack of calcium in the nutrient solution. *Science*, N.S., **68**, 426, 1928.

DELEANO, N. T., Studien über den Atmungsstoffwechsel abgeschnittener Laubblätter. *Jahrb. f. wiss. Bot.*, **51**, 541–92, 1912.

DELF, E. MARION, Studies of protoplasmic permeability by measurement of rate of shrinkage of turgid tissues. I. The influence of temperature on the permeability of protoplasm to water. *Ann. Bot.*, **30**, 283–310, 1916.

DENEKE, H., Über den Einfluss bewegter Luft auf die Kohlensäure-assimilation. *Jahrb. f. wiss. Bot.*, **74**, 1–32, 1931.

DETMER, W., *Vergleichende Physiologie des Keimungsprocesses*. Jena, 1880.

DEVAUX, H., Porosité du fruit des Cucurbitacées. *Rev. gén. bot.*, **3**, 49–56, 1891.

DILLEWIJN, C. VAN, Die Lichtwachstumsreaktionen von *Avena*. *Rec. trav. bot. Néerlandais*, **24**, 307–581, 1927.

DIMITRIEVICZ, *Haberlandt's wissenschaftlich-praktische Untersuchungen auf dem Gebiete des Pflanzenbaues*. Bd. 1, p. 75, Wien, 1875. (Ref. in V. Grafe : ' Ernahrungsphysiologisches Praktikum höherer Pflanzen '. Berlin, 1914.)

DIXON, H. H., and ATKINS, W. R. G., On osmotic pressure in plants. IV. On the constituents and concentration of the sap in the conducting tracts, and on the circulation of carbohydrates in plants. *Proc. Roy. Dublin Soc.*, N.S., **14**, 374–92, 1915.

DIXON, H. H., and GIBBON, M. W., Bast sap in plants. *Nature*, **130**, 661, 1932.

DIXON, H. H., and JOLY, J., On the ascent of sap. *Phil. Trans. Roy. Soc. London*, B, **186**, 563–76, 1895.

DOLK, H. E., Über die Wirkung der Schwerkraft auf Koleoptilen von *Avena sativa*. *K. Akad. Wetens. Amsterdam Proceedings*, **32**, 40–47, 1929.

—— Über die Wirkung der Schwerkraft auf Koleoptilen von *Avena sativa*, II. *K. Akad. Wetens. Amsterdam Proceedings*, **32**, 1127–1140, 1929.

—— *Geotropie und groeistof*. Diss. Utrecht, 1930.

—— The movements of the leaves of the compass-plant Lactuca sariola. *Amer. Journ. Bot.*, **18**, 195–204, 1931.

DOLK, H. E., and THIMANN, K. V., Studies on the growth hormone of plants. *Proc. Nat. Acad. Sci.*, **18**, 30–46, 1932.

DONNAN, F. G., and HARRIS, A. B., The osmotic pressure and conductivity of aqueous solutions of congo-red, and reversible membrane equilibria. *Journ. Chem. Soc., Trans.*, **99**, 1554–77, 1911.

DORSEY, M. J., A study of sterility in the plum. *Genetics*, **4**, 417–88, 1919.

DOSTÁL, R., Die Korrelationsbeziehung zwischen dem Blatt und seiner Axillarknospe. Vorläufige Mitteilung. *Ber. deut. bot. Ges.*, **27**, 547–54, 1909.

DOX, A. W., and NEIDIG, R. E., Enzymatische Spaltung von Hippursäure durch Schimmelpilze. *Zeitschr. f. Physiol. Chem.*, **85**, 68–71, 1913.

—— Cleavage of pyromucuric acid by mold enzymes. *Biochem. Bull.*, **2**, 407–9, 1913.

DU BUY, H. G., and NUERNBERGK, E., Ueber das Wachstum der Koleoptile und des Mesokotyls von Avena sativa unter verschiedenen Bedingungen (III Mitteilung). *K. Acad. Wetens. Amsterdam Proceedings*, **33**, 542–86, 1930.

DUTROCHET, R., *Recherches anatomiques et physiologiques sur la structure intime des animaux et des végétaux.* Paris, 1824.

—— *Mémoires pour servir à l'histoire anatomique et physiologique des végétaux et des animaux.* Bruxelles, 1837.

—— Recherches sur la volubilité des tiges de certains végétaux et sur la cause de ce phénomène. *Ann. sci. nat., Bot.*, Sér. 3, **2**, 156–67, 1844.

ECKERSON, SOPHIA, A physiological and chemical study of after-ripening. *Bot. Gaz.*, **55**, 286–99, 1913.

ELFVING, F., Ueber eine Wirkung des galvanischen Stromes auf wachsende Wurzeln. *Bot. Zeit*, **40**, 257–64, 273–8, 1882.

—— Ueber die Wasserleitung im Holz. *Bot. Zeit.*, **40**, 707–23, 1882.

EMBDEN, G., DEUTIGE, H. J., and KRAFT, G., Über die intermediären Vorgänge bei der Glykolyse in der Muskulatur. *Klin. Wochenschr.*, **12**, 213–15, 1933.

ERMAN, C., Thermowachstumsreaktionen bei den Koleoptilen von *Avena sativa*. *Ber. deut. bot. Ges.*, **44**, 432–9, 1926.

ERNEST, ELIZABETH C. M., Suction-pressure gradients and the measurement of suction pressure. *Ann. Bot.*, **45**, 717–31, 1931.

—— Studies in the suction pressure of plant cells II. *Ann. Bot.*, **48**, 293–305, 1934.

—— The water relations of the plant cell. *Journ. Linn. Soc., Bot.*, **49**, 495–502, 1934.

ERRERA, L., Die grosse Wachsthumsperiode bei den Fruchtträgern von Phycomyces. *Bot. Zeit.*, **42**, 497–503, 513–22, 529–37, 545–52, 561–6, 1884.

—— Une expérience sur l'ascension de la sève chez les plantes. *Comp. rend. soc. roy. bot. Belgique*, Bull. **25**, 24–32, 1886.

ESCHENHAGEN, F. Ueber den Einfluss von Lösungen verschiedener Koncentration auf das Wachsthum von Schimmelpilze. *Diss.* Leipzig, 1899. Cited from Pfeffer (1900).

EULER, H. v., and MYRBACK, K., Garungs-Co-Enzyme (Co-Zymase) der Hefe. I. *Zeitschr. f. physiol. Chem.*, **131**, 179–203, 1923.

—— Garungs-Co-Enzyme (Co-Zymase) der Hefe. III. *Zeitschr. f. physiol. Chem.*, **136**, 107–29, 1924.

—— Co-Zymase. XV. *Zeitschr. f. physiol. Chem.*, **177**, 237–47, 1928.

—— Co-Zymase. XVI. Weitere Isolierungsversuche. *Zeitschr. f. physiol. Chem.*, **184**, 163–8, 1929.

EWART, A. J., The ascent of water in trees. *Phil. Trans. Roy. Soc., London*, B, **198**, 41–85, 1905.

—— The ascent of water in trees. (Second paper.) *Phil. Trans. Roy. Soc.*, B, **199**, 341–92, 1908.

—— On the longevity of seeds. *Proc. Roy. Soc. Victoria*, **21** (N.S.), 1–210, 1908.

FABER, F. C. v., Über Transpiration und osmotische Druck bei den Mangroven (Vorläufige Mitteilung). *Ber. deut. bot. Ges.*, **31**, 277–81, 1913.

—— *Biophytum apodiscias*, eine neue sensitive Pflanze auf Java. *Ber. deut. bot. Ges.*, **31**, 282–5, 1913.

—— Physiologische Fragmente aus einem tropischen Urwald. *Jahrb. f. wiss. Bot.*, **56**, 197–220, 1915.

—— Zur Physiologie der Mangroven. *Ber. deut. bot. Ges.*, **41**, 227–34, 1923.

FALK, K. G., Studies on enzyme action VI. The specificity of lipase action. *Journ. Amer. Chem. Soc.*, **35**, 616–24, 1913.

FALK, K. G., and SUGUIRE, K., Studies on enzyme action XII. The esterase and lipase of castor beans. *Journ. Amer. Chem. Soc.*, **37**, 217–30, 1915.

FARMER, J. B., On the quantitative differences in the water-conductivity of the wood in trees and shrubs. Part I. The evergreens. *Proc. Roy. Soc.*, B, **90**, 218–32, 1918.

—— On the quantitative differences in the water-conductivity of the wood in trees and shrubs. Part II. The deciduous plants. *Proc. Roy. Soc.*, B, **90**, 232–50, 1918.

FERNANDEZ, D. S., Aerobe und anaerobe Atmung bei Keimlingen von *Pisum sativum*. *Rec. trav. bot. Néerlandais*, **20**, 107–256, 1923.

FISCHER, A., Wasserstoff- une Hydroxylionen als Keimungsreize. *Ber. deut. bot. Ges.*, **25**, 108–22, 1907.

FISCHER, E., *Untersuchungen über Amino-säuren, Polypeptide und Proteine*. Berlin, 1906.

FITTING, H., Untersuchungen über den Haptotropismus der Ranken. *Jahrb. f. wiss. Bot.*, **38**, 545–634, 1903.

—— Untersuchungen über den geotropischen Reizvorgang. Teil I: Die geotropische Empfindlichtkeit der Pflanzen. *Jahrb. f. wiss. Bot.*, **41**, 221–330, 1905.

—— Untersuchungen über den geotropischen Reizvorgang. Teil II: Weitere Erfolge mit der intermittierden Reizung. *Jahrb. f. wiss. Bot.*, **41**, 331–98, 1905.

—— Untersuchungen über die Aufnahme von Salzen in die lebende Zelle. *Jahrb. f. wiss, Bot.*, **56**, 1–64, 1915.

—— Untersuchungen über isotonische Koeffizienten und ihre Nutzen für Permeabilitätsbestimmungen. *Jahrb. f. wiss. Bot.*, **57**, 553–612, 1917.

—— Untersuchungen über die Aufnahme und über anomale osmotische Koeffizienten von Glyzerin und Harnstoff. *Jahrb. f. wiss. Bot.*, **59**, 1–170, 1919.

FRANK, A. B., *Die naturliche wagerechte Richtung von Pflanzenteilen*. Leipzig, 1870.

FRANK, B. Ueber die aut Wurzelsymbiose beruhende Ernährung gewisser Bäume durch unterirdische Pilze. *Ber. deut. bot. Ges.*, **3**, 128–45, 1885.

FREY, A., Die submikroskopische Struktur der Zellmembran. *Jahrb. f. wiss. Bot.*, **65**, 195–223, 1925.

FRÖSCHEL, P., Untersuchungen über die heliotropische Präsentationszeit. I. Mitteilung. *Sitzungsber. K. Akad. Wiss. Wien*, Abt. 1, *Math. Nat. Kl.*, **117**, 235–56, 1908.

—— Untersuchungen über die heliotropische Präsentationszeit. II. Mitteilung. *Sitzungsber. K. Akad. Wiss. Wien*, Abt. 1, *Math. nat. Kl.*, **118**, 1247–94, 1909.

GARDNER, V. R., Studies in the Nutrition of the Strawberry. *Univ. Missouri Agric. Exper. Sta., Bull.*, **57**, 1923.

GARNER, W. W., and ALLARD, H. A., Effect of the relative length of day and night and other factors of the environment on growth and reproduction in plants. *Journ. Agric. Res.*, **18**, 553–606, 1920.

—— Further studies in photoperiodism, the response of the plant to relative length of day and night. *Journ. Agric. Res.*, **23**, 871–920, 1923.

Garreau, Recherches sur l'absorption et l'exhalation des surfaces aeriennes des plantes. *Ann. sci. nat., Bot.,* Sér. 3, **13**, 321–46, 1850.

—— De la respiration chez les plantes (I), *Ann. sci. nat. Bot.*, Sér. 3, **15**, 5–36, 1851.

—— Nouvelles recherches sur la respiration des plantes, *Ann. sci. nat. Bot.*, Sér. 3, **16**, 271–92, 1851.

Gassner, G., Der Galvanotropismus der Wurzeln. *Bot. Zeit.*, Abt. 1, **64**, 149–222, 1906.

—— Über die keimungsauslösende Wirkung der Stickstoffsalze auf lichtempfindliche Samen. *Jahrb. f. wiss. Bot.*, **55**, 259–342, 1915.

—— Beiträge zur Frage der Lichtkeimung. *Zeitschr. f. Bot.*, **7**, 609–61, 1915.

—— Fruhtreibversuche mit Blausäure. *Ber. deut. bot. Ges.*, **43**, 132–7, 1925.

—— Untersuchungen über die Wirkung von Temperatur und Temperaturkombinationen auf die Keimung von *Poa pratensis* und anderen *Poa-Arten*. *Zeitschr. f. Bot.*, **23**, 767–838, 1930.

Genevois, L., Über Atmung und Gärung in grünen Pflanzen. II. *Biochem. Zeitschr.*, **191**, 147–57, 1927.

Gerber, C., Recherches sur la formation des réserves oléagineuses des graines et des fruits. *Comp. rend. acad. sci.*, **125**, 732–5, 1897.

Giesenhagen, K., Ueber innere Vorgänge bei der geotropischen Krümmung der Wurzeln von *Chara*. *Ber. deut. bot. Ges.*, **19**, 277–85, 1901.

Godlewski, E., Abhängigkeit der Sauerstoffausscheidung der Blätter von dem Kohlensäuregehalt der Luft. *Arb. bot. Inst. Würzburg*, **1**, 343–70, 1873.

—— Zur Theorie der Wassebewegung in den Pflanzen. *Jahrb. f. wiss. Bot.*, **15**, 569–630, 1884.

Godwin, H., and Bishop, L. R., The behaviour of the cyanogenetic glucosides of cherry laurel during starvation. *New Phyt.*, **26**, 295–315, 1927.

Goebel, K., Beiträge zur Morphologie und Physiologie des Blattes. *Bot. Zeit.*, **38**, 753–60, 769–78, 785–95, 801–15, 817–26, 833–45, 1880.

—— *Pflanzenbiologische Schilderungen*. Teil 2. Marburg, 1891–3.

—— Archegoniatenstudien I. *Flora*, **76**, 92–116, 1892.

—— *Einleitung in die experimentelle Morphologie der Pflanzen*. Leipzig und Berlin, 1908.

Gore, H. C., Studies on fruit respiration. *U.S. Dept. Agric. Bur. Chem., Bull.* **142**, 1911.

Gorup-Besanez, E. v., Ueber das Vorkommen eines diastatischen und peptonbildenden Ferments in den Wickensamen. *Ber. deut. chem. Ges.*, **7**, 1478–80, 1874.

—— Weitere Beobachtungen über diastatische und peptonbildende Fermente im Pflanzenreiche. *Ber. deut. chem. Ges.*, **8**, 1510–14, 1875.

Gorup-Besanez, E. v. and Will, H. Fortgesetzte Beobachtungen über peptonbildende Fermente im Pflanzenreiche (Dritte Mitteilung). *Ber. deut. chem. Ges.*, **9**, 673–8, 1876.

Graser, Marie, Untersuchungen über das Wachstum und die Reizbarkeit der Sporangienträger von *Phycomyces nitens*. *Beitr. z. bot. Centr.*, **36**, Abt. 1, 414–93, 1919.

Green, J. R., On the germination of the tuber of the Jerusalem artichoke. *Ann. Bot.*, **1**, 223–36, 1888.

—— On the germination of the seeds of the castor oil plant (*Ricinus communis*). *Proc. Roy. Soc.*, B, **48**, 370–92, 1890.

—— Researches on the germination of the pollen grain and the nutrition of the pollen tube. *Phil. Trans. Roy. Soc.*, B, **185**, 385–409, 1894.

Grew, N., *The Anatomy of Plants with an Idea of a Philosophical History of Plants*. London, 1682.

Gurjar, A. M., Carbon/Nitrogen Ratio in Relation to Plant Metabolism. *Science*, **51**, 351–2, 1920.

GUTTENBERG, H. R. v., Über die Verteilung der geotropischen Empfindlichkeit in der Koleoptile der Gramineen. *Jahrb. f. wiss. Bot.*, **50**, 289–327, 1912.

—— Studien über den Phototropismus der Pflanzen. *Beitr. z. all. Bot.*, **2**, 139–247, 1922.

HAAS, R. H. DE, On the connection between the geotropic curving and elasticity of the cell wall. *K. Akad. Wetens. Amsterdam Proceedings*, **32**, 371–3, 1929.

HABERLANDT, G., *Das Reizleitende Gewebesystem der Sinnpflanze.* Leipzig, 1890.

—— Ueber die Perception des geotropischen Reizes. *Ber. deut. bot. Ges.*, **18**, 261–72, 1900.

—— Die Perzeption des Lichtreizes durch das Laubblatt. *Ber. deut. bot. Ges.*, **22**, 105–19, 1904.

—— *Physiological Plant Anatomy.* English trans. London, 1914.

—— Zur Physiologie der Zelltheilung, VI. Über Auslösung von Zellteilungen durch Wundhormone. *Sitzungsber. preuss. Akad. Wiss.*, 221–34, 1921.

HALES, S., *Statical Essays.* London, 1731.

—— *Vegetable Staticks.* London, 1769.

HANES, C. S., and BARKER, J., The physiological action of cyanide. I. The effects of cyanide on the respiration and sugar content of the potato at 15° C. *Proc. Roy. Soc.*, B, **108**, 95–118, 1931.

HANSTEEN-CRANNER, B., Ueber das Verhalten der Kulturpflanzen zu den Bodensalzen. III. Beiträge zur Biochemie und Physiologie der Zellwand lebender Zellen. *Jahrb. f. wiss. Bot.*, **53**, 536–98, 1914.

—— Beiträge zur Biochemie und Physiologie der Zellwand und der plasmatischen Grenzschichten (Vorläufige Mitteilung). *Ber. deut. bot. Ges.*, **37**, 380–91, 1919.

—— Zum Biochemie und Physiologie der Grenzschichten lebender Pflanzenzellen. *Meldinger fra Norges Landsbrukheiskole, Kristiania*, **2**, 1–160. 1922.

—— Untersuchungen über die Frage ob in der Zellwand lösliche Phosphatide vorkommen. *Planta*, **2**, 438–45, 1926.

HANSTEIN, J., Versuche über die Leitung des Saftes durch die Rinde und Folgerungen daraus. *Jahrb. f. wiss. Bot.*, **2**, 392–467, 1860.

HARDER, R., Kritische Versuche zu Blackmans Theorie der 'begrenzender Faktoren' bei den Kohlensäureassimilation. *Jahrb. f. wiss. Bot.*, **60**, 531–71, 1921.

HARDY, W. B., On the structure of cell protoplasm. *Journ. Physiol.*, **24**, 158–210, 1899.

HARRIS, J. A., GORTNER, R. A., HOFMAN, W. F., and VALENTINE, A. T., Maximum values of osmotic concentration in plant tissue fluids. *Proc. Soc. Exper. Biol. and Med.*, **18**, 106–9, 1921.

HARRIS, J. A., GORTNER, R. A., and LAWRENCE, J. V., The relationship between the osmotic concentration of leaf sap and height of leaf insertion in trees. *Bull. Torrey Bot. Club*, **44**, 267–86, 1917.

—— On the relationship between freezing-point lowering, Δ, and specific conductivity, *K*, of plant fluids. *Science*, N.S., **52**, 494–5, 1920.

—— On the differentiation of the leaf tissue fluids of ligneous and herbaceous plants with respect to osmotic concentration and electrical conductivity. *Journ. Gen. Physiol.*, **3**, 343–5, 1921.

—— The osmotic concentration and electrical conductivity of the tissue fluids of ligneous and herbaceous plants. *Journ. Phys. Chem.*, **25**, 122–46, 1921.

HARRIS, J. A., and LAWRENCE, J. V., On the osmotic pressure of the tissue fluids of Jamaican Loranthaceae parasitic on various hosts. *Amer. Journ. Bot.*, **3**, 438–45, 1916.

HARRIS, J. A., and LAWRENCE, J. V., Cryoscopic determination of tissue fluids of plants of Jamaican coastal deserts. *Bot. Gaz.*, **64**, 285–305, 1917.

—— The osmotic concentration of the tissue fluids of Jamaican rain-forest vegetation. *Amer. Journ. Bot.*, **4**, 268–98, 1917.

—— with the co-operation of GORTNER, R. A., The cryoscopic constants of expressed vegetable saps as related to local conditions in the Arizona deserts. *Physiol. Res.*, **2**, 1–49, 1916.

HARVEY, E. M., and MURNEEK, A. E., The relation of carbohydrates and nitrogen to the behaviour of apple spurs. *Oregon. Agric. Coll. Exper. Stat. Bull.*, **176**, 1921.

HASSELBRING, H., The relation between the transpiration stream and the absorption of salts. *Bot. Gaz.*, **57**, 72–3, 1914.

HAWKER, LILIAN E., A quantitative study of the geotropism of seedlings with special reference to the nature and development of their statolith apparatus. *Ann. Bot.*, **46**, 121–57, 1932.

—— Experiments on the perception of gravity by roots. *New Phyt.*, **31**, 321–8, 1932.

—— The effect of temperature on the geotropism of seedlings of *Lathyrus odoratus*. *Ann. Bot.*, **47**, 503–15, 1933.

HAWKINS, L. A., and MAGNESS, J. R., Some changes in Florida grape-fruit in storage. *Journ. Agric. Res.*, **20**, 357–73, 1920.

HAWORTH, W. N., *The Constitution of Sugars*. London, 1929.

—— The Molecular Structure of Carbohydrates. Pres. Add. Section B, *Rep. Brit. Ass.*, Norwich Meeting, 1935, Publ. London, 1935.

HEIDMANN, Über Richtungsbewegungen hervorgerufen durch Verletzungen und Assimilationshemmung. *Sitzungsber. K. K. Akad. Wiss. Wien*, Abt. 1, *Math. nat. Kl.*, **122**, 1227–55, 1913.

HEILBRUNN, L. V., Protoplasmic viscosity of *Amoeba* at different temperatures. *Protoplasma*, **8**, 58–64, 1929.

—— The absolute viscosity of *Amoeba* protoplasm. *Protoplasma*, **8**, 65–9, 1929.

HEINRICHER, E., Die grünen Halbschmarotzer. I. *Odontites*, *Euphrasia* und *Orantha*. *Jahrb. f. wiss. Bot.*, **31**, 77–124, 1898.

—— Die grünen Halbschmarotzer. II. *Euphrasia*, *Alectorolophus* und *Odontites*. *Jahrb. f. wiss. Bot.* **32**, 389–452, 1898.

—— Die grünen Halbschmarotzer. III. *Bartschia* und *Tozzia*, nebst Bemerkungen zur Frage nach der assimilatorischen Leitungsfähigkeit der grünen Halbschmarotzen. *Jahrb. f. wiss. Bot.*, **36**, 665–752, 1901.

—— Die grünen Halbschmarotzer. IV. Nachträge zu *Euphrasia*, *Odontites* und *Alectorolopus*. Kritische Bemerkungen zur Systematik letzterer Gattung. *Jahr. f. wiss. Bot.*, **37**, 264–337, 1902.

—— Die grünen Halbschmarotzer. V. *Melampyrum*. *Jahrb. f. wiss. Bot.*, **46**, 273–376, 1909.

—— Die grünen Halbschmarotzer. VI. Zur Frage nach der assimilatorischen Leitungsfähigkeit der grünen, parasitischen Rhinanthaceen. *Jahrb. f. wiss. Bot.*, **47**, 539–87, 1910.

HELLRIEGEL, H., and WILFARTH, H., Untersuchungen über die Stickstoffnahrung der Gramineen und Leguminosen. *Zeitschr. d. Vereins f. d. Rubenzucker-Industrie*, 234 pp., 1888. (Ref. in E. J. Russell, ' Soil Conditions and Plant Growth '. London, 1932.)

HENDERSON, F. Y., On the effect of light and other conditions upon the rate of water loss from the mesophyll. *Ann. Bot.*, **40**, 507–33, 1926.

HEPBURN, J. S., Biochemical studies of the pitcher liquor of Nepenthes. *Proc. Amer. Phil. Soc.*, **57**, 112–29, 1918.

HEPBURN, J. S., ST. JOHN, E. Q., and JONES, F. M., The absorption of nutrients and allied phenomena in the pitchers of the Sarraceniaceae. *Journ. Franklin Inst.*, **189**, 147–84, 1920.

HERBERT, D. A., Anaesthesia in plants. *Philippine Agric.*, **11**, No. 5, 1922.

HERZOG, W., Über die Verteilung der geotropischen Empfindlichkeit in negativ-geotropen Pflanzenorganen. *Planta*, **1**, 116–44, 1925.

HESS, K., Zur Frage des Aufbaues pflanzlicher Membran. *Biochem. Zeitschr.*, **203**, 409–20, 1928.

HEYL, F. W and HEPNER, F. E., Some constituents of the leaves of *Zygadenus intermedius*. III. *Journ. Amer. Chem. Soc.*, **35**, 803–11, 1913.

HEYL, F. W., HEPNER, F. E., and LOY, S. K., Analyses of some Wyoming larkspurs. I. *Journ. Amer. Chem. Soc.*, **35**, 880–5, 1913.

HEYN, A. N. J., On the relation between growth and extensibility of the cell wall. *K. Akad. Wetensch. Amsterdam, Proceedings*, **33**, 1045–58, 1930.

—— Der Mechanismus der Zellstreckung. *Rec. trav. bot. Néerlandais*, **28**, 113–244, 1931.

—— Further investigations on the mechanism of cell elongation and the properties of the cell wall in connection with elongation. I. The load extension relationship. *Protoplasma*, **19**, 78–96, 1933.

—— Weitere Untersuchungen über den Mechanismus der Zellstrecknung und die Eigenschaften der Zellmembran. II. Das Röntgendiagramm von jungen wachsenden Zellwänden und parenchymatischen Geweben. *Protoplasma*, **21**, 299–305, 1934.

—— Weitere Untersuchungen uber den Mechanismus der Zellstrecknung und die Eigenschaften der Zellmembran. III. Die Änderungen der Plastizität der Zellwand bei verschiedenen Organen. *Jahrb. f. wiss. Bot.*, **79**, 753–89, 1934.

HILL, A. C., Reversible zymohydrolysis. *Journ. Chem. Soc. Trans.*, **73**, 634–58, 1898.

HILL, G. R., Respiration of fruits and growing plant tissues in certain gases, with reference to ventilation and fruit storage. *Cornell Univ. Agric. Exper. Sta. Bull.*, **330**, 1913.

HOAGLAND, D. R., and DAVIS, A. R., The intake and accumulation of electrolytes by plant cells. *Protoplasma*, **6**, 610–26, 1929.

HOAGLAND, D. R., DAVIS, A. R., and HIBBARD, P. L., The influence of one ion on the accumulation of another by plant cells with special reference to experiments with *Nitella*. *Plant. Physiol.*, **3**, 473–86, 1928.

HÖFLER, K., Die plasmolytisch-volumetrische Methode und ihre Andwendbarkeit zur Messung des osmotischen Wertes lebender Pflanzenzellen. *Ber. deut. bot. Ges.*, **35**, 706–26, 1917.

—— Permeabilitätsbestimmung nach der plasmometrischen Methode. *Ber. deut. bot. Ges.*, **36**, 414–22, 1918.

—— Ein Schema für die osmotische Leistung der Pflanzenzelle. *Ber. deut. bot. Ges.*, **38**, 288–98, 1920.

HÖFLER, K., and STIEGLER, A., Permeabilitätsverteilung in verschiedenen Geweben der Pflanze. *Protoplasma*, **9**, 469–512, 1930.

HOFMEISTER, W., *Die Lehre von der Pflanzenzelle.* Leipzig, 1867.

HOLLE, H., Untersuchungen über Welken, Vertrocknen und Wiederstraffwerden. *Flora*, **108**, 73–126, 1915.

HONERT, T. H. VAN DEN, Carbon dioxide assimilation and limiting factors. *Rec. trav. bot. Néerlandais*, **27**, 149–286, 1930.

HONING, J. A., De invloed van het licht op het kiemen van de zaden van verschillende variëteiten van *Nicotiana tabacum*. (With English summary.) *Bull. Deli Proefstation*, No. 7, 1–14, 1916.

—— De invloed van een behandeling met warm water op het keimprocent van de zaden van *Albizzia moluccana* Miq., *Pithecolobium saman* Bth., *Mimosa invisa* Mart. en *Crotalaria striata* D.C. (With English summary.) *Bull. Deli Proefstation*, No. 7, 15–24, 1916.

HOOKER, H. D., Hydrotropism in roots of *Lupinus albus*. *Ann. Bot.*, **29**, 265–83, 1915.

HOOKER, H. D., Physiological observations on *Drosera rotundifolia*. *Bull. Torrey Bot. Club*, **43**, 1–27, 1916.

—— Mechanics of movement in *Drosera rotundifolia*. *Bull. Torrey Bot. Club*, **44**, 389–404, 1917.

—— Seasonal changes in the chemical composition of apple spurs. *Univ. Missouri Agric. Exp. Stat., Bull.* **40**, 1920.

HOOKER, J. D., Address to the Department of Botany and Zoology. *Rep. Brit. Ass.*, Belfast Meeting, 1874, Trans. of Sections, 102–16, Publ. London, 1875.

HORNBERGER, R., MUTSCHLER, C., and HAMMERBACHER, F., Analysen der Asche der Früchte von *Lithospermum officinale* und des Holzes von *Calamus Rotang* und *Bambusa arundinacea*. *Ann. Chem.*, **176**, 84–8, 1875.

HOUWINK, A. L., The conduction of excitation in *Mimosa pudica*. *Rec. trav. bot. Néerlandais*, **32**, 51–91, 1935.

HUBER, B., Zur Physik der Spaltöffnungstranspiration. I. Das maximale Diffusionsvermögen von Porenmembranen (Vorläufige Mitteilung). *Ber. deut. bot. Ges.*, **46**, 610–20, 1928.

—— *Aster linosyris*, ein neuer Typus der Kompasspflanzen (Gnomonpflanzen). *Flora*, **129**, 113–19, 1934.

ILJIN, W. S., Über Offnen der Stomata bei starkem Welken der Pflanzen. *Jahrb. f. wiss. Bot.*, **77**, 220–51, 1932.

INAMDAR, R. S., and SINGH, B. N., Studies in the respiration of tropical plants. I. Seasonal variations in aerobic and anaerobic respiration in the leaves of *Arctocarpus integrifolia*. *Journ. Indian Bot. Soc.*, **6**, 133–212, 1927.

INGEN-HOUSZ, J., *Experiments upon vegetables, discovering their great power of purifying the common air in sunshine and of injuring it in the shade and at night; to which is joined a new method of examining the accurate degree of salubrity of the atmosphere.* London, 1779.

IVANOV, S., Über Ölsynthese unter Vermittlung der pflanzlichen Lipase. *Ber. deut. bot. Ges.*, **29**, 595–602, 1911.

JACQUES, A. G., and OSTERHOUT, W. J. V., The kinetics of penetration. II. The penetration of CO_2 into *Valonia*. *Journ. Gen. Physiol.*, **13**, 695–713, 1930.

—— The accumulation of electrolytes. III. Behaviour of sodium, potassium and ammonium in *Valonia*. *Journ. Gen. Physiol.*, **14**, 301–14, 1930.

—— The accumulation of electrolytes. IV. Internal versus external concentrations of potassium. *Journ. Gen. Physiol.*, **15**, 537–50, 1932.

—— The accumulation of electrolytes. VI. The effect of external *p*H. *Journ. Gen. Physiol.*, **17**, 727–50, 1934.

JAGER, L. DE, Erklärungsversuch über die Wirkungsart der ungeformten Fermente. *Arch. f. path. Anat. u. Physiol. u. f. klin. Med.*, **121**, 182–7, 1890.

JAMES, W. O., The effect of variations of carbon dioxide supply upon the rate of assimilation of submerged water plants. *Proc. Roy. Soc.*, B, **103**, 1–42, 1928.

—— Studies of the physiological importance of the mineral elements in plants. I. The relation of potassium to the properties and functions of the leaf. *Ann. Bot.*, **44**, 173–98, 1930.

—— The dynamics of photosynthesis. *New Phyt.*, **33**, 8–40, 1934.

JAMES, W. O., and BAKER, H., Sap pressure and the movements of sap. *New Phyt.*, **32**, 317–43, 1933.

JAMES, W. O., and CATTLE, MARGARET, Studies of the physiological importance of the mineral elements in plants. VI. The influence of potassium chloride on the rate of diastatic hydrolysis of starch. *Biochem. Journ.*, **27**, 1805–9, 1933.

JANNSEN, G., and BARTHOLOMEW, R. P., The translocation of potassium in tomato plants and its relation to their carbohydrate and nitrogen distribution. *Journ. Agric. Res.*, **38**, 447–64, 1929.

JANSE, J. M., Die Mitwirkung der Markstrahlen bei Wasserbewegung im Holze. *Jahrb. f. wiss. Bot.*, **18**, 1–69, 1887.

—— Der Aufsteigende Strom in der Pflanze. I. *Jahrb. f. wiss. Bot.*, **45**, 303–50, 1908.

—— Der Aufsteigende Strom in der Pflanze. II. *Jahrb. f. wiss. Bot.*, **52**, 509–602, 1913.

—— Die Wirkung des Protoplasten in den Zellen, welche bei der Wasserbewegung beteiligt sind. *Jahrb. f. wiss. Bot.*, **52**, 603–21, 1913.

JEFFREYS, H., Some problems of evaporation. *Phil. Mag.*, **35**, 270–80, 1918.

JENSEN, P. BOYSEN, Über die Leitung des phototropischen Reizes in *Avena*keimpflanzen (Vorläufige Mitteilung). *Ber. deut. bot. Ges.*, **28**, 118–20, 1910.

—— Über die Leitung des phototropischen Reizes in der *Avena*koleoptile. *Ber. deut. bot. Ges.*, **31**, 559–66, 1913.

—— Studies on the production of matter in light and shadow plants. *Bot. Tidssk.*, **36**, 219–62, 1918.

—— Studien über die genetischen Zusammenhang zwischen der normalen und intramolekularen Atmung der Pflanzen. *Kgl. Danske Videnskabernes Selskab. Biol. Med.* IV, **1**, 34 pp., 1923.

—— The connexion between the oxybiotic and anoxybiotic respiration in plants. Fifth International Botanical Congress, Cambridge, 1930. *Report of Proceedings*, 421–2, Publ. Cambridge, 1931.

—— Über Wachstumregulatoren bei Bakterien. *Biochem. Zeitschr.*, **236**, 205–10, 1931.

—— Über Bildung eines Wachstumregulators durch *Aspergillus niger*. *Biochem. Zeitschr.*, **239**, 243–9, 1931.

—— *Die Stoffproduktion der Pflanzen.* Jena, 1932.

—— Die Bedeutung des Wuchsstoffes für das Wachstum und die geotropische Krümmung der Wurzeln von *Vicia faba*. *Planta*, **20**, 688–98, 1933.

—— Über Wuchsstoff in Wurzeln, die mit Erythrosin vergiftet sind. *Planta*, **22**, 404–10, 1934.

JESENKO, F., Über das Austreiben im Sommer entblätterter Bäume und Sträucher. *Ber. deut. bot. Ges.*, **30**, 226–32, 1912.

—— Einige neue Verfahren, die Ruheperiode der Holzgewächse abzukurzen. II. Mitteilung. *Ber. deut. bot. Ges.*, **30**, 81–93, 1912.

JOHANNSEN, W., *Das Aether-Verfahren beim Frühtreiben mit besonderer Berücksichtigung der Fliedertreiberei.* Jena, 1900.

JOHNSON, H., The unsatisfactory nature of the theories proposed to account for the descent of the radicles in the germination of seeds, shewn by experiments. *Edinburgh New Philosophical Journal.* **6**, 312–17, 1829.

JOHNSTON, E. S., Boron : its importance in plant growth. *Journ. Chem. Educ.*, **5**, 1235–42, 1928.

JOHNSTON, E. S., and DORE, W. H., The influence of boron on the chemical composition and growth of the tomato plant. *Plant. Physiol.*, **4**, 31–62, 1929.

JOHOW, F., Die chlorophyllfreien Humuspflanzen nach ihren biologischen und anatomisch-entwicklungs-geschichten Verhältnissen. *Jahrb. f. wiss. Bot.*, **16**, 415–49, 1885.

JÖNSSON, B., Der richtende Einfluss strömenden Wassers auf wachsende Pflanzen und Pflanzentheile (Rheotropismus). *Ber. deut. bot. Ges.*, **1**, 512–21, 1883.

JÖRGENSEN, I., and KIDD, F., Some photochemical experiments with pure chlorophyll and their bearing on theories of carbon assimilation. *Proc. Roy. Soc.*, B, **89**, 342–61, 1916.

JOST, L., Ueber Dickenwachsthum und Jahresringbildung. *Bot. Zeit.*, **49**, 485–95, 501–10, 525–31, 541–7, 557–63, 573–9, 589–95, 605–11, 625–30, 1891.

—— Ueber Beziehungen zwischen der Blattentwickelung und der Gefässbildung in der Pflanze. *Bot. Zeit.*, **51**, 89–138, 1893.

—— *Pflanzenphysiologie.* Band II. Formwechsel und Ortwechsel, Jena, 1923.

JUEL, H. O., Untersuchungen über den Rheotropismus der Wurzeln. *Jahrb. f. wiss. Bot.*, **34**, 507–38, 1900.

KABSCH, Anatomische und physiologische Untersuchungen über einige Bewegungserscheinungen im Pflanzenreiche. *Bot. Zeit.*, **19**, 345–50, 353–8, 361–6, 369–75, 1861.

KAHHO, H., Ueber die Beeinflussung der Hitzekoagulation des Pflanzenplasmas durch Neutralsalze. I. Mitteilung. *Biochem. Zeitschr.*, **117**, 87–95, 1921.

—— Zur Kenntnis der Neutralsalzwirkung auf das Pflanzenplasma. II Mitteilung. *Biochem. Zeitschr.* **120**, 125–42, 1921.

—— Ein Beitrag zur Giftwirkung der Schwermetallsalze auf das Pflanzenplasma. III. Mitteilung. *Biochem. Zeitschr.*, **122**, 39–42, 1921.

—— Ein Beitrag zur Permeabilität des Pflanzenplasmas für Neutralsalze. IV Mitteilung. *Biochem. Zeitschr.*, **123**, 284–303, 1921.

KANITZ, A., *Temperatur und Lebensvorgänge.* Berlin, 1915.

KARSTEN, G., Über Kompasspflanzen. *Flora*, N.F., **11–12**, 48–59, 1918.

KASTENS, E., Beiträge zur Kenntnis der Funktion der Siebröhren. *Mitt. Inst. allg. Bot. Hamburg.* **6**, 33–70, 1924.

KEEBLE, F., NELSON, M. G., and SNOW, R., The integration of plant behaviour I. Separate geotropic stimulation of tip and stump in roots. *Proc. Roy. Soc.*, B, **105**, 493–8, 1929.

—— A wound substance retarding growth in roots. *New Phyt.*, **29**, 289–93, 1930.

—— The integration of plant behaviour IV. Geotropism and growth substance. *Proc. Roy. Soc.*, B, **108**, 537–45, 1931.

KEEBLE, F., and PELLEW, C., The mode of inheritance of stature and time of flowering in peas. *Journ. Gen.*, **1**, 47–56, 1910.

KEILIN, D., Cytochrome and respiratory enzymes. *Proc. Roy. Soc.*, B, **104**, 206–52, 1929.

KELLEY, W. P., and CUMMINS, A. B., Composition of normal and mottled citrus leaves. *Journ. Agric. Res.*, **20**, 161–91, 1920.

KERR, T., and BAILEY, I. W., Structure, optical properties and chemical composition of the so-called middle lamella. *Journ. Arnold Arbor.*, **15**. 327–49, 1934. (Ref. in Anderson, 1935.)

KIDD, F., The controlling influence of carbon dioxide in the maturation, dormancy and germination of seeds. Part I. *Proc. Roy. Soc.*, B, **87**, 408–21, 1914.

—— The controlling influence of carbon dioxide. Part III. The retarding effect of carbon dioxide on respiration. *Proc. Roy. Soc.*, B, **89** 136–56, 1915.

KIDD, F., and WEST, C., The controlling influence of carbon dioxide IV. On the production of secondary dormancy in seeds of *Brassica alba* following treatment with carbon dioxide, and the relation of this phenomenon to the question of stimuli in growth processes. *Ann. Bot.*, **31**, 457–87, 1917.

KIDD, F., WEST, C., and BRIGGS, C. E., A quantitative analysis of the growth of *Helianthus annuus*. Part I. The respiration of the plant and of its parts throughout the life cycle. *Proc. Roy. Soc.*, B, **92**, 368–84, 1921.

KIESEL, A., Untersuchungen über Protoplasma. Über die chemischen Bestandteile des Plasmodiums von *Reticularia lycoperdon*. *Zeitschr. f. physiol. Chem.*, **150**, 149–75, 1925.

KINZEL, W., *Frost und Licht als beeinflussende Kräfte bei den Samenkeimung.* Stüttgart, 1913.

KITE, G. L., Studies on the physical properties of protoplasm. I. The physical properties of protoplasm of certain animal and plant cells. *Amer. Journ. Physiol.*, **32**, 146–64, 1913.

—— Studies on the internal cytoplasm of animal and plant cells. *Amer Journ. Physiol.*, **37**, 282–99, 1915.

KLEBS, G., Über die Organisation der Gallerte bei einigen Algen und Flagellaten *Unters. bot. Inst. Tübingen*, **2**, 333–418, 1886.

—— Zur Physiologie der Fortpflanzen von *Vaucheria sessilis.* *Verh. Naturf. Ges. Basel*, **10**, 45–72, 1895.

—— *Die Bedingungen der Fortpflanzen bei einigen Algen und Pilzen.* Jena, 1896.

—— Einige Ergebnisse der Fortpflanzungsphysiologie. *Ber. deut. bot. Ges.*, **18**, (201)–(215), 1900.

—— Über das Treiben der einheimischen Bäume, speziell der Buche. *Abhandl. Heidelberger. Akad. Wiss., Math. nat. Kl.*, Abh. 3, 1914.

—— Zur Entwicklungsphysiologie der Farnprothallien, I–III. *Sitzungsber. Heidelberger Akad. Wiss., Math. nat. Kl., B*, 1916–17.

—— Über das Verhaltnis von Wachstum und Ruhe bei den Planzen. *Biol. Zeutr.*, **37**, 373–415, 1917.

—— Über die Blutenbildung von Sempervivum. *Flora*, **111–112**, 128–51, 1918.

KNIEP, H., Ueber die Lichtperzeption der Laubblätter. *Biol. Centr.*, **27**, 97–106, 129–42, 1907.

KNIGHT, L. I., Physiological aspects of self-sterility of the apple. *Proc. Amer. Soc. Hort. Sci.*, **14**, 101–5, 1917.

KNIGHT, L. I., and CROCKER, W., Toxicity of smoke. *Bot. Gaz.*, **55**, 337–71, 1913.

KNIGHT, R. C., Transpiration. *Science Progress*, **13**, 561–6, 1919.

—— Further observations on the transpiration, stomata, leaf water-content, and wilting of plants. *Ann. Bot.*, **36**, 361–83, 1922.

KNIGHT, T. A., On the direction of the radicle and germen during the vegetation of seeds. *Phil. Trans. Roy. Soc.*, 99–108, 1806.

KNOP, W., Quantitativ-analytische Arbeiten über den Ernährungsprocess der Pflanzen. *Landw. Versuchs-Stat.*, **3**, 295–324, 1861.

—— Quantitativ-analytische Arbeiten über den Ernährungsprocess der Pflanzen. *Landw. Versuchs-Stat.*, **4**, 173–87, 1862.

KNUDSON, L., Tannic acid fermentation. *Journ. Biol. Chem.*, **14**, 159–202, 1913.

—— Non-symbiotic germination of orchid seeds. *Bot. Gaz.*, **73**, 1–25, 1922.

—— Further observations on non-symbiotic germination of orchid seeds. *Bot. Gaz.*, **77**, 212–19, 1924.

—— Physiological study of the symbiotic germination of orchid seeds. *Bot. Gaz.*, **79**, 345–79, 1925.

—— Flower production by orchid grown non-symbiotically. *Bot. Gaz.*, **89**, 192–9, 1930.

KNY, L., Ueber den Einfluss des Lichtes auf das Wachsthum der Bodenwurzeln. *Jahrb. f. wiss. Bot.*, **38**, 421–46, 1903.

KOBEL, M., and SCHEUER, M., Über den Kohlenhydratumsatz im Tabakblatt. Nachweis von Methylglyoxal als Zwischenprodukt im Stoffwechsel grüner Blätter. *Biochem. Zeitschr.*, **216**, 216–23, 1929.

KÖGL, F., On plant growth hormones (Auxin *A* and auxin *B*). *Rep. Brit. Ass. Adv. Sci., Leicester Meeting*, 600–9, Publ. London, 1933.

KÖGL, F., and HAAGEN SMIT, A. J., Über die Chemie des Wuchsstoffs. *Kon. Akad. v. Wetenschappen te Amsterdam, Proc.*, **34**, 1411–16, 1933.

KOHL, F. G., *Die Transpiration der Pflanzen und ihre Einwirkung auf die Ausbildung pflanzlicher Gewebe.* Braunsweig, 1886.

KOSTYCHEV, S. (Kostytschew, S.), Der Einfluss des Substrates auf die anaërobe Athmung der Schimmelpilze. *Ber. deut. bot. Ges.*, **20**, 327–34, 1902.

—— Über die normale und die anaërobe Athmung bei Abwesenheit von Zucker. *Jahrb. f. wiss. Bot.*, **40**, 563–92, 1904.

—— Über die Alkoholgärung von *Aspergillus niger.* *Ber. deut. bot. Ges.*, **25**, 44–50, 1907.

KOSTYCHEV, S., TSCHESNOKOV, W., and BAZYRINA, K., Untersuchungen über den Tagesverlauf der Photosynthese an der Küste des Eismeeres. *Planta*, **11**, 160–8, 1930.

KRABBE, G., Zur Kenntnis der fixen Lichtlage der Laubblätter. *Jahrb. f. wiss. Bot.*, **20**, 211–60, 1889.

KRASNOSSELSKY-MAXIMOV, T. A., Daily variations in the water-content of leaves (in Russian). *Trav. Jard. Bot. Tiflis*, **19**, 1–22, 1917. (Cited by Maximov and Krasnosselsky-Maximov, 1924.)

KRAUS, E. J., and KRAYBILL, H. R., Vegetation and reproduction, with special reference to the tomato. *Oregon Agric. Coll. Exper. Sta. Bull*, **149**, 1918.

KRAUS, G., Ueber die Wasservertheilung in der Pflanze. I. *Festsch. Naturf. Ges. Halle*, 71 pp., 1879.

—— Ueber die Wasservertheilung in der Pflanze. II. Der Zellsaft une seine Inhalte. *Abhandl. Naturf. Ges. Halle*, **15**, 72 pp., 1880.

—— Ueber die Bluthenwarme bei *Arum italicum.* *Abhandl. Naturf. Ges. zu Halle*, **16**, 1882.

—— Über die Wasservertheilung in der Pflanze. IV. Die Acidität des Zellsaftes. *Abhandl. Naturf. Ges. zu Halle*, **16**, 66 pp., 1884.

KRAYBILL, H. R., Plant metabolism studies as an aid in determining fertiliser requirements. I. *Ind. Eng. Chem.*, **22**, 275–6, 1930.

KRONES, F. E., Einfluss des Lichtes auf den Geotonus. *Sitzungsber. k. Akad. Wiss. in Wien*, **123**, Abt. 1, 801–35, 1914.

KÜHNE, W., *Untersuchungen über das Protoplasma*, Leipzig, 1864.

—— Erfahrungen und Bemerkungen über Enzyme und Fermente. *Unters. a. d. physiol. Inst. Heidelberg*, **1**, 291–326, 1878.

KÜMMER, H., Fett und Fettsäuregehalt bei Gramineensamen in Beziehung zur Lichtbedürftigkeit bei der Keimung. *Ber. deut. bot. Ges.*, **50**, 300–3, 1932.

KÜSTER, E., Über die Aufnahme von Anilinfarben in lebende Pflanzenzellen. *Jarhb. f. wiss. Bot.*, **50**, 261–88, 1911.

LAIDLAW, C. G. P., and KNIGHT, R. C., A description of a recording porometer and a note on stomatal behaviour during wilting. *Ann. Bot.*, **30**, 47–56, 1916.

LAINE, T., On the absorption of electrolytes by the cut roots of plants and the chemistry of plant exudation sap. *Acta Bot. Fennica*, **16**, 64 pp., 1934.

LAMARCK, J. B. A. P. M. DE, *Flore Française.* 3 vols., Paris, 1778.

LANGE, S., Die Verteilung der Lichtempfindlichkeit in der Spitze der Haferkoleoptile. *Jahrb. f. wiss. Bot.*, **67**. 1–51, 1927.

LATSHAW, W. L., and MILLER, E. C., Elemental composition of the corn plant. *Journ. Agric. Res.*, **27**, 845–60, 1924.

LAWES, J. B., GILBERT, J. H., and PUGH, E., On the sources of the nitrogen of vegetation; with special reference to the question whether plants assimilate free or uncombined nitrogen. *Phil. Trans. Roy. Soc.*, **151**, 431–577, 1861.

LEACH, W., and DENT, K. W., Researches on plant respiration. III. The relationship between the respiration in air and in nitrogen of certain seeds during germination. (*a*) Seeds in which fats constitute the chief food reserve. *Proc. Roy. Soc.*, B, **116**, 150–69, 1934.

LECLERC DU SABLON, Sur la germination du Ricin. *Comp. rend. acad. sci.*, **117**, 524–7, 1893.

LECLERC DU SABLON, Recherches sur la germination des graines oléagineuses. *Rev. gén. bot.*, **7**, 145–65, 205–15, 258–69, 1895.

—— Sur la formation des réserves non azotées de la noix et de l'amande. *Comp. rend. acad. sci.*, **123**, 1084–6, 1896.

LECOMTE, H., Contribution à l'étude du liber des Angiospermes. *Ann. sci. nat., Bot.*, Sér. 7, **10**, 193–324, 1889.

LEE, E., The morphology of leaf-fall. *Ann. Bot.*, **25**, 52–106, 1911.

LEHMANN, E., Zur Keimungsphysiologie und -Biologie von *Ranunculus sceleratus* L. und einigen anderen Samen. *Ber. deut. bot. Ges.*, **27**, 476–94, 1909.

—— Über die Beeinflussung der Keimung lichtempfindlicher Samen durch die Temperatur. *Zeitschr. f. Bot.*, **4**, 465–529, 1912.

—— Ueber die minimale Belichtungskeit welche die Keimung der Samen von Lythrum Salicaria auslöst. *Ber. deut. bot. Ges.*, **35**, 157–62, 1918.

—— Über die keimfördernde Wirkung von Nitrat auf lichtgehemmte Samen von *Veronica Tournfortii. Zeitschr. f. Bot.*, **11**, 161–79, 1919.

LEHMANN, E., and OTTENWÄLDER, A., Über katalytische Wirkung des Lichtes bei den Keimung lichtempfindlicher Samen. *Zeitschr. f. Bot.*, **5**, 337–64, 1913.

LEITCH, I., Some experiments on the influence of temperature on the rate of growth in *Pisum sativum. Ann. Bot.*, **30**, 25–46, 1916.

LEPESCHKIN, W. W., Über die chemische Zusammensetzung des Protoplasmas des Plasmodiums. *Ber. deut. bot. Ges.*, **41**, 179–87, 1923.

—— *Kolloidchemie des Protoplasmas.* Berlin, 1924.

—— My opinion about protoplasm. *Protoplasma*, **9**, 269–330, 1930.

LIDFORSS, B., Zur Biologie des Pollens. *Jahrb. f. wiss. Bot.*, **29**, 1–138, 1896.

—— Weitere Beiträge zur Biologie des Pollens. *Jahrb. f. wiss. Bot.*, **33**, 232–312, 1898.

—— Ueber den Chemotropismus der Pollenschläuche. (Vorläufige Mitteilung.) *Ber. deut. bot. Ges.*, **17**, 236–42, 1899.

—— Untersuchungen über die Reizbewegungen der Pollenschläuche. *Zeitschr. f. Bot.*, **1**, 443–96, 1909.

LILIENFELD, M., Über den Chemotropismus der Wurzel. *Beih. z. bot. Centr.*, Abt. 1, **19**, 131–212, 1906.

LINSBAUER, K., Zur Kenntnis der Reizbarkeit der Centaurea-Filamente. *Sitzungsber. k. Akad. Wiss. Wien*, Abt. 1, *Math. nat. Kl.*, **114**, 809–22, 1905.

—— Zur Kenntnis der Reizbarkeit der Centaurea-Filamente. *Sitzungsber. k. Akad. Wiss. Wien*, Abt. 1, *Math. nat. Kl.*, **115**, 1741–56, 1906.

LIPMAN, C. B., and MACKINNEY, G., Proof of the essential nature of copper for higher green plants. *Plant. Physiol.*, **6**, 593–9, 1931.

LIVINGSTON, B. E., Physiological properties of bog water. *Bot. Gaz.*, **39**, 348–55, 1905.

—— The relation of desert plants to soil moisture and to evaporation (Publ. Carnegie Inst. No. 50), Washington, 1906.

LIVINGSTON, B. E., and BROWN, W. H., Relation of the daily march of transpiration to variations in the water content of foliage leaves. *Bot. Gaz.*, **53**, 309–30, 1912.

LIVINGSTON, B. E., and FREE, E., The effect of deficient soil oxygen on the roots of higher plants. *Johns Hopkins Univ. Circ.*, **293**, 182–5, 1917.

LLOYD, F. E., *The Physiology of Stomata*, Washington, 1908.

—— The types of entrance mechanisms of the traps of Utricularia (including Polypompholyx). Presidential Address to Section K, Botany, *Rep. Brit. Ass., Leicester Meeting*, pp. 183–218, 1933.

—— The carnivorous plants—a review with contributions. *Trans. Roy. Soc. Canada*, 3rd Ser., **27**, Appendix A, 1–67, 1933.

—— Is *Roridula* a carnivorous plant? *Canadian Journ. Res.*, **10**, 780–6, 1934.

LOEB, J., The physiological basis of morphological polarity in regeneration. I. *Journ. Gen. Physiol.*, **1**, 337–62, 1919.

LOEB, J., The physiological basis of morphological polarity in regeneration. II. *Journ. Gen. Physiol.*, **1**, 687–715, 1919.

—— Quantitative laws in regeneration. I. *Journ. Gen. Physiol.*, **2**, 297–307, 1920.

—— The nature of the directive influence of gravity on the arrangements of organs in regeneration. *Journ. Gen. Physiol.*, **2**, 373–86, 1920.

—— Quantitative laws in regeneration. III. The quantitative basis of polarity in regeneration. *Journ. Gen. Physiol.*, **4**, 447–61, 1922.

LOEW, O., Ueber die physiologische Functionen der Calcium und Magnesiumsalze im Pflanzenorganismus. *Flora*, **75**, 368–94, 1892.

LOFTFIELD, J. V. G., *The Behaviour of Stomata*, Washington, 1921.

LUNDEGÅRDH, H., Das geotropische Verhalten der Seitensprosse. *Lunds. Univ. Årskr.*, **14**, No. 27, 1917.

—— Ecological studies in the assimilation of certain forest-plants and shore-plants. *Svensk Bot. Tidsk.*, **15**, 46–95, 1921.

—— Die Kohlensäureassimilation der Zuckerrübe. *Flora*, **121** (N.F. 21), 273–300, 1927.

LUNDSGAARD, E., Die Monojodessigsäurewirkung auf die enzymatische Kohlenhydratspaltung. *Biochem. Zeitschr.*, **220**, 1–7, 1930.

LYON, C. J., The effect of phosphates on respiration. *Journ. Gen. Physiol.*, **6**, 299–306, 1924.

MᶜCLENAHAN, F. N., The development of fat in the black walnut (*Juglans nigra*). *Journ. Amer. Chem. Soc.*, **31**, 1093–8, 1909.

—— The development of fat in the black walnut, II. *Journ. Amer. Chem. Soc.*, **35**, 485–93, 1913.

MACDOUGAL, D. T., The mechanism of movement and transmission of impulses in *Mimosa* and other ' sensitive ' plants : a review with some additional experiments. *Bot. Gaz.*, **22**, 293–300, 1896.

MCDOUGALL, W. B., Symbiotic saprophytism. *Ann. Bot.*, **13**, 1–47, 1899.

MACFARLANE, J. M., Contributions to the history of *Dionaea muscipula*, Ellis. *Cont. Bot. Lab. Univ. Pennsylvania*, **1**, 7–44, 1902.

MACGILLIVRAY, J. H., Effect of phosphorus on the composition of the tomato plant. *Journ. Agric. Res.*, **34**, 97–127, 1927.

MCHARGUE, J. S., The effect of manganese on the growth of wheat ; a source of manganese for agricultural purposes. *Journ. Ind. Eng. Chem.*, **11**, 332–5, 1919.

—— Effect of different concentrations of manganese sulphate on the growth of plants in acid and neutral soils and the necessity of manganese as a plant nutrient. *Journ. Agric. Res.*, **24**, 781–94, 1923.

MCHARGUE and CALFEE, R. K., Effect of manganese, copper and zinc on growth and metabolism of *Aspergillus flavus* and *Rhizopus nigricans*. *Bot. Gaz.*, **91**, 183–93, 1931.

MACK, W. B., The relation of temperature and the partial pressure of oxygen to respiration and growth in germinating wheat. *Plant Physiol.*, **5**, 1–68, 1930.

MCLENNAN, E., The endophytic fungus of *Lolium*, II. The mycorrhiza on the roots of *Lolium temulentum* L., with a discussion on the physiological relationships of the organism concerned. *Ann. Bot.*, **40**, 43–68, 1926.

MAGNESS, J. R., Investigations in the ripening and storage of Bartlett pears. *Journ. Agric. Res.*, **19**, 473–500, 1920.

MAGNUS, W., Studien an der endotrophen Mycorrhiza von *Neottia Nidus avis* L. *Jahrb. f. wiss. Bot.*, **35**, 205–72, 1900.

MAIGE, A., and NICOLAS, G., Recherches sur l'influence des solutions sucrées de divers degrès de concentration sur la respiration, la turgescence et la croissance de la cellule. *Ann. sci. nat., Bot.*, Sér. 9, **12**, 315–68, 1910.

MAIGE, G., Recherches sur la respiration des différentes pièces florales. *Ann. sci. nat., Bot.*, Sér. 9, **14**, 1–62, 1911.

MANGHAM, S., The translocation of carbohydrates in plants. *Sci. Prog.*, **5**, 256–85, 1910; 457–79, 1911.

—— On the detection of sugars in the tissues of certain Angiosperms. *New Phyt.*, **10**, 160–6, 1911.

MANN, C. E. T., The antagonism between dyes and inorganic salts in their absorption by storage tissue. *Ann. Bot.*, **38**, 753–77, 1924.

MAQUENNE, L., and DEMOUSSY, E., Sur la valeur des coefficients chlorophylliens et leur rapports avec les quotients respiratoires réels. *Comp. rend. acad. sci.*, **156**, 505–12, 1913.

MASKELL, E. J., Experimental researches on vegetable assimilation and respiration. XVII. The diurnal rhythm of assimilation in leaves of cherry laurel at 'limiting' concentrations of carbon dioxide. *Proc. Roy. Soc.*, B, **102**, 467–87, 1928.

—— Experimental researches on vegetable assimilation and respiration. XVIII. The relation between stomatal opening and assimilation.—A critical study of assimilation rates and porometer rates in leaves of cherry laurel. *Proc. Roy. Soc.*, B, **102**, 488–533, 1928.

MASKELL, E. J., and MASON, T. G., Studies on the transport of nitrogenous substances in the cotton plant. I. Preliminary observations on the downward transport of nitrogen in the stem. *Ann. Bot.*, **43**, 205–31, 1929.

—— Studies on the transport of nitrogenous substances in the cotton plant. II. Observations on concentration gradients. *Ann. Bot.*, **43**, 615–52, 1929.

—— Studies on the transport of nitrogenous substances in the cotton plant. III. The relation between longitudinal movement and concentration gradients in the bark. *Ann. Bot.*, **44**, 1–29, 1930.

—— Studies on the transport of nitrogenous substances in the cotton plant. IV. The interpretation of the effects of ringing, with special reference to the lability of the nitrogen compounds of the bark. *Ann. Bot.*, **44**, 233–67, 1930.

—— Studies on the transport of nitrogenous substances in the cotton plant. V. Movement to the boll. *Ann. Bot.*, **44**, 657–88, 1930.

MASON, T. G., and MASKELL, E. J., Studies on the transport of carbohydrates in the cotton plant. I. A study of diurnal variation in the carbohydrates of the leaf, bark and wood, and of the effects of ringing. *Ann. Bot.*, **42**, 1–65, 1928.

—— Studies on the transport of carbohydrates on the cotton plant. II. The factors determining the rate and the direction of movement of sugars. *Ann. Bot.*, **42**, 571–636, 1928.

—— Further studies on transport in the cotton plant. I. Preliminary observations on the transport of phosphorus, potassium and calcium. *Ann. Bot.*, **45**, 125–73, 1931.

—— Further studies on transport in the cotton plant. II. An ontogenetic study of concentrations and vertical gradients. *Ann. Bot.*, **48**, 119–41, 1934.

MASON, T. G., and PHILLIS, E., Studies on the transport of nitrogenous substances in the cotton plant. VI. Concerning storage in the bark. *Ann. Bot.*, **48**, 315–33, 1934.

MATTHAEI, GABRIELLE, L. C., Experimental researches on vegetable assimilation and respiration. III. On the effect of temperature on carbon dioxide assimilation. *Phil. Trans. Roy. Soc.*, B, **197**, 47–105, 1904.

MAXIMOV, N. A., The physiological significance of the xeromorphic structure of plants. *Journ. Ecol.*, **19**, 273–82, 1931.

MAXIMOV, N. A., and KRASNOSSELSKY-MAXIMOV, T. A., Wilting of plants in its connection with drought resistance. *Journ. Ecol.*, **12**, 95–110, 1924.

MAZÉ, P., Influences respectives des éléments de la solution minérale sur le développement du maïs. *Ann. Inst. Pasteur*, **28**, 1–48, 1914.

—— Détermination des éléments minéraux rares nécessaires au développement du maïs. *Comp. rend. acad. sci.*, **160**, 211–14, 1915.

—— Recherche d'un solution purement minérale capable d'assurer l'évolution complète du maïs cultivé à l'abri des microbes. *Ann. Inst. Pasteur*, **33**, 139–73, 1919.

MEISSNER, R., Beiträge zur Kenntnis der Assimilationstätigkeit der Blätter. *Diss., Bonn*, 1894 (Abst. in *Bot. Centralbl.*, **60**, 206–7, 1894).

MEURER, R., Über die regulatorische Aufnahme anorganischen Stoffe durch die Wurzeln von *Beta vulgaris* und *Daucus Carota*. *Jahrb. f. wiss. Bot.*, **46**, 503–67, 1909.

MEYER, A., Ueber die Assimilationsproducte der Laublätter angiospermer Pflanzen. *Bot. Zeit.*, **43**, 417–23, 433–40, 449–57, 465–72, 481–91, 497–507, 1885.

MEYER, K. H., Die Chemie der Micelle und ihre Anwendung auf biochemische und biologische Probleme. *Biochem. Zeitschr.*, **208**, 1–31, 1929.

MEYERHOF, O., Über den Einfluss des Sauerstoffs auf die alkoholische Gärung der Hefe. *Biochem. Zeitschr.*, **162**, 43–86, 1925.

MEYERHOF, O., and KIESSLING, W., Über die phosphorylierten Zwischenprodukte und die letzten Phasen der alkoholischen Gärung. *Biochem. Zeitschr.*, **267**, 313–48, 1933.

MEYERHOF, O., LOHMANN, K., and MEIER, R., Über die Synthese des Kohlenhydrate im Muskel. *Biochem. Zeitschr.*, **157**, 459–91, 1925.

MICHAELIS, L., Die Permeabilität von Membranen. *Naturwissensch.* **14**, 33–42, 1926.

MICHAELIS, L., and MENTEN, M. L., Die Kinetik der Invertinwirkung. *Biochem. Zeitschr.*, **49**, 333–69, 1913.

MINDEN, M. v., Reizbare Griffel von zwei Arctotis-Arten. *Flora*, **88**, 238–42, 1901.

MILLARDET, A., Note sur la fausse hybridation chez les ampélidées. *Rev. de Viticulture*, **16**, 677–80, 1901.

MIYAKE, K., Über das Wachstum des Blutenschaftes von *Taraxacum*. *Beih. z. bot. Centr.*, **16**, 403–14, 1904.

MIYOSKI, M., Ueber Chemotropismus der Pilze. *Bot. Zeit.*, **52**, Abt. 1, 1–28, 1894.

—— Ueber Reizbewegungen der Pollenschläuche. *Flora*, **78**, 67–93, 1894.

MOLISCH, H., Zur Physiologie des Pollens mit besonderer Ruchsicht auf die chemotropischen Bewegungen der Pollenschläuche. *Sitzungsber. k. Akad. Wiss. Wien*, Abt. 1, **102**, 423–50, 1893.

—— *Das Warmbad als Mittel zum Treiben der Pflanzen*, Jena, 1909.

—— *Die Eisenbakterien*, Jena, 1910.

—— Ueber den Einfluss der Tabaksrauche auf die Pflanze, Tiel I. *Sitzungsber. Akad. Wiss. Wien*, Abt. 1, *Math.-nat. Kl.*, **120**, 3–30, 1911.

—— Ueber den Einfluss der Tabaksrauche auf die Pflanze, Tiel II. *Sitsungsber. Akad. Wiss. Wien*, Abt. 1, *Math.-nat. Kl.*, **120**, 813–38, 1911.

—— Das Offen- und Geschlossensein der Spaltöffnungen, veranschaulicht durch eine neue Methode (Infiltrations methode). *Zeitschr. f. Bot.*, **4**, 106–22, 1912.

—— Über das Treiben von Pflanzen mittels Radium. *Sitzungsber. k. Akad. Wiss. Wien*, **121**, Abt. 1, *Math.-nat. Kl.*, 121–38, 1912.

—— Über das Treiben ruhender Pflanzen mit Rauch. *Sitzungsber. k. Akad. Wiss. Wien*, Abt. 1, *Math.-nat. Kl.*, **125**, 141–62, 1916.

MONTFORT, C., Die aktive Wurzelsaugung aus Hochmoorwasser im Laboratorium und am Standort und die Frage seiner Giftwirkung. *Jahrb. f. wiss. Bot.*, **60**, 184–256, 1921.

MORINAGA, T., Germination of seeds under water. *Amer. Journ. Bot.*, **13**, 126–40, 1926.

—— Effect of alternating temperatures upon the germination of seeds. *Amer. Journ. Bot.*, **13**, 141–58, 1926.

—— The favourable effect of reduced oxygen supply upon the germination of certain seeds. *Amer. Journ. Bot.*, **13**, 159–66, 1926.

MUENSCHER, W. L. C., A study of the relation of transpiration to the size and number of stomata. *Amer. Journ. Bot.*, **2**, 487–504, 1915.

MÜLLER, D., Ein neues Enzym—Glycoseoxydase—aus *Aspergillus niger*. *Den Kgl. Veterinaer- og Landbohøjskole Aarskrift*, 329–31, 1925.

—— Studien über ein neues Enzym Glykoseoxydase. I. *Biochem. Zeitschr.*, **199**, 136–70, 1928.

MÜLLER, N. J. C., Untersuchungen über die Krümmungen der Pflanzen gegen das Sonnenlicht. *Bot. Unters. Heidelberg*, **1**, 57–83, 1872.

MÜLLER, O., Die Ortsbewegung der Bacillariaceen. V. *Ber. deut. bot. Ges.*, **15**, 70–86, 1897.

—— Kammern und Poren in der Zellwand der Bacillariaceen. II. *Ber. deut. bot. Ges.*, **17**, 423–52, 1899.

MÜLLER-THURGAU, H., Beitrag zur Erklärung der Ruheperioden der Pflanzen. *Landw. Jahrb.*, **14**, 851–907, 1885.

MÜNCH, E., *Die Stoffbewegungen in der Pflanze.* Jena, 1930.

MURNEEK, A. E., The nature of shedding of immature apples. *Univ. Missouri Coll. Agric., Agric. Exper. Stat., Res. Bull.*, **201**, 34 pp., 1933.

NAGAI, I., Some studies on the germination of the seed of *Oryza sativa*. *Journ Coll. Agric. Imp. Univ. Tokio*, **3**, 109–58, 1916.

NÄGELI, C., Ueber die Siebrohren von *Cucurbita*. *Sitzungsber. k. bayr. Akad. Wiss. München*, 212–38, 1861.

—— Ueber die inneren Bau der vegetabilischen Zellmembranen. *Sitzungsber. Bayer. Akad. Wiss. München*, **1**, 282–323 ; **2**, 114–71, 1864.

NANJI, D. R., PATON, F. J., and LING, A. R., Decarboxylation of polysaccharide acids : its application to the establishment of the constitution of pectins and their determination. *Journ. Soc. Chem. Ind., Trans.* **44**, 253–8, 1925.

NATHANSOHN, A., and PRINGSHEIM, E., Über die Summation intermittierender Lichtreize. *Jahrb. f. wiss. Bot.*, **45**, 137–90, 1907.

NAVEZ, A. E., Galvanotropism of roots. *Proc. Soc. Exp. Biol. Med.*, **24**, 341–3, 1927.

—— Respiration and geotropism in *Vicia faba*. I. *Journ. Gen. Physiol.*, **12**, 641–67, 1929.

NĚMEC, B., Die Perception des Schwerkraftreizes bei den Pflanzen. *Ber. deut. bot. Ges.*, **20**, 339–54, 1902.

—— Die Symmetrieverhältnisse und Wachstumsrichtungen einiger Laubmoose. *Jahrb. f. wiss. Bot.*, **43**, 501–79, 1906.

—— Ueber die Art der Wahrnehmung des Schwerkraftreizes den Pflanzen. *Ber. deut. bot. Ges.*, **18**, 241–5, 1900.

NEUBERG, C., Von der Chemie der Gärungserscheinungen. *Ber. deut. chem. Ges.*, **55**, 3624–8, 1922.

NEUBERG, C., and COHEN, CLARA, Über die Bildung von Acetaldehyd und die Verwirklichung der zweiten Vergärungsform bei verschiedenen Pilzen. *Biochem. Zeitschr.*, **122**, 204–24, 1921.

NEUBERG, C., and GORR, G., Über die Mechanismus der Milchsäurebildung bei Phanerogamen. *Biochem. Zeitschr.*, **171**, 475–84, 1926.

NEUBERG, C., and GOTTSCHALK, A., Beobachtungen über den Verlauf der anaeroben Pflanzenatmung. *Biochem. Zeitschr.*, **151**, 167–8, 1924.

—— Über den Nachweis von Acetaldehyd als der anaeroben Atmung höherer Pflanzen. *Biochem. Zeitschr.*, **160**, 256–60, 1925.

NEUBERG, C., and REINFURTH, ELSA, Die Festlegung der Aldehydstufe bei der alkoholischen Gärung. Ein experimenteller Beweis der Acetaldehyd-Brenztraubensäuretheorie. *Biochem. Zeitschr.*, **89**, 365–414, 1918.

NEUKIRCHEN, J., Über die Beeinflussung der tropischen Reizverhaltens von Gramineenkoleoptilen durch chemische Vorbehandlung des Saatguts. *Planta*, **12**, 505–31, 1929.

NEWCOMBE, F. C., The rheotropism of roots. *Bot. Gaz.*, **33**, 177–98, 1902.

—— Gravitation sensitiveness not confined to apex of root. *Beih. bot. Centr.* Abt. 1, **24**, 96–110, 1909.

NEWCOMBE, F. C., and RHODES, ANNA L., Chemotropism of roots. *Bot. Gaz.*, **37**, 23–35, 1904.

NICOLAS, M. G., Anthocyane et échanges respiratoires des feuilles. *Comp. rend. acad. sci.*, **167**, 130–3, 1918.

NIELSEN, N., Untersuchungen über einen neue Wachstumsregulierenden Stoff: Rhizopin. *Jahrb. f. wiss. Bot.*, **73**, 125–91, 1930.

NIETHAMMER, A., Die Beeinflussung der Pollenkeimung unserer Nutz und Ziergewächse durch die verschiedensten Giftstoffe, die im Pflanzenschutzdienste angewendet werden. *Gartenbauwiss.*, **1**, 471–87, 1929.

NIGHTINGALE, G. T., Light in relation to growth and chemical composition of some horticultural plants. *Proc. Amer. Soc. Hort. Sci.*, 18–29, 1922.

—— Ammonium and nitrate nutrition of dormant Delicious apple trees at 48° F. *Bot. Gaz.*, **95**, 437–52, 1934.

NIGHTINGALE, G. T., ADDOMS, R. M., ROBBINS, W. R., and SCHERMERHORN, L. G., Effects of calcium deficiency on nitrate absorption and on metabolism in tomato. *Plant Physiol.*, **6**, 605–30, 1931.

NIGHTINGALE, G. T., SCHERMERHORN, L. G., and ROBBINS, W. R., Some effects of potassium deficiency on the histological constituents of plants. *New Jersey Agric. Exp. Sta., Bull.* **499**, 36 pp., 1930.

—— Effect of sulphur deficiency on metabolism in tomato. *Plant Physiol.*, **4**, 565–95, 1932.

NIUS, E., Untersuchungen über den Einfluss des Interzellularvolumens und der Öffnungsweite der Stomata auf die Luftwegigkeit der Laubblätter. *Jahrb. f. wiss. Bot.*, **74**, 33–126, 1931.

NOACK, K., Die Bedeutung der schiefen Lichtrichtung für die Helioperzeption parallelotropen Organe. *Zeitschr. f. Bot.*, **6**, 1–79, 1914.

NOBBE, F., *Handbuch der Samenkunde.* Berlin, 1876.

NOLL, F., Die Orientirungsbewegungen dorsiventraler Organe. *Flora*, **76**, 265–89, 1892.

—— Ueber Geotropismus. *Jahrb. f. wiss. Bot.*, **34**, 457–506, 1900.

NORDHAUSEN, M., Über die Bedeutung der papillösen Epidermis als Organ für die Lichtperception des Laubblattes. *Ber. deut. bot. Ges.*, **25**, 398–410, 1907.

—— Blattepidermis und Lichtperception. *Zeitschr. f. Bot.*, **9**, 501–6, 1917.

NOWINSKI, M., L'influence des conditions exterieurs sur l'amidonité du pollen des fleurs. *Bull. internat. acad. Polon. sci. et let.*, 215–49, 1928.

OHGA, I., The germination of century-old and recently harvested Indian lotus fruits. *Amer. Journ. Bot.*, **13**, 754–9, 1926.

—— A comparison of the life activity of century-old and recently harvested Indian lotus fruits. *Amer. Journ. Bot.*, **13**, 760–5, 1926.

—— A report on the longevity of the fruit of *Nelumbium. Journ. Bot.*, **64**, 154–7, 1926.

ÖHOLM, L. W., Über die Hydrodiffusion der Elektrolyte. *Zeitschr. f. physik. Chem.*, **50**, 309–49, 1905.

—— A new method for determining the diffusion of dissolved substances. *Meddel. från Vetenskapsakad. Nobelinstitut*, **2**, No. 22, 23 pp., 1912.

ÖHOLM, L. W., The diffusion of an electrolyte in gelatine. *Meddel. från Vetenskapsakad. Nobelinstitut*, **2**, No. 30, 8 pp., 1913.

OLIVER, F. W., On the sensitive labellum of *Masdevallia muscosa*, Rchb. f. *Ann. Bot.*, **1**, 237–53, 1888.

OLNEY, A. J., Temperature and respiration of ripening bananas. *Bot. Gaz.*, **82**, 415–26, 1926.

OLTMANNS, F., Ueber positiven und negativen Heliotropismus. *Flora*, **82**, 1–32, 1897.

ONSLOW, MURIEL W., Oxidising enzymes. I. The nature of the 'peroxide' naturally associated with certain direct oxidising systems in plants. *Biochem. Journ.*, **13**, 1–9, 1919.

—— Oxidising enzymes. II. The nature of the enzymes associated with certain direct oxidising systems in plants. *Biochem. Journ.*, **14**, 535–40, 1920.

—— Oxidising enzymes. IV. The distribution of oxidising enzymes among the higher plants. *Biochem. Journal.*, **15**, 107–12, 1921.

ONSLOW, MURIEL W., and ROBINSON, MURIEL E., Oxidising enzymes. IX. On the mechanism of plant oxidases. *Biochem. Journ.*, **20**, 1138–45, 1926.

OSBORNE, T. B., *The Proteins of the Wheat Kernel.* Washington (Carnegie Inst. Publ. No. 84), 1907.

—— *The Vegetable Proteins.* London, Second edition, 1924.

OSTERHOUT, W. J. V., On the importance of physiologically balanced solutions for plants. I. Marine plants. *Bot. Gaz.*, **42**, 127–34, 1906.

—— On nutrient and balanced solutions. *Univ. California Publ., Bot.*, **2**, 10 pp., 1906.

—— Extreme toxicity of sodium chloride and its prevention by other salts. *Journ. Biol. Chem.*, **1**, 363–9, 1906.

—— On the importance of physiologically balanced solutions for plants. II. Fresh water and terrestrial plants. *Bot. Gaz.*, **44**, 259–72, 1907.

—— The antagonistic action of magnesium and potassium. *Bot. Gaz.*, **45**, 117–24, 1908.

—— On plasmolysis. *Bot. Gaz.*, **46**, 53–5, 1908.

—— Die Schutzwirkung des Natriums für Pflanzen. *Jahrb. f. wiss. Bot.*, **46**, 121–36, 1908.

—— On the penetration of inorganic salts into living protoplasm. *Zeitschr. f. physik. Chem.*, **70**, 408–13, 1910.

—— The permeability of living cells in pure and balanced solutions. *Science*, N.S., **34**, 187–9, 1911.

—— Plants which require sodium. *Bot. Gaz.*, **54**, 532–6, 1912.

—— Decrease in permeability and antagonistic effects caused by bile salts. *Journ. Gen. Physiol.*, **1**, 405–8, 1919.

—— Antagonism between alkaloids and salts in relation to permeability. *Journ. Gen. Physiol.*, **1**, 515–19, 1919.

—— Some aspects of selective absorption. *Journ. Gen. Physiol.*, **5**, 225–30, 1922.

OSTERHOUT, W. J. V., and DORCAS, M. J., The penetration of CO_2 into living protoplasm. *Journ. Gen. Physiol.*, **9**, 255–67, 1925.

OSTERHOUT, W. J. V., and HAAS, A. R. C., On the dynamics of photosynthesis. *Journ. Gen. Physiol.*, **1**, 1–16, 1918.

—— The temperature coefficient of photosynthesis. *Journ. Gen. Physiol.*, **1**, 295–8, 1919.

OTTENWALDER, A., Lichtintensität und Substrat bei der Lichtkeimung. *Zeitschr. f. Bot.*, **6**, 785–848, 1914.

OTTO, R., and KOOPER, W. D., Beiträge zur Abnahme bezw. Rückwanderung der Stickstoffverbindungen aus den Blättern wahrend der Nacht, sowie zur herbstlichen Rückwanderung von Stickstoffverbindungen aus den Blättern. *Landw. Jahrb.*, **39**, 167–72, 1910.

OVERBEEK, J. VAN, An analysis of phototropism in dicotyledons. *K. Akad. Wetens. Amsterdam Proceedings*, **35**, 1325–35, 1932.

OVERTON, E., Über die osmotischen Eigenschaften der lebenden Pflanzen und Thierzelle. *Vierteljahrschr. d. Naturf.-Ges. in Zurich*, **40**, 159–201, 1895.

—— Über die allgemeinen osmotischen Eigenschaften der Zelle, ihre vermuthlichen Ursachen und ihre Bedeutung für die Physiologie. *Vierteljahrschr. d. Naturf.-Ges. in Zurich*, **44**, 88–135, 1899.

—— Studien über die Aufnahme der Anilinfarben durch die lebende Zelle. *Jahrb. f. wiss. Bot.*, **34**, 669–701, 1900.

PAÁL, Á., Analyse des geotropischen Reizvorgänge mittels Luftverdünnung. *Jahrb. f. wiss. Bot.*, **50**, 1–20, 1911.

—— Über phototropische Reizleitung. *Jahrb. f. wiss. Bot.*, **58**, 406–58, 1919.

PALLADIN, V. I., Recherches sur la correlation entre la respiration des plantes et les substances azotées actives. *Rev. gén. Bot.*, **8**, 225–48, 1896.

—— Die Verbeitung der Atmungschromogene bei den Pflanzen. *Ber. deut. bot. Ges.*, **26a**, 378–89, 1908.

—— Über die Bildung der Atmungschromogene in den Pflanzen. *Ber. deut. bot. Ges.*, **26a**, 389–94, 1908.

—— *Plant Physiology*, Second American Edition. Philadelphia, 1923.

PANTANELLI, E., Assorbimento elettivo di ioni nelle piante. *Bull. Orto Bot. R. Univ. Napoli*, **5**, 1–54, 1915.

PARKER, F. W., and TRUOG, E., The relation between the calcium and the nitrogen content of plants and the function of calcium. *Soil Sci.*, **10**, 49–56, 1920.

PARR, ROSALIE, The response of *Pilobus* to light. *Ann. Bot.*, **32**, 177–205, 1918.

PASTEUR, L., Mémoire sur la fermentation de l'acide tartrique. *Comp. rend. acad. sci.*, **46**, 615–18, 1858.

—— Animalcules infusoires vivant sans gaz oxygène libre et determinant des fermentations. *Comp. rend. acad. sci.*, **52**, 344–7, 1861.

PAYEN, A., and PERSOZ, J., Mémoire sur la diastase, les principaux produits de ses réactions et leurs applications aux arts industriels. *Ann. chim. et phys.*, **53**, 73–92, 1833.

PEIRCE, J. G., On the structure of the haustoria of some phanerogamic parasites. *Ann. Bot.*, **7**, 291–327, 1893.

—— The liberation of heat in respiration. *Bot. Gaz.*, **53**, 89–112, 1912.

PELLETIER, J., and CAVENTOU, J. B., Sur la matière verte des feuilles. *Ann. Chim. et Phys.*, Sér. 2, **9**, 194–6, 1818.

PFEFFER, W., *Physiologische Untersuchungen.* Leipzig, 1873.

—— *Die periodischen Bewegungen der Blattorgane.* Leipzig, 1875.

—— *Osmotische Untersuchungen : Studien zur Zellmechanik.* Leipzig, 1877.

—— Locomotorische Richtungsbewegungen durch Chemischereize. *Unters. bot. Inst. Tübingen*, **1**, 363–482, 1884.

—— Zur Kenntnis der Kontaktreize. *Unters. bot. Inst. Tübingen*, **1**, 483–535, 1885.

—— Über Aufnahme von Anilinfarben in lebende Zellen. *Unters. a. d. bot. Inst. Tübingen*, **2**, 179–332, 1886.

—— Über chemotaktische Bewegungen von Bakterien, Flagellaten und Volvocineen. *Unters. bot. Inst., Tübingen*, **2**, 582–662, 1888.

—— Zur Kenntnis der Plasmahaut und der Vacuolen. *Abh. Sächs. Ges. Wiss.*, **16**, 187–343, 1890.

PFLÜGER, E., Beiträge zur Lehre von der Respiration. I. Ueber die physiologische Verbrennung in den lebendigen Organismen. *Arch. f. d. ges. Physiol.*, **10**, 251–367, 1875.

PHILLIS, E., and MASON, T. G., Studies on the transport of carbohydrates in the cotton plant. III. The polar distribution of sugar in the foliage leaf. *Ann. Bot.*, **47**, 585–634, 1933.

PICCARD, A., Neue Versuche über die geotropische Sensibilität der Wurzelspitze. *Jahrb. f. wiss. Bot.*, **40**, 94–102, 1904.

PISEK, A., and CARTELLIERI, E., Zur Kenntnis des Wasserhaushaltes des Pflanzen. II. Schattenpflanzen. *Jahrb. f. wiss. Bot.*, **75**, 643–78, 1932.

PLAETZER, HILDA, Untersuchungen über die Assimilation und Atmung von Wasserpflanzen. *Verhandl. phys. med. Ges., Wurzburg*, **45**, 31–102, 1917.

POLLARD, A., CHIBNALL, A. C., and PIPER, S. H., The wax constituents of forage grasses. I. Cocksfoot and perennial ryegrass. *Biochem. Journ.*, **25**, 2111–22, 1931.

—— The isolation of n-octacosanol from wheat wax. *Biochem. Journ.*, **27**, 1889–93, 1933.

POPOVICI, A. P., Der Einfluss der Vegetationsbedingungen auf Länge der wachsenden Zone. *Bot. Centrabl.*, **81**, 33–40, 87–97, 1900.

PORODKO, T. M., Über den Chemotropismus der Pflanzenwurzeln. *Jahrb. f. wiss. Bot.*, **49**, 307–88, 1911.

—— Untersuchungen über den Chemotropismus der Pflanzenwurzeln. *Jahrb. f. wiss. Bot.*, **64**, 450–508, 1925.

PRANKERD, T. L., Preliminary observations on the nature and distribution of the statolith apparatus in plants. *Rep. Brit. Ass.*, Manchester Meeting, 1915, p. 722, Publ. London, 1916.

—— On the irritability of the fronds of *Asplenium bulbiferum*, with special reference to graviperception. *Proc. Roy. Soc.*, B, **93**, 143–52, 1922.

—— The ontogeny of graviperception in *Osmunda regalis*. *Ann. Bot.*, **39**, 709–20, 1925.

—— Studies in the geotropism of Pteridophyta. IV. On specificity in graviperception. *Journ. Linn. Soc., Bot.*, **48**, 317–36, 1929.

—— Studies in the geotropism of the Pteridophyta. V. Some effects of temperature on growth and geotropism in *Asplenium bulbiferum*. *Proc. Roy. Soc.*, B, **116**, 479–93, 1935.

PRIANISCHNIKOW, D., Eiweisserfall und Eiweissrückbildung in den Pflanzen. *Ber. deut. bot. Ges.*, **17**, 151–5, 1899.

—— Zur Frage der Asparaginbildung. *Ber. deut. bot. Ges.*, **2**[illegible], 35–43, 1904.

—— Über den Aufbau und Abbau des Asparagins in den Pflanzen. *Ber. deut. bot. Ges.*, **40**, 242–8, 1922.

—— Sur le rôle de l'asparagine dans le transformations des matières azotées chez les plantes. *Rev. gén. bot.*, **36**, 108–22, 1924.

PRICE, S. R., Some studies on the structure of the plant cell by the method of dark ground illumination. *Ann. Bot.*, **28**, 601–32, 1914.

PRIESTLEY, J., Observations on different kinds of air. *Phil. Trans. Roy. Soc., London*, **62**, 147–264, 1772.

—— Experiments and observations on different kinds of air. Vols. I–IV. London, 1774–9.

PRIESTLEY, J. H., Light and growth, I. The effect of brief light exposure upon etiolated plants. *New Phyt.*, **24**, 271–83, 1925.

—— Light and growth, II. On the anatomy of etiolated plants. *New Phyt.*, **25**, 145–70, 1926.

PRIESTLEY, J. H., and ARMSTEAD, DOROTHY, Physiological studies in plant anatomy. II. The physiological relation of the surrounding tissue to the xylem and its contents. *New Phyt.*, **24**, 62–80, 1922.

PRIESTLEY, J. H., and EWING, J., Physiological studies in plant anatomy. VI. Etiolation. *New Phyt.*, **22**, 30–44, 1923.

PRINGSHEIM, E., Einfluss der Beleuchtung auf die heliotropische Stimmung. *Beitr. z. Biol. Pflanzen*, **9**, 263–306, 1909.

PRINGSHEIM, E., *Die Reizbewegungen der Pflanzen.* Berlin, 1912.

RAISTRICK, H., and Collaborators, Studies in the Biochemistry of micro-organisms. *Phil. Trans. Roy. Soc.*, B, **220**, 1–367, 1931.

RAMANN, E., and GOSSNER, B., Aschenanalysen der Esche (*Fraxinus excelsior* L.). *Landw. Versuchs-Stat.*, **76**, 117–24, 1912.

RAMSDEN, W., Separation of solids in the surface layers of solutions and suspensions (Observations on surface-membranes, bubbles, emulsions and mechanical coagulation). Preliminary account. *Proc. Roy. Soc.*, **72**, 156–64, 1903.

RAWITSCHER, F., Bewegungstudien an *Asparagus plumosus. Zeitschr. f. Bot.*, **23**, 537–69, 1930.

RAYNER, M. C., The nutrition of mycorrhiza plants : *Calluna vulgaris. Brit. Journ. Exper. Biol.*, **2**, 265–92,

REDFERN, GLADYS M., On the absorption of ions by the roots of living plants. I. The absorption of the ions of calcium chloride by pea and maize. *Ann. Bot.*, **36**, 167–74, 1922.

—— On the course of absorption and the position of equilibrium in the intake of dyes by discs of plant tissue. *Ann. Bot.*, **36**, 511–22, 1922.

REED, H. S., The value of certain nutritive elements to the plant cell. *Ann. Bot.*, **21**, 501–43, 1907.

—— The nature of the growth rate. *Journ. Gen. Physiol.*, **2**, 545–61, 1920.

REED, H. S., and HOLLAND, R. H., The growth rate of an annual plant *Helianthus. Proc. Nat. Acad. Sci.*, **5**, 135–44, 1919.

REINHARDT, M. O., Das Wachsthum der Pilzhyphen. Ein Beitrag zur Kenntnis des Flächenwachsthums vegetabilischer Zellmembranen. *Jahrb. f. wiss. Bot.*, **23**, 479–566, 1892.

REINKE, J., and RODEWALD, H., Studien über das Protoplasma. I. Die chemischen Zuzammensetzung des Protoplasma von *Aethalium septicum. Unters. a. d. bot. Lab. Univ. Göttingen*, Heft 2, 1–75, 1881.

REMER, W., Der Einfluss des Lichtes auf die Keimung bei *Phacelia tanacetifolia*, Benth. *Ber. deut. bot. Ges.*, **22**, 328–39, 1904.

RENNER, O., Beiträge zur Physik der Transpiration. *Flora*, **100**, 451–547, 1900.

—— Theoretisches und Experimentalles zur Kohäsionstheorie der Wasserbewegung. *Jahrb. f. wiss. Bot.*, **56**, 617–67, 1915.

—— Versuche zur Mechanik der Wasserversorgung. *Ber. deut. bot. Ges.*, **36**, 172–9, 1918.

RICCA, U., Soluzione d'un problema di fisiologia. La propagazione di stimulo nella ' Mimosa '. *Nuovo giorn. bot. ital.*, **23**, 51–171, 1916.

RICHARDSON, W. D., The ash of dune plants. *Science, N.S.*, **51**, 546–51, 1920.

RICHTER O. Über den Einfluss verunreinigter Luft auf Heliotropismus und Geotropismus. *Sitzungsber. k. Akad. Wiss. Wien*, Abt. 1, *Math.-nat. Kl.*, **115**, 265–352, 1906.

—— Konzentrierte Schwefelsäure, konzentrierte Kalilauge als Treibmittel und andere Erfahrungen über Pflanzentreiberei. *Ber. deut. bot. Ges.*, **40**, (43)–(52,) 1922.

ROBERTSON, T. B., Further remarks on the normal rate of growth of an individual, and its biochemical significance. *Arch. f. Entwickelungsmechanik*, **26**, 108–18, 1908.

—— On the nature of the autocatalyst of growth. *Arch. f. Entwickelungsmechanik*, **37**, 497–508, 1913.

ROTHERT, W., Über Heliotropismus. *Beitr. z. Biol. Pflanzen*, **7**, 1–224, 1894.

ROUGE, E., Recherche des premiers produits de l'assimilation chlorophyllienne du carbone. *Journ. suisse pharm.*, **59**, 1–9, 1924.

RUHLAND, W., Zur Frage der Ionenpermeabilität. *Zeitschr. f. Bot.*, **1**, 747–63, 1909.

RUHLAND, W., Studien über die Aufnahme von Kolloiden durch die Plasmahaut. *Jahrb. f. wiss. Bot.*, **51**, 376–431, 1912.

—— Weitere Beiträge zur Kolloidchemie und physikalischen Chemie der Zelle. *Jahrb. f. wiss. Bot.*, **54**, 391–447, 1914.

RUHLAND, W., and HOFFMANN, C., Die Permeabilität von *Beggiatoa mirabilis*. *Arch. f. wiss. Bot.*, **1**, 1–83, 1925.

RUHLAND, W., and WETZEL, K., Zur Physiologie der organischen Säuren in grünen Pflanzen. I. Wechselbeziehungen im Stickstoff- und Säurestoffwechsel von *Begonia sempervirens*. *Planta*, **1**, 558–64, 1926.

—— Zur Physiologie der organischen Säuren in grünen Pflanzen. III. *Rheum hybridum*, Hort. (Vorläufige Mitteilung.) *Planta*, **3**, 765–9, 1927.

—— Zur Physiologie der organischen Säuren in grünen Pflanzen. V. Weitere Untersuchungen an *Rheum hybridum*, Hort. (Vorläufige Mitteilung.) *Planta*, **7**, 503–7, 1929.

RUTGERS, A. A. L., The influence of temperature on the geotropic presentation time. *Rec. trav. bot. Néerlandais*, **9**, 1–123, 1912.

RUTTEN-PEKELHARING, C. J., Untersuchungen über die Perzeption des Schwerkraftreizes. *Rec. trav. bot. Néerlandais*, **7**, 241–348, 1910.

SACHS, J., Ueber die Erziehung von Landpflanzen in Wasser. *Bot. Zeit.*, **18**, 113–17, 1860.

—— Wirkungen farbigen Lichts auf Pflanzen. *Bot. Zeit.*, **22**, 353–8, 361–7, 369–72, 1864.

—— Ablenkung der Wurzel von ihrer normalen Wachsthumsrichtung durch feuchte Körpen. *Arb. bot. Inst., Würzburg*, **1**, 209–22, 1872.

—— Ueber das Wachsthum der Haupt- und Nebenwurzeln. *Arb. bot. Inst., Würzburg*, **1**, 385 474, 1873.

—— Ueber orthotrope und plagiotrope Pflanzentheile. *Arb. bot. Inst., Würzburg*, **2**, 226–84, 1879.

—— Ein Beitrag zur Kenntnis der Ernährungsthätigkeit der Blätter. *Arb. bot. Inst. Würzburg*, **3**, 1–33, 1884.

—— *Lectures on the Physiology of Plants.* English Ed. trans. by H. M. Ward. Oxford, 1887.

SAHAI, P. N., and CHIBNALL, A. C., Wax metabolism in the leaves of brussels sprout. *Biochem. Journ.*, **26**, 403–12, 1932.

SALISBURY, E. J., On the causes and ecological significance of stomatal frequency, with special reference to the woodland flora. *Phil. Trans. Roy. Soc.*, B, **216**, 1–65, 1927.

SAMMET, R., Untersuchungen über Chemotropismus und verwandte Erscheinungen bei Wurzeln, Sprossen und Pilzfäden. *Jahrb. f. wiss. Bot.*. **41**, 611–49, 1905.

SANDS, C. E., The process of ripening in the tomato, considered especially from the commercial standpoint. *U.S. Dept. Agric. Bull.* **859**, 38 pp., 1920.

SAPOZNIKOW, W., Eiweissstoffe und Kohlenhydrate der grünen Blätter als Assimilationsprodukte. Ref. in *Bot. Centralbl.*, **63**, 246–51, 1895.

SAUSSURE, T. DE, *Recherches chimiques sur la végétation.* Paris, 1804.

—— De l'action des fleurs sur l'air, et de leur chaleur propre. *Ann. chim. et phys.*, **21**, 279–303, 1822.

SAXTON, W. T., Autonomous movements in *Eleiotis soraria*. *Journ. Ind. Bot.*, **3**, 72–8, 1922.

SCARTH, G. W., Stomatal movement: its regulation and regulatory role. A review. *Protoplasma*, **2**, 498–511, 1927.

—— Mechanism of the action of light and other factors on stomatal movement. *Plant Physiology*, **7**, 481–504, 1932.

SCHIMPER, A. F. W., Ueber Bildung und Wanderung der Kohlenhydrate in den Laubblättern. *Bot. Zeit.*, **43**, 737–43, 753–63, 769–87, 1885.

SCHIMPER, A. F. W., Ueber Kalkoxalatbildung in den Laubblättern. *Bot. Zeit.*, **46**, 65–9, 81–9, 97–107, 113–23, 129–39, 145–53, 1888.

—— Zur Frage der Assimilation der mineral Salze durch die grüne Pflanze. *Flora*, **73**, 207–61, 1890.

—— Plant geography upon a physiological basis. English edition. Oxford, 1903.

SCHLEY, EVA O., Geo-presentation and geo-reaction. *Bot. Gaz.*, **70**, 69–81, 1920.

SCHNEIDER-ORELLI, O., Versuche über die Widerstandstähigkeit gewisser *Medicago* Samen (Wollkletten) gegen hohe Temperaturen. *Flora*, **100**, 305–11, 1910.

SCHULTZE, E., Über Zersetzung und Neubildung von Eiweissstoffen in Lupinenkeimlingen. *Landw. Jahrb.*, **7**, 411–44, 1878.

—— Über den Eiweissumsatz im Pflanzenorganismus. *Landw. Jahrb.*, **9**, 689–748, 1880.

—— Über den Eiweissumsatz im Pflanzenorganismus. II. *Landw. Jahrb.*, **12**, 909–20, 1883.

—— Über den Eiweissumsatz im Pflanzenorganismus. III. *Landw. Jahrb.*, **14**, 713–29, 1885.

—— Über den Eiweissumsatz im Pflanzenorganismus. IV. *Landw. Jahrb.*, **21**, 105–30, 1892.

—— Studien über die Proteinbildung in reifenden Pflanzensamen. II. Mitteilung. *Zeitschr. f. physiol. Chem.*, **71**, 31–48, 1911.

SCHULTZE, E., and UMLAUFT, W., Untersuchungen über einige chemische Vorgänge bei der Keimung der gelben Lupine. *Landw. Jahrb.*, **5**, 821–58, 1876.

SCHULTZE, E., and WINTERSTEIN, E., Studien über die Proteinbildung in reifenden Pflanzensamen. *Zeitschr. f. physiol. Chem.*, **65**, 431–76, 1910.

SCHWARZ, F., Die Wurzelhaare der Pflanzen. Ein Beitrag zur Biologie und Physiologie dieser Organe. *Unters. aus dem Bot. Inst. zu Tübingen*, **1**, 135–88, 1883.

SCHWENDENER, S., and KRABBE, G., Ueber die Beziehung zwischen dem Maas der Turgordehnung und der Geschwindigkeit der Langenzunahme wachsender Organe. *Jahrb. f. wiss. Bot.*, **25**, 323–69, 1893.

SEIDEL, K., Versuche über die Reizleitung bei *Mimosa pudica*. *Beitr. z. all. Bot.*, **2**, 557–75, 1923.

SEIFRIZ, W., Observations on the structure of protoplasm by aid of microdissection. *Biol. Bull.*, **34**, 307–24, 1918.

—— Viscosity values of protoplasm as determined by microdissection. *Bot. Gaz.*, **70**, 360–86, 1920.

—— Observations on some physical properties of protoplasm by aid of microdissection. *Ann. Bot.*, **35**, 269–96, 1921.

—— An elastic value of protoplasm, with further observations on the viscosity of protoplasm. *Brit. Journ. Exper. Biol.*, **2**, 1–13, 1924.

—— Elasticity as an indicator of protoplasmic structure. *Amer. Nat.*, **60**, 124–32, 1926.

—— The contractility of protoplasm. *Amer. Nat.*, **63**, 410–34, 1929.

—— The alveolar structure of protoplasm. *Protoplasma*, **9**, 177–208, 1930.

—— The origin, composition and structure of cellulose in the living plant. *Protoplasma*, **21**, 129–59, 1934.

—— The structure of protoplasm. *Bot. Rev.*, **1**, 18–36, 1935.

SENEBIER, J., *Mémoires physico-chimiques, sur l'influence de la lumière solaire pour modifier les êtres des trois regnes de la nature et surtout ceux du règne végétal.* 3 vols. Genève, 1783.

—— *Expériences sur l'action de la lumière solaire dans la végétation.* Genève, 1788.

—— *Physiologie végétale, contenant une description des plantes, et une exposition des phénomènes produits par leur organisation.* 5 vols. Genève, 1800.

SEN GUPTA, J., Untersuchungen über Rheotropismus. *Zeitschr. f. Bot.*, **21**, 353–98, 1929.

SHREVE, E. B., An analysis of the causes of variations in the transpiring power of cacti. *Physiol. Researches*, **2**, 72–127, 1916.

SHULL, C. A., Semipermeability of seed coats. *Bot. Gaz.*, **56**, 169–99, 1913.

—— Temperature and rate of moisture intake in seeds. *Bot. Gaz.*, **69**, 361–90, 1920.

SIERP, H., Ein Beitrag zur Kenntnis des Einflusses des Lichts auf das Wachstum der Koleoptile von *Avena sativa*. *Zeitschr. f. Bot.*, **10**, 641–729, 1918.

SIERP, H., and SEYBOLD, A., Untersuchungen uber die Lichtempfindlichkeit der Spitze und des Stumpfes in der Koleoptile von *Avena sativa*. *Jahrb. f. wiss. Bot.*, **65**, 592–610, 1926.

SILBERSCHMIDT, K., Untersuchungen über die Thermowachstumsreaktion. *Ber. deut. bot. Ges.*, **43**, 475–82, 1925.

SMITH, A. M., On the application of the theory of limiting factors to measurements and observations of growth in Ceylon. *Ann. Roy. Bot. Gard. Peradeniya*, **3**, 303–75, 1906.

SMITH, J. A. B., and CHIBNALL, A. C., The glyceride fatty acids of forage grasses. I. Cocksfoot and perennial ryegrass. *Biochem. Journ.*, **26**, 218–34, 1932.

—— The phosphatides of forage grasses. I. Cocksfoot. *Biochem. Journ.*, **26**, 1345–57, 1932.

SNOW, R., The conduction of geotropic excitation in roots. *Ann. Bot.*, **37**, 43–53, 1923.

—— Conduction of excitation in stem and leaf of *Mimosa pudica*. *Proc. Roy. Soc.*, B, **96**, 349–74, 1924.

—— Conduction of excitation in the leaf of *Mimosa spegazzinii*. *Proc. Roy. Soc.*, B, **98**, 188–201, 1925.

—— The young leaf as an inhibiting organ. *New Phyt.*, **28**, 345–58, 1929.

—— Experiments on growth and inhibition I. The increase of inhibition with distance. *Proc. Roy. Soc.*, B, **108**, 209–23, 1931.

—— Experiments on growth and inhibition II. New phenomena of inhibition. *Proc. Roy. Soc.*, B, **108**, 305–16, 1931.

—— Experiments on growth and inhibition III. Inhibition and growth promotion. *Proc. Roy. Soc.*, B, **111**, 86–105, 1932.

—— Growth regulators in plants. *New Phyt.*, **31**, 336–54, 1932.

—— The nature of the cambial stimulus. *New Phyt.*, **32**, 288–96, 1933.

—— Activation of cambial growth by pure hormones. *Nature*, **135**, 867, 1935.

SNOW, R., and LE FANU, B., Activation of cambial growth. *Nature*, **135**, 149, 1935.

SÖDING, H., Wachstum und Wanddehnbarkeit bei den Haferkoleoptile. *Jahrb. f. wiss. Bot.*, **74**, 127–51, 1931.

—— Über das Wachstum der Infloreszenzschäfte. *Jahrb. f. wiss. Bot.*, **77**, 627–56, 1933.

—— Über die Wachstumsmechanik der Hafeskoleoptile. *Jahrb. f. wiss. Bot.*, **79**, 231–55, 1934.

—— Ist der Wuchsstoff unspezifisch? *Zesde Internat. Bot. Congres Amsterdam. Proc.* Vol. II, 272–3. Leiden, 1935.

SOMMER, A. L., Copper as an essential for plant growth. *Plant Physiol.*, **6**, 339–45, 1931.

SOMMER, A. L., and LIPMAN, C. B., Evidence of the indispensable nature of zinc and boron for higher plants. *Plant Physiol.*, **1**, 231–49, 1926.

SOMMER, A. L., and SOROKIN, HELEN, Effects of the absence of boron and of some other essential elements on the cell and tissue structure of the root tips of *Pisum sativum*. *Plant Physiol.*, **3**, 237–60, 1928.

SORBY, H. C., On comparative vegetable chromatology. *Proc. Roy. Soc.*, **21**, 442–83, 1873.

SÖRENSEN, S. P. L., Studies on proteins. I–V. *Comp. rend. trav. lab. Carlsberg*, **12**, 1–372, 1917.

SOROKIN, HELEN, and SOMMER, ANNA L., Changes in the cells and tissues of root tips induced by the absence of calcium. *Amer. Journ. Bot.*, **16**, 23–39, 1929.

SPERLICH, A., Über Krümmungsursachen bei Keimstengeln und beim Monocotylenkeimblatte nebst Bemerkungen über den Phototropismus der positiv geotropischen Zonen des Hypocotyls und über das Stemmorgan bei Cucurbitaceen. *Jahrb. f. wiss. Bot.*, **50**, 502–653, 1912.

SPONSLER, O. L., Orientation of cellulose space lattice in the cell wall. *Protoplasma*, **12**, 241–54, 1930.

STAHL, E., Ueber sogenannte Kompasspflanzen. *Jen. Zeitschr. f. Naturw.*, **15**, 382–9, 1882.

—— Einfluss des Lichtes auf den Geotropismus einiger Pflanzenorgane. *Ber. deut. bot. Ges.*, **2**, 383–97, 1884.

—— Einige Versuche über Transpiration und Assimilation. *Bot. Zeit.*, **52**, 117–46, 1894.

—— Zur Physiologie und Biologie der Exkrete. *Flora*, **113**, 1–132, 1919.

STÅLFELT, M. G., Ljuset i frukträdkronorna. *Pomologisk Aarskrift*, 125–36, 1920.

—— Till Kännedomen om Förhållandet mellan Solbladens och Skuggbladens Kolhydratsproduktion. *Medd. f. Statens Skogsförsöksanstalt, Stokholm*, 221–80, 1921.

STAMEROFF, K., Zur Frage über den Einfluss des Lichtes auf das Wachsthum der Pflanzen. *Flora*, **83**, 135–50, 1897.

STARK, P., Experimentelle Untersuchungen über das Wesen und die Verbreitung der Kontaktreizbarkeit. *Jahrb. f. wiss. Bot.*, **57**, 189–320, 1916.

—— Beiträge zur Kenntnis der Traumatotropismus. *Jahrb. f. wiss. Bot.*, **57**, 461–552, 1917.

—— Studien über traumatotrope und haptotrope Reizleitungsvorgänge mit besonderer Beruchsichtigung auf fremde Artem und Gattungen. *Jahrb. f. wiss. Bot.*, **60**, 67–135, 1921.

—— Weitere Untersuchungen über das Resultantende setz beim Haptotropismus (Mit besonderer Berucksichtigung physiologisch nicht radiärer Organe). *Jahrb. f. wiss. Bot.*, **61**, 126–67, 1922.

—— Geotropische Reizleitung bei Unterbrechung des organischen Zusammenhangs. *Ber. deut. bot. Ges.*, **42**, 125–30, 1924.

STEFAN, J., Ueber die Verdampfung aus einen kreisförmig oder elliptisch begrenzten Becken. *Ann. Phys. u. Chem.*, **17**, 550–60, 1882.

STERN, K., Über die Fluoreszenz des Chlorophylls und ihre Bedeutung beim Assimilationsprozess. *Zeitschr. f. Bot.*, **13**, 193–230, 1921.

STEWARD, F. C., The absorption and accumulation of solutes by living plant cells. I. Experimental conditions which determine salt absorption by storage tissue. *Protoplasma*, **15**, 29–58, 1932.

STEWARD, F. C., and PRIESTLEY, J. H., Movement of organic materials in plants ; a note on a recently suggested mechanism. *Plant Physiol.*, **7**, 165–71, 1932.

STICH, C., Die Athmung der Pflanzen bei verminderter Sauerstoffspannung und bei Verletzungen. *Flora* (N.R. 49), **74**, 1–57, 1891.

STILES, W., The absorption of salts by storage tissues. *Ann. Bot.*, **38**, 617–33, 1924.

—— Plant respiration. *Biol. Rev.*, **1**, 249–68, 1935.

STILES, W., and ADAIR, G. S., The penetration of electrolytes into gels. III. The influence of the concentration of the gel on the coefficient of diffusion of sodium chloride. *Biochem. Journ.*, **15**, 620–8, 1921.

STILES, W., and JØRGENSEN, I., Studies in Permeability. II. On the permeability of plant cells to the hydrogen ion. *Ann. Bot.*, **29**, 611–18, 1915.

—— Studies in permeability. V. The swelling of plant tissue in water and its relation to temperature and various dissolved substances. *Ann. Bot.*, **31**, 415–34, 1917.

STILES, W., and KIDD, F., The influence of external concentration on the position of the equilibrium attained in the intake of salts by plant cells. *Proc. Roy. Soc.*, B, **90**, 448–70, 1919.

STILES, W., and LEACH, W., Researches on plant respiration. I. The course of respiration of *Lathyrus odoratus* during germination of the seed and the early development of the seedling. *Proc. Roy. Soc.*, B, **111**, 338–55, 1932.

—— Researches on plant respiration. II. Variations in the respiratory quotient during germination of seeds with different food reserves. *Proc. Roy. Soc.*, B, **113**, 405–28, 1933.

STILES, W., and REES, W. J., Studies on toxic action. IX. The toxicity of organic acids of the formic acid series. *Protoplasma*, **22**, 518–38, 1935.

STILES, W., and STIRK, M. L. L., Studies on toxic action. II. The toxicity of normal aliphatic alcohols towards potato tuber. *Protoplasma*, **13**, 1–20, 1931.

STOKLASA, J., Ueber die Entstehung und Umwandlung der Lecithins der Pflanze. *Zeitschr. f. physiol. Chem.*, **25**, 398–405, 1898.

STOKLASA, J., and CZERNY, F., Isolierung des die anaerobe Atmung der Zelle der hoher organisierten Pflanzen und Thiere bewirkenden Enzyms. *Ber. deut. chem. Ges.*, **36**, 622–34, 1903.

STOKLASA, J., and MATOUŠEK, A. (unter Mitwirkung VON SENFT, E., ŠEBOR, J., and ZDOBNICKY, W.), *Beiträge zur Kenntnis der Ernährung der Zuckerrube. Physiologische Bedeutung des Kalium-Ions im Organismus der Zuckerrube*, Jena, 1916.

STOKLASA, J. (unter Mitwirkung VON ŠEBOR, J., TÝMICH, F., and CWACHA, J.), Über die Resorption des Aluminium-Ions durch das Wurzelsystem der Pflanzen. *Biochem. Zeitschr.*, **128**, 35–47, 1922.

STOLL, A., Zusammenhänge zwischen der Chemie des Chlorophylls und seiner Funktion in der Photosynthese. *Naturwiss.*, **24**, p. 53 (reprint 16 pp.), 1936.

STOPPEL, R., Die Abhängigkeit der Schlafbewegungen von *Phaseolus multiflorus* von verschiedenen Aussenfaktoren. *Zeitschr. f. Bot.*, **8**, 609–84, 1916.

STOUT, A. B., Self- and cross-pollinations in *Cichorium Intybus* with reference to sterility. *Mem. New York Bot. Gard.*, **6**, 333–454, 1916.

STRASBURGER, E., *Ueber den Bau und Verrichtungen der Leitungsbahnen in den Pflanzen*, Jena, 1891.

SUMNER, J. B., and HAND, D. B., Crystalline urease, II. *Journ. Biol. Chem.*, **76**, 149–62, 1928.

SUTHERLAND, W., A dynamical theory of diffusion for non-electrolytes and the molecular mass of albumin. *Phil. Mag.* (Ser. 6), **9**, 781–5, 1905.

SUZUKI, U., The formation of asparagin in plants under different conditions. *Imp. Univ. Tokyo Coll. Agric.*, **2**, 409–57, 1897.

—— An important function of leaves. *Imp. Univ. Tokyo Coll. Agric.*, **3**, 241–52, 1897.

—— On the formation of proteins and the assimilation of nitrates by phanerogams in the absence of light. *Imp. Univ. Tokyo Coll. Agric.*, **4**, 488–587, 1898.

SZÜCS, J., Studien über Protoplasmapermeabilität. Ueber die Aufnahme der Anilinfarben durch die lebende Zelle und ihre Hemmung durch Elektrolyte. *Sitzungsber. kais. Akad. Wiss. in Wien, Mat.-nat. Kl.*, **119**, Abt. 1, 737–73, 1910.

Szücs, J., *A növényi sejtek elektrolyt felvétele és az adsorptio.* (Reprint, 26 pp.) Budapest, 1911.

—— Experimentelle Beiträge zu einer Theorie der antagonischen Ionenwirkung. I. Mitteilung. *Jahrb. f. wiss. Bot.*, **52**, 85–142, 1912.

—— Ueber einige charakteristiche Wirkungen des Aluminiumions auf das Protoplasma. *Jahrb. f. wiss. Bot.*, **52**, 269–332, 1913.

Talma, E. G. C., The relation between temperature and growth in the roots of *Lepidium sativum*. *Rec. trav. bot. Néerlandais*, **15**, 366–422, 1918.

Thimann, K. V., and Skoog, F., On the inhibition of bud development and other functions of growth substance in *Vicia faba*. *Proc. Roy. Soc.*, B, **114**, 317–39, 1934.

Thimann, K. V., and Went, F. W., On the chemical nature of the root-forming hormone. *Kon. Akad. Wetensch. Amsterdam, Proceedings*, **37**, 456–9, 1934.

Thoday, D., Experimental researches on vegetable assimilation and respiration. V. A critical examination of Sachs' method for using increase of dry weight as a measure of carbon dioxide assimilation in leaves. *Proc. Roy. Soc.*, B, **82**, 1–55, 1909.

—— Experimental researches on vegetable assimilation and respiration. VI. Some experiments on assimilation in the open air. *Proc. Roy. Soc.*, B, **82**, 421–50, 1910.

Thomas, Nesta, and Ferguson, A., On evaporation from a circular water surface. *Phil. Mag.*, **34**, 308–21, 1917.

Thunberg, T., Zur Kenntnis der Einwirkung tierischer Gewebe auf Methylenblau. *Skand. Arch. Physiol.*, **35**, 163–95, 1917.

Tincker, M. A. H., The effect of length of day upon the growth and reproduction of some economic plants. *Ann. Bot.*, **39**, 721–54, 1925.

Tischler, G., Untersuchungen an Mangrove- und Orchidaceenwurzeln mit spezieller Beziehung auf die Statolithentheorie des Geotropismus. *Ann. Jard. Bot. Buitenzorg*, **3**, 131–86, 1910.

True, R. H., On the influence of sudden changes of turgor and of temperature on growth. *Ann. Bot.*, **35**, 365–402, 1895.

Tswett, M., Zur Kenntnis der Phaeophyceenfarbstoffe. *Ber. deut. bot. Ges.*, **24**, 235–44, 1906.

—— Physikalisch-chemische Studien über das Chlorophyll. Die Adsorptionen. *Ber. deut. bot. Ges.*, **24**, 235–44, 1906.

—— Adsorptionsanalyse und chromatographische Methode. Anwendung auf die Chemie des Chlorophylls. *Ber. deut. bot. Ges.*, **24**, 384–93, 1906.

Ubisch, G. v., Betrachtungen und Versuche zur Statolithenhypothese. *Biol. Zentrabl.*, **48**, 172–90, 1928.

Umrath, K., Über die Erregungsleitung im Blatte von *Mimosa pudica*. *Sitz. Akad. Wiss. Wien, Math.-nat. Kl.*, Abt. 1, **134**, 21–44, 1925.

—— Über die Erregungsleitung bei *Mimosen*. *Sitz. Akad. Wiss. Wien, Math.-nat. Kl.*, Abt. 1, **134**, 189–208, 1925.

Ursprung, A., Über die Kohäsion des Wassers im Farnannulus. *Ber. deut. bot. Ges.*, **33**, 153–62, 1915.

—— Zweiter Beitrag zur Demonstration des Flüssigkeitskohäsion. *Ber. deut. bot. Ges.*, **33**, 253–65, 1915.

—— Dritter Beitrag zur Demonstration der Flüssigkeitskohäsion. *Ber. deut. bot. Ges.*, **34**, 475–88, 1916.

—— Über die Stärkebildung im Spektrum. *Ber. deut., bot. Ges.*, **35**, 44–69, 1917.

—— Zur Kenntnis der Saugkraft, VII. Eine neue vereinfachte Methode zur Messung der Saugkraft. *Ber. deut. bot. Ges.*, **41**, 338–43, 1923.

URSPRUNG, A., and BLUM, G., Zur Kenntnis der Saugkraft. *Ber. deut. bot. Ges.*, **34**, 539–54, 1916.

—— Zur Kenntnis der Saugkraft, II. *Ber. deut. bot. Ges.*, **36**, 577–99, 1918.

—— Zur Kenntnis der Saugkraft, III. 4. *Hedera Helix.* Abgeschnittenes Blatt. *Ber. deut. bot. Ges.*, **37**, 453–62, 1919.

—— Zur Kenntnis der Saugkraft, IV. Die Absorptionszone der Wurzel. Der Endodermissprung. *Ber. deut. bot. Ges.*, **39**, 70–9, 1921.

—— Zur Kenntnis der Saugkraft, V. Eine Methode zur Bestimmung des Widerstandes, den der Boden der Wasserabsorption durch die Wurzel entgegensetzt. *Ber. deut. bot. Ges.*, **39**, 139–48, 1921.

URSPRUNG, A., and HAYOZ, C., Zur Kenntnis der Saugkraft. VI. Weitere. Beiträge zur Saugkraft normalen und abgeschnittenen Hederablattes. *Ber. deut. bot. Ges.*, **40**, 368–73, 1922.

UYLDERT, INA E., *De invloed van groeistof op planten met intercalaire groei.* (Diss. Utrecht) Santpoort, 1931.

VERKAIK, C., Ueber das Enstehen von Zygophoren von *Mucor Mucedo* (+) unter Beeinflussung eines von *Mucor Mucedo* (—) abgeschieden Stoffes. *K. Akad. Wetens. Amsterdam, Proceedings,* **33**, 656–8, 1930.

VERSCHAFFELT, E., Le traitement chimique des graines à imbibit on tardive. *Rec. trav. bot. Néerlandais,* **9**, 401–35, 1912.

VERWORN, M., *Die Biogenhypothese. Eine kritisch-experimentelle Studie über die Vorgänge in der lebendigen Substanz.* Jena, 1903.

VESQUE, J., Recherches sur le mouvement de la sève ascendante. *Ann. sci. nat. Bot.*, sér. 6, **19**, 159–99, 1884.

VINES, S. H., The proteolytic enzymes of Nepenthes. *Ann. Bot.*, **11**, 563–84, 1897.

—— The proteolytic enzymes of Nepenthes (II). *Ann. Bot.*, **12**, 545–55, 1898.

—— The proteolytic enzymes of Nepenthes (III). *Ann. Bot.*, **15**, 563–73, 1901.

—— Proteolytic enzymes in plants. *Ann. Bot.*, **17**, 237–64, 1903.

—— Proteolytic enzymes in plants (II). *Ann. Bot.*, **17**, 597–616, 1903.

—— The proteases of plants. *Ann. Bot.*, **18**, 289–317, 1904.

—— The proteases of plants (II). *Ann. Bot.*, **19**, 149–62, 1905.

—— The proteases of plants (III). *Ann. Bot.*, **19**, 171–87, 1905.

—— The proteases of plants (IV). *Ann. Bot.*, **20**, 113–22, 1906.

—— The proteases of plants (V). *Ann. Bot.*, **22**, 103–13, 1908.

—— The proteases of plants (VI). *Ann. Bot.*, **23**, 1–18, 1909.

—— The proteases of plants (VII). *Ann. Bot.*, **24**, 213–22, 1910.

VISSER, A. W., Reaktionsgeschwindigkeit und chemisches Gleichgewicht in homogenen Systemen und deren Anwendung auf Enzymwirkungen. *Zeitschr. f. physik. Chem.*, **52**, 257–309, 1905.

VÖCHTING, H., Ueber die Lichtstellung der Laubblätter. *Bot. Zeit.*, **46**, 501–14, 1888.

—— Ueber den Einfluss des Lichtes auf die Gestaltung und Anlage der Blüthen. *Jahrb. f. wiss. Bot.*, **25**, 149–208, 1893.

—— Ueber die Bedeutung des Lichtes für die Gestaltung blättförmiger Cacteen. Zur Theorie der Blattstellungen. *Jahrb. f. wiss. Bot.*, **26**, 438–94, 1894.

—— Ueber den Einfluss niedriger Temperatur auf die Sprossrichtung. *Ber. deut. bot. Ges.*, **16**, 37–52, 1898.

—— Zur Physiologie der Knollengewächse. Studien über vicarirende Organe am Pflanzenkörper. *Jahrb. f. wiss. Bot.*, **34**, 1–148, 1900.

—— Ueber die Keimung der Kartoffelknollen. *Bot. Zeit.*, **60**, 87–114, 1902.

—— Ueber den Sprossscheitel der *Linaria spuria.* *Jahrb. f. wiss. Bot.*, **38**, 83–118, 1902.

VOGT, E., Über den Einfluss des Lichts auf das Wachstum der Koleoptile von *Avena sativa.* *Zeitschr. f. Bot.*, **7**, 193–270, 1915.

VRIES, H. DE, Ueber einige Ursachen der Richtung bilateralsymmetrischer Pflanzentheile. *Arb. bot. Inst. Würzburg*, **1**, 223–77, 1872.

—— Ueber den Antheil der Pflanzensäuren an der Turgorkraft wachsender Organe. *Bot. Zeit.*, **41**, 849–54, 1883.

—— Eine Methode zur Analyse der Turgorkraft. *Jahrb. f. wiss. Bot.*, **14**, 427–60, 1884.

VRIES, MARIE S. DE, Der Einfluss der Temperatur auf den Phototropismus. *Rec. trav. bot. Néerlandais*, **11**, 195–290, 1914.

WÄCHTER, W., Untersuchungen über den Austritt von Zucker aus den Zellen der Speicherorgane von *Allium Cepa* und *Beta vulgaris*. *Jahrb. f. wiss. Bot.*, **41**, 165–220, 1905.

WAIGHT, F. M. O., On the presentation time and latent time for reaction to gravity in fronds of *Asplenium bulbiferum*. *Ann. Bot.*, **37**, 55–61, 1923.

WALTER, H., Wachstumsschwankungen und hydrotropische Krümmungen bei *Phycomyces nitens*. *Zeitschr. f. Bot.*, **13**, 673–718, 1921.

—— Plasmaquellung und Wachstum. *Zeitschr. f. Bot.*, **16**, 353–417, 1924.

—— Die Wassersättigung der Pflanze und ihre Bedeutung fur das Pflanzenwachstum. *Zeitschr. f. Pflanzenernahrung und Dungen.* Teil A, **6**, 65–88, 1925.

WARBURG, O., Über die Geschwindigkeit der photochemischen Kohensäurezersetzung in lebenden Zellen. I. *Biochem. Zeitschr.*, **100**, 230–70, 1919.

—— Über die Geschwindigkeit der photochemischen Kohlensäurezersetzung in lebenden Zellen. II. *Biochem. Zeitschr.*, **103**, 188–217, 1920.

—— Physikalische Chemie der Zellatmung. *Biochem. Zeitschr.*, **119**, 134–66, 1921.

—— Über die Grundlagen der Wielandschen Atmungstheorie. *Biochem. Zeitsch.*, **141**, 518–23, 1923.

WARBURG, O., and NEGELEIN, E., Über den Energieumsatz bei der Kohlensäureassimilation. *Zeitschr. f. physikal. Chem.*, **102**, 235–66, 1922 ; *Naturwiss*, **10**, 647–53, 1922.

—— Über den Einfluss der Wellenlänge auf den Energieumsatz bei der Kohlensäureassimilation. *Zeitschr. f. physikal. Chem.*, **106**, 191–218, 1923.

WARRINGTON, KATHERINE, The effect of boric acid and borax on the broad bean and certain other plants. *Ann. Bot.*, **37**, 629–72, 1923.

—— The changes induced in the anatomical structure of *Vicia Faba* by the absence of boron from the nutrient solution. *Ann. Bot.*, **40**, 27–42, 1926.

WARNER, T., Über den Einfluss der geotropischen Reizung auf den Zucker und Säuregehalt von Sprossen. *Jahrb. f. wiss. Bot.*, **68**, 431–97, 1928.

WEBER, F., Über die Abkürzung der Ruheperiode der Holzgewächse durch Verletzung der Knospen, beziehungsweise Injektion derselben mit Wasser (Verletzungsmethode). *Sitzungsb. K. Akad. Wiss. Wien*, Abt. 1, **120**, 179–94, 1911.

—— Über ein neues Verfahren, Pflanzen zu treiben. Azethylenmethode. *Sitzungsber. K. Akad. Wiss. Wien*, Abt. L, *Math.-nat. Kl.*, **125**, 189–216, 1916.

—— Über die Winterruhe des Holzgewächse. *Ber. deut. bot. Ges.*, **39**, 152–6, 1921.

—— Fruhtreiben durch Quetschen. *Ber. deut. bot. Ges.*, **40**, 148–52, 1922.

—— Ruhesperiode und Fruhtreiben. *Ber. deut. bot. Ges.*, **42**, 109–12, 1924.

WEBER, G., and WEBER, F., Wirkung der Schwerkraft auf die Plasmaviskosität. *Jahrb. f. wiss. Bot.*, **57**, 129–88, 1916.

WEBER, U., Untersuchungen über Wachstum und Krümmung unverletzter und halbierter Koleoptilen nach geotropischer Reizung. *Jahrb. f. wiss. Bot.*, **66**, 35–108, 1926.

WEBER. U., Wachstum und Krümmung einzelner Zonen geotropisch gereizter Gerstenkeimlinge. *Jahrb. f. wiss. Bot.*, **75**, 312–76, 1931.

WEEVERS, T., Betractungen und Untersuchungen über die Nekrobiose und die letale Chloroformeinwirkung. *Rec. trav. bot. Néerlandais*, **9**, 236–80, 1912.

—— Die letale Einwirkung einiger organischen Giftstoffe auf die Pflanzenzelle. *Rec. trav. bot. Néerlandais*, **11**, 321–41, 1912.

—— The first carbohydrates that originate during the assimilatory process. A physiological study with variegated leaves. *Kon. Akad. Wetensch. Amsterdam, Proc.* (English Version), **27**, 1–11, 1924.

WEIJ, H. G. VAN DER, Die quantitative Arbeitsmethode mit Wuchsstoff. *K. Akad. Wetens. Amsterdam, Proceedings*, **34**, 875–92, 1931.

—— Der Mechanismus des Wuchsstofftransportes. *Rec. trav. bot. Néerlandais*, **29**, 379–496, 1932.

—— Der Mechanismus des Wuchsstofftransportes II. *Rec. trav. bot. Néerlandais*, **31**, 810–57, 1934.

WEIS, F., Sur la rapport entre l'intensité lumineuse et l'énergie assimilatrice chez des plantes appartenant à des types biologiques differents. *Comp. rend. acad. sci.*, **137**, 801–4, 1903.

WENT, F. A. F. C., The investigations on growth and tropisms carried on in the botanical laboratory of the University of Utrecht during the last decade. *Biol. Rev.*, **10**, 187–207, 1935.

WENT, F. W., Concerning the difference in sensibility of the tip and base of *Avena* to light. *K. Akad. Wetens. Amsterdam, Proceedings*, **29**, 185–91, 1925.

—— Wuchsstoff und Wachstum. *Réc. trav. bot. Néerlandais*, **25**, 1–116, 1928.

—— On a substance causing root-formation. *K. Akad. Wetens. Amsterdam, Proceedings*, **32**, 35–6, 1929.

—— Ein botanisches Polaritätstheorie. *Jahrb. f. wiss. Bot.*, **76**, 528–57, 1932.

—— Auxin, the plant growth-hormone. *Bot. Rev.*, **1**, 162–82, 1935.

—— Hormones, involved in root formation. The phenomenon of inhibition. *Zesde Internat. Bot. Congres Amsterdam. Proc.*, Vol. II, 267–9, Leiden, 1935.

WEST, C., BRIGGS, G. E., and KIDD, F., Methods and significant relations in the quantitative analysis of plant growth. *New Phyt.*, **19**, 200–7, 1920.

WESTERMEIER, M., Zur Kenntniss der osmotischen Leistungen des lebenden Parenchym's. *Ber. deut. bot. Ges.*, **1**, 371–83, 1883.

WHELDALE, M. (*see also* ONSLOW, M. W.), On the direct guaiacum reaction given by plant extracts. *Proc. Roy. Soc.*, B, **84**, 121–4, 1911.

WHITAKER, D. M., On the rate of oxygen consumption by fertilised and unfertilised eggs. I. *Fucus vesiculosus*. *Journ. Gen. Physiol.*, **15**, 167–82, 1931.

WHITE, JEAN, The influence of pollination on the respiratory activity of the gynaecium. *Ann. Bot.*, **21**, 487–99, 1907.

WIECHULLA, O., Beiträge zur Kenntnis der Lichtwachstumreacktion bei *Phycomyces*. *Beitr. z. Biol. Pflanzen*, **19**, 371–419, 1932.

WIEDERSHEIM, W., Studien über photonastische und thermonastische Bewegungen. *Jahrb. f. wiss. Bot.*, **40**, 230–78, 1904.

WIELAND, H., Zur Verbrennung des Kohlenoxyds. *Ber. deut. chem. Ges.*, **45**, 679–85, 1912.

—— Studien über den Mechanismus der Oxydationsvorgänge. *Ber. deut. chem. Ges.*, **45**, 2606–15, 1912.

—— Über den Mechanismus der Oxydationsvorgänge. *Ber. deut. chem. Ges.*, **46**, 3327–42, 1913.

—— Über den Verlauf der Oxydationsvorgänge. *Ber. deut. chem. Ges.*, **55**, 3639–48, 1923.

WIELER, A., Die Beeinflussung des Wachsens durch verminderte Partiarpressung des Sauerstoffs. *Unters. bot. Inst. Tübingen*, **1**, 189–232, 1883.

WIESNER, J., Die heliotropischen Erscheinungen im Pflanzenreiche. Eine physiologische Monographie. *Denkschr. Akad. Wiss. Wien*, **39**, 143–209, 1879.

—— Grundversuche über den Einfluss der Luftbewegung auf die Transpiration der Pflanzen. *Sitzungsber. k. Akad. Wiss, Wien*, Abt. 1, **96**, 182–214, 1887.

—— Ueber die Ruheperiode und über einige Keimungsbedingungen der Samen von *Viscum album*. *Ber. deut. bot. Ges.*, **15**, 503–16, 1887.

—— *Der Lichtgenuss der Pflanzen*. Leipzig, 1907.

WILLSTÄTTER, R., Über Isolierung von Enzymen. *Ber. deut. chem. Ges.*, **55**, 3601–23, 1922.

WILLSTÄTTER, R., and PAGE, H. J., Untersuchungen über Chlorophyll XXIV. Über die Pigmente der Braunalgen. *Ann. Chem.*, **404**, 237–71, 1914.

WILLSTÄTTER, R., and POLLINGER, A., Über Peroxidase. III. *Ann. Chem.*, **430**, 269–319, 1923.

WILLSTÄTTER, R., and STOLL, A., *Untersuchungen über Chlorophyll*. Berlin, 1913.

—— *Untersuchungen über die Assimilation der Kohlensäure*. Berlin, 1918.

WILLSTÄTTER, R., and WALDSCHMIDT-LEITZ, E., Über Ricinus lipase. *Zeitschr. f. physiol. Chem.*, **134**, 161–223, 1924.

WISSELINGH, V. VAN, Sur la paroi des cellules subéreuses. *Arch. Néerlandais des Sci. exactes et nat.*, **22**, 253–98, 1899.

—— Sur la lamelle subéreuse et la suberine. *Arch. Néerlandais des Sci. exactes et nat.*, **26**, 305–53, 1893.

WOLFF, E., *Aschen-Analysen von landwirtschaftlichen Produkten, Fabrik-Abfällen und wildwachsenden Pflanzen*. Theil 1. Berlin, 1871.

—— *Aschen-Analysen von land- und forstwirthschaftlichten Produkten, Fabrik-Abfällen und wildwachsenden Pflanzen*. Theil 2. Berlin, 1880.

WOLFF, H., Zur Physiologie des Wurzelpilzes von *Neottia Nidus avis*, Rich. und einigen grünen Orchideen. *Jahrb. f. wiss. Bot.*, **66**, 1–34, 1926.

WOO, M. L., Chemical constituents of *Amaranthus retroflexus*. *Bot. Gaz.*, **68**, 313–44, 1919.

WOOD, J. G., The physiology of xerophytism in Australian plants. The stomatal frequencies, transpiration and osmotic pressures of sclerophyll and tomentose-succulent leaved plants. *Journ. Ecol.*, **22**, 69–87, 1934.

WORTMANN, J., Ueber die Beziehungen der intramolecularen zur normalen Athmung der Pflanzen. *Arb. bot. Inst. Würzburg*, **2**, 500–20, 1880.

—— Ueber den Einfluss der strahlenden Wärme auf wachsende Pflanzentheile. *Bot. Zeit.*, **41**, 457–70, 473–80, 1883.

—— Studien über geotropische Nachwirkungs-erscheinungen. *Bot. Zeit.*, **42**, 705–13, 1884.

WURTZ, A., Sur la papaine. Contribution à l'histoire des ferments solubles. *Comp. rend. acad. sci.*, **90**, 1379–85, 1880.

—— Sur la papaine. Nouvelle contribution a l'histoire des ferments solubles. *Comp. rend. acad. sci.*, **91**, 787–91, 1880.

YAPP, R. H., and MASON, U. C., The distribution of water in the shoots of certain herbaceous plants. *Ann. Bot.*, **46**, 160–81, 1931.

YASUDA, S., Physiological researches on the fertility in *Petunia violacea*. *Bot. Mag. Tokyo*, **41**, 17–27, 1927.

YEMM, E. W., The respiration of barley plants II. Carbohydrate concentration, and carbon dioxide production in starving leaves. *Proc. Roy. Soc.*, B, **117**, 504–25, 1935.

YOSHIDA, H., Chemistry of lacquer (urushi). Part I. *Journ. Chem. Soc.*, **43**, 472–86, 1883.

ZALESKI, W., Zur Kenntnis der Eiweissbildung in den Pflanzen. *Ber. deut. bot. Ges.*, **15**, 536–42, 1897.

—— Beiträge zur Kenntnis der Eiweissbildung in den Pflanzen. *Ber. deut. bot. Ges.*, **19**, 331–9, 1901.

—— Über den Aufbau der Eiweissstoffe in den Pflanzen. *Ber. deut. bot. Ges.*, **25**, 360–7, 1907.

ZALESKI, W., and MARX, E., Über die Carboxylase bei höheren Pflanzen. *Biochem. Zeitschr.*, **47**, 184–5, 1912.

—— Über die Rolle der Carboxylase in den Pflanzen. *Biochem. Zeitschr.*, **48**, 175–80, 1913.

ZOLLIKOFER, CLARA, Über die Wirkung der Schwerkraft auf die Plasmaviskosität. *Ber. deut. bot. Ges.*, **35**, 291–8, 1917.

—— Über das geotropische Verhalten entstärkter Keimpflanzen und den Abbau der Stärke in Gramineen-Koleoptilen. *Ber. deut. bot. Ges.*, **36**, 30–8, 1918.

INDEX